EQUINOX
BOOK OF SCIENCE

EQUINOX: THE EARTH
Anna Grayson with Douglas Palmer, Paul Simons, David Jackson and Karl P N Shuker

EQUINOX: THE BRAIN
Jack Challoner

EQUINOX: SPACE
Jack Challoner

EQUINOX: WARFARE
Peter Harclerode

First published in 2001 by Channel 4 Books, an imprint of Macmillan Publishers Ltd, 25 Eccleston Place, London SW1W 9NF,
Basingstoke and Oxford
Associated companies throughout the world.
www.macmillan.co.uk

ISBN 0 7522 6136 3

1 3 5 7 9 8 6 4 2

A CIP catalogue record for this book is available from the British Library.

Design by Jane Coney
Typeset by Ferdinand Pageworks
Printed in Great Britain by Mackays of Chatham plc

ACKNOWLEDGEMENTS

First and foremost I would to thank fellow authors, David Jackson, Douglas Palmer, Karl Shuker and Paul Simons for their contributions and help. Many other have helped in the writing of this book – my husband, Desmond Clark, and my sons Nicholas and Christopher Clark. I am also deeply grateful to Caroline Davidson for her support. Many scientists have generously given of their time to help me check for factual accuracy, but particular thanks are due to Mike Benton, Norm MacLeod, Vincent Courtillot, Ian Gilmour, Dave Rothery, Mike Widdowson, Bob Spicer, Roger Musson, Brian Upton, Paul Wignall, Ed Stephens, Roger Smith and the late Harald Drever. I would like to thank Emma Tait of Channel 4 Books and my editor Christine King for being so delightful and efficient to work with. Thanks is also due to the teams who made the *Equinox* programmes, and to Charles Furneaux and Sara Ramsden of Channel 4 for facilitating these programmes which make a major contribution to the wider understanding of the workings of our planet.

Anna Grayson

Many people have helped me to write this book. Friends and family – in particular Paula James, Carolyn McGoldrick, Daniel Brookman, Karen Darling and Jilly Duckworth – have been supportive as ever. I would like to thank Emma Tait at Channel 4 Books, and my editor Christine King, for their professionalism and enthusiasm. I would also like to acknowledge the makers of the *Equinox* series for communicating issues of contemporary science and contributing to the public understanding of the subject.

Jack Challoner

The following very kindly provided me with considerable technical assistance during my research: Lorna Arnold, former historian at AERE; Igor Kudrick of the Bellona Foundation; Anne Aldis of the Conflict Studies Research Centre at the RMA Sandhurst; Duncan Lennox, editor of Jane's Strategic Weapons Systems; Christopher Foss, editor of Jane's Armour & Artillery; weapons expert Ian Hogg; Peter Donaldson, technical editor of Defence Helicopter; Alan Warnes, editor of Air Forces Monthly; Keith Atkin, editor of Jane's Electro-Optic Systems; and Bob Kemp of BAE Systems Avionics Group. I would like to express my sincere thanks to them all.

I am most grateful to the producers who supplied me with post-production scripts and additional research material originally used in the production of the documentaries on which Warfare is based: Andy Patterson and Patrick Forbes of Oxford Films & Television Company; William Woollard of Inca Films; and David Dugan of Windfall Films.

Peter Harclerode

PRODUCTION CREDITS

When Pigs Ruled the Earth
accompanies the *Equinox* programme of the same name made by The Mission Film and Television Company for Channel 4.
First broadcast: 10 November 1998

Killer Earth
accompanies the *Equinox* programme of the same name made by Pioneer Film and Television Productions for Channel 4.
First broadcast: 22 September 1998

Resurrecting the Mammoth
accompanies the *Equinox* programme of the same name made by Cicada Films for Channel 4.
First broadcast: 6 November 1999

Ice Warriors
accompanies the *Equinox* programme of the same name made by RDF Productions for Channel 4.
First broadcast: 21 September 1997

Lethal Seas
accompanies the *Equinox* programme of the same name made by Northlight Productions Limited for Channel 4.
First broadcast: 16 October 2000

A Sense of Disaster
accompanies the programme of the same name made by Granada Productions for Channel 4.
First broadcast: 10 May 1999

Mind Readers
accompanies the *Equinox* programme of the same name made by Illuminations Television Limited for Channel 4.
First broadcast: 1 December 1997

Natural-born Genius
accompanies the *Equinox* programme of the same name made by John Gau Productions for Channel 4.
First broadcast: 3 November 1997

Phantom Brains
accompanies the *Equinox* programme of the same name made by Oxford Television Limited for Channel 4.
First broadcast: 7 June 1999

Living Dangerously
accompanies the *Equinox* programme of the same name made by TVF for Channel 4.
First broadcast: 12 April 1999

Thin Air
accompanies the *Equinox* programme of the same name made by WGBH/*NOVA*
Research cited in this chapter is drawn from 'Alive on Everest', a *NOVA* Online Adventure.
www.pbs.org/wgbh/nova/everest/expeditions/97/index.html
First broadcast: 11 August 1998

Lies and Delusions
accompanies the programme of

the same name made by Rosetta Pictures for Channel 4.
First broadcast: 28 February 2000 and 6 March 2000

Day Return to Space
accompanies the *Equinox* programme of the same name made by TV6 (Scotland) Limited for Channel 4.
First broadcast: 19 July 1999

What Shall we do with the Moon?
accompanies the *Equinox* programme of the same name made by Wall to Wall Productions for Channel 4.
First broadcast: 19 July 1999

Space – The Final Junkyard
accompanies the *Equinox* programme of the same name made by Eagle & Eagle Limited for Channel 4.
First broadcast: 13 September 1999

Sun Storm
accompanies the *Equinox* programme of the same name made by Eagle & Eagle Limited for Channel 4.
First broadcast: 31 July 1999

Black Holes
accompanies the *Equinox* programme of the same name made by Pioneer Film & TV Productions for Channel 4.
First broadcast: 8 September 1997

The Rubber Universe
accompanies the *Equinox* programme of the same name made by Union Pictures Limited for Channel 4.
First broadcast: 5 June 1996

A Very British Bomb
accompanies the *Equinox* programme of the same name made by Oxford Films & Television Company for Channel 4.
First broadcast: 15 September 1997

Dismantling the Bomb
accompanies the *Equinox* programme of the same name made by Inca Limited for Channel 4.
First broadcast: 20 June 1996

Russian Roulette
accompanies the *Equinox* programme of the same name made by Windfall Films for Channel 4.
First broadcast: 14 July 1998

After Desert Storm
accompanies the *Equinox* programme of the same name made by Inca Limited for Channel 4.
First broadcast: 16 January 1993

Dawn of the Death Ray
accompanies the *Equinox* programme of the same name made by Windfall Films for TLC and first broadcast in the USA.
First broadcast on Channel 4: 4 August 1998

CONTENTS

EQUINOX: THE EARTH
The Introduction

The last century saw a complete revolution in the Earth sciences – from the old philosophy of a static unchanging Earth, in which life and landscapes had evolved gradually, according to a constant and unwavering pattern, to a new image of a dynamic ever-changing Earth. An Earth whose surface is constantly on the move, and whose interior is always churning and moving, providing heat energy that drives the workings of oceans and continents. The old view was shattered by the slow realization that the continents were not static but were in fact mobile, had drifted around the surface of the globe, and travelled thousands of miles.

The understanding of the mechanism of continental drift in the 1960s – the great theory of plate tectonics – was the climax to this change and revolutionized the way scientists understand the Earth. The 1970s were almost dizzy with a series of unfolding realizations that plate tectonics affected so many aspects of our interaction with the Earth's crust. It became known that really explosive and dangerous volcanoes and many cities at risk from earthquakes tended to be sited at plate boundaries. The Canadian geophysicist Tuzo Wilson realized that it was possible for an ocean basin to be transformed into a mountain chain, and for that mountain chain to be eroded and to split and turn back into an ocean basin. Even the highest mountains are transient objects in the timescale of the planet.

I remember being asked as a student to give a seminar on how plate tectonics had affected the formation of economic resources, such as metal ores, oil and gas. In researching this, I was quite astounded to find that plate tectonics had actually played a role in just about everything on which modern society depends. For example, oil is formed in basins where the crust is stretching and sagging, ores are formed by volcanic processes and in mountain-building where the crust is being compressed, even the water cycle follows the cycle of the rocks, in that rain falls on chains of mountains and washes their eroded products into the sea. At the rims of some oceans, water-laden sediment is carried down, and back into the Earth's interior on great slabs of lithosphere (Earth's outer solid layer, the crust, and the top of the layer below, the mantle). Deep inside, the water acts as a flux, making the rocks melt and shoot back to the surface as volcanoes. The truth was beginning to unfold – we are what we are because we live on an active planet, an active planet that is also covered in liquid water. It was a lively seminar, at a time when senior academic staff and students were on a learning curve together. Plate tectonics was becoming a unifying theory in the Earth sciences.

At about the same time, I was working on another type of volcanic rock in north-west Scotland for a dissertation. These were rocks of the basalt family which made up volcanoes that were once active in and around the Western Isles. At the time of my student days, these black satanic rocks were not fashionable, as they had not been part of a plate margin. They were dismissed as a minor oddity. But to me there was something powerful about them, as one kind of once-molten rock (called a picrite) had clearly been emplaced at a much higher temperature than anything I had ever seen before, so hot that the surrounding sandstones through which it passed had actually melted. The now-cold lavas were full of the mineral olivine which is olive-green in colour and a rough version of the gem peridot. Also there were mahogany-red crystals of spinel, rich in the metal chromium. Even then, we knew these minerals implied a source deep in the Earth's interior, way beneath the crust in the region called the mantle. But we did not know just how deep the ultimate origins of these lavas were. Nor did I, or anyone else at the time, suspect their role in the story of the planet's continental configuration and in the history of life.

Now, well over twenty years later, in writing the first two chapters of this book, the significance of these dark volcanic rocks is all too clear. My black outcrops in the Hebrides had once formed a small part of the plumbing for the most prolific and formidable kind of volcanism that ever occurs on the Earth.

These are flood basalts, so called because they quite literally flood vast expanses – thousands of square kilometres of landscape – with thick layers of lava. They have occurred periodically in different parts of the globe, throughout geological history, but fortunately for us there are no flood basalts active today. Even our largest active volcano, Hawaii, is a pimple compared to flood basalts.

But in the past their effects on the planet, way beyond the landscape they submerge, have been significant. It would appear that they are capable of splitting continents apart, and sending chunks of crust sailing in opposite directions around the globe. Also, modern dating techniques are showing that there is a remarkable coincidence between the occurrence of these flood basalts and a series of mass-extinction events that have occurred throughout the history of life on this planet.

Most remarkable of all, to me, is that modern geophysical techniques are now revealing that these basalts probably owe their origins to the fact that the Earth is still cooling, following a hot birth 4,500 million years ago. Convection currents are rising from the deep interior, causing melting and the expelling of searing hot lavas, heat energy and gases. Small wonder, then, that some of these magmas had melted sandstone on the Hebrides, 58 million years ago. At that time, it turns out, the Hebrides were welded to Greenland, and the Highlands to the Appalachians. It was the power that I had sensed in the Hebridean basalts and picrites that tore the mountains apart, created the rift that was to become the North Atlantic and predestined the pattern of our recent human history. My unprepossessing black volcanic outcrops were no minor oddity, but part of the big picture of the history of this very active planet, and tell of heat energy within the Earth that drives activity at the surface.

Earth science is currently full of such surprises, and reveals epic stories of the planet's past, and its probable future, which are far stranger and more awesome than science fiction. This book tells six stories, each a piece of remarkable science in its own right, but weaving together to give a new view of the Earth's working patterns and of life's interaction with its host planet.

The first chapter, 'When Pigs Ruled the Earth', examines the largest of these extinction events, 250 million years ago, when something like 95 per cent of life on Earth was wiped out. The pigs in our story are not the pigs we know today, but a species called *Lystrosaurus* – a type of reptile with mammal-like features that looked rather pig-like and shared their ability to root for whatever food was available. This animal survived the extinction and found itself almost alone on the great supercontinent of Pangaea. These 'pigs' had no predators and multiplied to occupy and dominate the world.

After the great continent of Pangaea split up (rifting that was most likely to have been initiated by the eruption of flood basalts) the fossils of *Lystrosaurus* were left on all the continents. When the pieces of the global jigsaw – Africa, the Americas, Eurasia, Australia, Antarctica – were fitted together by disbelieving scientists in the first half of the last century, it was this that helped prove that continental drift really had happened.

Study of these ancient 'pigs' illustrates another, quieter, revolution in Earth sciences, in the way that fossils are studied. They are no longer just collected and classified like stamps, or even used just for the relative dating of rocks, but are studied in much more detail to reconstruct the lives and environments of the organisms. Palaeontology now has a much more detailed, almost forensic dimension to it. Sediments in which the fossils are found are collected and examined; footprints are studied to recreate the gait and speed of their maker. From these studies, *Lystrosaurus* and its close relatives are shown to have features that we would recognize in mammals and they were probably the first group of animals to which the word 'cute' could apply today.

Our understanding of the patterns of past ecosystems is growing year by year, and this is not just of academic interest, or even of interest to those who make money with scary dino-movies – they have many lessons to teach us about our future on this planet. These 'pigs' are a metaphor for our own existence and future, for we are the only other creatures in the history of the world to have achieved such dominance.

The next chapter, 'Killer Earth', examines the possible causes of mass extinctions. From the press and media reports of the last twenty years you would be forgiven for believing that it was an asteroid whodunnit – end of story. But if you examine the evidence, the asteroid idea does not fully explain mass extinctions at all (though it really did happen, and clearly would have made a tremendous impact). It is philosophically far more comfortable for us to cite an external factor for death and destruction on the Earth, and absolve ourselves of any environmental blame today.

But it would be very wrong not to examine other theories of extinction, in particular to question whether climate change alone could have been brought about by the massive outpourings of flood basalts that occurred in India at the same time as the last dinosaurs died.

At the time of writing, however, science is still far from the end of the extinction story. As we go to press, new dating and new data from an Open University team are casting doubt on the extent to which the Indian lavas contributed to the demise of the dinosaurs. There is a big unanswered question in my mind, too – for although many outpourings of flood basalts do coincide throughout Earth's history with mass extinctions, there are some that do not, including the basalts I studied in the Hebrides. There is no evidence whatsoever for any death, destruction or wiping out of species at the time they were emplaced. There is clearly more of the extinction story to be told.

We go on in the next chapter to look at another prehistoric extinct animal, the mammoth. Actually, to say *pre*historic is not strictly true, for it has emerged in the last five years that the last of these magnificent hairy creatures died only 3,700 years ago, which does in fact take us well into the times of recorded history – two-thirds through the history of the Pharaohs as illustrated in the hieroglyphs and tomb paintings of Egypt, in fact. In the scale of geological time, we have missed seeing the mammoth by the blink of an eye.

The chapter mentions the wonder with which children view the mammoth. How true that is. I remember seeing my first one in the Natural History Museum in London at the age of seven, and being so awestruck that I ate the picnic my mother had prepared in absolute silence – not a common state of affairs with me. I don't know whether it is that childhood wonderment in us all, or a kind of collective guilt that it might have been our hunting that killed the mammoth, but there seems to be something inside us that would like to see the great hairy elephant alive again. With modern genetic technology might it be possible to extract frozen mammoth sperm from one of the carcasses preserved in Siberian ice, and to use it to recreate the mammoth?

If we did resurrect the mammoth, there is the question of where we would put it in order to find the 100 kilograms (over 200 pounds) of fresh grass it would need every day. Its habitat is no longer with us, partly because of human expansion into every niche dry land has to offer, and partly because of climate change – the mammoth's real heyday was in the last cold period of the Ice Age.

This brings us to 'Ice Warriors', which as well as going into the science behind the most famous shipping disaster of all time, examines aspects of another revolution in Earth sciences – the study of climatic change. Starting with the iceberg that hit the *Titanic* in 1912, we look into the icy giants, as into a crystal ball, to read messages that give clues to our climatic future. The tale that icebergs have to tell is not just confined to the poles, but to the whole circulation of the planet's oceans and to global climate patterns.

The glib soundbites of news reports suggest that man alone produces carbon dioxide to change climate, that the 'greenhouse effect' is entirely of our making – and that CO_2 from our burning of fossil fuels will raise global temperatures so that we can grow large glossy aubergines outside in the north of England. As the story unfolds, it is clear that this is far too simplistic a view – CO_2 levels have changed quite naturally by themselves in the past, with no human intervention. Other factors have affected climate too, and icebergs have been the heralds of some highly unexpected effects of past global warming.

'Lethal Seas' starts with the story of a legacy from the Ice Age – the Corryvreckan whirlpool off the western coast of Scotland. Ice has carved the mountains of Scotland into their present form, and has also shaped parts of the present sea floor. Near the island of Jura, submerged beneath the waves, is a shallow basin, where it is presumed that ice has gouged out a comparatively soft layer of rock. Nearby is a pinnacle, presumably made of harder rock, that has been more resistant to the scouring of sand-laden ice. We are given the history, folklore and physics of this whirlpool, as well as a tour of whirlpools around the world – by their very nature they are not the easiest of phenomena to study and understand. Like any areas of Earth science an interdisciplinary approach is required, bringing together geology, physics and biology.

'A Sense of Disaster', embraces views far wider than conventional Earth science in investigating whether people, animals and even some plants have the ability to sense an impending earthquake. Reading the account of Californian 'earthquake sensitives', you might be forgiven for feeling that this is para-science, or unproven mumbo-jumbo. But read on and you may be surprised to find a critical and open analysis of phenomena that deserve further investigation. There is a tantalizing glimpse into possible electromagnetic disturbances that might be associated with seismic activity.

When the programme *A Sense of Disaster* was originally broadcast, it was greeted positively by many Earth scientists who recognized that an open-minded approach was required to such observations. The Earth does contain mobile fluid conductors – iron in the outer core and water circulating in the crust – so we should not be surprised to find electromagnetic effects associated with the movement of these fluids.

I hope that this book will lead you to ask more questions about the Earth – particularly the areas about which little is known, such as earthquake prediction. Although there are so many phenomena explained by the great unifying theory of plate tectonics, many things are not yet fully understood, no matter how much some individual scientists may claim to have cracked the problem. Climate change and extinctions are important examples of areas where more research needs to be done, and where the simple answer to their cause is that there is no simple answer.

Half this book is devoted to examining extinction, and questioning whether an internal cause from the Earth itself, such as sulphurous gas from flood basalts, is the smoking gun behind times in Earth history when huge numbers of species disappeared. This book also looks at an overall cooling of our climate in the past few million years culminating in some pretty extreme and quite sudden climatic swings in the past half million years or so. Yet there have been no flood basalts to bring on this change, and no asteroid impact. Some scientists believe we are in the middle of the most traumatic extinction event to date, so we owe it to ourselves to try to understand the workings of the planet better, and to observe what we see around us with an open mind. Only then will we be able to understand what the future holds for our species, and to plan accordingly.

WHEN PIGS RULED THE EARTH

Anna Grayson

Two hundred and fifty million years ago, fifty million years before the start of the Jurassic period – that great age in Earth history when dinosaurs ruled the world – another set of weird and wonderful reptiles held sway. This was the world of the Permian period, when landscapes were a mosaic of lush green forests and orange desert, and when all the continents had joined together to make one super-continent, the vast land mass of Pangaea.

Some of the Permian reptiles were the ancestors of the dinosaurs, and showed glimpses of the giants to come in their bodies and behaviours. Another large group of reptiles was in a way more interesting, for they were the group of animals that gave rise to the mammals, and so eventually to us. These reptiles are termed the mammal-like reptiles and showed tantalizing glimpses of mammalian features inside an essentially reptilian framework. Within this group was a highly successful creature called *Lystrosaurus,* which looked and behaved rather like a modern pig. *Lystrosaurus* was not our direct ancestor, more of a great-great-aunt on a side chain, whose progeny eventually died out with no issue. But, despite it not being a direct forebear, its success illustrates, and to a large extent explains, two of the great phenomena of Earth's history – continental drift and survival after a mass extinction.

Lystrosaurus and its relatives, a whole group of mammal-like reptiles called therapsids, are known from fossils all over the world, and most particularly from the Karoo desert of South Africa where their fossilized remains are found in large numbers, and in many varieties. The British vertebrate palaeontologist Mike Benton of Bristol University has studied them as the first group of animals that lived in complex ecosystems parallel to those we see in the world today: 'There were small ones the size of a mouse, through dog-sized ones, right up to some animals the size of a rhinoceros.' Over in the USA, at Washington State University, Peter Ward is interested in their relationship to us and the lessons we can learn from them; as he says, 'Mammal-like reptiles are our distant ancestors – the precursors to mammals.'

The finding and interpretation of such fossils is the nearest thing science has to a time machine, enabling palaeontologists to revisit and reconstruct an ancient world. But it is more than that – the messages read from fossils of the past can tell stories of how the Earth works, and how animals and plants have responded to natural climate changes and global disasters. So, scientists can read messages and signs, locked within stone, which may tell us something about our future on this planet.

Prehistoric pigs?

Pigs are mammals and, as George Orwell so eloquently pointed out in his novel *Animal Farm*, they are pretty close relations to humans in terms of behaviour and evolutionary heritage. But there is a strange phenomenon in the story of life on Earth: convergent evolution, in which totally separate groups of animals, frequently many millions of years apart, on drastically different evolutionary branches, develop a similar appearance. The most famous example is the resemblance of the great sea reptiles of the Jurassic period, the ichthyosaurs, to modern-day dolphins. The ichthyosaurs were reptiles and the dolphins are mammals. Both adapted, quite independently, to swimming in water (which provided a rich source of food), popping up to the surface periodically to breathe air, taking it into their lungs in the same manner as their land-based relatives. Both modern dolphins and ancient sea reptiles suffer confusion with fish – the word 'ichthyosaur' actually means fish-lizard.

So, in a similar way, the modern pig family has developed a similar shape and way of rooting for food that the essentially reptilian *Lystrosaurus* did 250 million years ago. *Lystrosaurus*'s rump was very porcine, but lacked the curly tail. Its body and legs were in similar proportions to a wild pig, but both males and females had tusks. Instead of a snout *Lystrosaurus* had a beak, made of the same sort of material as the shell of a tortoise, which served the function of a snout and teeth combined – it could browse for low-lying vegetation, and tear it off with the cutting edge of the beak.

While bearing a resemblance to pigs, *Lystrosaurus* was not a true mammal – more a halfway house between reptiles and mammals, living at an evolutionary watershed between the dawn of primitive animals and the present day.

Drifting continents

Anyone who has ever looked at a map of the world will have noticed that Africa and the Americas seem to fit into one another like pieces of a jigsaw puzzle. The first scientist to write about this was a Frenchman, Antonio Snider, who published maps showing a united Africa, Europe and the Americas in a book entitled *La Création et ses mystères dévoilés* in 1858. No one took the slightest notice – the idea was considered absurd.

It was not until 1915 that a (literally) ground-breaking book was published suggesting that there may indeed once have been one large continent that split up and drifted apart. The author of that paper was Alfred Wegener, a meteorologist and astronomer rather than a geologist – and his being an 'outsider' may well have contributed to the extreme hostility encountered by his theory. Yet now, Wegener is considered probably the most significant contributor to the Earth sciences of the twentieth century.

It was not just the shape of the continents that drew Wegener to the conclusion that they had once all been joined together in one large supercontinent. There was fossil evidence too in the form of the seed-fern *Glossopteris* which was found on all continents. Rocks matched up on different continents, in particular a very distinctive rock-type called a till which is deposited by ice sheets. Finding tills that fitted like the pictures printed on jigsaw pieces was very compelling evidence: not only did this add evidence to the fit, but it also suggested that when the continents had been joined, they had been in a different position, near the South Pole. Clearly, it seemed to Wegener, the continents had split up and drifted northwards. Wegener went so far as to name this supercontinent Pangaea (the name we still use today), and to work out that it had split up in two stages – South America from Africa during the age of the dinosaurs; Australia from Antarctica and Europe from North America some time later.

The opposition and hostility to which Wegener was subjected must have been very hard for him. Only a handful of geologists took him seriously, and came up with other explanations for the distribution of fossils, the favourite being land-bridges that had long since disappeared. The fact that there was no evidence for land-bridges, nor any explanation for their disappearance, did not subdue the hostility; neither did the fact that land-bridges failed to explain the distribution of glacial tills in the tropics and equatorial regions. The geologists had closed ranks against him – a stance that remains a permanent scar on the profession.

The big problem for Wegener was his failure to find a mechanism that could move continents. The best he could come up with was some kind of magnetic 'flight from the poles' or a slow nudging formed by oceanic tides. Wegener died during a meteorological expedition to Greenland in 1930 and never knew that his theory would be proved right by a series of courageous and more open-minded geologists.

The first of these was Arthur Holmes, a remarkably astute British geologist working in Edinburgh, who the year before Wegener's death suggested a workable mechanism. Holmes, realizing that the Earth had been cooling down since its creation, suggested that convection currents – carrying heat from the Earth's interior towards the surface, and then (having cooled) sinking back down again – could provide enough force to move mountains.

Then in 1937 a South African geologist, Alexander Logie du Toit, published a work dedicated to Wegener, and entitled *Our Wandering Continents*. He cited many more fossil plants, vertebrates and insects that were common to all the continents Wegener had cited as parts of Pangaea. In particular, du Toit noted occurrences of *Lystrosaurus* all over the globe – in 1934 they were recorded as far away as China. How could a squat, pig-like herbivore have crossed oceans and colonized corners of the globe as far apart as Antarctica and Russia?

The only rational answer was for the continents to have once nestled together so that *Lystrosaurus* could walk from its presumed birthplace of South Africa immediately across to all its other current resting-places.

Of all the fossil evidence, *Lystrosaurus*'s conquering of the continents was the most compelling. Undoubtedly other scientists had to take this evidence seriously. Nonetheless, according to the eminent South African palaeontologist Colin McCrae, writing in his book *Life Etched in Stone*, du Toit's support of Wegener earned him 'much censure and hostile comment'. McCrae describes his diligent and thorough field-working style: 'He would send his donkey wagon on ahead on the few passable roads and then cover the area on foot or by bicycle ... his observations and maps were of an outstanding standard and many of his maps cannot be improved today despite all the modern geological aids available.' Now, of course, du Toit is considered a hero of South African geology, and he was eventually made a Fellow of the Royal Society of London, which was a great honour for someone living and working outside Britain.

Du Toit died in 1948, living long enough to see Arthur Holmes's work on convection currents gain respectability and a body of support after Holmes's publication in 1944 of the famous textbook, *Principles of Physical Geology.* In this book Holmes eloquently argues that the idea of continental drift should be taken seriously. In the final paragraph of this book, however, Holmes inserts a caveat to his proposal of a convection-current mechanism: '... many generations of work may be necessary before the hypothesis can be adequately tested'. How wonderful, then, it must have been for Holmes to find that proof came within his own lifetime – in the 1960s. And how tragic that Wegener and du Toit did not live to see their work vindicated.

In 1963 a young research student, Fred Vine, together with his supervisor, Drummond Matthews, discovered a phenomenon called sea-floor spreading. The idea had been put in Vine's mind a year earlier by an American scientist and ex-naval officer, Harry Hess. During voyages at sea, Hess had put some thought to Holmes's ideas of convection. He wondered whether the mid-ocean ridges – which were just revealing themselves to those who were mapping the ocean floors – were places where Holmes's convection currents rose. This indeed turned out to be the case.

All ocean-floor rocks are the same – black basalt, a volcanic rock that comes up from the Earth's interior as hot molten lava. As it cools and solidifies, magnetic minerals form within it. These 'mini-magnets' retain a record of Earth's magnetic field at the time of cooling. For the Earth's magnetic field is not static – it wanders about, and every so often completely flips over, so that the North Pole becomes a magnetic South Pole, and the South Pole a magnetic North Pole.

Reading magnetic traces recorded by instruments on board ship crossing over the mid-ocean ridges, Fred Vine and Drummond Matthews found a remarkable symmetry. Either side of the ridges were identical 'stripes' of magnetic polarity – matching alternate bands of north–south then south–north polarity. Also the age of the basalts became progressively older as they moved away from the ridges: at the edges of continents the basalts were of the order of 50 million years old, but by the ridges they were real youngsters – erupted yesterday in geological terms.

So what Vine and Matthews had in front of them was the basaltic equivalent of a tape recording – a magnetically coded recording of the Earth's magnetic history. As there were two of these rocky recordings, sitting symmetrically either side of the mid-ocean ridges, getting older further away, Vine and Matthews concluded that the mid-ocean ridges must be the source of these basalts.

What was happening – and is still happening – was that basalt was erupting along the mid-ocean ridges, and then falls to the sides, pushing older basalt to one side. So, to think of an ocean as a whole, it gradually gets wider as more basalt is formed at its middle. As the ocean widens, so the continents on either side of it get pushed further apart – and there you have it, a mechanism for continental drift. Vine and Matthews had proved Wegener's theory to be right all along. The great continent of Pangaea had been a reality. *Lystrosaurus* did indeed walk its way round to colonize the supercontinent.

Continents are not always moving away from each other, of course. The whole point is that they had once moved closer and joined up to form the one massive continent – Pangaea, where *Lystrosaurus* and its friends lived 250 million years ago. Oceans cannot continue to get wider for ever. At the margins of some oceans, such as the Pacific today, the black basalt ocean floor takes a nose-dive back into the Earth – because it has cooled and become heavier. So the ocean floor gets swallowed up and, as it does so, the ocean gets progressively narrower. All round the 'Ring of Fire' surrounding the Pacific, the ocean floor is returning to the bowels of the Earth – the Pacific is shrinking, and in around another 100 million years, the west coast of the Americas will collide with Kamchatka, Japan, the Philippines and Australia to form another supercontinent. Whether we humans, or our evolutionary descendants, will be here to see it is a question to which we shall return later. No doubt *Lystrosaurus* will provide us with some clues and guidance.

The origins of the vertebrates

The evolution of our line, the animals with backbones, goes back a very long time – way, way before the dinosaurs, or even the pig-like *Lystrosaurus*. Our ultimate undisputed ancestor – that is, a fossil with a distinctive backbone that can unquestionably be defined as a vertebrate – is a 550-million-year-old fish found in South China called *Haikouichthys*.

A timescale that encompasses hundreds of millions of years may seem rather meaningless in the timescales of our own rushed lives, which stretch to a mere three-score years and ten or so. It is hard to imagine one million years, let alone the 4,600 million years that have elapsed since the world began. A commonly used aid to coping with geological timescales is to imagine geological time crushed into a year. Thus the world began at 12.01 a.m. on 1 January and we, *Homo sapiens*, arrived to crack a champagne cork just in time for midnight the following 31 December. On this scale, simple, primitive, single-celled life began on Earth in February, and the dinosaurs reigned between 13 and 27 December. Our story of vertebrates began in early November, and our mammal-like reptile ancestors arrived just in time to celebrate the feast of St Nicholas on 6 December.

But even before the first fish (and going back to early November or between 600 and 500 million years ago) there were precursors to vertebrates that are assumed to be ancestral to our line. In Canada, in black shales laid down in the sea around 520 million years ago, is the imprint of a worm-like organism called *Pikaia.* Even older than that is a 2-centimetre (under an inch) Chinese creature called *Cathaymyrus*, which appears to have a head end and a back end and some kind of linear structure between.

Another intriguing 600-million-year-old find was discovered in the 1990s next to a brewery on the Isle of Islay in Scotland – a row of tiny faecal pellets. Faecal pellets could be extruded only by an animal with a mouth and an anus, and some form of gut that moves food down in a series of muscle contractions (peristalsis). We human beings are such animals, as are all vertebrates, so the inference is that the pellets were dropped (underwater, not on land) by some small precursor of the vertebrates.

The pellets demonstrate another palaeontological truism that has major significance much later on in our story about the lives of the mammal-like reptiles: if an animal has left traces of itself behind in the form of faecal pellets, or footprints, that constitutes evidence that the animal actually lived there. Dry fossil bones, however, are evidence of death rather than life, and indeed the environment that forged their host rocks may have been so hostile as to have been responsible for that death.

Nonetheless, dry fossil bones do yield copious amounts of information, and the early fishes are no exception. These fishes were jawless, and in many respects resembled the modern lamprey rather than a cod or mackerel. But the most important thing about them was that they were the blueprint for all vertebrates. Over millions of years of evolution the basic skeletal building blocks of a jawless fish became adapted to make the body parts of creatures such as *Lystrosaurus* and ourselves. So *Lystrosaurus*'s four legs were adapted fins, and the inner mechanism of our ears has been adapted from fish gills. Most intriguing of all is that these early fish had three eyes, and a vestige of that third eye is still with us today as the pineal gland – an organ that functions as a 'body clock' dispensing hormones according to the time of day, or available sunlight.

Early fishy vertebrates lived in the ancient waters for millions and millions of years, evolving into new varieties on the basic design to exploit all sources of nutrition available. They coexisted with many shell creatures that would not seem out of place in the sea today, and also with some weird forms that are now extinct – including the trilobites, woodlouse-like arthropods that reached up to half a metre in length. The vertebrates and arthropods have coexisted on Earth ever since – sometimes happily and sometimes in full-scale competition and battle. So today we have a friendly attitude towards bees and the garden butterflies, but there is mutual hostility with wasps.

Life in the early oceans did not always run smooth. Not only was there competition for food and ecological niches, but evolution was interrupted by episodes of death and apparent destruction, or mass extinctions. There were no fewer than five mass extinctions during the Cambrian period alone – the first 50 million years of early vertebrate history. Yet each time life fought back and filled the seas, and the proto-fish survived. Then, 443 million years ago, at the end of the Ordovician period, 70 per cent of all life in the sea was wiped out – there was no life on land at this point – and there is no clear reason why this should have happened. However, once again life recovered, and new species evolved to fill the voids left by those that had died. The fish in particular bounced back, and Scottish rocks from early in the following era – the Silurian – have yielded a wide range of fish.

Towards the end of the Silurian, about 430 million years ago, plants started to colonize the land. Many people assume that vertebrates were the first to follow plants. But this assumption is not only incorrect – it is a form of arrogance on the part of higher vertebrates to assume that we must always have been best at everything. The first animals to live on the land were very small arthropods – creepy-crawlies that would not look out of place in the grass today.

It was not until another 60 million years later that a fleshy-finned ('lobe-finned') fish started to develop limb-like fins and a stronger backbone, with interlocked vertebrae. (Think of a fish steak compared to a lamb chop – the fishmonger can cut clean between the vertebrae of a fish, but the butcher has to cut through the bone when cutting a lamb into chops.) These fish, which ate small arthropods, had lungs as well as gills, so they could take gulps of air to get oxygen.

Momentous and important though it was, the move of fish to land did not happen in one sudden leap. Per Ahlberg of the Natural History Museum in London can cite a whole series of intermediate forms. But what seems to have happened is that these four-legged fish used their limbs to move through weeds by the shores of lakes and rivers, searching for bugs to eat. Gradually they leaned further out over the shore to reach more appetizing mouthfuls until, after a few million years, they were able to walk on land and breathe air into primitive lungs, returning to the water at will to breed.

These four-legged ex-fish are termed tetrapods, and are the ancestral form of all four-limbed vertebrate creatures such as ourselves, dogs, dinosaurs and the mammal-like reptiles of the Permian.

The amphibious tetrapods survived another mass extinction around 355 million years ago and life moved into the Carboniferous period, which in equatorial regions (where Britain and North America were at the time) was characterized by the deposition of limestones and coals. (This is why the Carboniferous was so named – because of layers of rock containing carbon, in the element form as coal and as carbonate in the calcium carbonate of limestone.) Meanwhile, in the southern hemisphere, Africa, India, South America and Antarctica were clustered around the South Pole in a single continent called Gondwana. They, like Antarctica today, were covered in a deep layer of ice. The amphibians continued to evolve and expand in the pools, watercourses and forests of the united Europe and America.

During the early Carboniferous, the Midland Valley of Scotland in particular seems to have been a centre for amphibian evolution – in a manner that mirrors the evolution of hominids in the African Rift Valley some 350 million years later. The Midland Valley was also a rift valley, and at the time also equatorial. Lush green forests were punctuated by rivers, lakes – and by volcanoes. One very small lake has been preserved near the town of Bathgate, just outside Edinburgh. It had been heated by nearby volcanoes, and its water became steamy and acidic. Any creature that fell in died and sank to the bottom; 348 million years later a professional fossil collector called Stan Wood discovered their petrified remains. He revealed a whole ecosystem of animals and plants hidden between the fine layers of rock, among them one of the most significant fossils ever discovered in Britain.

'Lizzie the Lizard', as she is affectionately known, is no spectacular museum dinosaur. About 16 centimetres (nearly 8 inches) long, she had no spikes, no claws and no scary teeth. Yet on her discovery, museums vied to possess her, and eventually the Royal Museum of Scotland paid £180,000 for her. Her scientific name is *Westlothiana lizziae*, and she is probably not a she – but the specimen has acquired that gender in popular culture because it is a representative of a group of vertebrates that is ancestral to all the dinosaurs, the birds, and to all mammals including ourselves.

Lizzie the Lizard was the first vertebrate to lay eggs on dry land. So that the eggs would not dry up, they would have been surrounded by a membrane, the amnion, which can still be found surrounding hens' eggs today – and which also surrounds our babies in the womb. All the other vertebrates in Lizzie's Carboniferous world returned to the water to lay jelly-eggs, like amphibians such as frogs and newts do today. Lizzie, although very lizard-like and reptilian in appearance, is termed an amniote.

Just what drove her to lay her eggs away from water is a story of how natural environmental change can shift ecological balance and catalyse an evolutionary step.

There is more than one reason, the first and most obvious being to shield them from predators. The Bathgate pool preserved a whole community of Carboniferous life, consisting in the main of a range of amphibious vertebrates and arthropods. The arthropods were not just the kind of harmless little creepy-crawlies you find in your garden – there were scorpions up to a metre (over 3 feet) long, and gigantic millipedes. Arthropods and vertebrates ate each other, and were evolving rapidly, finding new ways to outsmart or outrun the enemy. Although there were plants around, none of these creatures was herbivorous – they all ate each other. Laying eggs away from everyone else was a sure way to protect them from becoming someone's breakfast.

There may also have been climatic pressure to find an alternative to water for egg-laying. The climate was fluctuating and becoming drier – this is shown by a change in vegetation in younger layers of the petrified pool. There was an increase in gymnosperms (the group of plants to which pine trees belong), and a decline in the tree ferns that liked to sit with their roots in water. If there was less water, and fewer pools to choose from for egg-laying, there was an obvious advantage in developing an egg that could be laid on dry land.

Although the Bathgate pool preserves a whole community of life, it was not the same kind of ecosystem we see in the world around us today. Apart from their being no herbivores, there is strong evidence that the atmospheric composition was different, and that levels of oxygen were significantly higher than they are today. Nonetheless, the Bathgate petrified pool is a window on the past at the very moment our vertebrate ancestors won the advantage over the arthropods.

It was not long before Lizzie's progeny evolved and diversified into the kind of ecological patterns we would recognize today, and the pattern was set for new types of vertebrate to evolve.

Diversity and deserts

Once creatures laying amniotic eggs on dry land were established, evolution moved on apace. New streamlined and faster animals evolved – these were the first true reptiles. Remains of such animals have been found in fossilized tree stumps in Nova Scotia, Canada. They are accompanied by fossil millipedes, insects and snails, so the inference is that these lizard-like reptiles either fell into the tree stumps and were trapped, or that they actually lived there, taking advantage of the creepy-crawly food supply.

No actual eggs from these early reptiles have been found – the earliest fossil egg laid on dry land dates from the period just after the Carboniferous, the Permian, and is about 270 million years old. But, although it is egg-like, not all scientists are convinced that it really is an egg. So why, if these early reptiles were busy laying, are rocks not littered with fossil eggs? There are two main reasons: first, early eggs were likely to have had a leathery coating rather than a hard shell; and second, eggs do not fossilize readily. Broken shells do not last long in the environment, so it is only on very rare occasions when unbroken eggs become smothered in sediment that they have any chance of preservation.

There is of course a third possibility – that these creatures gave birth to live young. That seems highly improbable, but it was these creatures that were to produce the line that led to mammals like us as well as to the reptiles and birds. In fact it was not long after the appearance of amniotes, such as Lizzie the Lizard, that the stock divided into two lines – one reptilian, and one essentially mammalian.

But as the Carboniferous drew to a close, the continents were moving northwards and, more significantly, were colliding to form Pangaea.

Pangaea stretched from the South Pole, straddling the equator and into what are now temperate latitudes of the Northern Hemisphere. Antarctica was fractionally north of its present position, with Australia wrapped around its eastern flank. To the north of Antarctica were South America, Africa and India (with the Island of Madagascar wedged between Africa and India). Round about the equator was the junction with North America and Eurasia, which stretched to the north.

Even today, large continental land masses tend to form their own climate systems, with desert conditions in the interior. Pangaea was no exception and, being that much larger and straddling the equator, the effect was exaggerated. Petrified red sand dunes are found in many parts of the globe, dating from the days of Pangaea. By pure fate, the new animals, the reptiles, were ideally built to adapt to these conditions. The chief advantage was being able to lay eggs in their own sealed environment on the driest of land. But reptiles also developed waterproof scaly skin, which does not lose water through evaporation. Most modern reptiles have the ability to produce almost solid urine, and there is every reason to suppose that this is a legacy of Pangaea's deserts.

But by this time, the reptiles had evolved into two distinct and vitally important groups: one gave rise to the dinosaurs, modern reptiles and birds, and the other to us. Scientists classify these two groups by the number of holes in their skulls. Apart from the eye sockets, the reptile/dinosaur/bird group have two holes in the back of the skull, so they are called the diapsids. The mammal ancestor group have just one set of holes in the skull behind the eye socket, and they are called synapsids. These holes are very hard to find in a living creature as they are full of muscles; ours are just behind the eye sockets, inherited from this early division of the reptiles.

These synapsid reptiles were less reptilian than the diapsid group, and they evolved into a group of intermediate creatures, the mammal-like reptiles. It is within this group that *Lystrosaurus* was to evolve.

Dimetrodon was one of these mammal-like reptiles. Every pack of plastic toy dinosaurs seems to contain a *Dimetrodon*, with the dramatic semicircular sail on its back. But *Dimetrodon* was not a dinosaur. It died out well over 50 million years before the first dinosaur was born. It was indeed every bit as exciting as the dinosaurs – wonderfully shaped and part of our mammalian ancestral stock too – but it was nothing to do with the dinosaurs.

Dimetrodon was one of a group of synapsid mammal-like reptiles that dominated the land of Pangaea. The sail on its back is probably representative of its desert way of life – although its habitat, the swampy areas surrounding the Pangaean desert, were very hot during the day, they could be very cold indeed at night with no cloud cover to act as an insulator to keep the heat in. So first thing in the morning *Dimetrodon* would stand with its side facing the Sun, so that the rays would fall on to the sail and heat it up. Blood vessels then carried the warmed blood to the rest of the body, to warm the animal in order to (quite literally) get it moving. Then in the heat of the day, with no shade in the desert, the *Dimetrodon* would stand with its back end facing the Sun, so that heat could radiate away from the sail, to cool it down.

The idea of temperature control in this early mammal-like reptile is supported by marks on the bones of the sail that look to have been left by blood vessels. Many other scientists have suggested that the sail might also have been used for display, to attract a mate. *Dimetrodon* has been depicted with bright colours on the sail, changing colour in the way that many reptiles can change the colour of their skin today, but there is no hard evidence for this. However, comparison with living animals is a vital tool of palaeontology, and so, given that many animals and humans today operate an 'if you've got it, flaunt it' approach to finding a mate, there is every reason for supposing that *Dimetrodon* did the same.

If the primary purpose of the sail was temperature control, that tells us something very important about the rate of transition from reptile synapsid to mammal synapsid. If *Dimetrodon* had been warm-blooded, like a mammal, it would not have needed fancy sails to regulate its temperature. So we can assume that it was still reptilian.

Also, from its teeth, it is clear that *Dimetrodon* was a meat eater. Yet it would have had a problem running fast to chase prey because of the way its legs were arranged. *Dimetrodon* stood with the classic reptilian gait – elbows out, with the upper limb (which had a huge bone to support the weight) at right angles to the length of the body, and then the 'forearms' going straight down to the ground. It is likely that it lay around sunning or cooling itself all day, suddenly pouncing if a meal walked within range. Its jaws were not adapted to chewing, and it could not breathe and eat at the same time, so it had to gulp down great chunks of meat and take snatches of air through its mouth. Clearly evolution had to improve on this design.

Dimetrodon was accompanied by a variety of other synapsid mammal-like reptiles – there were smaller insect-eating creatures, some of which developed a taste for flesh as well. But there were others which for the first time started eating food hitherto untried and untested by the vertebrate line. Some vertebrates became plant eaters – and there were some problems for the first vegetarians. They had to eat a great deal of leaves, stems and seeds to get the kind of calorific value they would have got from meat. Also, being unable to chew their food in the way that modern mammalian herbivores can, it took a very long time to digest – they had to develop huge bellies to contain large quantities of fermenting food. It is perhaps a blessing that the fossil record does not say whether flatulence was a common early-Permian problem.

Some of these first herbivores had tiny heads and stumpy, sticking-out legs to go with their huge bodies. One genus, *Edaphosaurus*, independently developed a sail, like *Demetrodon*, presumably for the same function.

The winners of the early-Permian mammal-like reptile scene were the much smaller insectivores. These animals were to give rise to a new group of mammal-like animals that replaced *Dimetrodon*, *Edaphosaurus* and their peers. The newcomers overcame some of the disadvantages of posture and digestion that had imposed limitations of lifestyle on the larger early-Permian animals. This new group of mammal-like synapsids were called the therapsids. One line of therapsids gave birth to *Lystrosaurus*. Another line, after many millions of years, gave rise to true mammals and eventually to us.

The rise of the therapsid empire

The therapsid empire was global – but nowhere are these animals better preserved than in the Karoo of South Africa. The Karoo is now desert – or as Roger Smith of the South African Museum in Cape Town puts it, 'vast areas of nothingness'. Yet the layers of rock which make up that nothingness are like the pages of a history book: each layer of rock tells a page in the history of the evolution of the mammal-like reptiles and the true mammals. The layers are piled up in a shallow basin, with the older ones at the outside, and progressively younger as Roger Smith and his colleagues drive their Land Rovers towards the interior.

About twenty years ago, Smith was called to a farm not far out of Cape Town, where a farmer had found some footprints preserved on a slab of rock. Alongside the five-toed prints were water-lain sand ripples, lines showing falling water levels, and desiccation cracks where water and mud had dried out completely. The rock shows the movements of real live animals over a mudflat between two floods 260 million years ago. 'The whole sequence must have occurred over a period of about eight weeks,' says Roger Smith, still excited about the find twenty years later. 'That is, from the original flood to the drying up, to the impression of the footprints to the burying again by a subsequent flood. Eight weeks in geological time – a mere instant!' It is no wonder that the site has been declared a national monument.

Even to an untrained eye, these footprints intuitively seem part reptilian and part mammalian. For that is what the therapsids were – quite literally mammal-like reptiles. From measuring the tracks, Roger Smith can see that they were made by an animal with a less sprawling gait than the reptiles, but with legs not as upright as mammals.

Driving further into the interior, to slightly younger rocks, Smith has been working on an area littered with actual skeletons. This is a classic example of footprints being evidence of life, and skeletons being evidence of a fatally hostile environment: 'It appears that animals migrated to this point before they died. That's why there's so many in the area, and the likely scenario is a shrinking waterhole where the animals came to drink but were drought-stricken.'

The skulls of these fossils show important differences from their predecessors, *Dimetrodon* and friends, and from true reptiles. First, it has an arched cheekbone and the jaw joint is placed far back in the skull, giving a much wider bite than true reptiles. This was a result of the shape of the synapsid window in the skull, and in fact the word 'therapsid' means arched beast, referring to the arched shape of the bone surrounding the window in the skull. Second, as Bristol University's Mike Benton has found, their method of eating had improved, and it was no longer a case of tearing off mouthfuls of food and swallowing them whole: 'They had jaws just like our jaws, that they could rotate, and they could chew. They would move them backwards and forwards and side to side and chew in a complex way.' This is a skill unique to the mammalian line – not even the most advanced of the later dinosaurs managed this one – and, in being able to chew, the therapsids developed another distinguishing feature: 'Many of them had differentiated teeth – incisors, longer canines and molars.' The various therapsids could between them cope with every kind of food the Permian world had to offer – plants, meat or insects.

The ears, too, were more developed. In reptiles the ear (evolved from the gill structure in fish) comprises just one bone and an eardrum. In the mammal-like reptiles, bones behind the lower jaw have joined the ear structure. (In true mammals, the ear has a chain of three bones, and the structure is now separate from the jaw.) It is tempting to speculate that a more sophisticated ear might reflect the need to hear noises made by each other as well as listening out for danger. As the evidence is mounting that they were moving between waterholes in groups, a more advanced system of aural communication would certainly have been useful.

Permian paradise – found and lost

Towards the end of the Permian period, a very complex ecosystem had evolved with communities of the various therapsid mammal-like reptiles as the dominant animals on land. Although the landscapes would have looked different, with no flowers or grasses, and the animals would have looked very weird, the niches they occupied and their interrelationships would not seem out of place in Africa today. Instead of lions, there were creatures called Gorgonopsians – named after the three sisters of Greek mythology who were so ugly that the sight of one could turn a man to stone. Gorgonopsians had large toothy heads, with two huge protruding incisors very like those of the sabre-toothed tiger that was to come 250 million years later in the Ice Age. 'They certainly would not have won a beauty competition,' comments Peter Ward: 'imagine a large lion with reptilian characteristics.' This fearsome beast was supported on comparatively long legs and would have been able to trot, if not run, after its prey.

Smaller predators included the – if anything even more ugly – *Titanosuchus*, a dog-sized carnivore with sabre teeth set in an outsized crushing jaw, and *Bauria*, which was more mammal-like in appearance and may have fed on smaller prey, such as very small lizard-like true reptiles that scuttled around the sand and undergrowth. Smaller still were various mammal-like insectivores including the cynodonts, which at this stage were insignificant – but it was very much a case of 'watch this space' for them in the future.

The largest herbivores, living a similar lifestyle to modern hippos, were ungainly creatures called *Moschops* (pronounced 'moss-chops' – a wonderfully graphic name, given a lifestyle that involved chomping poor-quality vegetation). *Moschops* had an extraordinarily thick skull, up to 10 centimetres (4 inches) of bone, which was presumably used in head-butting contests of the sort seen among some herding herbivores today, such as deer and wildebeest.

Another herbivorous group, much smaller than *Moschops*, were the dicynodonts, dog- or pig-sized grazers found in large numbers. Instead of teeth, dicynodonts had a horny sharp beak attached to the front of the face which was used to cut off low-lying leaves. The males had two tusks either side of the beak. Dicynodonts were evolving and a new form was occasionally found. This was *Lystrosaurus*.

There would have been no birdsong in the Permian world, for birds had not evolved, and would not do so for another 100 million years. Instead the skies were dominated by insects – some were giants with wingspans of over 20 centimetres (8 inches) – and by a graceful gliding reptile, *Coelurosaurarus*. This was the first vertebrate to attempt flight and did so by means of a membrane stretched over long extensions of its ribs. The ribs were jointed so that it could fold the 'wings' back while at rest. Fossils of *Coelurosaurarus* have been found all over the world – from Madagascar, Canada and northern England – so it was a successful animal. But its success in the Permian paradise was short-lived, for disaster was around the corner.

The Permian period came to a close very abruptly, geologically speaking, nearly 250 million years ago. Exactly what happened is uncertain and the possibilities are discussed in the next chapter. But as Peter Ward puts it: 'Something horrible happened.' Ninety per cent of all known species – plants and animals – were made extinct. Some groups of animals, such as the ammonites in the sea, and reptiles like *Coelurosaurarus* on land, were wiped out altogether, while other families dwindled down to a few survivors. Such was the case with the mammal-like reptiles: *Moschops* had gone, so had the sabre-toothed predators. *Dicynodon* had gone – all that remained was its relative and descendant *Lystrosaurus* along with some of the much smaller carnivores and insectivores such as *Bauria* and the cynodonts. These latter animals were our direct ancestors and so, as Peter Ward points out, this extinction put the whole mammalian line at risk: 'I think really the most dangerous moment in the history of humanity, strange as that may seem, happened 250 million years ago.'

He explains: 'The greatest mass extinction in the history of the Earth happens. The number of mammal-like reptiles – our ancestors – dwindles to almost nothingness. If the last of them go extinct, no mammals exist. If no mammals exist, then no humans evolve ... perhaps the most dangerous of links in our long history – dangerous in the sense that were this link broken at this time: no humanity.' But, by the skin of our teeth, our line did survive, or perhaps it was our skin and our teeth that helped us survive. An examination of the mammal-like reptiles that did make it, should reveal some clues for our present survival too.

The Lystrosaurus Zone

Many of the best clues come from the rocks that make up the magnificent scenery of the Karoo scrubland in South Africa. As Roger Smith drives further into the Karoo, he passes a point where there are no more coal seams (coal being the product of plants, so the absence of enough plants means no coal is formed). The line marking the extinction is clear. 'The boundary interval is recognizable from a distance as a dark red band of laminated mudrocks. The impression I get is that this interval represents an arid landscape subjected to extended drought.'

A little further on is a layer of rock termed the *Lystrosaurus* zone, so called because fossils of this species are abundant and well preserved. So abundant, in fact, that 90 per cent of the fossils taken out of this zone are of *Lystrosaurus*. Unlike the rocks below, from before the extinction, which yield a whole community, these rocks mark a time when the land was dominated by just one species of animal. It is not just in South Africa that this is the case, for rocks of the same age in China, Russia and Antarctica reveal the same pattern. Only one other time in the whole of geological history has one species been so ubiquitous – our time, and we human beings are dominant in just about every environment on the Earth. Why?

The Karoo Lystrosaurs were first discovered in the 1850s by a Scottish engineer, Andrew Geddes Bain. He described finds of many skeletons together as 'Charnel Houses', and was very puzzled as to what these creatures were and how they died. Bain was followed by another Scot, Robert Broom, who found and named many dozens of South African species, despite the fact that he conducted his field work done up in a tail coat and starched white collar and tie. But the science of palaeontology has moved on and changed in more than dress code. In the early days the priorities were to collect as much as possible, using hammer and pick, and then to sort them into families and name them. It was a job that needed to be done to ascertain what there was, and what there had been, on Earth.

Roger Smith, Mike Benton and Peter Ward are examples of a new breed of palaeontologist that arose in the last decades of the twentieth century. They are as likely to be seen with a broom or a dentist's pick as with a hammer, and their study of fossils is more akin to forensic science than the art of the gentleman-collector. Using a far more detailed and enquiring approach, these modern scientists can put flesh back on the bones, paint landscapes and leaves back into ancient environments and then, most importantly, answer some crucial questions about past events on Earth, to establish likely patterns for the future of our planet. There are several mysteries to be solved about *Lystrosaurus* – more about what it looked like, how it lived, whether warm-blooded and viviparous or cold-blooded and egg-laying, whether it lived in herds, and what it ate. More importantly are the questions of how it survived the Permian extinction event and just how it came to be quite so dominant.

Roger Smith and his colleagues in Cape Town have acquired a new specimen, found 'with eleven sub-adults' in a shallow depression in flood-plain sediment. Smith interprets the find as 'a drought accumulation of herd-living grazers'. While his fossil preparator is starting the painfully slow but rewarding task of moving the rock, bit by bit, from the bones using a compressed air-driven engraving tool, Peter Ward over in Washington has an important appointment at the local hospital.

Ward has been granted an hour on a CAT-scan to examine a *Lystrosaurus* skull. He could crack open the petrified skull, but that would ruin the valuable fossil for ever. The CAT-scan enables images of the bones of the interior of the skull to be examined. All warm-blooded creatures have a series of bones inside their noses called turbinate bones, which help to warm up air as it enters the nasal passages and before it goes down into the lungs. So if the CAT-scan reveals evidence of these bones, that would show that *Lystrosaurus* really was warm-blooded. The evidence is faint: 'It's almost impossible to see the actual bones themselves – they're made of cartilage, and that never preserves,' reports Ward. 'But you can see here a contact point where these turbinates were attached – and that's an indication that it is a warm-blooded animal.' Being able to control its own body temperature would be a distinct advantage in a period of fluctuating climate following an extinction.

So it seems possible that *Lystrosaurus* was warm-blooded. Roger Smith's interpretation of groups of Lystrosaurs wandering to and from waterholes is backed up by a new find from Australia. A slab of footprints from the roof of a coal mine in Australia has trackways which seem to be those of *Lystrosaurus.* (It is significant that the tracks were found in the roof of the mine – that is, above where the coal seams are, and therefore younger, probably just after the extinction event.) The trackways (which are so clear they could have been made yesterday rather than 250 million years ago) are evidence of living animals going about their daily business. They show sets of five-fingered prints walking down towards rocks where a river would have been, and tracks returning to a forested bank, marked by a fossil soil in the rocks. Mathematics can be used to analyse trackways to bring out a remarkable amount of information. These Australian trackways show animals of just under a metre (3 feet or so) from nose to tail, and walking at a leisurely pace (which suggests that there was no threat from predators). With only about 12 centimetres (4½ inches) between left and right prints, the animal must have had a fairly upright gait, and have been an efficient mover.

Beside the *Lystrosaurus* tracks are the marks made by an insect, probably a cockroach, which had survived the mass extinction too. But, most intriguingly, each footprint is surrounded by a halo of fine lines. It is possible that these are wrinkles in a thin algal mat on top of the mud, but it is equally possible that they are hairs on the animal's foot. If this were the case, and the Lystrosaur's body were covered in hair, it would be the most compelling evidence yet that they were warm-blooded and well on the way to becoming true mammals.

So far, no fossil eggs have been found. This could be because they were laid in soft shells, or simply because no one has found them yet. Or it could mean that *Lystrosaurus* was like a true mammal, and gave birth to live young. But no fossilized pregnant female, with the skeletons of young in her belly, has been found either. So the jury is still out on the question of eggs versus live babies.

Roger Smith's examination of *Lystrosaurus* skulls provides more evidence of an animal adapted to living in groups. The ears have the same advances in structure as all the therapsids, suggesting that they may have been tuned to hear alarm calls from its companions as well as any noises made by predators. Smith thinks the alarm calls would have been made by 'expelling air rapidly through the nose'.

Examining the skull and face of the fossil gives more clues as to why the animal was such a survivor. *Lystrosaurus* had no teeth, apart from its two tusks. Instead it cut its food with the beak-like structure at the front of its face. This structure at first glance appears to be broken, but on further examination it is clear that the beak is flexible, and can be used like a pig's snout to grub away at low-lying plants, or even buried tubers which would be highly nutritious and a food source unavailable to any other herbivores. This idea of rooting for food is supported by the fact that in many specimens the tusks are worn. On top of the skull is a shallow crest that would have supported some pretty hefty jaw muscles – suggesting that *Lystrosaurus* was capable of chewing tough plant material.

Another crucial clue to survival comes from the position in which some specimens have been found in the *Lystrosaurus*-zone rocks where Roger Smith has been working: tightly curled-up skeletons at the end of tubular sandstone structures. If these are burrows, this could be the most exciting find and the most telling clue yet: 'These are interpreted as aestivation burrows to overcome the extreme heat of the day. This behaviour would put this animal at an advantage when the temperatures soared as a result of greenhouse gases emitted into the atmosphere from volcanoes or oceanic overturn, or perhaps – as Peter Ward would hypothesize – a meteorite impact.' Whatever the cause of the surge of carbon dioxide and the extinction (which we shall discuss in the next chapter), these curled-up buried bodies must surely be the smoking gun that explains their survival. Their posture adds to the evidence that they were warm-blooded and able to control their body temperature.

All this information is helping Mike Benton to brief a model-maker to reconstruct *Lystrosaurus* and the landscape it shared with other creatures. Scientist and artist (Neil Gorton) can work together to create a remarkable likeness of the live animals. The only stumbling block is Gorton's question about what colours to give to the animals, for in spite of the detailed information that can be read from fossils and footprints, colours do not preserve. The only answer is an educated guess based on animals today – so a dark green-brown back and lighter belly seems a possibility.

The range of animals that lived with *Lystrosaurus* immediately after the Permian extinction was limited – and this provides the final clue to its worldwide dominance. With very few animals living alongside *Lystrosaurus* after the extinction, *Lystrosaurus* had no predators to eat it and no competitors for its food.

The only other plant eater was a small, primitive reptile called *Procolophon*. The skull had nothing of the adaptations of *Lystrosaurus*, and a reptilian gait would make long-distance walking across Pangaea rather slow. There was another true reptile called *Proterosuchus*, which like *Procolophon* had two extra holes in the skull behind the eye sockets and so was a diapsid, or true, reptile. *Proterosuchus* was a meat eater and, being rather like a crocodile in appearance, it is safe to assume that it lived the same kind of lifestyle, lurking in rivers and eating mainly fish. A *Lystrosaurus* would have presented an unmanageable meal. But *Proterosuchus* was an important animal for another reason: this was the animal that was to found the stock that led to all the crocodiles we see today, and – on a separate line – to all the dinosaurs that were to take over 30 million years later.

The only other flesh eaters in *Lystrosaurus*'s world were other therapsids – and even these ate insects or much smaller amphibians and reptiles. These were the cynodonts, which ranged from rat-like to dog-like in appearance. One of their number, *Thrinaxodon*, has been found to have tiny pits in the bone around the nose, and these have been interpreted as attachment points for whiskers. This implies very strongly that *Thrinaxodon* was furry, which is a very clear indication of warm-blood and of mammalian lifestyle. It too has been found curled up in burrows, which may also have been its technique for surviving the great extinction. This little animal, living in the lonely Permian world alongside its therapsid cousin *Lystrosaurus*, seems to be the parent of all the mammals including ourselves.

The lessons of *Lystrosaurus*

Mike Benton, Roger Smith and Peter Ward are all examples of a species that has colonized widely dispersed and very different corners of the globe. On the land, they and *Lystrosaurus* are the only species known to have achieved such widespread dominance. But there is one clear difference – *Lystrosaurus* conquered the globe *after* a major mass extinction, whereas Benton, Smith and Ward are all convinced that we are in the *middle* of a mass extinction now. More than that – the evidence suggests that the radiation of human beings is the *cause* of the current mass extinction. No one could accuse *Lystrosaurus* of actually causing the end-Permian event.

'If you look back through history,' says Mike Benton, 'you can see that humans – we – have wiped out many, many species. Something like two or three species of bird and mammal disappear every year or so – we all know examples like the dodo, the passenger pigeon and the great flightless birds of New Zealand.' Nowhere is this more graphically illustrated, however, than in the island of Madagascar, the small chunk of continent that once lay landlocked in Pangaea between Africa and India. It began to break away 100 million years ago, and to develop its own isolated evolution of animals and plants. A recent survey in Madagascar suggested that in recent years three species of bear, two species of hippo and 125 other animals and birds had become extinct. A hundred years ago, tens of thousands of ploughshare tortoises lived on the island. That number is now down to 200 – and they are far smaller than previous generations.

No asteroid has hit Madagascar, neither has it been swamped by ash and noxious fumes from a volcano. The obvious cause of the extinctions is loss of habitat following deforestation by human beings. Mike Benton has no illusions as to the implications: 'If you calculate from this the rate of loss you might work out that all birds and all mammals will have disappeared within a few thousand years as a result of human activity ... Many people would interpret that to mean that we are definitely in an extinction event as big as any of the big five in the past – and this one is unique because it's being caused by human beings.'

Roger Smith describes the current rate of extinction as being of the order of the end-Permian when at least 90 per cent of all species were wiped out. Peter Ward blames another human factor as well as deforestation: the burning of fossil fuels to increase the global greenhouse effect. 'I think we are in the middle of a tragedy ... I believe that what we are doing to the atmosphere, creating greenhouse gas emissions, is in every way comparable to what happened at the end of the Permian. The end-Permian event is to me, in my mind, a greenhouse event – it is caused by excess carbon dioxide. We, industrialized humans, are doing the exact same thing to the atmosphere.'

The big question is – will we survive the extinction? Will we continue to dominate the Earth, or will we die out to be replaced by another vertebrate such as rats? Or even an invertebrate such as the cockroach? Peter Ward paints a gloomy picture either way: 'Well, if you love humans, as I do, the worst case is that we could go extinct, but I don't think that will happen. I believe humans are the least endangered species on the planet – and I think that's maybe the great curse, that we may end up a thousand years from now like *Lystrosaurus*, in a very empty world.'

So like the Lystrosaurs that made the trackways in Australia, we could find ourselves surviving with only a cockroach for company. Just what might cause such a mass extinction and leave us in that position is the subject of the next chapter.

KILLER EARTH

Anna Grayson

We all love dinosaurs. We all know that they were big and some of them were scary, that they lived millions and millions of years ago and that they were wiped out when an asteroid hit the Earth ... Or so we are told by popular culture and by the media. Dinosaurs did indeed live a long time ago and were the dominant land animals from about 225 million years ago until 65 million years ago. They did indeed go extinct, that much is true. It is also true that around the same time, 65 million years ago, a huge asteroid hit the Earth, landing on the Yucatán peninsula in Mexico. But whether the impact caused the extinction is far from proven.

The huge fiery asteroid crashing to Earth and wiping out the huge *Tyrannosaurus rex*, baring his teeth for the last time as he falls to the ground, is a good story, and one that fits into our dino- and space-loving, disaster-movie culture very well. It is not surprising, in a society that boldly goes where no man has been before in sci-fi movies, and which sees dinosaurs as fictional cartoon figures as well as scientific realities, that this easy explanation and exciting story caught on. So well did it catch on that, just a few years after the asteroid theory was first published in 1980, the then American President, Ronald Reagan, instigated the expensive Star Wars programme. This aimed to destroy any asteroid coming too close for comfort to the Earth, so that we humans might be spared the same fate as the dinosaurs.

Many scientists became uneasy with what was really only a hypothesis being promoted as fact. The only proven fact was that there had been an extinction 65 million years ago. It is known by scientists as the K-T extinction, after the German *Kreide*, meaning chalk, and from Tertiary, which is the geological era following the extinction event. There were many other threads to the story: it was not just the dinosaurs that died – many other species of animals and plants were killed both on land and in the sea. This extinction event was not the only one the Earth had suffered – there have been many more, including the end-Permian extinction (the subject of the previous chapter), which wiped out up to 95 per cent of all species – yet there was no evidence for asteroid impacts then. Nonetheless there are a good many scientists, particularly in America, who still promote the view that it was an asteroid impact that wiped out an estimated 60 per cent of all species, including the dinosaurs, 65 million years ago.

There are other scientists who have never been satisfied with this explanation. One of them is Vincent Courtillot from the Paris Institute of Earth Physics who, with others, discovered that huge quantities of basalt lavas in the Deccan region of India were erupted at the same time as the K-T mass extinction. He suggested that this volcanism was a more plausible cause. A decade of debate followed, in which Courtillot was a key player: 'Many people believe that the debate on what caused mass extinctions is over, but I believe this is far from true.' There are two main philosophies on what might have happened: the first is that nothing spectacular happened, it was just something ingrown in many species that made them survive for a certain amount of time, and then die out, according to the normal ways of evolution. At certain times in history more species just happened to die out, maybe because of changing sea levels or natural climatic fluctuations. Vincent Courtillot is not of this 'gradualist' persuasion and believes it is a minority view.

'The other camp, if you want to use warlike terms, has a catastrophe happening at the time of a mass extinction,' explains Courtillot – and his use of the word 'warlike' is not much of an exaggeration; emotions in Earth science can run rather high. 'Again, what the catastrophe was divides people into two camps. Some people believe it was a meteorite or a comet coming from outside the Earth that hit and changed the climate, destroyed species ...' But this is not Courtillot's view.

'Some people think it was actually from inside the Earth, that enormous volcanism as seen in flood basalts was the cause of it.' When Courtillot says 'some people', he is of course referring to himself, for it is his belief that a cataclysmic series of volcanic eruptions coincided in time with the extinction event of 65 million years ago. These gigantic outpourings of hot black lava are called flood basalts. A growing body of scientists now agrees with Vincent Courtillot that there is a strong causal link between these eruptions and the death of the dinosaurs. It is possible that the asteroid impact may have had an effect on the dinosaurs, but many scientists believe that other extinction events in the Earth's past may have been due to flood-basalt eruptions.

The concept of extinction

You will not find any reference to mass extinctions in Darwin's seminal work on evolution, *The Origin of Species*, published in 1859. Darwin was very much of the view that evolution happened gradually and steadily, and that any apparent gaps or jumps in the fossil record were due to the chance way in which some dead organisms become fossilized and others do not. In fact the first person to recognize that there might have been some form of catastrophe in the past was another Frenchman, Georges Cuvier (1769–1832), who is famous for working out what fossil animals would have looked like by comparing them with similar living animals of today.

Cuvier recognized that animals such as the mammoth had gone extinct, and he went on to argue that there had been several episodes of extinction in the past. Whole populations had been swept away and replaced by populations of new species. Cuvier's work was published in 1812 – the year of the eponymous overture reflecting the Napoleonic Wars. So it was perhaps hardly surprising that France's enemy, the English, ignored Cuvier and developed their own ideas. Darwin was very much influenced by another Englishman, Charles Lyell, who coined one of the most frequently used phrases among geologists: 'The present is the key to the past.'

By this, Lyell meant that what we see around us in the world today can explain everything we see preserved in the rocks of the past. So ever since the dawn of time, Lyell argued, rivers have flowed in the same way, volcanoes have erupted, and mountains have worn down and weathered. This theory was called uniformitarianism, and it fitted perfectly the Darwinian model of continuous gradual change. It was accepted without question by many generations of geology students, and the fact that there had indeed been extinction events on the Earth was effectively masked until the last few decades of the twentieth century. It meant that any major changes from one set of rocks to another were explained by there being 'gaps' in the fossil record.

Various minor papers on the idea of extinctions were published in the 1950s and 1960s, but it was not until the 1980s that the topic became a priority for science, with the emergence of a clear idea of what extinction events really were. Two scientists working at the University of Chicago, David Raup and Jack Sepkoski, drew a graph plotting the number of families of organisms against time. The results were shocking. Over geological time the graph showed five major dips in the numbers of families of organisms in existence at any one time. From this the number of species lost in each of the five events was estimated as shown in the table:

Event	Years ago	Percentage of species lost
End-Ordovician	440 million	85
Late Devonian	367 million	83
End-Permian	250 million	95
End-Triassic	210 million	80
End-Cretaceous (K-T)	65 million	76

Those were just the big five; Sepkoski recognized another twenty-three smaller extinctions, going right back to a few million years after animals and plants as we would recognize them (composed of many cells, rather than single cells) evolved. Raup and Sepkoski even went as far as to suggest that there was a regular pattern in extinctions through the fossil record, in that they seem to come round regularly every 25–28 million years. This sparked off all kinds of theories about their being a 'death star' circling the Solar System – but independent evidence for such a star (or a mechanism for its causing extinction) was never found. Nonetheless, the name 'Nemesis Star' has remained part of folk-science culture.

But it was the end-Cretaceous, K-T event that caught everyone's imagination. This was partly because of the popularity of dinosaurs, but also because in 1980 the asteroid impact theory had been published. Perhaps surprisingly, science has its fashions and, almost overnight, the question of what happened to the dinosaurs was the height of fashion in Earth science circles. The extinction fashion even seemed to overtake the revolution in Earth sciences that followed the discovery of plate tectonics, discussed in the previous chapter.

The origins of the Earth and its structure

In order to understand the history of life, with all its vicissitudes, it is essential to consider the origins of this planet and how forces have acted on it both from outside and from within the planet itself. The Earth and all the other bodies that make up the Solar System, including the Sun itself, were formed at the same time – around 4,500 million years ago. So the Earth is not just a sphere of isolated rock, but part of an interacting family of other bodies circling a moderately sized star.

Just over 4,500 million years ago there was a cloud of dust somewhere near the edge of the Milky Way galaxy. By this stage the Universe was about 1,000 million years old, and the galaxy of stars just over 500 million years old. The dust cloud had been formed from the debris of several stars that had exploded in the vicinity. Exploding stars – or supernovae – are not uncommon, for all stars have limited life: they are formed, they exist, creating energy and new elements by nuclear fission, and then they die. A supernova is just one way a star may die, ejecting dust made up of a whole mix of chemical elements into surrounding space to form a cloud of stardust, a nebula.

The nebula started to swirl round and round and it became flattened and disc-shaped. Gravity pushed matter to the centre where it became hotter and hotter and eventually ignited, triggering nuclear fusion, to form a star – our Sun. The rest of the dust started to clump together, forming mini-planets (planetesimals). The mini-planets collided to form bigger mini-planets and eventually the nine planets we know: Mercury, Venus, Earth, Mars, Jupiter, Saturn, Uranus, Neptune and Pluto. Between Mars and Jupiter is the asteroid belt, which comprises thousands of small bodies that did not manage to accrete into a planet. The asteroid belt is considered to be the source of most meteorites that hit the Earth today. (Although there are also much-hyped meteorites from Mars and a few from the Moon.)

Way beyond the orbit of Pluto (a whole light year beyond) is a region from which most comets appear to come. No one has actually seen this cometary source through a telescope, yet it has been given a name – the Oort

Cloud. Comets from this supposed Oort Cloud whizz by the Earth from time to time, including some famous visitors such as comets Halley (which last visited in 1986) and Hale-Bopp (which was clearly visible in 1997). Comets are composed mainly of ice, with a small core in the middle.

The Earth was not immune from collisions. Most scientists now think that very soon after its formation 4,500 million years ago, the Earth was hit by a planet the size of Mars; this impact knocked a great chunk off – which became the Moon. In the process the Earth became tilted on its axis, so we have the seasons, and the Moon's gravitational pull made tides – both of which have affected the pattern of life on Earth.

Comets too would have collided with the early Earth, and there is a strong body of thought that suggests that much of the water in our oceans may have come from the melted ice of comets that impacted on the Earth's surface. Impacts of comets, meteorites and whole asteroids would have been very frequent until about 3,500 million years ago. Since then they have been less frequent, but nonetheless recurrent, visitors.

Some scientists would go as far as to suggest that life – the life that evolved into the creatures that suffered extinction – was brought to Earth by a comet or by a meteorite from Mars. But there was also much activity within the Earth itself.

Not long after its formation the whole Earth melted and the aggregated stardust became liquid. Gravity caused heavier elements such as iron to sink towards the centre, and lighter substances such as air and water to float to the surface. Eventually the planet settled down to a layered structure with a heavy iron/nickel core in the centre, and a light rocky crust on the outside, all blanketed in water and air.

Between the core and the crust was a vast layer of solid iron- and magnesium-rich silicate rock, the mantle. The top of the mantle and the crust together form a solid, brittle layer, the lithosphere. The mantle itself is layered, with a marked discontinuity layer at a depth of about 700 kilometres (435 miles), where the high pressures change the nature of the material from crystals we would recognize at the surface (such as garnet and a rough form of peridot called olivine) to much more tightly packed, dense structures. The core was, and still is, in two parts – a molten layer on the outside and a central solid core. So the whole Earth is something like an onion in its structure. We live on the brown skin – the outer layer. Occasionally we get glimpses of the upper mantle brought up in volcanoes. The only time we get a glimpse of what the core might look like is from iron meteorites, which are assumed to have once formed the core of another planetary body or asteroid.

Long after its formation, the Earth still continued to out-gas the very light materials such as water and gases in volcanoes. Convection currents conveyed heat from the core, through the mantle to the surface. Indeed, the Earth is still cooling and out-gassing today, and very slow currents carry heat away from the Earth's core.

A group of scientists based at Glasgow University believe that life started around an out-gassing vent deep on an ancient ocean floor. Hot water and other chemicals coming up from inside the Earth mixed with cold sea water, and a chemical reaction took place. Very small 'cells', or small hollow spheres of iron sulphide, were formed. Inside them was the chemical soup needed to make life. Exactly how that chemical soup transformed itself into DNA and cytoplasm remains a mystery, but it is now accepted by many scientists that volcanism played an important part in the start of life, some 4,000 million years ago. Vincent Courtillot is just one of a growing band of scientists who believe that volcanism may also have been responsible for destroying life during the mass extinctions. To examine the evidence either way, we first need to look at the evidence for mass extinction in the rocks.

Evidence for the K-T extinction event

The evidence we know, perhaps only too well, is that dinosaur fossils are found from the end of the Triassic period (around 225 million years ago), on through the Jurassic (about 210 million to 145 million years ago) and right through the Cretaceous (up to 65 million years ago). But they are never found in younger rocks (the Tertiary Era, 65 million to 2 million years ago).

But dinosaurs were not the only lifeforms to die out. On land, many flowering plants disappeared – they had not been around for long, having made their first appearance in the Cretaceous. Many of the larger tree-ferns and club-mosses, which were such an integral part of the dinosaurs' landscape, no longer grew.

There were major changes in the sea, too. If you visit Lyme Regis on the English coast on the borders of Devon and Dorset, you cannot help but see the elegant coiled shells of ammonites on the beach. With them are bullet-shaped fossils called belemnites, which are the remains of squid-like organisms that also lived in the sea. Lyme Regis is also famous for its fossil 'sea-dragons' – the ichthyosaurs and plesiosaurs that were first discovered in the early nineteenth century by the fossil collector Mary Anning. Just a few tens of miles to the east on the Hampshire coast at Barton-on-Sea, where Tertiary rocks are crumbling into the sea, there are no ammonites, belemnites or sea-dragons to be found, just fossil shells that look for all the world like modern clams and whelks. Many creatures that had been abundant in the Jurassic and Cretaceous seas were not found after the K-T boundary.

Norm MacLeod of the Natural History Museum in London feels that it is equally, if not more, telling to look at which organisms survived the K-T extinction – something to which he has devoted a great deal of his work. For example, on land 100 per cent of placental mammals survived (that is, mammals, like us, whose young are fed and receive oxygen in the uterus from the mother's bloodstream via a placenta and umbilical cord), but only 10 per cent of marsupial mammals made it. All frogs and salamanders survived, yet only 30 per cent of lizards survived. In the sea, while the coiled ammonites were snuffed out, the nautilus lived on, and survives much unchanged today. Fish did very well: not only did they survive, but they radiated and thrived. Any hypothesis for the cause of the K-T extinction has to explain the success of the survivors, as well as the extinction of the victims.

Nautilus and fish may have been able to move to deeper water to survive, something the ammonites could not do. Or maybe their breeding patterns were less risky – we just do not know. Placental mammals may have been able to burrow, in the same way as *Lystrosaurus* at the end of the Permian. Mammals would almost certainly have had fur by the Cretaceous. Frogs and salamanders still hibernate under rocks and stones to survive the British winter. All these facts provide clues, but not an explanation. There is another important, but seldom acknowledged, fact about the K-T extinction: one group of dinosaurs did survive – the birds. All birds we see around us today descend from a group of dinosaurs that existed during the Jurassic. Birds are just dinosaurs with wings and feathers, which were clearly a strong evolutionary advantage in surviving beyond the Cretaceous.

Evidence in the USA shows that the first plants to emerge after the extinction were ferns. Plant populations can be studied in the fossil record by microscopic grains of pollen which preserve well. Grains of pollen taken from layers of rock are studied, classified and counted by specialist palaeontologists called palynologists. In rocks just above the K-T boundary in North America they find an abundance of fern pollen – referred to as the fern spike, which is nothing to do with the shape of the ferns themselves but to the sharp and tall spike they find if they plot their results on a graph.

After forest fires, ferns are often the first plants to recolonize burnt and damaged land. This has been taken by many scientists as evidence for a catastrophic disaster at the K-T boundary.

The rocks themselves

The rocks in which fossils are found can themselves yield a great deal of evidence. These are sedimentary rocks – made from consolidated sediment, which may itself comprise grains of older rocks, chemical accretions, or fragments of the hard parts of creatures (limestones). An experienced sedimentologist can use the composition of the rock, and structures found within it, to reconstruct environments of the past.

There are several places in the world where the last rocks of the Cretaceous are overlain directly by the first rocks of the Tertiary. One of these places is about an hour's drive from Tunis in North Africa – a place called El Kef. This is the scientists' yardstick, the world standard for the K-T boundary. Such boundaries are called 'golden spikes' – but sadly, no gleaming golden dagger is hammered into the rock. In fact the El Kef section is now barely visible at all, the land having been ploughed up by a local farmer.

Fortunately there are other places in the world where the boundary can be seen. Typically there are two layers of rock sandwiched together by a layer of clay. In his book *On Methuselah's Trail*, Peter Ward describes a section in northern Spain, with dark reddish-brown finely layered rocks containing ammonites below the boundary, then thicker lighter layers above with no ammonites. Between them is the layer of clay. Differences in the appearance of rocks in a sequence reflect different conditions in the past, and the striking changes in the rocks at Zumaya show that something pretty drastic had happened. (Red is an interesting colour to find in rocks – it is caused by iron oxides, different combinations of iron and oxygen leading to different hues.

The presence of oxides in the lower Cretaceous rocks suggest that there was plenty of oxygen around, yet at the boundary the red colour disappears. This is an important clue – it would suggest that something may have happened to levels of oxygen at the K-T boundary.)

Oxygen plays a big part in the story of another famous K-T boundary site – Stevns Klint in Denmark. Here the lower set of rocks is pure white chalk, and the boundary layer black clay. Walter Alvarez described his first view of Stevns Klint as showing that: 'Something unpleasant had happened to the Danish sea bottom ... the rest of the cliff was white chalk ... full of fossils of all kinds, representing a healthy sea floor teeming with all kinds of life. But the clay bed was black, smelled sulphurous and had no fossils except for fish bones.' This layer is a clear indication of oxygen starvation. 'The healthy sea bottom had turned into a lifeless, stagnant, oxygen-starved graveyard where dead fish slowly rotted.'

The oxygen in the lower, older, set of rocks from K-T boundary sites has another tale to tell if it is analysed in a mass spectrometer. It can act as a palaeo-thermometer, giving information about the temperature of ancient sea water. Oxygen comes in two varieties, O-16 and O-18. The O-18 simply has two extra particles (neutrons) in its nucleus. This makes it slightly heavier than O-16. If you imagine a body of sea water evaporating in the Sun, it would be easier for the lighter O-16 to 'escape' into the atmosphere. So the warmer the water gets, the more O-16 escapes and the more O-18 gets left behind.

So, if the ratio of O-18 to O-16 is measured, the temperature of the water can be estimated. This method has shown that during the end of the Cretaceous there was a gradual cooling around the Earth. But it wasn't exactly a continuous cooling, for within the general trend were wild fluctuations from hot to cold and back again. This climatic instability and general cooling has been verified by

Professor Bob Spicer of the Open University, who has used the shapes of fossil leaves as a climate indicator. (For example, in the modern world, temperate leaves such as beech and hazel have serrated edges, whereas leaves of tropical plants, like the rubber plant, have smooth edges – there would have been parallel differences in the Cretaceous.) During the whole of the age of the dinosaurs there was no permanent ice at the poles, and plants could grow all over the world. Dinosaurs probably migrated to the polar regions in the summer months, while other animals may have hibernated during the months of darkness. Nonetheless, cooling and climate fluctuations would have put pressure on all ecosystems. But what could have caused this climate change? How does a gradual climate change tie in with the idea of a catastrophe? There are no easy answers to these questions, but there is no shortage of theories as to what might have happened 65 million years ago.

The asteroid impact theory

It is rare for a single scientific paper about a hypothesis to shake the world. But the paper in the Journal *Science* in 1980 on a possible asteroid impact as a cause for the K-T extinction did just that. In fact, father and son team Luis and Walter Alvarez had the same impact on the scientific world as their proposed asteroid turned out to have had on the Earth. The most remarkable thing about the paper was that at the time there was no evidence for an impact crater of the right age and size – yet as time went on the Alvarez team were proved right. The key to the success of the work and thought that went into the paper was probably twofold – an interdisciplinary approach (rather than relying purely on geology) and the enquiring mind and remarkable intuition of the Alvarezes.

In fact Walter Alvarez's mother must be given some credit too – for it was she who encouraged his interest in geology. His father Luis was a Nobel Laureate physicist who worked on subatomic particles. It was not until Walter was an adult that father and son started to work together – and it was Walter who inspired his father to apply his particle physics to a geological problem.

In his book *T-Rex and the Crater of Doom* Walter Alvarez describes how he was fascinated by the K-T boundary at Gubbio in Italy, and how the narrow layer of clay between the Cretaceous and Tertiary rocks intrigued him. There are places on Earth where a narrow boundary between two sets of rock can represent a huge chasm of time where little happened, and deposition was slow to the point of being almost non-existent. The influence of Charles Lyell made this 'gap' assumption the obvious one – it allowed enough time to elapse for gradual evolution to have made the changes in fossils from the lower beds to the upper ones. Walter Alvarez was interested in measuring just how much time – and asked his father for advice.

After much discussion, Luis thought of a way to measure the accumulation of meteorite dust that, in theory, should have fallen slowly and steadily on the Earth. He chose to measure the quantities of the element iridium, which is found in meteorites and the Earth's mantle and core but not usually in the Earth's crust. Luis Alvarez was not expecting to find vast amounts of iridium, but tiny quantities in the order of parts per billion – maybe 0.1 ppb. When the results came back they got the shock of their lives: the clay layer at Gubbio contained 90 times that amount – 9ppb. Father and son were astounded.

Scientists always repeat their results to eliminate the possibilities of experimental error or fluke. They decided to sample the black sulphurous fish clay at Stevns Klint in Denmark. Again they obtained the anomalous high level of iridium. A few years previously, a palaeontologist and a physicist (Dale Russell and Wallace Tucker) had suggested that an exploding star or supernova might have killed the dinosaurs.

It had been a series of supernova explosions that had created all the matter of the Solar System in the first place – all the elements that make up the chemists' periodic table and all the matter we see around us. One of the elements created by a supernova is plutonium. But, being radioactive, plutonium decays – after 83 million years half of it has gone. So in the 4,500 million years since the Solar System was formed, all the original plutonium has long since decayed. So if plutonium were to be found in the clay, it would have to have come from a much more recent supernova, such as one that exploded not too far away from the Solar System 65 million years ago. A supernova would also explain the iridium neatly – for iridium is a daughter product of decaying plutonium.

Walter and Luis Alvarez fully expected to find plutonium in the Gubbio clay, and thus prove the earlier supernova hypothesis. But science is full of surprises, twists and turns. There was no plutonium, and no evidence for a K-T supernova. Walter and his father had to find another explanation for the iridium anomaly. After racking their brains they came up with the idea of a giant meteorite, or even a whole asteroid, having impacted with the Earth 65 million years ago.

The problem was that there was no evidence for a crater visible on the Earth. This was odd because a meteorite or asteroid large enough to scatter iridium around the globe would have made an enormous hole. Yet there was no hole to be seen. Impact craters had been recognized on Earth – perhaps the most famous example being Barringer Crater in Arizona which is just over a kilometre (more than half a mile) wide and 50,000 years old. It is probably one of the most photographed landforms in the world, with its clear basin-shape and crisply upturned rim. But that very clarity and crispness of the Barringer crater is because it is only a few tens of thousands of years old – youthful compared to the tens of millions of years that have elapsed since the dinosaurs died. The Earth's surface is active, and subject to erosion, so any crater from 65 million years ago may well have eroded away.

On the other side of the coin, impacts very much more ancient than 65 million years have left substantial traces on the Earth: in Sudbury, Ontario, is a huge crater 200 kilometres (125 miles) wide and an astounding 1,800 million years old. The crater has been pretty much levelled by erosion and squashed by the forces of the Earth's moving crust into an oval shape. But it is there – concentric bands of rock where both Earth-rock and meteorite melted, to form one of the richest sources of nickel in the world. If there is such compelling evidence for an impact crater as ancient as Sudbury, surely one thirty times younger would have left some form of mark?

Despite the lack of firm evidence for a crater, Walter and Luis Alvarez went on to publish their paper, going so far as to calculate the size of the impacting body – 6 kilometres (3.7 miles) in diameter. This courage and vision was remarkable and, even if the impact did not annihilate the dinosaurs, the Alvarez work has to go down in the annals of scientific history as a first-class piece of detective work and accurate prediction. For recent history has proved them right: the Earth was hit by a massive body 65 million years ago. That it caused the extinction, however, is not proven.

The Alvarez team went on in their paper to analyse exactly how the dinosaurs and their peers might have met their deaths. They assumed that the clay layer in Italy and Denmark was formed from the dust that would have circled the Earth after the meteorite collided and broke up. Interestingly, they used for comparison not another known meteorite impact but the dust cloud that had resulted from the explosion of the Indonesian volcano, Krakatoa, in 1883. Before the explosion Krakatoa had been an island, 9 kilometres (5.5 miles) long; after the explosion two-thirds of the island had blown away. Twenty cubic kilometres (4.8 cubic miles) of ash had been shot 5 kilometres (3 miles) into the air where it circled the globe, and caused bright sunsets for two years as far away as London. The atmosphere was cooled for several years, affecting climate worldwide. Tsunamis, great tidal waves up to 40 metres (130 feet) high, killed 36,000 people living in coastal villages in Indonesia.

Scaling the Krakatoa effects up to the kind of dust cloud that a 6-kilometre (3¾-mile) wide asteroid would make as it impacted, Luis Alvarez thought that it would get dark all around the world. With no light, he thought, plants would no longer be able to photosynthesize and they would die. There would be nothing for animals to eat, so they would die too – the whole food chain would collapse. It is this aspect of the Alvarez theory with which some scientists now take issue. But no one can take away from Luis and Walter Alvarez their vision in suggesting that an asteroid hit the Earth 65 million years ago.

The Chicxulub impact crater

After the publication of the 1980 Alvarez paper, more evidence for a huge impact at the K-T boundary arose. A Dutch geologist, Jan Smit, independently confirmed the iridium anomaly. Then, in 1984, Bruce Bohor of the United States Geological Survey found crystals of 'shocked' quartz – tiny fragments of rock crystal containing internal fractures that could only have resulted from having had tremendous sudden pressure applied to them.

In 1985 a team of scientists from the University of Chicago were working on a remote exposure of the K-T boundary clay called Woodside Creek, on the South Island of New Zealand. They were taking samples, hoping to find tiny bubbles of noble gases (gases such as neon and argon) that might have an extraterrestrial origin. One of the team, Wendy Wolbach, took a closer look under the microscope at some black material she had found in the clay. It turned out to be soot. Just as there is no smoke without fire, there is no soot without fire either, and the team investigated further.

In fact the soot turned up at K-T boundary sites all over the world. This suggested that there had been a global conflagration, with fires raging on every continent, at the very time the last dinosaurs died. The Chicago team put forward the idea that the fires were started by material ejected by a gigantic impact falling back to Earth and burning up in the atmosphere. There was no other way of explaining fire storms on such a global scale. It was extremely compelling evidence, and could not be ignored.

Finally in 1991 the impact site was located – Chicxulub in the Yucatán peninsula in Mexico. (In fact the site had been noted ten years earlier, but had not been connected with the K-T boundary.) This seemed to be the impact predicted by the Alvarezes. Geophysical measurements revealed that, buried beneath a kilometre of younger sediments, lay an impact crater consisting of concentric rings 130 kilometres (80 miles) and 195 kilometres (120 miles) wide. Evidence emerged of sediments dropped by gigantic tsunami.

In 1992 one of the Chicago team who had discovered the soot, Ian Gilmour, found diamonds at K-T boundary sites in Mexico and in Montana, USA. Diamond is the hardest known form of the element carbon, and it can be formed only at high temperatures and under the most extreme pressure. So it must have been an almighty smash. A paper in the journal *Nature*, published at the end of 1997, estimated that there must have been an asteroid (or comet) of around 12 kilometres (7.5 miles) diameter (twice the Alvarez estimate). Around 50,000 cubic kilometres (12,000 cubic miles) of material would have been ejected and at least 6.5 billion tonnes of sulphur would have been released into the atmosphere. (The latter phenomenon was not from the impacting body itself, but from calcium sulphate – gypsum – in the rocks that were hit.)

There is now no doubt that a massive asteroid or comet did indeed hit Chicxulub 65 million years ago – it would be a very brave scientist indeed to challenge that. The discovery of the crater was one of the most exciting pieces of science to emerge in the twentieth century. But the question remains: is there any definite proof that the impact actually caused a sudden worldwide mass extinction?

Problems with the impact theory

One of the main problems with blaming the asteroid impact is that not all species died actually at the boundary. Many species of dinosaur had died well before the end of the Cretaceous, and in the sea the ammonites were dwindling. A now extinct form of colonial molluscs called rudists reached their peak of success 7 million years before the end of the Cretaceous. Yet, by 1 million before the end, they had declined to only four or five species. So the fossil record does seem to show a gradual decline as much as a sudden devastation. There is also no really good explanation as to why some creatures, particularly frogs and salamanders, should have survived.

Evidence has recently come to light suggesting that the Alvarez idea of plants being wiped out and the food chain having broken down is flawed. Margaret Collinson of Royal Holloway College, University of London, has studied plants across the K-T boundary in Wyoming. She has found that although plants were not producing many spores immediately after the extinction, they were growing and photosynthesizing. In other words, there was food for animals to eat, so food chains did not break down completely.

But perhaps the biggest flaw in connecting the Chicxulub impact with the K-T extinction is that asteroid or cometary impact cannot be cited for all the other mass extinctions that have taken place on the Earth. There is no evidence whatsoever for impact 250 million years ago when 95 per cent of all species were wiped out, leaving the *Lystrosaurus* of the previous chapter wandering unaccompanied in an empty world. Towards the end of the

Triassic period 80 per cent of species went, leaving niches on the land empty for the newly evolving dinosaurs to fill – yet there is no hint of an impact structure.

It could be argued that the impacts are there but scientists have not found them yet – and it is just a matter of time before they find other buried impact craters like Chicxulub. It could also be argued that being further back in time some other impact craters have been lost by erosion, or have been swallowed back into the Earth's interior by the forces of plate tectonics. Or alternatives could be considered. Palaeontologist Norm MacLeod of the Natural History Museum in London feels that the impact theory puts off the explanation of what is actually seen happening to animals and plants in the fossil record: 'The catastrophic model essentially comes down to a claim that a rock fell out of the sky, changed the environment and everything died, except those things that didn't die. And more and more palaeontologists are becoming aware that this isn't very satisfying as an explanation.'

Many others took Norm MacLeod's view and, while all the excitement and vigorous debate was going on, several scientists were quietly developing alternative theories. One of them was Vincent Courtillot, who had been taking a long hard look at a massive volcanic event that happened in India at exactly the same time, 65 million years ago.

Other extinction theories

Besides Vincent Courtillot's ideas of volcanism for the extinction 65 million years ago, there are many other ideas, some plausible and others just plain daft, which offer an alternative to the Alvarez impact theory. Included in the latter category is the idea that the diet of vegetarian dinosaurs was so coarse and indigestible that it brought about such extreme flatulence as to be fatal. (Methane is a greenhouse gas – so there was in this case a serious message beneath the silliness.) Another daft idea, that the dinosaurs were too big and cumbersome to achieve satisfactory sex, falls down simply because dinosaurs survived for so long and clearly reproduced generation after generation very successfully.

The idea of a 'death star' lurking in the Oort Cloud, which was alleged to come round every 25 to 28 million years, wreaking death and destruction, no longer has any plausibility. Although, paradoxically, scientists working separately in Italy and at the Open University in England have detected signs that there may indeed be a mysterious body such as a giant planet or dead star (brown dwarf) circling the Solar System at those distances.

There are, however, other extinction theories that need to be taken very seriously before we consider Vincent Courtillot's volcanic theory. The main theory concerns changes in sea level that are detailed in the rock record. Towards the end of the Cretaceous there is evidence that sea levels fell, exposing the continental shelves as dry land and decreasing the area of shallow oxygenated water in the oceans. If a slight rise in sea level followed, anoxic water would flow back on to the continental shelves and suffocate organisms living there. This could explain extinctions of plankton and other marine fauna, and the increased area of low-lying land with longer rivers must have benefited the frogs, salamanders, crocodiles and turtles that survived.

It is possible that the sea level change was driven by plate tectonics as the Earth's internal forces pushed Australia away from Antarctica. This would have channelled colder water towards the equator, and may therefore explain the climatic changes of the time. But this does not explain the end of the dinosaurs. Many of them were well used to the cooler conditions of high latitudes, and a fall in sea level would increase the land available to them, rather than taking habitat from them. It is time to explore another explanation – the idea put forward by Vincent Courtillot, and others, that it was a volcanic catastrophe that wiped out the dinosaurs.

Flood basalts

The island of Staffa in the Inner Hebrides is one of Britain's most precious natural monuments, an inspiration to the composer Mendelssohn for his *Hebrides* overture, and now owned by the National Trust. The lower part of the black cliffs of Staffa, into which the famous Fingal's Cave has been etched by the waves, is formed of tall, tightly packed, vertical polygonal columns. The upper part has a rougher appearance, formed from narrower contorted and twisted columns. The whole island is made of one rock type – smooth black basalt. It is a chunk of lava left from major volcanic eruptions that occurred between 62 million and 58 million years ago over north-west Scotland, Northern Ireland and Greenland. That might sound like a large area, but it is modest compared to the volumes of basalt that have been erupted in other areas during the past. The Deccan traps, for example, cover an area of 500,000 square kilometres (193,000 square miles) – or about a third of the Indian subcontinent. There is no volcano on Earth today that could produce anything like that quantity of lava.

The word 'trap' comes from an ancient Scandinavian term meaning 'staircase', because successive sheets of lava lying on top of one another do look like a staircase on the landscape. This effect is particularly visible on the skyline of some Hebridean islands such as Mull, where whole mountains are formed from lava flows, with gentle slopes on one side and jagged stairways, several hundreds of metres high, on the other. These gigantic outpourings of lava are now called 'flood basalts', and many episodes of them have been identified in the past, all over the globe. All of them dwarf the largest active volcano on the Earth today – Hawaii.

As well as being the largest, Hawaii is also the most studied volcano on Earth. Steve Self, a professor at the

University of Hawaii, spends much of his time on the active lava fields of Kilauea – one of two currently active volcanoes on the main island of the Hawaiian chain. This might sound a very dangerous occupation, but these volcanoes are not in the habit of exploding without warning like the huge ash cones around the edge of the Pacific Rim. These basalt volcanoes are quite flat, and regularly extrude rivers of fast-flowing, very runny lava. Often as not the lava is extruded not from a crater as you might imagine, but from a long fissure in the ground. Quite often the lava flows beneath the surface, through lava tubes underneath a surface crust that has formed earlier during the eruption. Steve Self and his colleagues can get very close to holes looking into one of these lava tubes, and point their instruments inside to measure the flow-rate and temperature. A laser thermometer measures the temperature of the lava, which can be as hot as 1200°C, and a radar gun measures speeds, which for the current eruptions at Kilauea are a few kilometres per day.

The ground surface on top of lava flows take two forms, each of which has been given a Polynesian word. First there is *pahoehoe*, which means coiled rope, for the cooled surface looks like a pile of thick ropes. Then there is *aa-aa*, which is sharp, jagged clinker – it is said that the name comes from the gasps of pain one would experience walking over the lava in bare feet. 'If you could go back and wander around many millions of years ago when a flood-basalt province was forming,' says Steve Self, 'I think you'd see a scene that is very similar to what we see today here on Kilauea.'

The difference would be that flood-basalt provinces would have been very much larger, as Self says: 'just immense in extent and thickness with individual flows about the size of England'. But the landscape would have been similar – large expanses of flat basalt surfaces. Older flows would have been colonized by plants and rich soils would have formed – again something you see on the older parts of Hawaii.

The scale of flood basalts is hard to imagine. The nearest thing to a flood-basalt eruption in modern times happened in Iceland in 1973, when cracks and fissures almost a kilometre long opened up and spewed out fountains of lava. Millions of tonnes of gases were poured into the atmosphere too. But, according to most vulcanologists, including Vincent Courtillot, this was tiny by comparison to the flood basalts of the past. 'Picture a crack in the Earth, well, in Iceland, a few hundred metres long – tens of metres of fire fountains – and blow that up by factors of a hundred or more,' says Courtillot, still apparently in awe himself by the immensity of flood-basalt eruptions. 'Then think of fissures that must have been 400 kilometres [250 miles] long – with fire fountains that may have reached maybe hundreds of metres, possibly more than a kilometre ... injecting gases and ash all the way to the upper layers of the atmosphere and then circulating them around the Earth. This is something that really must have been a frightening sight.'

In his book *Evolutionary Catastrophes* Courtillot also cites a much larger eruption in Iceland that took place between June 1783 and March 1784. Gases emitted from the eruption destroyed grassland and crops and led to famine on the island, with the death of livestock and a quarter of the human population. Temperatures in the northern hemisphere were the lowest in more than two centuries. There were fogs and a haze that extended as far as China. Twelve cubic kilometres (about 3 cubic miles) of basalt was extruded in this eruption – yet this is small by flood-basalt standards of the past.

Every day scientists measure the gas given off on Kilauea in Hawaii, and there is estimated to be about 1,000 tonnes of sulphur emitted per day. The Icelandic eruption is estimated to have produced 1.7 million tonnes of sulphur per day. The flood basalts of the past must have produced far more.

Ancient flood basalts have left their mark on a number of the world's landscapes. Steve Self from the University of Hawaii and Steve Reidel from the Pacific Northwest National Laboratory have both studied the Columbia River basalts that erupted over Oregon and Washington State 15 million years ago. More than 200 vast lava flows, some over 200 kilometres (125 miles) long, poured out and flooded 160,000 square kilometres (about 62,000 square miles) of the landscape in less than a million years. Steve Reidel describes conditions during one of these eruptions as 'the closest thing we can know to Hell'. The lava would have inundated everything in its path, and would have continued to flow for many years. It is estimated that just one flow in the Columbia River flood basalts could have produced 12,000 million tonnes of sulphur dioxide. That is about 6,000 times more than the whole of the 1783–4 Icelandic eruption.

The volume of lava extruded in the Deccan traps was a whole order of magnitude larger than the Columbia River – they cover an area three times the size and are about ten times thicker. They must have sent hundreds of millions of tonnes of sulphur dioxide – and other poisonous gases – shooting up into the atmosphere. Hillside after hillside in the Deccan is made up of flow upon flow upon flow. Even the great temple at Ellora, carved directly into two lava flows, is completely dwarfed by the trap landscape. The time gap between two flows is marked by a thin white line passing through the trunks of carved elephants – it is a weak point and many elephants have lost their trunks. But where did this vast outpouring of magma come from?

The formation of flood basalts

You may recall that the Alvarez team used the explosion of Krakatoa as a mini-analogue for what might have happened with an explosion following an asteroid impact. In living memory we have seen great volcanic explosions like Mount St Helens (1980) and Pinatubo (1991), with their gigantic and spectacular plumes of ash. On television pictures they look so much more poisonous and dangerous than the rivers of glowing Hawaiian basalt. Several eminent geologists have been killed by ash explosions – so why is it that basalts are invoked as a cause of extinction?

Explosive volcanoes have a very different structure, chemistry, source and eruption pattern. They are formed in places where slabs of the Earth's crust are colliding, where one slab of crust is swallowed up and sinks down into the Earth's interior under another slab. So their starting material is made of crustal rocks (very wet crustal rocks, usually, as these slabs of rock sink around the edges of oceans, such as the Pacific Rim). Part of the lower crustal material melts, steam and gases build up pressure, and eventually a whole mountain or oceanic island literally blows its top. Ash is indeed blown everywhere, very visibly. There is a great deal of gas. But it only happens once in a while. After a few months – a few days, even – it all goes quiet again and the ash and dust disperse.

With large basalt volcanoes, however, the eruptions can go on relentlessly for years. They have a very different source – for basalt is formed way down in the Earth's mantle. For several decades it has been possible to track the lava on Hawaii coming up from a depth of 100 kilometres (60 miles) – but recent advances in geophysics have allowed a new view into the Earth's interior, and have revealed a source for basalts, far deeper than anyone had imagined.

It is all down to the fact that the Earth is still cooling and losing heat from deep in its interior. New techniques in geophysics, akin to the various methods of medical imaging where doctors can 'see' inside the body, are giving some interesting pictures of the Earth's interior. The origin of flood basalts is now believed to be as deep as the core–mantle boundary, 3,000 kilometres (1,860 miles) inside the Earth.

Although the mantle is solid, it can convect, very slowly – at about the speed your fingernails grow. That is, it can carry hot material away from the centre to the crust and thus transfer heat energy away from the centre to the crust. A rising current of hot material in the mantle is termed a plume or a hot spot (technically a hot spot is where the plume hits the Earth's crust). Vulcanologist Steve Sparks likens the process to that wonderful kitsch icon, the lava lamp: 'You have two oils in a lamp, and you heat one oil up at the bottom and it gets a little lighter and it forms a blob that rises up through the other oil. That seems to be what happens with flood basalts.'

If you watch the initial heating of a lava-lamp, you would notice that the rising plume of oil has a big blob at the top, and a long narrow tail of smaller elongate blobs behind. As it reaches the top of the lamp the initial large blob spreads out somewhat. The same thing happens with the rising plumes in the mantle. As the head of the plume reaches the base of the lithosphere (the rigid layer comprising the crust and the very top of the mantle), it spreads out to the side, and pushes on the lithosphere causing it to dome upwards.

Also, as the plume rises, the pressure becomes less, and eventually the hot material starts to melt. A film of molten material forms around the solid crystals of rising rock. Molten films join up and very quickly there is a flow of material to the surface. This is molten basalt. Once the melt is in contact with the floor of the lithosphere it will utilize any weaknesses to spurt onwards and upwards to the surface. The result is flood-basalt eruptions – on a massive scale.

But the lithosphere plays another part in the flood-basalt story. The lithosphere is on the move, and great slabs of rock (plates) move from the ocean ridges to the ocean rims, shunting the continents around. This means that a plate of lithosphere can move over a stationary hot spot, creating a line of basaltic eruptions. Nowhere is this clearer than in Hawaii, where a string of islands has been formed as the Pacific plate moves north-westwards carrying older edifices of basalt with it to form the string of Hawaiian islands. Although what we are seeing now on Hawaii is not the big fat head of the plume, but literally the tail end – which is why Hawaii, despite its size, represents a mere pimple when compared to ancient flood basalts such as Columbia River and the Deccan traps.

The Deccan traps clearly represent the head of a plume that brought forth flood basalts 65 million years ago. At that time, India was further south, where the island of Réunion is now. The current volcanism on Réunion is caused by the dying tail of that plume. At the time India was attached to the Seychelles, but the rising plume not only engendered the basalts, it caused the crust to crack and split, so Madagascar and India parted. Over the 65 million years that have elapsed, India has migrated northwards to collide with Asia and form the Himalayas. In its trail are islands – the Maldives and Mauritius, all formed from the tail of the great mantle plume.

The Deccan traps – a smoking gun for the K-T?

As we have seen, the Deccan traps cover an area a third the size of India. Their formation resulted in a continental parting of India from what is now a submerged piece of continent surrounding the Seychelles, as well as a doming effect on the crust. That in itself would have contributed to the changes in sea level that have been recorded in late-Cretaceous sedimentary rocks around the world. But there is also a very powerful, almost literal, smoking gun in the form of the gases that would have been given off. Steve Self describes the effect these gases would have had on the atmosphere: 'Gases like sulphur dioxide, chlorine and fluorine ... All these can combine with water in the atmosphere to form various acids.'

It is the way the sulphur dioxide combines with water that provides the main mechanism for extinction: 'Sulphuric acid stays in the atmosphere to form little round droplets or aerosols – and these droplets are what interfere with the incoming radiation from the Sun. They both absorb it and back-scatter it. So if there are a lot of aerosols in the atmosphere after an eruption, then less of the Sun's radiation reaches the Earth's surface, and you generally have a cooling of the Earth's surface and thus change climate and alter weather patterns.' So Steve Self can explain the cooling of climate recorded in Cretaceous rocks by the two isotopes of oxygen.

But what about the fluctuations observed within that cooling? Steve Sparks of Bristol University blames gases too: 'We also get carbon dioxide – a greenhouse gas. So if you put huge amounts of carbon dioxide into the atmosphere over a few years then you might have warming effects. So, you can get both cooling and warming.' This would put huge strains on ecosystems and life in general. It is the key to Vincent Courtillot's thesis of the Deccan traps and the K-T extinction: 'Through this cycle in years, decades, centuries you would completely harm the environment, destroy plants, generate fires, generate winter – global winter; and then animals that feed on plants would die and then animals that feed on animals would die and you break up the entire chain.' Adds Courtillot, 'Then think that you get another flow and another one, and another one. The flows of flood-basalt hit and hit and hit again over hundreds and thousands of years. So you can imagine that by adding up these effects you can in some cases reach a critical stage in which you really generate not just a small harmful event, but a mass extinction.' For in the sea, too, sulphur would 'poison' the top layers of the water and cause anoxia (oxygen shortages) of the sort found in the sulphurous K-T clay.

The repetitive nature of flood-basalt flows over a period of time may explain why some species died out before the

K-T layer itself. For although Vincent Courtillot and colleagues estimated the Deccan traps to have been formed in only a million years or so, new dates calculated by Mike Widdowson at the Open University (using the relatively new argon-argon method of dating) suggest they may have taken as long as 5 million years to form.

The K-T boundary itself is found within the Deccan traps as a clay horizon between lava flows, complete with the anomalous iridium. Below that layer, dinosaur nests have been found in fossil soils between flows. Above that layer, no dinosaurs. Surely, Vincent Courtillot must accept that the asteroid cannot have done those dinosaurs much good? 'I believe we have field evidence that the two happened,' he states, quite categorically. 'The big question is how much did each one contribute to whatever happened? I would say off the top of my head, volcanism was probably responsible for something like two-thirds of the extinctions ... The additional stress put on the environment by the impact may have pushed overboard another third of them.'

So Vincent Courtillot accepts a possible double-whammy for the K-T extinction 65 million years ago – but the K-T is just one of many extinctions in the long history of life on Earth.

A pattern of extinctions

Perhaps the most compelling evidence for the Deccan traps having played a major role in the K-T extinction is the fact that it has now been found by state-of-the-art radiometric dating techniques that many other flood basalts coincide with extinctions – both large and small. According to Vincent Courtillot, eight out of ten episodes of flood basalts are associated with an extinction. This is a marked contrast to the impact theory where only one impact coincides conclusively with an extinction.

Most excitingly, there is a flood-basalt that coincides exactly with the largest extinction of all – the end-Permian, discussed in the previous chapter. These are the Siberian traps that form the vast plateaux that make the landscape of that part of Russia so monotonous. They were extruded 250 million years ago.

Very recently a correlation has been found between an extinction event at the end of the Triassic (about 210 million years ago) and flood basalts in West Africa and America. The original size of these basalts had not been appreciated until now – they had been erupted on to the great continent of Pangaea, which split up (probably as a result of these basalts) and, in their fragmented and eroded form, their significance was easy to miss.

The regularity of flood basalts coinciding with minor extinctions is beginning to recall to some scientists the 25- to 28-million-year pattern noticed by Raup and Sepkoski. Applying mathematics to the natural world is making a few people wonder if there is not some form of cyclicity or chaotic behaviour (in the mathematical sense) that governs heat loss deep in the interior of the Earth – and what happens at the base of the mantle could therefore dictate the pattern of life at the surface.

A catastrophe for Darwin's evolutionary model?

The Triassic extinction of 210 million years ago was the one that cleared the way for dinosaurs to radiate and take over niches left by extinct animals. In other words, without the Triassic extinction the dinosaurs would never have flourished as they did. Scientists such as Norm MacLeod, Steve Sparks and Vincent Courtillot are keen to emphasize this positive side that extinction has to play in creating opportunities for new species to evolve. If it were not for the radiation of mammal-like reptiles after the Permian, and for the death of the dinosaurs that allowed true mammals to radiate and evolve, I would not be writing this and you would not be reading it.

For Vincent Courtillot, the concept of extinction and renewal must surely be a kind of vindication for his fellow countryman, Georges Cuvier, who essentially founded the school of catastrophism. Is Lyell and Darwin's idea of gradualism dead? Courtillot is something of a diplomat: 'It seems to me that Darwin is right most of the time and wrong at some key times in the Earth's history.'

The Darwinian way sees evolution as an interaction between species and environments leading to the survival of the fittest. 'Most of the time evolution proceeds in the normal Darwinian way,' says Courtillot. 'But in the times of those short catastrophes when either an asteroid hits or a flood-basalt is emplaced, conditions are changing so much that you cannot say that animals were not adapted – they could not be adapted to something that almost never happens.' Courtillot suggests a slight alteration to Darwin's most famous phrase: 'At those times that completely reorientate the course of evolution you should rather speak of the survival of the *luckiest*!'

One day, a new giant plume will rise up from the bowels of the Earth, and pour flood basalts out on to the surface. It may not happen for millions of years, but as the Earth is still cooling, it certainly will happen one day. When it does, whatever life forms are living on our planet – ourselves or our progeny – will face the prospect of extinction. Once again, it may be a case of the survival of the luckiest.

RESURRECTING THE MAMMOTH

Douglas Palmer

The mammoth is one of the most intriguing and iconic of the extinct 'monsters' of prehistory. It is perhaps no accident that the word 'mammoth' is still entrenched in many languages as a word for awesome hugeness, despite the fact that many dinosaurs were vastly bigger than mammoths. But then our relationship with the mammoth has a more intimate and complex basis than our relationship with the dinosaurs we have encountered only as fossils. Our ancestors lived alongside mammoths; they painted and carved beautiful images of them on cave walls, pieces of bone, ivory and stone. Perhaps it is because we empathize so much with elephants and their renowned intelligence, sociability and strength that we can also relate to their ancient mammoth relatives.

I have personal experience of the power that the image of the mammoth still has over people. In 1991 I organized a major exhibition on 'Mammoths and the Ice Age' at the National Museum of Wales in Cardiff. The exhibition included a full-size reconstruction of an Ice Age scene showing a mother mammoth protecting its calf from wolves. The cow and wolves were robotic and their movements were synchronized with sound and lighting to create a very dramatic effect. The power of the mammoth was so great that whole parties of raucous schoolchildren were reduced to silence when they came to it. And, after some tearful scenes, we had to put up a notice warning parents that young children might be frightened by what they saw. In the eighteen months that the exhibition was on, some 300,000 people came to see it – which is about the same level of attendance attracted by dinosaur exhibitions.

These awesome hairy relatives of living elephants roamed the northern hemisphere in vast numbers during the Ice Age between 250,000 and 10,000 years ago, when most, but not all, of them became extinct. The last mammoths finally died out a mere 3,700 years ago, stranded on islands in the Arctic Ocean. This was long after the construction of Ancient Egyptian pyramids, with just 350 years to go before Tutankhamun's brief reign as pharaoh.

Did our ancestors have a hand in the extinction of the mammoths, just as we, their descendants today, are threatening the survival of some populations of elephants? A number of scientists think that modern biotechnology and our romantic nostalgia for the past can be combined in an effort to resurrect the extinct mammoth.

What is a mammoth?

The name 'mammoth' is thought to originate from the word *mammut* in a number of ancient northern Eurasian languages. In Estonian the word *maa* means earth and *mutt* means mole. The *maamutt* was originally seen as an earth-mole because its remains were found buried in the ground and people thought that it lived there. Not until the eighteenth century did the anglicized version 'mammoth' become associated with the notion of enormity.

Giant bones

Fossil bones, teeth and occasionally tusks of elephant-like animals have been found scattered over a vast area of the northern hemisphere for many hundreds of years. Their occurrence was puzzling to the scholars of the time and often required fanciful explanations. An ancient English chronicler, Ralph of Coggeshall, recounts how in 1171 a river bank collapsed, revealing the bones of a 'man' who 'must have been fifty feet high'. Similarly, a huge thighbone found in 1443 by workmen digging the foundation of St Stephen's Cathedral in Vienna was thought to be that of a giant. The bone was chained to a cathedral door, which became known as the Giant's Door, but it was almost certainly the leg bone of a mammoth.

In the last 300 years or so, similar finds of large bones have been recognized for more or less what they are – the remains of some sort of elephant. However, the implied occurrence of such beasts so far north was difficult to account for. Scholars searched historic records in an effort to trace their origin. They found that the Carthaginian general Hannibal had brought elephants to Europe as part of his armoured attack on Rome in 218 BC, and that the Roman Emperor Claudius also used the big beasts in his invasion of Britain in AD 43. So there was some historic precedent for their occurrence in Europe but the explanations were not particularly satisfactory, especially as the bones were often found buried deep in the ground.

The other most important historical source of mammoth remains was the frozen tundra of Siberia. Chinese merchants began buying ivory from Siberia over 2,000 years ago. The tribal hunters told the Chinese that the tusks belonged to giant rats or mole-like animals, which used their tusks to tunnel through the rock-hard, frozen ground. The same explanation for the mammoth remains was still being given to inquisitive foreigners at the end of the seventeenth century.

Evert Ysbrant Ides was a Dutch diplomat working for the Russian Czar, Peter the Great. On his way to China in 1692, he asked hunters of the Yakut, Ostiak and Tungus tribes about the origin of the beautiful white Siberian ivory. Ides noted in his journal how they spoke of the 'mammut' which lived in tunnels beneath the ground and how 'if this animal comes near the surface of the frozen earth so as to smell or discern the air, he immediately dies. This is the reason that they are found dead on the high banks of the rivers, where they accidentally come out of the ground.'

Ides also recounted how mammut legs and tongues could be found especially on the banks of the great rivers that drained into the Arctic Ocean, such as the River Lena. 'In spring when the ice of this river breaks, it is driven in such vast quantities, and with such force by high swollen waters, that it frequently carries very high banks before it, and breaks the tops off hills, which, falling down, reveal these animals whole, or their teeth only, almost frozen to the earth.' Furthermore, the 'teeth' (that is, tusks) are 'placed before the mouth as those of the elephants are'.

Mammoth bones and corpses have been occasionally exposed in exactly the same way for centuries, and the Siberian hunters have searched the river banks for their tusks. Huge quantities have been harvested for sale. Records show that throughout the nineteenth century an average of 50,000 pounds (about 23,000 kilograms) of ivory were traded each year in the Siberian town of Yakutsk. Even in 1872, 1,630 mammoth tusks were sold by London ivory dealers and the following year another 1,140 tusks were up for auction.

At least those 'elephants' were not being killed by poachers although, as we shall see, there is a question over whether or not humans had a hand in their extinction. The sheer volume tells us something about the ubiquity of the beast in Eurasia during the Ice Age. There was no way that all these remains could be attributed to Hannibal's few elephantine chargers, nor could the discovery of mammoths in North America. As the New World was being opened up at the end of the eighteenth century, the discovery of mammoths in America added a new dimension to our understanding of the great beasts.

The North American mammoth

During the eighteenth century a prolific source of fossil bones was found at a place called Big Bone Lick in Kentucky. In 1765 George Croghan, a wealthy Irish trader whose hobby was collecting fossils, noted in his diary that he had found elephant bones 'in vast quantities ... five or six feet underground ... and two tusks above six feet long'. He sent some of his specimens to the American diplomat and scientist Benjamin Franklin, who was living in London at the time.

Franklin's interest was aroused and on 5 August 1767 he wrote to Croghan, '... many thanks for the box of elephants' tusks and grinders. They are extremely curious on many accounts; no living elephants have been seen in any part of America by any of the Europeans settled there ... The tusks agree with those of the African and Asiatic elephant in being nearly of the same form and texture ... but the grinders differ, being full of knobs, like the grinders of a carnivorous animal; when those of the elephant, who eats only vegetables, are almost smooth. But then we know of no other animal with tusks like an elephant, to whom such grinders might belong.'

Franklin went on to point out the different geographical and climatic distribution of living elephants compared with the fossil forms. He explained the discrepancies 'as if the earth had anciently been in another position, and the climates differently placed from what they are at present'. Franklin was an acute observer and had picked up an important distinction in the teeth of the North American fossil elephants. They were indeed different from those of both living elephants and the Eurasian mammoths.

Croghan had also sent specimens to an English scientist, Peter Collinson, who lectured on the problem of their identity to the Royal Society in London on 10 December 1767. Collinson explained the peculiar distribution of the fossil elephants in the cold regions of North America,

Europe and Asia as being due to the torrential currents of the Noachian Deluge – the biblical Flood – which had swept them north from their normal habitats. And he concluded that the unusual features of the North American teeth showed that they 'belong to another species of elephant, not yet known'.

The catastrophic action of the Deluge, as described in the Old Testament, had long been the generally accepted explanation for the occurrence of fossilized remains of life being trapped in rock strata. For some people with fundamentalist views it is still a viable explanation but, by the middle of the nineteenth century, it had become untenable for an increasing number of scientists. Also, there was the associated problem of extinction. How could a benevolent God allow any of his creations to fail and become extinct? In the eighteenth century, with so much land and ocean still unexplored, there was still the possibility of apparently extinct creatures lurking somewhere in the far reaches of the Earth.

Another eminent American, Thomas Jefferson, who like Franklin was a man of many parts, was greatly interested in the problem of the fossil elephants. He recognized a distinction between elephants and mammoths but did not realize that the 'knobbly' teeth of the North American fossil elephant belonged to yet another kind of elephant-related animal. Jefferson thought that the American fossils belonged to the same kind of elephant as the Siberian fossils, namely the mammoth. He also thought it possible that living mammoths would be found in the unexplored forests and mountains of the great North American continent. Indeed, when he was President, Jefferson instructed Meriwether Lewis and William Clark to search for live mammoths while on their famous expedition to the interior from 1804 to 1806. Jefferson insisted that 'in the present interior of our continent there is surely space enough' for such huge creatures. Unfortunately he was wrong – the last North American relatives of mammoths had died out some 10,000 years before he gave the instruction.

Naming the beast

Meanwhile, one of Germany's greatest naturalists and experts on fossils in the eighteenth century, Johann Friedrich Blumenbach, of the University of Göttingen in Germany, had made a particular study of the fossil elephants of Europe. In 1799 he published the results of his study, writing that the European bones were certainly those of an elephant but were sufficiently different from those of the African and Asian elephant to warrant being named as a new species: *Elephas primigenius*, meaning 'the first born of the elephants'.

At the same time Georges Cuvier, (the even more famous French scientist whose work we have already come across in 'Killer Earth') was considering the problem of extinction and the identity of these fossil elephants. Cuvier subscribed to the catastrophic biblical Flood story as an explanation for the occurrence of a variety of fossil animals he was finding buried in sedimentary rock strata around Paris. But he also realized that there had to have been more than one flood to distribute the fossils at various levels within the rock strata.

Cuvier paid particular attention to the form of the jaws and cheek teeth of the elephant. From the differing jaw form and patterns of ridges on the grinding surfaces, he showed that the African and Indian elephants were actually sufficiently different to necessitate being placed in separate genera and species, namely *Loxodonta africana* and *Elephas maximus*. Then he showed that the fossil elephants were in turn different enough from either of the living species to be a separate and extinct species, for which he accepted Blumenbach's name of *Elephas primigenius*. Cuvier went on to distinguish the North American fossil elephant with 'knobbly' teeth as a separate genus which he called *Mastodon*, meaning 'breast-shaped tooth'.

A body in the Siberian freezer

Also in 1799 Ossip Shumakov, chief of the Siberian Tungus people, found a curious body embedded in the icy banks of the Lena River, while he was searching for mammoth tusks. When he returned to the spot in 1803 he found that the icy ground had melted away, leaving a huge mammoth exposed on the river bank. He told an ivory dealer, Roman Boltunov, of his find and in 1804 they removed the beast's tusks and Boltunov made a rough sketch of the animal. Another two years passed before a Russian scientist, Mikhail Ivanovich Adams, who was passing through Yakutsk, heard about the mammoth and saw Boltunov's drawing.

Adams set out to recover what he could of the beast. By the time he got there much of the flesh had been scavenged by wolves and foxes. However, the skull was still covered in skin, and one eye and an ear remained. There was still some skin and flesh under the animal, which Adams removed along with about 16 kilograms (37 pounds) of long reddish hair. Adams recovered all the skeleton and as much of the skin as he could. In addition he managed to buy the spectacular and beautifully preserved tusks from Boltunov and sent all the remains back to the St Petersburg Academy of Sciences, where he taught botany.

The skeleton of the magnificent bull mammoth was mounted in the Academy's museum of zoology, along with the remaining head skin from around the eye and ear. As a result, the identity and characteristics of the Siberian mammoths became quite clear. The mounted skeleton and bits of skin are still on display there, nearly 200 years later.

Mammoth data

Strangely, despite the association of the name with enormous size, the mammoth was no bigger than an African elephant.

Mammoths were between 2.7 and 3.4 metres (9 and 11 feet) high, weighed 4 to 6 tonnes and generally looked rather like living elephants. There were some important differences but only a few of these showed up in the skeletal remains. Their forelegs were longer than their hindlegs so that the back sloped down towards the rear. It would have been difficult to ride the back of a mammoth – living elephants have much more horizontal backbones. Mammoth tusks were often spectacularly long and curved; indeed a tusk recovered from the Kolyma river in Siberia measures 4.2 metres (13½ feet) along the curve from tip to root and weighs 84 kilograms (185 pounds). By comparison the largest African elephant tusks rarely reach more than 3 metres (10 feet) in length or weigh more than 60 kilograms (130 pounds). Female mammoths, like female elephants, were generally smaller than the males and had smaller and lighter tusks.

The most obvious differences between elephants and mammoths were realized fully only when soft tissue had been recovered from the frozen mammoth corpses of Siberia and Alaska. Most striking of mammoth characteristics was their long hairy coat, which consisted of a coarse outer layer of hair that in places was up to a metre (3 feet) long. Below was a thick and shorter woolly underlayer, about 2.5 to 8 centimetres (1 to 5 inches) long. The overall appearance of the coat would have been like that of the living musk ox, which is now restricted to Arctic Canada. Living elephants are born with a covering of hair over much of the body but they soon lose most of it.

Mammoth hair is clearly an adaptation for body insulation and is accompanied by several other features that helped protect the animals from the cold. Below the skin lay a fat layer 8 to 10 centimetres (3 to 4 inches) thick, which is not seen in elephants but is similar to that found in marine coldwater mammals. The mammoth had small ears, which were only about 38 centimetres (15 inches) long, about one-fifteenth the size of the African elephant's ear. The tail was also shorter, with between seven and twelve fewer vertebrae. Also, mammoths had a distinct fatty hump on top of their shoulders and a curious topknot of fat and long hair right on top of the skull. Finally, one other external distinction was the tip of the trunk, which in mammoths had two long prehensile, finger-like projections used for fairly delicate manipulation or selection of plant material for eating. By comparison, the living elephants have only one short projection at the end of the trunk, and tend to curl the whole end of the trunk when plucking plant food.

When the first cave paintings and engravings of the animals of the Ice Age 'game park' were discovered in the nineteenth century, especially in south-west France and northern Spain, mammoths were among the most common beasts to be portrayed. All the distinct features of the mammoth body were accurately depicted by the Cro-Magnons, early modern humans, who clearly knew the beasts intimately.

The elephant family

Over the last two centuries, large numbers of elephant-like fossils have been found in rock strata which are now known to have accumulated over some tens of millions of years. Scientists have come to realize that the living elephants and their extinct mammoth relatives are only some of the most recent members of a much larger and ancient family of elephant-like relatives. A few million years ago there were at least half a dozen different genera and many more species of elephant-like animals roaming the landscapes of the Earth. The surviving African and Asian elephants are but a remnant of this wonderful diversity our early australopithecine ancestors lived alongside.

The ancestry of the elephant family stretches back some 55 or more million years to a time when modern mammals were beginning to dominate the landscapes of the Earth after the demise of the dinosaurs. Indeed, the early elephants 'inherited' the habitats vacated by the large plant-eating dinosaurs, but none of the elephants reached the extraordinary dimensions of the biggest sauropod dinosaurs.

Three major groups of these elephant relatives are recognized: the primitive proboscideans, the mastodontids and the elephantids. Collectively they are known as proboscideans (proboscis being the Greek word for the trunk of an elephant) and, strictly speaking, they constitute an 'order' rather than a family in zoological terms. The proboscideans seem to share an ancestry with some unlikely-looking relatives such as the living sea cows (sirenians) and the hyraxes which look more like rabbits than elephants. But recent gene mapping supports this association. Together, they are thought to have shared a common ancestor in early Tertiary times some 55 million years ago.

However, the real story of mammoth and elephant evolution begins in Africa between 40 and 35 million years ago with a small, hippo-like animal, a metre (3 feet) long, called *Moeritherium*, which lived in freshwater. Two of its incisor teeth were already somewhat tusk-like and it had no lower canines. The deinotheres were the first proboscideans to look more elephant-like with elongate trunks. Some of them were the biggest members of the order, growing to 4 metres (13 feet) high. The striking difference from later elephants was that their tusks were developed from the lower incisor teeth and curved downwards. The exact function of this somewhat bizarre arrangement is not really known. The tusks may have been used for stripping bark from trees – but whatever it was, it was certainly successful because over 20 million years the beasts spread throughout Europe, Asia and Africa. They became extinct only about 2 million years ago.

The major group of elephant-like animals originated about 20 million years ago and initially included several distinct groups – the mastodonts, gomphotheres and stegodontids, often collectively known as the mastodontids. The primitive mastodonts were characterized by teeth with the distinctive 'knobbly' surfaces that had caused so much confusion when they were first found in America in the eighteenth century. The mastodonts probably arose in Asia and spread into Africa, Europe and North America. The gomphotheres, with their four short tusks, were equally, if not more, successful and spread further south in the Americas, down into South America. The stegodontids looked very like the true elephants and had a single pair of long tusks. They were once thought to be intermediate between mastodonts and true elephants, but are now regarded by experts as a kind of mastodont and quite separate from the elephants.

It was about 6 million years ago that the family of true elephants, the elephantids, evolved in Africa. The surviving African and Asian elephants belong in this group, along with the extinct mammoths and some other extinct elephant species. Their distinctive features include their large cheek teeth with ridged grinding surfaces and tusks made of solid dentine, the hard material that forms the bulk of our cheek teeth, but no enamel coating. Around 5 million years ago, the African elephants diversified into three different groups. One group has remained in Africa ever since and survives as *Loxodonta africana*, the African elephant. Another, the *Elephas* group, produced several species, some of which stayed in Africa but one migrated north into India and south-east Asia and survives there as the Asian elephant *Elephas maximus*.

A third group that split away while still in Africa went on to form the diverse *Mammuthus* branch. Some of their skeletal features suggest that they are more closely related to the Asian elephant, but genetic analysis of their DNA shows a closer affinity with the African branch. For the 4 million years of the life of the group they lived alongside the elephants but evolved separately; they were not ancestral to the elephants but contemporaries.

The earliest-known true mammoth fossils, *Mammuthus subplanifrons*, were found in Africa in the 1920s and were distributed throughout southern and eastern Africa between 4 and 3 million years ago. Then between 3 and 2.5 million years ago, the first mammoths appeared in Europe and soon spread widely, with their fossil remains being found from Italy to southern England. Their migratory route to north-western Europe, like that of the humans who were to follow them out of Africa, was around the eastern end of the Mediterranean, through today's Turkey and Greece. At over 6,000 kilometres (3,700 miles), the journey might seem extraordinarily long but it did not happen overnight. Some living elephants migrate enormous distances annually and even at a rate of only 5 kilometres (3 miles) a year, the journey could have been accomplished in 1,200 years or sixty generations, which is relatively short on the scale of 'evolutionary' time.

The early European mammoth is known as *Mammuthus meridionalis.* It was considerably bigger than living elephants, standing about 4 metres (13 feet) high and weighing 10 tonnes. It fed by browsing on typically mild climate woodland vegetation of oak, ash, beech and hickory. It probably looked very like the living elephants and did not need to be cold-adapted as the European climate was still mild (but getting cooler).

From 2 million years ago, the world climates cooled and descended into the Ice Age. The cooling path was not smooth but consisted of periods of fluctuating climate, which had drastic effects on the plant life and was 'knocked on' through the plant-eating animals to the meat-eaters that depended on the herbivores as a source of food. In the cold winters of Europe and Asia, only those that could adapt to the cold survived. By about 750,000 years ago, a new mammoth species, *Mammuthus trogontherii,* had appeared in Europe and Asia and may well have been cold-adapted. Certainly there is evidence from changes in their tooth structure that their diet had shifted to include tough grasses, which were then abundant on the cold northern steppes of Asia and eastern Europe. This steppe mammoth can be seen as something of an intermediate stage between the older woodland mammoths and the true woolly mammoth, *Mammuthus primigenius*.

The true woolly mammoth seems to have appeared in Europe around 250,000 years ago, when the Neanderthal people were also emerging for the first time. The climate was in fact entering a relatively warm interglacial phase. Early fossils of the mammoth (around 200,000 years old) found in 1992 at Stanton Harcourt, near Oxford, were associated with plant and insect fossils that reflect the mild climate of the time. By 100,000 years ago, when the last glacial phase was well under way, fully cold-adapted mammoths had spread from Siberia all the way across Europe to Britain.

With so much of the global water supply locked up in polar ice sheets, sea levels dropped over 100 metres (more than 300 feet). The English Channel became land and allowed mammoths, along with the other creatures of the Ice Age (including the Neanderthal people), to gain access to Britain. The vast kingdom of the woolly mammoth was finally established and stretched from the Atlantic eastwards over Asia to the Pacific and then across the Bering Strait into North America. The total numbers of beasts that made up the vast migratory herds must have been an awesome sight. They were seen by both the Neanderthal people and the later Cro-Magnon modern humans who replaced them. Both groups used mammoth materials for tools and certainly the Cro-Magnons actively hunted them and left us their superb wall paintings as silent witness of the great beasts. But sadly, the reign of the mammoth was all too brief; by 10,000 years ago most of them had disappeared.

Making a mammoth

Today, the idea of recreating a mammoth is not as fanciful as it would have been even ten years ago. Unlike

Dr Frankenstein, scientists no longer have to steal 'fresh' bodies from graveyards or cut down criminals from gallows in the dead of night in order to attempt to regenerate living tissues. Now, theoretically all a scientist needs is the genetic code from a single body cell to recreate the rest of the animal.

DNA and its fossil remains

It is now nearly fifty years since James Watson and Francis Crick first unravelled the genetic code of life and revolutionized the science of molecular genetics. They discovered the double-helix structure of DNA, the main constituent of the chromosomes, which allows replication of those chromosomes during cell division. Not until the 1980s was the very difficult technology available for the extraction, amplification and sequencing of the DNA molecule. Even more difficult has been the recovery of fossil DNA. Thanks to the global interest in dinosaurs and the huge success of the film *Jurassic Park*, much effort has been spent trying to recover fossil DNA.

There were claims made in the early 1990s that fossil DNA had been recovered from insects trapped in amber, and from dinosaur bone. But when other scientists tried to replicate those results in the late 1990s they found that they could not do so. The apparently ancient DNA was in fact modern contamination. The complex DNA molecule is fairly fragile and soon deteriorates, especially in the presence of water or oxygen and generally within hours or days of a cell's death.

For there to be any chance of preserving fossil DNA it has to be freeze-dried as soon after death as possible, and there are not too many natural circumstances where that can easily occur, except in the permafrost. Even then, after thousands of years only tiny amounts of the original DNA can be recovered. The best chance of recovering DNA is by taking it from tissues such as skin and bone – but only when amplified (repeatedly copied) by the polymerase chain reaction

(PCR) method can these tiny bits of ancient genetic codes be read.

Svante Pääbo, a Finnish biologist, successfully recovered bits of DNA from the skin of a quagga, a recently extinct horse-like animal, and a mummified Egyptian human in the 1980s. His results were replicated by other laboratories and were further verified by the similarity of the DNA to that of living horses and modern Egyptians. Pääbo's success encouraged others to try and recover even older DNA. Since then fossil DNA has been recovered from the bones of Neanderthals around 50,000 years old and a 70,000-year-old frozen mammoth .

The conceit of the *Jurassic Park* story was based on the idea of using fossil DNA to resurrect the dinosaurs. The DNA was to be recovered from bloodsucking insects that had originally fed on dinosaurs. But in the real world, even if dinosaur DNA could be recovered and sequenced, it would form only a minuscule part of the total dinosaur genome and be quite inadequate for cloning the beasts. The same applies to mammoth DNA, recovered by Russian scientists from frozen specimens in Siberia. Although the tissue seemed fresh enough, it was very fragmentary and the longest fragment contained only 545 base pairs. The complete mammoth sequence contains many millions of base pairs.

The best analogy for this process is to imagine an organism's DNA code as a long book, which only makes sense when the letters, words, sentences and pages are in order. Tear some of the pages out of the book, throw the rest away, then tear the pages up into little pieces, mix them up and pull a few of them out at random and try to reconstruct the story line from the bits. Not so easy, especially if you do not even know what the story is about to begin with.

However, two Japanese scientists think that if only they can get hold of one half of the double-stranded DNA helix they can create a new mammoth. Kazufumi Goto and

Akira Iritani both have considerable international reputations in reproductive biology and genetic engineering, and they have the technology to overcome many of the biological problems involved. They believe that they will never be able to find a complete sequence of mammoth DNA from the normal frozen mammoth tissue – but what if they could find some deep-frozen mammoth sperm?

Goto and Iritani's professional expertise lies in their pioneering application of cryotechnology to reproduction through artificial insemination. In the 1990s their research team at Kagoshima University was the first to use 'dead' sperm to reproduce a mammal – a calf from an endangered breed of cattle. Although frozen sperm is regularly used in artificial insemination techniques, it is normally defrosted to make it viable. Goto's team 'killed' the bovine sperm (that is, immobilized them), and froze them for several months. He goes on to explain, 'We inject the spermatozoa into a bovine oocyte [egg cell] and then we obtain a live calf.' He needs the egg because sperm only contains half the DNA code needed to create the mammoth. He makes it sound easy, but it involved a great deal of hard work by his team. As he points out, the most important thing is that it 'indicates that the sperm DNA is very strong against freezing'.

The main reason for this seems to be that the genetic material of a sperm is more tightly bound and 'tougher' than that of any other cells. Evolution has ensured that sperm DNA is robust because it has to survive some pretty rough treatment in its passage from the male testes to meet up with a female egg cell.

Male mammoths and elephants are unusual among mammals in that their testes are not carried in an external scrotum but lie within the body cavity close to the kidneys. The reason for this is not known, but it may well be that they are afforded greater protection within the body. Huge quantities of sperm, estimated at around 250 quadrillion (250×10^{24}) are stored in the 50-metre (160-foot) long sperm duct; immobile at the far end, they can easily be remobilized by a weak salt solution.

The cow elephant's genital opening is quite different from that of other quadrupedal mammals in being placed in front of the hind legs. Its opening is also forward pointing; consequently the bull's penis has to be S-shaped to enter the opening when he mounts the female from behind. The penis enters the elongate urino-genital canal but never reaches the vagina. As a result the sperm has a very long journey before it reaches the egg cell in the uterus. In recent years, sperm has been successfully collected from drugged wild bull elephants and frozen for insemination of zoo elephants. Bull elephants are too dangerous to keep in captivity once they become mature. The sperm has to be quick-frozen and stored at below –20°C. Elephant sperm survives thawing just as well as cattle sperm, and artificial insemination is now routine in zoos.

Presumably, if a mammoth was originally frozen quickly enough in the Ice Age permafrost, it is just possible that some of the tough sperm cells might have survived with their DNA intact. The Siberian permafrost is still maintained today at temperatures of –30°C even though it is over 10,000 years since the end of the Ice Age when most of the Siberian mammoths died out.

It is a pity that mammoth testes lay within the body cavity; if only they were external as in most other terrestrial mammals, then they would certainly have frozen very quickly upon death. As it is, their internal position may have delayed the freezing process with unfortunate results for the potential preservation of the DNA. It remains to be seen, because nobody has yet recovered any mammoth testes let alone tried to extract DNA from them. First Goto and his team have to find a frozen post-adolescent bull mammoth.

Gathering DNA

As we saw earlier, scientists have recovered mammoth DNA

from specimens deep-frozen in the Siberian permafrost. Natural freezing processes have shown that animal and plant tissue can be very well preserved by freezing. Post-mortem decay can be arrested. The process is so successful that we have replicated it in our domestic and commercial deep freezers. Indeed, some foods can be kept frozen almost indefinitely and still be edible when defrosted.

Normally, upon death, decay sets in pretty quickly with the soft tissues of the body breaking down into simpler chemical components, often aided by microbial activity. As part of this process, the complex and lengthy DNA protein strands within individual body cells soon begin to break up. Water is the main natural agent which leads to the destruction of body proteins whereas desiccation can preserve many tissues and proteins for substantial periods of time.

Bodies of animals that have died in deserts often preserve some skin and muscle tissues. Historically many different groups of humans, especially South American Indians of the Andes, have learned from this natural process and used it as a means of preserving the bodies of their dead.

Natural freezing processes are even more effective at preserving body organs, soft tissues and even the most delicate of cell contents, the genetic material. The popular concept of the process is that everything just freezes solid in a block of ice, but it is more complicated than that and actually involves dehydration of much of the body tissue. Freezing meat in a domestic freezer produces quite a lot of ice on the surface of the tissue and small pieces of meat lose a significant amount of weight as the water is drawn out of them during the freezing process.

The most famous of naturally frozen cadavers to have been found recently is Ötzi, the 5,200-year-old Neolithic 'iceman', found high in the Alps near the border between Austria and Italy. When alive he probably weighed about 50 kilograms (110 pounds), but his frozen body had been reduced by dehydration during the freezing process to just 20 kilograms (44 pounds). His corpse is now kept at –6°C to prevent any further deterioration. However, to achieve preservation, freezing has to be very rapid and subsequently maintained at temperatures of –30°C or lower. Small and medium-sized animals, including humans, can be quite quickly frozen solid but large animals like mammoths with a proportionally high ratio of body volume to surface area may take much longer to be completely frozen. The only place where this may have happened in the past is within the frozen terrain of the Arctic permafrost.

By studying the circumstances under which frozen bodies have been found in the Siberian permafrost, scientists have been able to reconstruct the sort of circumstances that led to their death. The remains of mammoth, bison, horse, woolly rhinocerous and wolverine have been found in the permafrost.

In the Siberian deep-freeze

The permafrost is the hidden face of the Arctic landscape. It forms wherever the ground has been at a temperature below 0°C for several years, irrespective of what the ground material is or how wet or dry it is. Some 26 per cent of the Earth's land surface is permafrost and most of it is in the northern hemisphere, covering about 7.6 million square kilometres (3 million square miles). This vast area has mean annual air temperatures below –8°C. In Siberia the thickness of the permafrost layer varies from over 600 metres (nearly 2,000 feet) in the coastal region of the Arctic Ocean to 300 metres (around 1,500 feet) at the southern margin.

But it is not all solidly frozen all year; even in the high Arctic, during the very brief summer – which lasts just a few weeks – the top layer thaws. Despite the low-angle sunlight, its persistence day and night raises the surface temperature above freezing. The depth of the thaw depends on many factors but generally varies between 15 and 100 centimetres (6 and 40 inches). Where there are soils, a remarkable variety of plant life can not only survive but thrive and reproduce.

The flowers produce an extraordinary burst of colour to attract insects by the myriad for pollination. Animal life is attracted by the plants and insects, especially birds and grazing mammals, and so are some top predators. The active layer can be very varied in form, ranging from marshy ponds to patterned ground broken up into large polygonal cracks, and hummocky terrain with small conical hills called pingos.

During cold Ice Age periods, the region of permafrost spread much further south into Eurasia and North America. Huge herds of cold-adapted mammals such as bison, reindeer, horses and mammoths, along with woolly rhinoceros, roamed these vast landscapes. They were particularly attracted by the nutritious grasses of the mammoth steppe (named after its chief inhabitant), which expanded and contracted as the climate changed. Virtually none of it is left in Eurasia now because the climate is wetter than previously.

Along the northern margins of the mammoth steppe many of these animals ventured into more treacherous areas of tundra, particularly along the coastal region of the Arctic Ocean, where the great rivers of Siberia drained into the sea. Here the permafrost conditions were probably much more like they are today. During summer, the active layer forms and plants flourish, particularly in the wetter parts such as ponds and water-filled cracks. These plants tend to be the more tender annuals rather than the tougher perennial woody shrubs that live on the drier tundra surface.

Not surprisingly, the lush growth of tender plants seems to have attracted mammoths and other grazers. Mammoths were like huge lawnmowers and had to eat enormous quantities of plant material each day to fuel their massive energy-expensive bulk. A 6-ton elephant requires about 90 kilograms (200 pounds) of forage a day and may have to feed for twenty hours daily to get this amount of plant material. They also produce a commensurate quantity of dung that helps fertilize further plant growth.

With this sort of food requirement, it is not surprising that from time to time hungry mammoths ventured too close to the edges of deep bog pools in summer. Organic rich mud is very glutinous and slippery and, if a pool was deep enough, it may have been impossible for a bulky animal to get any foothold to extricate itself from the cold water. The chill factor would eventually cause hypothermia and death even if the mammoth didn't drown.

What is yet unknown is whether any of the mammoths that perished in this sort of way were frozen quickly enough to preserve their sperm and DNA intact. The circumstances surrounding the discovery of a specimen called the Beresovka mammoth in 1900 suggests that it might just be possible.

The Beresovka mammoth

Reports of the discovery of a frozen mammoth in Siberia reached the Russian Imperial Academy of Sciences in St Petersburg in 1900. The following year the academicians sent an expedition, led by Austrian scientists Otto Herz and Eugen Pfizenmeyer, to recover the frozen cadaver. The expedition took four months to travel the 9,600 kilometres (6,000 miles) to reach the beast, still lying frozen on the bank of the Beresovka River, a tributary of the Kolyma which flows into the Arctic Ocean.

Originally, the mammoth was spotted by a Siberian deer hunter, who cut away the tusks and sold them in Kolyma. Most of the flesh had gone from the head and the trunk had been eaten by wolves but the rest of the animal, its long hair and flesh, was frozen solid. There was still grass clenched between its massive grinding cheek teeth. The scientists had to thaw the body so that they could cut it up and freeze it again for transportation. They managed to recover quite a lot of skin, flesh and even some of the stomach contents. Four months later, it was all back in St Petersburg. On the return journey, the air temperature fell to –48°C (–54°F). Eventually the preserved remains were reconstructed and mounted and are still on show in the Zoological Museum of the St Petersburg Academy of Sciences.

The animal was a 35- to 40-year-old bull mammoth that died somewhere between 33,000 and 29,000 years ago. Most of its stomach content was grass so evidently it had not died of starvation. The animal was found on its haunches, perhaps originally mired in mud, and had probably died of hypothermia. If such a large mammoth can be frozen to death quickly enough to preserve the grass still clenched between its teeth, there is some hope for preservation of the testes and sperm. The scientists did in fact recover the mammoth's penis – but of course, Goto and his team were not around at the time.

Reasons for extinction

Extinction events have happened throughout the history of life on Earth. There have been many different causes, as were discussed in 'Killer Earth', all long before humans arrived on the scene. The most recent extinction has been that associated with the end of the last Ice Age. The big question here has been whether it was climate change or the arrival of modern human hunters that was responsible.

During the Ice Age, the zone of permanently frozen ground spread south from the North Pole across northern Asia, Europe and North America; glaciers developed in the mountains and sea-ice spread south into the Atlantic. The large mammal faunas had to retreat southwards, but many of them soon adapted to the cold and were able to survive on the windswept tundra and grassland steppe of Asia and North America. When the ice retreated during the warmer periods, the cold-adapted plants and animals moved north again.

We know that woolly mammoths, rhinoceros, bears, wolves, wolverines, giant deer, horses, big cats and many other mammals occupied these cold regions because their fossils, including their butchered remains, have been found. These animals shared their habitat with one of the most effective predators of all time, humans, who have left a pictorial record of the great Ice Age 'game park'.

Mammoths are not the only members of the Ice Age bestiary to become extinct: some thirty-three different kinds of big game from Australia to Siberia, Ireland to South America, all became extinct within a few thousand years of the end of the last glaciation (between 12,000 and 10,000 years ago). Did they go quietly or were they pushed? Two main suspects have been identified – climate change and modern humans.

Ice Age climate change

Recently, remarkable evidence from a variety of sources has become available about the details of climate change through the latter part of the Ice Age. Drill cores of sediment layers retrieved from the ocean floor have provided a measure of such change. The shells of certain microscopic plankton preserve a chemical signature of changes in the composition of sea water and the size of the oceans. When more freshwater was locked up in the growing ice sheets, the oceans shrank and their composition changed slightly but measurably. When the climate warmed, the oceans increased in size and again there was a slight change in the sea-water composition.

Moreover, by analysing the fossil shells preserved in successive layers of seabed sediments, which have accumulated over many hundreds of thousands of years, scientists can obtain a measure of fluctuating climates through the Ice Ages. Some fourteen swings of climate between relatively cold glacial phases and warmer interglacial phases have been measured over the last 1.8 million years.

Independent but supporting evidence has been obtained from drill cores recovered from holes bored through the ice caps and sheets of Greenland and Antarctica. They too have provided indirect measures of climate change, virtually on a year-to-year basis, since the ice is built up by annual accumulation at the surface. Each surface layer is then covered and progressively buried deeper and deeper.

The cores have penetrated over 250,000 years' worth of annual layers, and analysis has shown some startling features of climate change. For instance, there were at least a dozen significant swings in climate between 60,000 and 25,000 years ago, which involved changes of between 5°C and 8°C. And some of these oscillations happened very quickly, over periods of hundreds or a few thousand years (see 'Ice Warriors'). Such changes can drastically affect plant life and subsequently cascade through the plant eaters and the meat eaters that prey upon them, right through the food chain.

Mammoth diets

As we have seen, mammoths were grazers that depended on forage from the vast swathes of cold, windswept steppe grassland that stretched across much of northern Eurasia. Analysis of the stomach contents of frozen mammoths from Siberia, such as the Shandrin mammoth found in 1972, shows that their main food was grass. The Shandrin mammoth's stomach contained 90 per cent grasses and sedges along with some twig tips of willow, larch, birch and alder.

With their massive grinding molar teeth the mammoths, like bison and horse, could use the coarsest end of the grazing spectrum, the poorer quality fibrous plant materials – such as coarse grasses, twig tips of woody herbs and shrubs which are relatively less defended by plant toxins. Nevertheless to avoid 'overdosing' on any one plant toxin these megaherbivores had to eat a variety of plants.

A 37,000-year-old frozen horse from Selerikan had a stomach content of 90 per cent herbaceous material, mainly grasses and sedges, along with smaller amounts of willow, dwarf birch and moss. Similarly the stomach of a woolly rhino from Yakutia in Siberia showed that it had been eating mostly grass.

Clearly, for the vast herds of grazers to thrive on the periglacial plains of Ice Age Eurasia and North America, there must have been extensive grasslands. Today, there are no such steppe grasslands in northern Eurasia – the climate is too wet and the grassland has been replaced with shrub tundra. The tough woody herbs that make up the tundra vegetation are not so nutritious as the grasses. With the disappearance of their main food supply, far fewer grazers would have been able to survive in these areas.

Undoubtedly, rapid climate and associated vegetation change at the end of the Ice Age could well have been responsible for their demise. However, in the northern hemisphere the pressure of human hunting on the dwindling population stocks probably had a significant effect as well.

Mammoth killers

We know only too well that humans are ruthless hunters, who will kill just for enjoyment and even deplete essential animal food reserves to our own long-term detriment, such as bison in North America and even fish and whale stocks. We have also threatened the populations of tiger, elephant and rhinocerous, panda, and our primate relations.

Paul Martin, an American academic, is convinced that humans drove the mammoth and other Ice Age beasts to extinction, especially in North America. He thinks that human hunters killed more animals than they really needed and that the resulting overkill had a disastrous effect, especially on the large plant eaters such as the mammoth. Certainly in North America some mammoth remains have been found closely associated with a particular kind of stone tool used in hunting – but large-scale mammoth slaughter is difficult to prove.

Computer modelling shows how populations can be affected by different combinations of factors such as changes in climate, vegetation and hunting. Stephen Mithen, an archaeologist at Reading University, has modelled the effects on North American mammoth populations of colonization by the Clovis hunters. Very little is known about these people, who lived around 11,000 years ago and belong to the earliest Palaeoindian tradition in North America. They are so called because their characteristic stone spear points were first found at Clovis in New Mexico. Since then the points have been found across the continent, along with the remains of early horses, tapirs, camels and, importantly for us, mammoths.

Mammoth reproduction is relatively slow with few offspring being produced by the females over long lifespans. As Mithen says, 'Start taking out some of those young females and you start seriously depleting the populations.' His model suggests that if just three in every 100 mammoths were killed each year, then it would take only a century to wipe out the mammoth.

It is a convincing argument – after all, in 1979 there were some 1.3 million elephants in Africa. Just a decade later there were only 600,000 left. Vance Haynes, another American academic, has studied one of the dozen known mammoth kill sites in North America, at Murray Springs. Here he found a Clovis point close to the skeletal remains of a mammoth. Vance Haynes is not convinced that humans alone could have wiped the mammoths out: 'Certainly they had a hand in it, and the fact that they killed bison and mammoth is a significant factor. But sloth, camel, sabre-tooth cat, Ice Age wolf, all these other members of the megafauna [in North America] went out at the same time. My feeling is that something else happened.'

That something else was probably a very sudden return to glacial conditions about 11,000 years ago (and this is something that will be re-examined in 'Ice Warriors'). As British mammoth expert Adrian Lister points out, there is good evidence 'that the world was plunged into a time of very intense cold that could have played a major part in the extinctions'. It may be possible to test this theory using a particular kind of 'stressometer' carried around by all mammoths. Mammoth tusks grew throughout life in an incremental fashion and, like tree rings, provide a record of times of plenty and times of famine. Since mammoths, like elephants, lived for at least sixty years any climate stress should show up in the tusk record. So far there is no clear signature in the tusks that have been analysed.

As with attempts to pin down the numerous earlier extinction events in the geological past, it is virtually impossible to point the finger of blame at a single cause. Climate change and our ancestors probably had a hand in pushing much of the Ice Age megafauna over the brink into extinction. The lesson is that it has happened time and time again in the past and will happen again in the future, whether or not it is aided and abetted by humans.

A mammoth task

If Professor Goto manages to recover some good-quality frozen mammoth sperm from Siberia, he then has the difficult bit to do – successfully breeding several generations of viable mammoth/elephant hybrids. This will be a lengthy and very complicated process, and some scientists have severe reservations about the whole operation. Nevertheless, Goto and his team are convinced that they can bring the mammoth back to life – and that the effort will be worthwhile.

Let us suppose that Goto's team do get some mammoth sperm: what then? They have chosen a female Asian elephant to act as the surrogate mother, and have to operate on her to remove an egg for *in vitro* (test tube) fertilization before being reimplanted in the elephant's uterus. They have to select sperm carrying the X chromosome to ensure breeding a female hybrid. The next hurdle will be the possibility of rejection of the egg by the elephant. Then there is a long wait while the embryo develops – if it does.

In the wild, a female elephant only comes into heat every three to five years, and once she has conceived it will be at least another three years before she mates again. Gestation in elephants is unusually long, lasting between 652 and 660 days. Some good news for Professor Goto is that birth is relatively straightforward and usually successful in elephants. This is mainly because calves are small at birth in relation to the mother, being about 3.5 per cent of the mother's weight; a human baby is 6 to 7 per cent of the mother's weight. Also, as the elephant's birth canal is nearly vertical, birth can be fairly quick – even so, labour may take several hours.

Many calves can walk within an hour of birth and so can start to feed from the mother's teats. Feeding continues for several years. So, despite the long life of an elephant, the maximum number of calves a female elephant can produce is about ten. Puberty in African elephants is not achieved until about the age of nine years, and sometimes as late as eighteen, in both sexes.

But the situation in resurrecting a mammoth is different. To begin with, the calf will be a hybrid, or chimera, because of the significant genetic distance between its parents. A hybrid elephant was born a few years ago in Chester Zoo as a result of an accidental mating between the more closely related living African and Indian elephants. The calf died ten days after it was born as a result of internal bleeding. Furthermore, the best-known domestic hybrid, the mule – produced by cross-breeding a horse with an ass – is sterile. Mules have been artificially bred by humans in Asia since at least the seventh century BC. All other cross-breeds between different kinds of horse relatives and cat relatives are also sterile, but some cattle species can be crossed.

Part of the reason for this is the different number of chromosomes in the different species. In normal cell division during an organism's growth and development, the exchange of genetic material is facilitated by the basic similarity of the chromosomes from the two parents. However, in hybrids, where there is a considerable difference between chromosomes, it is more difficult for successful exchange of genetic material (crossing over) to take place. However, in Dubai recently, a camel and a llama were successfully crossbred using artificial insemination. But with the mammoth/elephant the two parents are separated by 30 million years of evolution. So mammoth/elephant hybridization may be possible after all. Nevertheless, it would take three generations and at least fifty years, after which the calf will still only be 88 per cent mammoth.

Mammoths are not the only animals being stored in the freezer awaiting future revival. Scientists have stored the sperm of other endangered species such as the tiger, panda, mountain gorilla and chimpanzee. And, these days, humans are being deep-frozen upon death in the hope that the technology will be available in the not too distant future to bring them back to life. For a mere 120,000 US dollars, your body can be stored indefinitely, using modern cryotechnology. But what is the morality of such endeavours as resurrecting the mammoth?

Some scientists, such as Andy Currant of the Natural History Museum in London, question the morality on the basis that they are 'trying to create an extinct mammal, which has no environment to live in, and at a time when we're busy doing our best to exterminate its two nearest relatives'. And Steve Mithen thinks that 'it would be a bit of an insult to these marvellous creatures if we were to produce some sort of bizarre chimera by combining bits of animals today with a bit of ancient DNA ... we respect the mammoths by letting them have their time and let them rest in peace.'

A mammoth hybrid would be a freak, out of time and place, with nowhere to go. Even if it were possible to create a hybrid, any viable breeding population would have to number hundreds of animals to avoid inbreeding. And where would they live? The cold steppe grasslands that maintained the vast herds of mammoths disappeared when most of their ancestors did, around 10,000 years ago. Sergei Zhimov, an ecologist and director of Siberia's Northeast Scientific Station in Duvannyi Yar, has planned a 'Pleistocene Park' for the region. He and his team have already introduced thirty-two Yakutian horses to the 160-square-kilometre (60-square-mile) preserve and want to build up herds of moose, reindeer and bison to recreate the kind of environment in which the mammoths lived. Zhimov hopes that in twenty years' time in the park 'the density of the animals will be the same as in the Serengeti Game Park in Africa'. Whether or not mammoths will be joining them is still very much a matter of luck.

Professor Goto believes that we can reverse the ravages we have inflicted on the animals with which we share the planet. He believes we can help protect endangered species even after they have gone over the brink. According to Goto, cryotechnology is already being used to preserve 'sperm, eggs and fertilized eggs of endangered species for future use'. He argues that all this effort will be wasted unless we know more about the long-term effects of freezing genetic material. The mammoth and the elephant are just the 'guinea-pigs' in Professor Goto's long-term conservation strategy.

Latest news

The Japanese team drew a blank in their attempt to find a frozen mammoth and recover its DNA in 1997. But meanwhile another international team was being assembled to try again. In March 2000 they hit the international headlines with a carefully orchestrated media blitz of pictures and stories about a mammoth encased in a 23-tonne block of frozen mud. Everywhere there were slightly absurd pictures of a Russian helicopter carrying a big brown block, which was supposed to contain the mammoth, with two superb mammoth tusks sticking out of one end.

This time Bernard Buigues, a French explorer and entrepreneur, was the moving spirit. Buigues lives in the Siberian town of Khatanga, 800 kilometres (500 miles) inside the Arctic Circle, and organizes expeditions in the region. In 1997 Simion Jarkov, a young Dolgan tribesman, claimed to have found the tusks sticking out of the ground while on a hunting trip and told Buigues of his find. In June 1998 a research trip was organized to the site, 400 kilometres (250 miles) north-west of Khatanga, near the banks of the Balskhnya River, and an attempt to excavate the mammoth was begun. Beneath the shallow defrosted surface layer of shrubs and mud they found that the upper part of the skull had already gone, but the lower jaw with its massive molars was still there along with strands of mammoth hair.

Buigues assembled a multinational team to try to assess whether the rest of the animal was present and to investigate the possibilities of recovering it. Hi-tech ground-penetrating radar suggested that there was something there and so Buigues returned with compressors and pneumatic drills, and a film crew. They eventually managed to excavate the block and took it to Khatanga, where it is stored in an underground ice cave. The plan is to defrost the mammoth using twenty-five scientists working in shifts and wielding hairdryers! The laboratory is being set up as I write.

Analysis of the remains that have already been recovered shows that the animal was aged about forty-seven when it died 23,000 years ago. In contrast to Professor Goto's ambition to 'resurrect' a mammoth, there is a more realistic intention to find some well-preserved DNA. Behind the scenes and the hype there are scientists from the American Museum of Natural History in New York and the University of London who hope to get some scientifically useful and interesting results.

As Adrian Lister of University College, London, who heads one of the laboratory teams, says, 'There is a chance that the new Jarkov mammoth does preserve some tissue but we will have to wait and see what the quality of preservation is. Our main hope is to retrieve some DNA that is less fragmentary than previous sample, and to improve our knowledge of mammoth genetics and the mammoth's relationship to living elephants. I do not believe that cloning or hybridization is a realistic proposition.'

The most recent report from one of the Russian members of the team, Alexei Tikhonov of the Zoological Institute in St Petersburg, is not promising. According to Tikhonov, earlier media reports suggesting that the mammoth is complete is speculative: 'In our opinion there's a lot of mammoth wool, probably some bones and a piece of skin.'

The good news is that even if the Jarkov mammoth does not provide the DNA 'goods', it is only a matter of time before a better-preserved carcass is found in the frozen Pleistocene park of Siberia. Next time, there will be plenty of scientists who will make sure that the wolves do not get it all.

In the end, the big question remains: even if some quality DNA is obtained, would cloning a mammoth actually work?

ICE WARRIORS

Paul Simons

On the evening of 14 April 1912, an iceberg slowly floated southwards in the cold waters off the coast of Newfoundland. It had already drifted nearly 3,000 kilometres (1,800 miles) in over three years after shearing off a glacier in west Greenland, and now it was nearing the end of its life before melting in the Gulf Stream. But before it disappeared this iceberg would become the most notorious chunk of ice in world history.

On its maiden voyage from Southampton to New York, the *Titanic* had received several warnings of icebergs in the Newfoundland area but despite these the crew continued to sail into treacherous cold waters. Then just before midnight the 66,000-ton liner collided with the 120-metre (400-foot) long iceberg and a tremendous judder went through the entire vessel. Yet the passengers had such faith in the reputation of the unsinkable *Titanic* that many of them went out on deck and actually played with the ice lodged next to the ship.

Now, recent research using a deep-sea submersible has revealed that the iceberg made only small tears in the front section of the ship's sixteen watertight compartments. Had the ship not been steaming so fast at 22 knots the damage might have been contained, but the gashes were just large enough to let in water and 39,000 tons of it surged into the hull. The huge pressure of water placed such unbearable strain on the mid-section of the vessel that it split in two as it sank, and by 2.20 a.m. the ship had vanished and over 1,500 passengers and crew died.

The whole tragedy started as snow that fell about 3,000 years ago. As the snow piled up it became squashed into granular snow called firn, and eventually turned into glacial ice. Arctic icebergs are born in Greenland, an island roughly the size of western Europe and almost completely covered in ice, slowly surging outwards at up to 20 metres (65 feet) a day. When a glacier meets the sea, the rising and falling tides, winds and warm spring temperatures break off huge slabs of ice which crash into the water, and icebergs are born – a process called calving. Each year an estimated 10,000 to 15,000 icebergs are calved, largely from Greenland's west-coast glaciers. They vary from 'growlers' (the size of a grand piano) to the tallest known Arctic iceberg spotted in 1967, which towered some 170 metres (550 feet) above the ocean, slightly less than half the height of the Empire State Building. The bulkiest Arctic iceberg measured 11 kilometres long by 5.8 kilometres wide (7 by 3½ miles) and was sighted near Baffin Island in 1882.

From its launch into the freezing seas around western Greenland, an iceberg has to escape the fjords and bays of the Greenland coast before hitching a ride north on the West Greenland Current into Baffin Bay on the east coast of Canada. Many get stuck there and can take up to four years to melt, but for those that break free it can be another three months to two years before they reach the open seas again. They sweep south past the Labrador coast in the greatest concentration of icebergs in the world, anything up to 2,000 a year, in what is known as 'Iceberg Alley', refrigerated by the cold water of the Labrador Current. Some of the icebergs run aground in the coastal shallows and melt away, but icebergs swept along in the main current drift further south and reach Newfoundland. Some hit shoals and stay out of harm's way, but the remainder are carried even further out beyond latitude 48°N where they meet the open North Atlantic, the busiest shipping lanes in the world.

Only when the Labrador Current crashes into the Gulf Stream's warm waters south of Newfoundland do the surviving icebergs finally melt away. Although by the time an iceberg reaches the waters near Newfoundland it has lost 90 per cent of its mass through melting, it still packs enough weight to be highly dangerous. During this elaborate journey iceberg mortality is extremely high – of the 10,000 or so bergs that start their journey in Greenland each year, only about 2,000 get past northern Labrador, and only an average of 466 a year make it into the open Atlantic and threaten shipping. A few exceptional icebergs have been sighted well over 2,000 miles from their origin; in June 1907 one got to within a few hundred miles of south-west Ireland, and the citizens of Bermuda twice got a surprise when icebergs paid them a visit in the last century.

Some ingenious uses have been found for icebergs. In the Second World War British scientists planned to carve aircraft carriers out of icebergs and tow them to the English Channel where they would be clad in iron. Winston Churchill ordered so-called Project Habbakut be given top priority, but despite that it was never implemented, probably because it was totally impracticable. For decades engineers have dreamt of towing icebergs to far-off lands short of water; after all, an iceberg of 40 million tons contains enough highly pure water to supply a city of half a million people for a year. The idea was first seriously considered in the 1970s to carry fresh drinking water to Saudi Arabia, but the technical hurdles are enormous: icebergs break up in heavy waves and even if you manage to get one to shore how do you get the water on to land? Perhaps the closest anyone has come to exploiting icebergs is a Canadian entrepreneur who has carved out chunks from icebergs and bottled the melted ice into drinking water or made it into vodka.

Patrolling the ice

Probably the first recorded mention of icebergs was by St Brendan, an Irish monk, who encountered a 'floating crystal castle' on the high seas. They have caused havoc to shipping ever since, and are still one of the biggest threats to navigation, especially the small growlers that are difficult to see by eye or by radar. Apart from the *Titanic*, hundreds of other ships were hit by icebergs during the twentieth century; the last major disaster was in 1959 when another supposedly iceberg-proof vessel, the *Hans Hedtoft*, a Danish passenger and cargo ship, hit an iceberg and sank with the loss of all ninety-five crew and passengers.

But it was the sinking of the *Titanic* that sent shockwaves around the world like no other shipping disaster. The maritime nations vowed it would never happen again and in 1914 an international conference of the major seafaring nations called for protection against icebergs. Through their efforts the International Iceberg Patrol was established, to cover about 1.3 million square kilometres (over half a million square miles) of the most treacherous iceberg-infested seas in the North Atlantic. However, their ultimate dream was extraordinary – to destroy icebergs. Over the next fifty years they tried to smash icebergs to oblivion with machine-guns, shellfire, torpedoes, bombs and high temperature explosives, but it was all futile. 'We tried everything we could think of short of a nuclear bomb,' remembers Captain Bob Dinsmore, commander of the patrol during the 1950s and 1960s, 'but there was almost no effect at all.' What no one realized is that icebergs are remarkably good at absorbing shocks.

One ingenious idea was to paint an iceberg black. The concept was wonderfully simple – normally the white colour of icebergs reflects the heat of the sun, but a black surface would help to soak up enough solar heat to melt the ice. The idea worked up to a point, but soon after the ice started to thaw the meltwater washed the paint away.

The Ice Patrol has now given up the idea of destroying icebergs, and their philosophy is 'if you can't beat them, steer clear of them'. It sounds simple, but although modern ships use radar to detect icebergs they are still difficult to spot – with a ship travelling at 15 knots the radar only gives about eight minutes' warning, and in heavy seas radar signals become confused.

The Ice Patrol flies scouting missions about three times a week during the iceberg season, from March to July, tracking icebergs by eye and radar. Satellites follow changes in ocean currents using buoys with radio transmitters, and computer models predict the speed and direction of icebergs and how fast they are melting. The data are drawn up into a map showing the limit of the ice and the results are faxed and broadcast to ships. As a result, the Ice Patrol has been a huge success – since it was set up no ship that took its advice has been struck by an iceberg. But avoiding icebergs comes at a price, ships often have to take a wide detour to avoid collisions, and for a ship travelling from northern Europe to New York this could involve a detour of 340 nautical miles. At an average speed of 20 knots that adds seventeen hours' sailing time to the journey, which in today's economic climate is a significant cost for shipping lines.

Now there is a new and vastly more difficult challenge. In the 1970s a large oilfield was discovered on the Grand Banks off the coast of Newfoundland in Iceberg Alley, and it was too much for the oil companies to resist. So they decided to take the biggest technological risk they have ever faced – to drill from offshore rigs in iceberg-strewn waters in the Atlantic Ocean.

So what exactly makes a an iceberg so dangerous? It sounds a perfectly easy question to answer, but when the oil companies investigated the threat it soon became clear that very little was known about icebergs – how they moved, how hard they were, how much damage they could do, and many other vital questions. Over the course of several years, their research showed that the underwater part of an iceberg is extraordinarily strong. They towed medium-sized icebergs weighing half a million tons into a rig lined with pressure pads on the sides to measure the force of blows from the berg. The results showed that an iceberg could pack the weight equivalent to 24,000 cars piled on top of one another. What's more, this is not the sort of ice you have floating round in a gin and tonic. Iceberg ice is incredibly strong because it is made from compressed snow, which is much denser than ice made from frozen fresh water or sea water. It behaves more like a rock that happens to be made of water, thanks to the pattern and size of its frozen water crystals, and also because the air bubbles trapped inside when it first fell as snow get squeezed smaller as the ice is compressed in glaciers. You can even hear the iceberg air bubbles make a fizzing noise as the ice melts, when the bubbles escape under pressure – what scientists quaintly call 'bergy seltzer'.

The oil companies have had to resort to desperate measures to keep their rigs safe – they tow icebergs as large as 2 million tons out of the way by running a lasso round them and pulling them with a high-powered tugboat. Larger icebergs simply drag the tugboats with them, so the oil companies also use mobile rigs, or 'semi-submersibles', which as a last resort up-anchor from the seabed and move off the oil field if an iceberg heads towards them. The problem is that the movements of icebergs can be highly erratic. Although only about one-seventh of an iceberg sticks out of the water, this is large enough to act like a sail and catch the wind, while the submerged part is dragged along by the ocean current. This means that they often take unpredictable turns, making oil-rig safety even more of a nightmare. In one case an iceberg headed towards one of six rigs drilling fairly close to each other on the Grand Banks, so the rig in danger was disconnected from its well and moved off. No sooner had that been achieved that the berg changed direction and headed straight for the next rig, and then went for all of the other rigs and drove them all off the oil field. The rigs were all saved but, of course, it was a big operation and the companies lost a hefty chunk of oil production.

Now for the first time ever, oil companies have pushed back the frontiers of technology and designed a drilling platform capable of resisting collision with an iceberg. The result is the *Hibernia* rig, protected from icebergs using an underwater 1-million-ton concrete wall 15 metres (50 feet) thick and 85 metres (280 feet) tall and ringed by sixteen huge teeth designed to absorb the impact of bergs of up to 6 million tons. It is also the largest single insurance risk the world has ever seen.

The *Hibernia* is an extraordinary feat of engineering and just getting it into place was an achievement in itself, taking nine of the world's most powerful tugboats ten days to make the 315-kilometre (195-mile) journey from its construction site, dodging several icebergs on the way. It is now producing oil in excess of forecasts and has yet to face a direct assault from a berg.

More icebergs on the way

However, the perils of drilling or shipping in iceberg-prone waters are likely to grow worse in the near future, because climate experts are forecasting many more icebergs in years to come. If there is any place on Earth showing unmistakable signs of climate change, it is the Arctic. The temperature in some parts in the Arctic Circle has shot up by a staggering 9°C in the last century, faster than anywhere else in the world. This region has become the planet's bellwether for climate change with widespread melting ice, thawing permafrost and huge shifts in plant life and wildlife.

For most of the past decade there has also been an alarming rise in the numbers of icebergs. Captain Bob Dinsmore of the International Ice Patrol has noticed these changes at 48°N, the traditional boundary where icebergs are considered a menace to transatlantic shipping. 'When I was on ice patrol from the 1950s to the 1960s, we would average about 400 icebergs a year,' he explains. 'In 1998 there were well over 900 icebergs.' According to the textbooks, a severe season is classified as recording more than 600 icebergs crossing 48°N, so these are worrying numbers.

Having said that, the numbers of icebergs carried into the Atlantic can be notoriously fickle and in 1999 only twenty-two icebergs passed 48°N, so few that the Ice Patrol's aerial reconnaissance flights were suspended by the end of May instead of finishing in July as normal. What made that iceberg season even more remarkable was that several thousand icebergs were stretched out along the northern Newfoundland and Labrador coasts in the spring and early summer, but very few of them moved into southern waters. The seas off Newfoundland were also unusually warm, 2 to 3°C higher than normal, and the Labrador Current was weaker than normal. These features were difficult to explain, and it just goes to show there is still a lot left to learn about icebergs and the currents that carry them.

Given that there has been a striking rise in iceberg numbers in most recent years, is this really a sign of global warming? You would imagine that the answer must be an emphatic 'yes' simply because global warming should make the ice sheets thinner and hence make them break off icebergs more easily. In fact, the higher temperatures in Greenland are actually making *more* ice, and on the west of the island, which bears the brunt of the snowstorms carried on the prevailing westerly winds, the ice has thickened on average by more than 15 centimetres (6 inches) a year. That is because more water evaporates from the sea and the warmer atmosphere also carries more moisture. So warmer, wetter winds are dumping more snow over Greenland; the ice sheet is growing thicker; more ice is moving outwards into the sea and that is breaking off more icebergs. This is what all computer models of global warming have predicted will happen.

Apart from threatening oil rigs and shipping off the Newfoundland coast, the rising numbers of icebergs are bringing a raft of other problems in their wake. As more icebergs melt in the sea they are feeding the Labrador Current with more freezing cold water and could push the current further south, potentially launching fleets of icebergs far into the North Atlantic. Perhaps one day there may be icebergs floating along the coast of Cornwall.

A bizarre sight like this would pale into insignificance compared to another surprise lurking in the North Atlantic that could potentially devastate north-western Europe. The first hint of something truly apocalyptic was dug up from the bottom of the ocean.

In 1988 oceanographer Hartmut Heinrich of the Hydrographic Institute in Hamburg was drilling into the seabed off Labrador, when to his amazement he unearthed six layers of light-coloured stones unlike anything else he had seen before buried in the ocean sediment. Even more remarkable, the same six layers of stones turned up on the other side of the Atlantic in the seabed west of France, and since then they have also been excavated from a dozen sites spanning the Atlantic from Labrador to Portugal.

Heinrich traced this rocky debris to the Hudson Bay in Canada, and dated its journey back to the Ice Age when Canada was crushed under a vast ice sheet. As the ice ripped and tore across the Canadian landscape it scraped up masses of rocks and stones, and when the ice sheet eventually cracked up it launched massive numbers of icebergs carrying the rocky cargo out to sea. When the ice melted in the North Atlantic the stones simply fell to the seafloor – the ghostly imprints of ancient icebergs.

To get such extraordinary amounts of stones stretching across the North Atlantic would have needed millions of icebergs ranged across the ocean in a vast flotilla. By dating the sediments the stones were buried in we know that these so-called Heinrich events happened roughly every 5,000 to 10,000 years during the Ice Age. What caused an even bigger stir in scientific circles is that the Heinrich events also carried a potent sting in their tail. As the icebergs melted they would have also dumped colossal amounts of freshwater in the North Atlantic. And not just the North Atlantic but ocean currents all over the world would have been affected, with enormous impacts on the climate of the Earth.

The North Atlantic is one of the most important dynamos in the world's weather and oceans. It behaves like a gigantic central-heating system, collecting heat from the tropics, carrying the warm water north on the Gulf Stream where the heat warms the atmosphere of western Europe. As the current reaches the Arctic some of it moves into the Arctic Ocean as a subsurface current while the rest turns southwards and mixes with cold water coming out of the Arctic Ocean itself. This forms a southward current down the east coast of Greenland. At about 75°N latitude some of this cold water is diverted by ocean ridges out into the centre of the Greenland Sea. There the surface evaporates and cools and forms sea ice in winter, leaving behind cold, salty water that eventually grows so dense it sinks down to the bottom of the ocean like water rushing down the plughole in a bath, sucking more warm surface water northwards from the Gulf of Mexico. Meanwhile, deep down in the ocean this current then turns round and heads back south, pushes into the South Atlantic, rounds the southern tip of Africa into the Indian Ocean and eventually reaches the Pacific Ocean. There the deep water wells up to the surface of the sea, where the water soaks up the heat of the hot tropical sun and then travels all the way back to the North Atlantic as a warm surface current. It is a round-the-world odyssey that can take a thousand years to complete. This global central-heating system helps balance out temperatures across the world by carrying warmth from the tropics towards the poles, and without it the planet's climate changes radically.

The icebergs melting during the Heinrich events would have stalled the heating system by flooding the North Atlantic with so much fresh water the surface currents would have been too light to sink. The deep water currents would slow or stop, heat would no longer spread around the globe as efficiently as before, and the world would be plunged into another bout of glaciation.

The vast armadas of icebergs would have driven temperatures even lower because, like ice cubes floating in a drink, they would have cooled the surface of the ocean. Their whiteness also helped reflect the Sun's radiation back into space. The increasing cold of the oceans in turn made the climate drier, as cold air holds less moisture than warm air. So less rain or snow fell over the continents, creating huge deserts with winds so ferocious they blew up gigantic dust storms. That in turn kicked up so much dust high into the atmosphere that it blocked out enough sunlight to shade the Earth, cooling the climate even further. And so the world descended into a vicious cycle of cold and drove glaciation even deeper. Just to prove how profoundly the Heinrich events shook the world, scientists in the southern hemisphere have discovered that glaciers in Chile and New Zealand advanced and retreated in synchrony with the surges of the North Atlantic icebergs. These were truly global events.

Climate flips

Shortly after the Heinrich events were discovered, another dark secret of the Ice Age was uncovered, this time from the vast ice sheet covering Greenland.

Ice covers nearly 2 million square kilometres (800,000 square miles) of Greenland. A team of twenty-five international scientists led by Danish glaciologists has been studying the heart of the ice sheet for several years in atrocious conditions, where even in the summer months temperatures average only –32°C. The remote camp site is called North Grip, supplied by the US military in huge C130 aircraft that can only land on skis and take off with the help of rockets. The North Grip team has been drilling out cores of ice about 3 kilometres (2 miles) deep all the way through the ice sheet down to the bedrock underneath at the highest point on the Greenland ice sheet. By analysing this prehistoric ice the scientists get annual 'weather reports' going back some 150,000 years.

As snow falls on the huge ice sheet each year it is slowly squashed into layers of ice, each layer representing a year's snowfall. Like counting tree rings, these layers reveal the age of the ice and also something about the past climate. The thickness of each year's layer of ice shows how much snow fell each year. Also, as each snowflake falls it traps a tiny piece of the atmosphere and eventually seals it into a time-capsule of ice. By unlocking the gases and dust trapped in the ice the past climate can be worked out, especially the ratio of different forms of oxygen in the ice, which indicates the ancient temperature at the time the original snowflake fell (in the way described in 'Killer Earth').

The results of the Greenland ice-core work have given scientists a fright. We used to think that the ice ages were all unending bitter cold and that the climate changed slowly over thousands of years, but the ice records revealed that the world went through several convulsions in climate with sudden bouts of warming or cold lasting hundreds to thousands of years before flipping back again. Even more astonishing, these changes happened staggeringly fast, sometimes in just two or three years.

So, for instance, we used to think that the last Ice Age gradually drew to a close 15,000 years ago when the world warmed up and the ice sheets gracefully melted away. The Greenland ice cores show that the huge continental ice sheets started to melt and disintegrate within a decade – just *ten* years! Then just as suddenly the cold returned again and this yo-yo of climate change happened a few times more before the warm climate won.

What made the world go through such violent spasms of climate? The Heinrich armadas of icebergs can explain some of the episodes of wild swings in temperature. As the climate warmed at the end of the last Ice Age, the vast ice sheets across eastern Canada and America disintegrated and released the masses of icebergs into the North Atlantic. This turned the climate so cold that ice sheets grew again and it eventually took a huge bout of solar heating before the Ice Age finally released its grip.

These astonishing flips in climate must have been catastrophic for life on Earth. As was considered in the previous chapter, maybe it led to the extinction of the mammoth and the other big mammals in the high latitudes that could not adapt quickly enough to the changing conditions. The climate must also have been diabolical for ancient man trying to adapt to a warmer climate, only to be plunged into the depths of another mini-Ice Age.

What we do know for sure is that sea creatures went though turmoil, because the tiny microscopic skeletons of plankton left buried on the sea floor show that warm-adapted species suddenly gave way to cold-adapted species and then back again. In Scotland, the remains of midges exhumed from lake beds and bogs are proving to be the most sensitive indicators yet of climate change. Swarms of warm-sensitive midges infested the Scottish mountains more than 10,000 years ago, then their numbers fluctuated wildly as the thaw went into reverse at least three times as the ice returned. The longest of these cold spells 1,100 years ago was dubbed the Younger Dryas period, named after the little Arctic Dryas flower in Scotland which typified the cold climate, during which icecaps regrew in North America and Europe. According to the midge remains in Scotland, summer temperatures crashed by about 10°C over just a few decades and stayed that way for about 1,500 years. Equally intriguing was an earlier sudden, shorter and so far unexplained freeze around 12,500 years ago when summer temperatures plummeted in Scotland by 11.5°C for about 150 years, the equivalent of Madrid taking on the climate of Reykjavik. These 'blips' mirror those recently found by midge researchers in Canada, suggesting that they were widespread.

Freeze or fry

This idea that the world's climate can suddenly perform somersaults has come as a slap in the face for scientists, who now realize that although the past 10,000 years have been relatively stable and warm times, the climate can suddenly swing violently at very short notice. 'We used to think climate changed gradually, like slowly turning up a dial on an oven,' says Jeff Severinghaus of the Scripps Institution of Oceanography in La Jolla, California. 'But it's more like a light switch.'

If the past climate was capable of dramatic rapid changes, the scientists now reason, then it could throw another fit in our own lifetime. The average world temperature has gone up about 0.6°C in the past century, which doesn't sound much except when you consider that in the last Ice Age average global temperatures were only 4 or 5°C lower than today. So the Earth is now warming at such a punishing rate the fear is that it could push the climate to some unknown critical threshold when it suddenly tips into a totally new mode more terrifying than anything since the last Ice Age. 'The consequences could be not just slow change but a rapid switch to something that we've never experienced in the last 11,000 years,' reckons Jack Dibb, who works on the Greenland Ice Sheet Project.

Even more disturbing, the Greenland ice cores also revealed that in the last big warm interglacial period called the Eemian (sometimes called the Ipswichian in Britain) just over 100,000 years ago, wild lurches in temperature were also matched by changes in levels of carbon dioxide. This is one of the key gases that keep the Earth warm by trapping much of the Sun's heat in the atmosphere – the so-called greenhouse effect. What presses the alarm bells with climate experts is that today's global warming is being fed by our own greenhouse pollution made from carbon dioxide, methane and all sorts of other waste gases. So are we going to repeat the ice melt of the Eemian? Many experts fear that there are already warning signs from the Arctic, and not just in the numbers of icebergs.

While the Greenland glaciers are growing and shedding more icebergs, the Arctic sea icecap is melting at a terrifying rate. The Arctic Ocean is an oblong of water nearly one and a half times the size of the United States, the surface frozen into ice sitting on top of icy waters. This pack ice is made from sea water, a thin layer between one and 30 metres (3 and 100 feet) thick, floating around with the ocean currents and winds like the skin on a bowl of soup. It grows and shrinks with the seasons and roughly 3,000 cubic kilometres (720 cubic miles) of ice float off each spring and summer and drift into the Greenland Sea where most of it melts before reaching the latitude of Iceland.

While the air temperature in the Arctic is rising alarmingly, the sea is also changing. In February 1997 vast stretches of the Arctic Ocean were found to have warmed by a remarkable 1°C or more since the late 1980s. Indeed, the climate change was so worrying that more than three dozen leading Arctic scientists wrote to the US National Science Foundation urging it to support a monitoring programme to find out what is going on. 'It is becoming increasingly clear that the Arctic is in the midst of a significant change,' they warned.

The rising temperatures are having spectacular consequences. A series of satellite pictures shows that in just sixteen years the Arctic Ocean icecap has retreated by 5 per cent, about twice the area of Norway, and over the past couple of decades the melt has been accelerating. The amount of ice drifting down from the Arctic to the Greenland Sea has fallen by nearly 40 per cent as the sea has grown warmer, although the Labrador Current to the west of Greenland has turned colder.

Further reports of changes in the Arctic have been coming in thick and fast. In April 1996 a team of US and British scientists made an epic voyage across the pole by icebreaker in a 3700-kilometre (2300-mile) voyage from Alaska to Iceland. They discovered that the North Pole itself is melting. They had expected to park the icebreakers amid floes measuring 3 metres (10 feet) thick, the kind of ice seen in the 1970s during the last major US Arctic initiative. Instead, the crew was shocked by what it encountered in 1997: 'When we went up there, the first problem we had was trying to find a floe that was thick enough. The thickest ice we could find was 1.5 to 2 metres [5 to 6.5 feet],' said Donald Perovich of the US Army Cold Regions Research and Engineering Laboratory in Hanover, New Hampshire. The rate of melt is so fast that they even went as far as to predict that the *entire* polar ice cap would disappear some time in the twenty-first century.

The team also found that the shallow waters of the Arctic Ocean were less salty than the previous expedition twenty-two years earlier, thanks to melting ice. Their results were backed up by a submarine expedition that travelled under the North Pole; by taking echo-soundings, this found that the ice was starting to melt before it had even left the Arctic Ocean.

Sea ice is only a thin skin over the Arctic Ocean, which makes it especially sensitive to climate change. But it also serves as the linchpin of the region's climate and perhaps that of the whole globe. Because ice is so bright it reflects more than half the sunlight that hits it during summer and refrigerates the Arctic Ocean. On the other hand, the naked

Arctic water *absorbs* 90 per cent of the incident sunlight. That is going to heat the atmosphere above and so more water is going to evaporate. Already the weather is changing over most parts of the Arctic Ocean, with atmospheric pressure sinking lower every year since 1988; wind patterns have changed as well.

Are we now starting a chain reaction that we will never be able to stop? The big fear is that the sea warms up so fast that the entire Arctic climate runs out of control. Ocean currents and weather patterns further south could be disturbed, sending world temperatures climbing even higher.

Added to that, there is another catastrophe hiding in the ground at the Arctic Circle. The permafrost on Arctic land is thawing out the old remains of plants, releasing huge amounts of methane, a greenhouse gas far more potent than carbon dioxide, and which could potentially send global temperatures soaring even higher. And so the greenhouse effect could spiral out of control in a methane-triggered chain reaction.

Having said all this, the interactions between sea, ice and atmosphere are so complex that all these apocalyptic scenarios are far from certain. What is clear is that the warming Arctic is already melting land glaciers and thawing permafrost across Canada, Scandinavia and Russia. Animals, trees and plants from the south are pushing northwards on land, while fish such as cod are penetrating deeper into the Arctic seas. Polar bears are facing starvation as the pack ice they depend on for their hunting grounds becomes too thin or disappears. The snow caves they rely on for rearing their young are collapsing in the warmer springs, exposing the cubs to the harsh Arctic weather too early in their lives. Polar bird life is also changing. This has been followed closely for decades by George Divoky at the University of Alaska, Fairbanks, who explains, 'We're certainly seeing the effect of climate change in the Arctic. The summer ice edge has retreated so far from the Arctic

Alaskan coast that coastal species that used to breed there and used to use the coast for feeding have greatly decreased.' In their place, guillemots have flocked in from the south as the climate has warmed, thriving in the open waters revealed by the melting ice.

The shrinking ice floes also affect the Inuit, such as the settlement at Barrow in Alaska. Marine mammals there use ice floes about 190 kilometres (120 miles) out to sea for rest and shelter during their seasonal voyage, but the animals have disappeared along with the ice, leaving the Inuit little to hunt for food and skins.

What now makes the hair stand up on the back of scientists' necks is that all these changes in the Arctic are setting the scene for another 'flip' in the Atlantic. The melting Arctic Ocean sea ice and lack of winter sea ice in the Greenland Sea are creating a flood of freshwater into the sea water around the Arctic, and making it so dilute that the Gulf Stream is losing power and may eventually grind to a halt. Peter Wadhams, director of the Scott Polar Research Institute in Cambridge, has found evidence from Greenland that one of the natural 'pumps' that drives the Gulf Stream has not worked for the past few years because the sea ice there has failed to form. It is called the Odden Feature – a tongue of ice that forms off the coast of Greenland. This is where water is sucked down from the surface to the seabed and draws the Gulf Stream north towards Iceland and Scandinavia. It also helps drive the vast transworld ocean current which takes heat from the tropics to the cold polar regions. Without the Odden Feature's sea ice the current has lost half its engine, and the Gulf Stream will weaken and be less effective in heating Europe.

'The changes are out of all proportion to anything that anyone has experienced in modern times,' says Wadhams, and he fears much worse. 'They're very rapid changes, and frighteningly rapid as far as changes in vegetation and crop yields are concerned.' Wadhams's research has many years to run, but the prime suspect for the disappearance of the Greenland ice is global warming.

The Gulf Stream sweeps up from the Gulf of Mexico and laps around the shores of the British Isles and north-western Europe with warm air, eventually petering out towards the Arctic Circle. This ocean heat is our free passport to mild winters – when you consider that London is on the same latitude as the Hudson Bay in Canada and Belfast lies equal to Novosibirsk in Siberia, you can appreciate that the Gulf Stream keeps people here about 5°C warmer in winter with the energy of some 20 million large power stations. It is calculated that the Gulf Stream brings a *third* as much heat as the Sun. Where the full force of the Gulf Stream hits County Kerry in the south-west corner of Ireland, it has nurtured a natural subtropical paradise of plants and trees that you would normally expect to see only on the southern tip of Portugal or Spain.

Computer simulations suggest that as soon as the world is a little warmer than it is now, the entire Gulf Stream will become unstable and flips erratically. 'According to the computer models the Gulf Stream will weaken so that the climate of north-west Europe will cool,' Wadhams explains. When exactly the Gulf Stream is due to shut down altogether is the crucial question scientists are trying to fathom, but if Wadhams is correct, time is running out.

It is ironic, then, that all the talk at the moment is of global warming turning Britain into a hot Mediterranean paradise with vineyards and a south coast like the French Riviera. Those predictions are based on temperatures set to increase by 2°C this century, but the local predictions for Britain will look pretty feeble if the Gulf Stream disappears from its shores in just a decade or so.

The paradox is that even though most of the world is warming up, Britain and much of north-west Europe could be thrown into terrifyingly cold winters and chilly summers. But what exactly would it be like living here? Probably the best recent comparison is to look back at the winter of 1962–3 when Britain was in the grip of the most ferocious cold spell for 200 years. In that winter, temperatures stayed so low that snow lay on the ground continuously from Boxing Day to March; the sea froze off many eastern coasts; and ice floes bobbed around in the English Channel. Transport ground to a halt in the snow, and power cuts left swathes of the country cold and dark as the national grid found itself unable to cope with the huge demand for electricity. Some forty-nine people were killed directly by the cold, and unemployment increased with 160,000 workers laid off. For farmers, thousands of sheep died and pneumatic drills were used to dig up root crops such as parsnips. Hundreds of thousands of British birds died but for other species, like the snowy owl from the Arctic, new opportunities opened up as they started flying further south into the Shetlands.

But even that winter hardly compared to periods in the 1600s and 1700s when Britain was in the grip of a cold epoch called the Little Ice Age. Some winters were so cold that the Thames froze over thick enough for 'frost fairs' with bonfires and booths, which became a regular feature for weeks on end. It was so cold that the ice at one time stretched from Iceland to the Faroe Islands, just 320 kilometres (200 miles) north of the Shetlands. People walked out on the frozen sea to ships trapped in the Firth of Forth, and Eskimos even visited Aberdeen. The dismal climate devastated a succession of harvests in Scotland, leading to famine and mass starvation.

The worst picture of a future in Britain without the Gulf Stream would be taking on the climate of Spitsbergen, 960 kilometres (600 miles) inside the Arctic Circle. During the summer temperatures could climb to 15°C, but winter temperatures could plummet to –13°C or lower. Each winter, London could be several feet under snow, with rivers frozen and glaciers forming in the mountains of Wales and Scotland. The climate would also turn drier because cold air carries less moisture – rainfall would average only a couple of centimetres (about an inch) a month. The views across the countryside would be spectacular because there would be few trees left to get in the way – only small polar willows and stunted dwarf birch would grow among the mosses and lichens. Birdwatchers would see snow buntings, ptarmigan, sandpipers and eider ducks, and instead of red deer and badgers there would be musk ox and polar bears. Seaports would need icebreakers to stay open; vast numbers of snowploughs would have to be ready for roads, railways and airports; farming would completely change and new power stations would have to be built. Britain already has one of the worst winter-cold death rates in Europe and if heating, clothing and even outdoor behaviour did not change radically the population might well decline.

The trouble is that the British people have been lulled into a false sense of security over the past two decades during a remarkable run of mild winters. Could they adapt to such a dramatic climate shock? 'If we have a change of five or six degrees in a decade it's almost inconceivable that society could adapt to that without huge disruptions,' predicts glaciologist Jack Dibb. The future on these islands rests on the Gulf Stream, and no one knows how much punishment it can take before wreaking horrible revenge for its mistreatment at the hands of global warming. All we can do is pray that it is resilient enough to ward off the onslaught of melting Arctic ice. Yet if the Gulf Stream does hold up we will steadily grow warmer, so perhaps the big question for the twenty-first century is: 'Are we going to freeze or fry?'

The other end of the world

So far we have looked only at the Arctic ice, but the largest icebergs and ice sheets in the world are in the Antarctic.

On the face of it, both polar regions look similar – but appearances are deceptive. Whereas the Arctic is mostly ocean, the Antarctic is a massive land mass with the Antarctic ice cap covering 13.5 square kilometres (5.2 million square miles), compared to Greenland's 1.7 square kilometres (just over half a million square miles). Its ice cap is on average about 1,800 metres (just over a mile) thick, containing an astonishing 70 per cent of the world's fresh water.

While the sea ice of the Arctic changes by 20 per cent between summer and winter, the Antarctic sea ice varies by 80 per cent and effectively doubles the size of the landmass. That huge expanse of sea ice helps cool the Antarctic because the snow-covered pack ice reflects so much sunlight that it delays the warming effect of the southern spring in September and October. It also helps to keep the air cold, and because cold air is dry the sea ice helps to make the Antarctic the largest desert in the world, with the equivalent of 5 centimetres (2 inches) of rain a year in the interior of the continent (although a lot more falls around the coast).

About half of Antarctica's coastline is bordered by ice shelves, a tenth of its total area. They are hundreds of feet thick, fed by huge inland glaciers flowing out under their own weight, which slip into the sea and float, underpinned by subterranean hills and mountains, and possibly helping to block other glaciers from rolling into the sea. Over 20 million million tons of ice each year falls into the Southern Ocean, which sounds like a catastrophic loss except that it is balanced by a similar amount from snow falling on the continent.

Antarctic icebergs are monsters compared to their puny northern cousins, weighing up to 400 million tons, ten storeys above the water; and occasionally truly colossal bergs break off – measuring 160 kilometres (about 100 miles) or more across. These icebergs can roam for years over large parts of the Southern Ocean up to about 1,600 kilometres (1,000 miles) farther north than the extent of sea pack ice, but mostly out of harm's way as they are swept along in the strong Southern Ocean currents south of about

60°S. But some do stray north and the furthest an Antarctic iceberg has been spotted was about 50 kilometres (30 miles) south of the Cape of Good Hope, South Africa, in 1850.

Because the Antarctic is so far from civilization its icebergs affect far less shipping than in the North Atlantic. But the most southerly shipping lanes still have to be on the lookout for the largest Antarctic icebergs, and these are tracked using satellites by the National Ice Center based in Suitland, Maryland, run by the US Navy, the National Oceanic and Atmospheric Administration and the US Coast Guard. When an iceberg is identified, the National Ice Center documents its point of origin and follows its progress, sometimes over several years as huge icebergs break up into flotillas of smaller bergs.

Anarchy in the Antarctic?

The Antarctic ice shelf is now going through some extraordinary changes. Vast icebergs have sheared off in recent years and it was all brought home in stark television pictures in 1997 when Greenpeace sent an expedition to Antarctica and filmed an ice shelf cracking up with a rupture in the ice as wide as a football pitch stretching for miles as far as the eye could see. The message it sent to the world was clear – global warming was finally hitting the world's largest ice sheets, with dire consequences for all of us if the Antarctic melted and raised sea levels.

The evidence from previous ice shelf incidents already seemed to be damning. In 1967 an iceberg sheared off an ice shelf facing the Indian Ocean, collided with another shelf and tore off one of the largest icebergs ever known so large that it was christened 'Trolltunga'. With an area of around 8,000 square kilometres (3,000 square miles), it survived twelve years before eventually breaking up.

In September 1986 a huge piece of the Filchner Ice Shelf facing the South Atlantic broke off to form three mammoth icebergs which, combined together, were almost as large as

Northern Ireland. Even more alarming for the scientists working there, the giant icebergs also carried off three of their research stations. Then, most recently in March 2000, an iceberg roughly the size of Connecticut broke free from the Ross Ice Shelf opposite the Pacific Ocean.

You could argue that we know so little about Antarctic ice shelves that these could all be part of a natural cycle of events. But five Antarctic ice shelves out of nine studied have disintegrated in the past fifty years, a loss equivalent in size to the area of Cyprus. By far the biggest change is happening in the Antarctic Peninsula which juts out towards South America like a provocative finger. Here the floating ice shelves are breaking up at such a startling rate that they are in danger of disappearing altogether and redrawing the map of Antarctica for ever.

In the late 1980s the Wordie Ice Shelf on the west side of the peninsula broke up. It covered 2,000 square kilometres (770 square miles) a couple of decades ago and was regularly crossed by scientific parties. Soon afterwards 1,300 square kilometres (500 square miles) of ice disappeared from the Larsen Ice Shelf which runs along the north-eastern tip of the Antarctic Peninsula. In January 1995 the northernmost part of the Larsen Ice Shelf suddenly collapsed during a storm, breaking off an iceberg the size of Oxfordshire. Scientists who witnessed it said they could not tell it was an iceberg because it filled the entire horizon.

The 1997 Greenpeace expedition had shown massive rifts in the remainders of the more southerly Larsen B Ice Shelf, and the following year satellite pictures confirmed the collapse of 195 square kilometres (75 square miles) of that shelf. The remaining ice is now the most northerly ice shelf surviving in Antarctica but, as pieces shear off, it is like bricks being taken out of a bridge, making the ice sheet weaker as it flexes in the sea's waves until eventually it too will completely crack up.

'Here are things we thought were permanent which

have just crumbled away. It's as if a godlike hammer has fallen on them,' commented glaciologist David Vaughan of the British Antarctic Survey.

Changes as drastic as these look irreversible in the foreseeable future, and on the Antarctic Peninsula land itself grasses are now growing on newly exposed bare rock, possibly for the first time in thousands of years.

The wildlife is also going through dramatic changes. The hardy little Adélie penguins have been studied by Bill Fraser at Montana University and he has found their numbers on the Antarctic island of Orbison have collapsed by 60 per cent in twenty years. Part of the reason is that more snow is falling on the ground, which makes it difficult for the small Adélie penguins to get through to their breeding areas. Also, in winter the penguins need ice to stand on before they dive through cracks to catch food, but when there is less sea ice the adult and young penguins struggle to survive. The Adélie shares its breeding grounds with sealions and elephant seals and their numbers are increasing, thus breaking up the Adélie colonies, leaving their eggs and chicks more exposed to attack from skua birds.

Meanwhile, the Adélie are being ousted by chinstrap penguins which seem be to thriving on the shrinking sea ice and warmer seas. The chinstraps are spreading further south and their breeding success rises abruptly after warmer winters – their population has more than tripled between the 1940s and the 1980s.

Elephant seals are suffering a drastic fall in numbers on the tiny sub-Antarctic Macquarie Island, half-way between Tasmania and Antarctica. Elephant seals are usually found everywhere along the island's coastline but scientists now estimate that the population has halved from 200,000 to 100,000. Harry Burton of the Australian Antarctic Division believes the elephant-seal populations are falling due to shifting ocean currents and climate changes: the average temperature there has risen from 4.5°C in 1912 to

5.4°C today. 'When a large mammal that can weigh five tons starts to disappear, that is a significant signal coming from the marine ecosystem,' he says.

A big problem in studying any climate change in Antarctica is that weather records are very brief because the first permanent scientific bases were not established until after the Second World War. Recently the archives were pushed back much further by a brilliant piece of historical detective work. Australian Bill de la Mare dusted down old whaling records going back to 1904 and from them revealed how the sea ice surrounding the Antarctic is shrinking. The old whalers knew that whales tended to congregate near the ice edge of Antarctica each spring and their ships' logs showed that the whaling fleets had to push further southwards each year because the ice was shrinking: between 1950 and 1970 the area covered by sea ice retreated by a staggering 25 per cent.

This loss of sea ice affects the climate in a self-driven cycle. Like the Arctic, the Antarctic is its own refrigerator because white sea ice reflects sunlight and insulates the ocean and so keeps it frozen. But after a warmer year the reverse can apply and dark ocean waters absorb sunlight and increase warming, so the ice retreats and continues to go on retreating.

The immediate culprit behind these astonishing changes is not too difficult to pin down. The temperature on the Antarctic Peninsula has shot up by a staggering 2.5°C in fifty years, while the rest of the world has warmed by about 0.3°C in the same period. It has been a fairly consistent trend – the longest weather records in Antarctica at Orcadas Station, South Orkney Island, on the edge of the sea ice limit, indicate warming there since the 1930s.

As global warming began to be taken seriously in the 1970s, all eyes turned with great apprehension to the white continent as the new climate began to bite. Fears were raised that the Antarctic icecap covering the land could collapse. The plot went like this: if the greenhouse effect was to warm the south polar region by just 5°C, the floating ice shelves surrounding the West Antarctic ice sheet further south of the peninsula would disappear. Robbed of these buttresses, the vast ice sheet on land might slip into the sea, disintegrate and send world sea levels surging by an estimated 5 or 6 metres (16 or 20 feet). Low-lying areas such as the Netherlands, parts of eastern England, Bangladesh, the Mississippi, Miami, the Nile, the Mekong Delta and many Pacific islands would face catastrophic floods.

It was a terrifying scenario and largely theoretical, but there was evidence that the West Antarctic ice sheet had melted at least once before. In the Eemian period between 110,000 and 130,000 years ago in the last big interglacial, the world warmed up to a much higher temperature than today. Sea level stood about 5 metres (16 feet) higher than it does now, submerging many of the world's coastlines and islands. If the ice sheet had collapsed before, the reasoning goes, then the present-day warming might repeat the same performance.

The theory sparked a group of Americans to look for signs of modern ice collapse, and their first reports were indeed ominous. The West Antarctic ice sheet moves in streams, creeping over the bedrock below, some of the ice pushing out into the sea as icebergs. The Americans saw five streams of ice pulling ice from the interior of West Antarctica into the sea and 'may be manifestations of collapse already under way'. That pressed the alarm bells and some experts warned of a global flood the like of which had not been seen since the days of Noah. The Antarctic ice could disappear in a domino effect, they said: as one piece of ice gives way the one behind it lurches forward, and so on until all the ice ends up in the sea as monstrous icebergs. If Antarctica melted entirely, sea levels across the globe would rise 60 metres (200 feet). The world as we know it would be doomed.

So it would seem that Greenpeace was right and Antarctica is finally cracking up and will trigger world flooding. But although the changes in the Antarctic look like the clues to imminent global catastrophe, there is another side to this story we rarely hear about in the media.

The connection between climate warming and the movement of West Antarctic ice streams has become increasingly tenuous. The ice seems to stop and start, and no one really knows why because it is incredibly difficult to find out what is going on underneath the ice as it rolls over the land buried below. The latest evidence from satellite radar suggests that the West Antarctic ice sheet is very stable and almost in equilibrium – the amount of new snow falling on it matches the amount of ice flowing into the sea as glaciers. It is neither growing nor shrinking, and there is no evidence of any collapse in the West Antarctic ice sheet. Scientists such as David Vaughan of the British Antarctic Survey caution against scaremongering: 'Nobody is saying at the moment that we are going to lose the West Antarctic land ice.' Even if the temperature rise doubles in the next century, ice sheets are remarkably stable lumbering giants, taking thousands of years to respond to changes in surface temperatures because it takes so long for those changes to penetrate close to the bottom of the ice where it slips along the ground on a bed of water.

Even the extraordinary changes in the Antarctic Peninsula look different when set in a wider picture. The peninsula is much further north than the rest of the continent so it is more sensitive to outside influences. Temperatures in the interior of the Antarctic continent are showing little sign of change – at Britain's Halley Station on the eastern side of the Weddell Sea there is no warming trend, and there has even been a slight *cooling* at the South Pole in recent years. It could well be a similar story in other places, but unfortunately climate records are few and far between in the Antarctic.

David Vaughan believes the upward temperature trend in the peninsula 'is almost certainly a result of local conditions there. We suspect an instability in the local climate – this could be linked to ocean circulation, the amounts of sea ice forming, routes of depressions, and many other factors that may have nothing to do with Antarctica.'

His point was rammed home during the last El Niño in 1997–8, when warm waters from the Pacific surged through the gap between Antarctica and the tip of South America and raised sea temperatures so high in the Southern Ocean that it killed off the local krill – a shrimp-like creature – leaving seals, albatrosses and penguins without enough food and their breeding levels collapsed. Ice shelves are much more sensitive to temperature than the glaciers because slight changes in the temperature of sea water can melt them, so they bore the brunt of the El Niño warming.

And even if the temperature across the entire Antarctic continent were to rise, a warmer climate should actually protect some ice shelves by thickening them. As we saw with the Greenland ice sheet, one of the great paradoxes of global warming is that as the climate warms it might make more ice, not less. Warmer seas evaporate more water, making more clouds, which drop more snow and that makes more ice. Computer models of greenhouse-induced warming at the Antarctic show no dramatic change in ice volume, and the volume of ice in the West and East Antarctica ice sheets are predicted to increase, not decrease in future warming predictions. The increased snowfall over the continent may even compensate for some of the melting of glaciers in other parts of the world. This is not the picture we have been getting from most television and press reports.

Bearing out these predictions, the rate of snowfall near the South Pole has mounted substantially in recent decades. Donna Roberts at the University of Tasmania has found another sign of increasing snowfall, in that salt water lakes in Antarctica have turned less salty in the past two centuries than at any time in the preceding 6,000 years as more snow dilutes the water. Even though the climate fluctuated between drier and wetter periods during this time, for the past two centuries it has been steadily getting much wetter – possibly because of global warming. 'What has been happening to the climate in Antarctica over the past 200 years is vastly different from what it was like in the preceding few thousand years,' Roberts explains.

Another nail in the coffin of the scaremongers is that the sea itself could save some ice shelves from collapse. The Filchner-Ronne Ice Shelf, the most massive of the southerly ice shelves at the foot of the Antarctic Peninsula, is fed by warm salty water under the shelf. But in warmer winters less of the salt water feeds underneath, so the bottom of the ice shelf cools and less of it melts.

Some of the media has also been drawn into thinking that melting ice shelves will raise world sea levels, but this is also completely wrong. Because the ice was already floating in the sea before breaking off, the melting ice shelves make no difference to world sea-level rises – it is the same as ice cubes melting in a drink of water. And dramatic as the retreat of the Peninsula land ice has been, it is only 1 per cent of the total ice volume of Antarctica.

In short, we have to be careful about jumping to conclusions about global warming and the ice caps. There are colossal changes going on in both the polar regions, and although some of the Antarctic ice looks seriously damaged it is the Arctic we should be more worried about at the moment. Unfortunately we are only just starting to understand how the poles affect the rest of the world's climate and we are in a race against time trying to find the answers we need to forecast the future climate. The consequences for us could be the greatest challenge to the human race since the last Ice Age ended in a violent bout of climate convulsions.

LETHAL SEAS

David Jackson

'The Sea begins to boil and ferment with the Tide of Flood,' wrote Martin Martin of his visit to the Gulf of Corryvreckan in the late 1680s. The natural wonder he saw lies just off the northern tip of the Isle of Jura and 'not above a Pistol shot distant from the coast of Scarba Isle'. As the rising tide comes through the strait, 'the boiling [turbulent water] increases gradually until in it appear many Whirlpools'. In seventeenth-century language Martin reveals the horror associated with its ancient name: 'They call it the *Kaillach,* i.e. an old Hag,' he wrote, 'and they say that when she puts on her *Kerchief,* i.e. the whitest Waves, it is then reckon'd fatal to approach her.'

The *Kaillach* appears in several different 'manifestations' which build up a truly awesome spectacle: eddies of swirling water, huge waves falling apart at the top, swells, standing waves – upwellings or bulges in the sea surface, a lot of white water and – if you are there at the right time – whirlpools. It is difficult to believe that some kind of mysterious force is not at work here. Add the science and you will become drawn into a marine phenomenon that has still not revealed all its secrets.

The area where whirlpools are created is found, as Martin Martin suggested, very close to the island of Scarba. However, the most accessible viewing point is on Jura. It is quite an expedition on foot – a 25-kilometre (16-mile) round trip from the tiny settlement of Lealt. But you do have to choose the weather and tide carefully: ideally, a strongish wind from the west confronting the fastest-moving tidal stream flowing from the east.

The tidal stream here builds up in speed to be one of the fastest in Europe. At its peak, water will be moving through the gulf between the islands at an amazing 4.5 metres (15 feet) per second. The traditional whirlpool site is about 300 metres (nearly 1,000 feet) off the cliffs on the western side of Camas nam Bairneach on Scarba – that is a small bay along its southern rocky shore. So remote and wild is this region that the bay's name isn't even on the map. Nor is it easy to get to the high ground on Scarba, where the best photographs from land have been taken. From there, the full length of the narrow Gulf of Corryvreckan separating the southern coast of Scarba from the northern coast of Jura is spread out below.

At the western entrance to the gulf, where the Corryvreckan spills into the Firth of Lorn, is an extensive patch of turbulent water known as the 'Great Race'. The name associates this area of unpredictable eddies and standing waves with its scientific designation of a 'tidal race'. There are several other tidal races found between the islands in this part of western Scotland.

Looking down on the Great Race from a high point on Scarba, the most ordered stretch of water is the main tidal stream running straight through the centre of the gulf. Like the backwash from a speedboat, the tidal stream creates a backwash of eddies (those swirling or circular movements of water that develop whenever there is a flow discontinuity). These appear to break off the stream, drift backwards and fan out to both the coasts of Jura and Scarba. They are clearly seen in calm conditions when the tidal stream is 'setting westward', that is, running with increasing speed from east to west. The eddies are not so visible if the west wind is blowing very strongly and there is a large swell and massive breakers at the entrance to the gulf.

Understanding the behaviour of tidal streams is very important to mariners. Each tidal stream is described in detail in the Admiralty's *West Coast of Scotland Pilot*. After leaving the gulf, the west-going tidal stream preserves its direction for some miles, although it gradually decreases in speed. It is moving a large body of water at fast speeds for several hours. The central core is 30 metres (about 100 feet) deep and at least 30 metres wide, if not more. That is a lot of energy to dissipate. Most of the energy is lost through friction between the moving water and the rocky seabed. Ultimately the energy goes to heat, which warms the water by an imperceptible fraction of a degree. But some frictional energy is lost by the eddies until the tidal stream's momentum runs out. The release of these eddies marks out the long westerly course of this tidal stream into the Firth of Lorn.

Tides and tidal streams

The tidal stream 'sets westward' through the Gulf of Corryvreckan with the flood tide, and returns eastward with the ebb tide. The stream reverses its direction. Roughly twice a day the tidal stream moves a huge body of water laterally: forth and back, forth and back. These tides are caused by the rotation of the Earth and the gravitational attraction of the Moon. At the same time as the Earth is rotating, the Moon is also moving in its orbit around the Earth. This means that the time interval between the Moon passing twice over any one meridian on Earth is slightly longer than a twenty-four-hour 'Earth' day. During this time the Earth experiences two tide cycles.

But why, in broad terms, are there two high tides and two low tides a day? Gravitational pull attracts a bulge of water in the hemisphere of the Earth closest to the Moon, and the centripetal force of the rotating Earth/Moon system produces a complementary bulge of water in the opposite hemisphere – farthest away from the Moon. These bulges create the two high tides and are compensated for by two depressions that separate them and experience low tides. During the rotation of the Earth, any one point experiences all four situations. In fact the bulges are not quite equal. A slightly greater bulge occurs when the Moon is on the same side as the Earth. Then the high water is slightly higher than the high water caused by the smaller bulge on the opposite side of the Earth. So when that bulge 'rotates' round to where we are and we get the second high tide of the day, it is a lower high tide than the previous one.

To find out when the highest tides occur, we need some more astronomy. The Moon orbits the Earth in 27.3 days. Because the Earth is moving too, it takes a little longer for us to observe two successive full Moons, and what we call the 'lunar month' ends up being a 29.5-day cycle. On two occasions within this lunar month Sun, Moon and Earth align and exert the maximum gravitational pull on the sea water. In turn, this produces the maximum tidal range – the highest high water and the lowest low water – that we experience at the shoreline. These two tides are *spring* tides. In contrast, twice a lunar month an arrangement of Sun, Earth and Moon occurs that produces a minimum tidal range. These two tides are *neap* tides. Obviously, throughout the lunar month there are intermediate situations where the tidal range builds up to a maximum at the two springs and decreases to a minimum at the two neaps.

Although in its slightly elliptical orbit around the Sun, the Earth is closest on 3 January, this is not the time when the greatest spring tides occur. These highest tides are spring tides nearest 21 March and 21 September – the spring and autumn equinoxes. The equinoxes separate the summer and winter seasons when the Earth's axis tilts to its fullest extent either towards or away from the Sun. Only at the equinoxes is the Sun exactly at right angles to the Earth's axis. As a result, at noon the Sun is directly overhead at the equator, making the length of day and night equal. Also, the combined influence of the Earth and Moon is at its greatest at the equinoxes – and so the highest spring tides occur. These two maximum tides of the year are called the equinoctial springs.

Therefore the tide table takes into account three tidal periodicities: daily, monthly and yearly. But it can be confusing. Does the table describe what is experienced on shore or out at sea? On shore, we are most aware of high water and low water and we are really concerned with the vertical difference in the height of the sea water. Out at sea, mariners are more concerned with the water flow, and their 'high tide' is when the flow is moving with maximum velocity. It so happens that on the west coast of Scotland – and this is just a coincidence – the time when the tidal current out at sea (and generally parallel to the shore) is running at its fastest is pretty much the same as the time of high water on the beaches. But the time the tide turns out at sea is roughly three hours earlier and three hours later than the time we would say the tide 'had turned' if we were on the shore. If you draw out the tide cycles in terms of current speed and water height this becomes clear.

The west-going tidal stream – and that is the one to go for to see the Corryvreckan in all its glory – 'runs' for six hours and the very best effect will of course be in the middle of that period. The *West Coast of Scotland Pilot* tells you that the time the tide runs is one hour before and five hours after the time of high tide at Dover. So, equipped with a standard UK tide table, you could answer the key question: what is the very worst situation a mariner might find in the Gulf of Corryvreckan?

Dougie MacDougall, a retired skipper living on Jura's neighbouring island, Islay, says, 'When a strong westerly wind meets a westerly flowing spring flood, the breakers at the overfalls can be 20 feet (6 metres) high. The Corryvreckan roar can be heard over a 20-mile (30-kilometre) stretch of coast.' He warns, 'The gulf is not to be tampered with, and never on a flood tide with a force 3, or you are indeed in very bad trouble.' Just imagine what it would be like during gale-force 8 or storm-force 10 ...'

The effect of high wind speed is obvious enough – but exactly why is the water so fast? This is not obvious from a map, but very clear from the Admiralty chart. The Corryvreckan's sea topography is a narrow straight trough, 1.6 kilometres (1 mile) long and 120 to 30 metres (about 400 feet) deep. It acts like a 'strait' channelling water between the two coasts of Scarba and Jura. During the flood part of the tidal cycle water is funnelled into it from the Sound of Jura. Constantly gaining speed, it approaches the 'strait', and as it is forced through it soon becomes a very fast moving tidal stream. Water moves in the opposite direction during the ebb part of the tide cycle. It is funnelled from the Firth of Lorn through the strait and at ever increasing speed it shoots water back into the Sound of Jura.

The west-going spring-flood tidal stream builds up to 8.5 knots – nearly 16 kilometres (10 miles) per hour. Some say this is an underestimate and the speed is more like 14 knots – 26 kilometres (16 miles) per hour. It is as fast as this because high water occurs thirty minutes earlier at the east end. There is a difference in water height of at least a metre between the east and west ends of the strait. This height difference over such a short length of sea adds a substantial force to the current until the hydraulic head (height difference) is obliterated. A simple analogy would be a river where the speed of water flow between two points can be related to the drop in height. The speed can be calculated by converting the potential energy associated with the height drop to the kinetic energy associated with the motion. Elsewhere in the world where hydraulic pressure is greater, tidal streams have reached speeds of 20 to 25 knots – 35 to 45 kilometres (23 to 29 miles) per hour.

The east-going spring-ebb tidal stream through the gulf does not run with quite the same rapidity and violence as the west-going stream – because it emerges into the comparatively tranquil waters of the Sound of Jura, which are sheltered from the Atlantic Ocean. Or, at least, that is the simplified explanation.

So here are two good tips from the *West Coast of Scotland Pilot*: If you are determined to sail through the gulf, choose the passage from west to east. If you must go in the opposite direction, batten down the hatches, steer for the middle of the gulf, hope to be carried through in the centre of the tidal stream, and note that the most violent breakers lie on each side – theoretically!

The whirlpool

As Martin Martin suggested, there is not just the one 'Corryvreckan whirlpool' –but several. Although the Ordnance Survey map shows only one, the site marked 'whirlpool' is the place where they hit the breakers when the tide is flowing in a westerly direction. This is moving very quickly, while the whirlpools move in a westerly direction and fan out at a slight angle to the main stream. The whirlpools on the northern side of the westerly flowing tidal stream generally rotate clockwise whereas those on the southern side rotate in the opposite direction.

Whirlpool formation

The Admiralty chart shows that a submerged pinnacle exists very close to where the whirlpools are heading. The pinnacle's top is roughly circular, about 20 metres (65 feet) in diameter and about 30 metres (nearly 100 feet) below the surface. To the south, east and west the steeply sloping sides of the pinnacle drop 100 metres (over 300 feet) to the sea bottom. To the north the pinnacle is connected to the Scarba massif by an underwater rocky ridge and this runs to the headland immediately west of Camas nam Bairneach. About 800 metres (2,600 feet) to the east of the whirlpool site is a pit in the seabed. The pit is about 250 metres (800 feet) long by 50 metres (160 feet) wide and 219 metres (a little over 700 feet) deep at its deepest point.

The most consistent explanation of how Corryvreckan's whirlpools form – and this also applies to tidal whirlpools in other parts of the world – involves a flow separation of a fast-moving stream through an essentially stationary body of water. Eddies form on either side of the stream at the boundaries between the fast and slow water. Small vortices can be seen in calm water. But to form substantial whirlpools, the energy conditions have to be high, and flow separation has to be stimulated by other factors as well. If the fast-flowing current is influenced by unusual sea-bed topography – in Corryvreckan's case, the pit and the pinnacle – then the eddies that form are so energetic that they become whirlpools. When flow separation occurs at depth, subsurface layers have to move at higher speeds than surface layers to keep up. This creates an unstable situation on the surface, and turbulence, and whirlpools form.

Some accounts suggest there is a huge upsurge caused by another current, said to gather momentum as it climbs up the sides of the pit. This submarine current is then supposed to follow the sea-bed topography up the ridge to the pinnacle beneath the whirlpool. But there is no corroborative evidence that this current really exists and it may just be another way of looking at the effects of the enhanced flow separation.

Boatman Duncan Phillips says: 'Sometimes the Corryvreckan is benign and fishermen go through it every day. Then again it turns horrendous in two minutes, and I'm sure that people must have lost their lives there.' Seeing a central funnel develop in a whirlpool depends on the forcing conditions exerted by the currents. When the flow is strong enough and the channel is constricted, the fluid dynamical frequencies of all the water particles involved come together. The water particles spin in harmony around an axis – and a whirlpool is created.

To be sucked into a whirlpool is surely a primordial fear – but could it actually happen? We are aware that there should be a strong downward force in its centre. But the laws of physics require the downdraught to be balanced by an upwelling so that the whirlpool remains in equilibrium. When there is a whirlpool, a surrounding or nearby upwelling would tend to repel any object approaching it. To experience the downdraught and actually be sucked into the centre would require a determined effort – or some external agency like a storm-force wind.

It is just possible to dive at the traditional 'whirlpool' site. Gordon Ridley's guide *Dive West Scotland* – written from a sport diver's perspective – says that when the pinnacle, 30 metres (95 feet) down, was dived for the first time in 1982, 'interesting experiences' were had by all. The 'safe window' is twenty minutes, but only when the conditions are very quiet during slack water at neap tides. A commercial scallop diver could not stop himself being dragged down: 'He therefore fired his ABLJ (Adjustable Buoyancy Life Jacket), and still found himself going down. He saw the bottom at 75 metres (nearly 250 feet)! It was going past very quickly, but at last his ABLJ finally pulled him towards the surface! Be warned.' Gordon Ridley emphasizes, 'The dangers of terrifying vertical down-eddies must never be underestimated. Many consider it to be the most serious undertaking in British waters.'

The downward flow within a whirlpool is very strong. It has been measured at half a metre per second even within only moderately powerful eddies. Divers who have ventured into these waters near slack tide talk of the bubbles of their exhaled air heading downwards. Even with fully inflated life jackets they have had to climb back up the underwater cliff face, then pull themselves up a fixed or 'shot' line to the surface. A shot line has a large weight at the bottom and a marker buoy at the top to indicate the dive site.

In an experiment at Corryvreckan, a dummy was

dragged under by the forces acting on the submerged portion of its body. Below 10 metres (over 30 feet) it lost its positive buoyancy due to the compression of the foam material from which it was made. So anyone pulled down to this depth would become negatively buoyant because the air in his or her lungs would have compressed. Incidentally, a depth gauge attached to the dummy went off the scale at 99.9 metres (nearly 330 feet).

During bad weather, rather more dangerous than the 'whirlpool' area is the Great Race, that much larger stretch of turbulence at the western entrance to the Gulf of Corryvreckan. A simpler example of a tidal race in a more gentle sea helps to understand what is going on here. Anyone who has ventured across a tidal race just off a headland will know how unsettling the experience can be – strong currents and white water. Because the sea is very shallow the skipper is often concerned as to whether there is enough water to make it across the race. Understanding how the tidal race forms explains unstable water flow and relates to all the phenomena we associate with turbulence, including whirlpools.

Imagine a body of water in a series of layers all moving past a headland from A to B. These layers extend downwards from the surface and in a sense are a mathematical invention to understand the fluid flow. The lower layers may have to travel a longer distance from A to B than the surface layers – for example if their route is bent to enable them to get over an underwater ridge. The lower layers have to travel much faster over the longer distance in an attempt to keep up with the surface layers and, to some extent, they are dragged along by frictional contact with them. Include the roughness of the seafloor, and all this makes the water in the upper and surface layers very unstable. In extreme circumstances frictional contact between all the layers is reduced and the water becomes chaotic and turbulent.

At the western entrance to the Gulf of Corryvreckan and alongside the Scarba and Jura coastlines there are shallow rocky shelves about 30 metres (95 feet) deep. Given the huge hydraulic force of the tidal currents involved, these shelves are shallow enough to stimulate the formation of a tidal race by much the same fluid mechanics as above. On entering and leaving the western end of the central trough, water on both sides of the tidal stream passes over the undulating topography of the shelves with their rugged bedrock reefs.

Troughs with flanking shelves are a common feature in the Firth of Lorn and are often associated with tidal races. For example, the 'Grey Dogs Race' is just off the north shore of Scarba (and sometimes referred to as 'Little Corryvreckan'). Quite often these races are called 'tidal rips' – a rip meaning a disturbed state of the sea with turbulence, breakers and overfalls. But the real cause of the Great Race is the interaction of the current with the seabed 30 metres (95 feet) down.

To be certain of all the phenomena associated with the Corryvreckan we have to be confident in the first-hand descriptions by those who have experienced them. Here is an eyewitness account of a mariner who was determined to brave the Corryvreckan in a small prawn-fishing boat in 1964. The skipper, Ronnie Johnson, starts off by attempting to sail west through the Gulf, and head on into the prevailing wind towards Colonsay.

'Occasionally the surface would erupt and water would come boiling up from below as if there was an enormous fire on the seabed, keeping the sea boiling. As we passed the north points of Jura the hatches were battened down ... Within a few minutes we would be in a wall of water turning white which we could see ahead of us. Suddenly this wall of water erupted and short, steep seas gathered ahead ... The seas became even shorter and steeper; the white-crested waves overfalling themselves into deep troughs ... Vicious currents leapt into the air and I was kept so busy at the time conning the boat that I had no time to think what was happening ...'

The boat was fighting wind, tide, turbulent water and swell. Johnson turned the boat around and made headway in the opposite direction by catching the edges of the eddies and using their energy to gain ground. He had survived – but on this occasion he did not make it through the Gulf. He continued his voyage by taking the longer and safer way round Scarba.

Ronnie Johnson also describes the loss of buoyancy in the white water: 'I had the impression something was seriously wrong. The boat seemed to be further down in the water than usual from where I was standing in the wheelhouse.' This buoyancy loss is explained by a fluidized bed experiment. A heavy object is supported on the surface of a container of sand and extreme agitation is applied to the sand. Each vibrating sand grain acquires a little pocket of air around it. The body of sand as a whole 'fluidizes' and can no longer support the object, which sinks down into the sand and eventually disappears below the surface. When the vibration stops, the body of sand returns to its former state but there is no trace of the object.

One of the features that Ronnie Johnson described in rough conditions was the 'wall of water' which he found stretching right across the Gulf. So what, beside the wind, are the combination of factors that produce unusual waves and overfalls? First, it is worth remembering that the waves commonly experienced at sea depend on the wind speed, the wind duration and the fetch – that is, the distance over which the wind blows without an impediment that might change its direction. Under storm conditions wind-produced waves out at sea could be expected to build up to a height of 15 metres (nearly 50 feet) and greater.

Second, the idea of making waves when you throw a stone into a pond is very familiar. It is also possible to make waves in a horizontal piece of string if one end is moved up and down while the other end is fixed. The waves move along the string, and spread outwards on the surface of the pond. But do they move or do they give the illusion of moving? Although the waves move across the entire length of string, it has not moved any distance: it is still in your hand. The same is true in the pond. A float on a fishing line will simply move up and down as the wave appears to pass by. In fact, a wave is merely a vibration and when we see it 'travelling' from A to B, we are really seeing the transfer of energy from A to B and not the material substance – not the string, nor the pond water.

But what occurs if the medium is moving – as it does when there is a strong tidal current acting in opposition to the wind-generated waves? It is rather like trying to run up a 'down escalator' where the runner is working hard against the motion but stays put. A similar situation can arise in the ocean when the current is attempting to move water in the opposite direction to the waves generated by the wind. If the water and wave speeds are the same, then the waves stay put and are called standing waves. The wind and current creating these standing waves are called forcing agents.

Mathematically, when more and more energy is added and the waves become larger and larger, they change shape and become steeper and narrower. And this is exactly what is experienced at sea too, except a steep narrow wave often cannot support the water it contains and it 'overfalls'. Standing waves of 3.6 metres (12 feet) high have been claimed even in calm conditions but other reports suggest half this height is more likely. Nevertheless, as a barrier standing waves are a pretty daunting feature in the Gulf of Corryvreckan and present the worst danger to yachts and small boats.

But just how are these enormous standing waves achieved? Because a body of restricted water has a natural mode of oscillation, resonance must be considered. A well-known example of resonance occurs when a guitar string is plucked and the wooden board vibrates with the same frequency. The waves that are produced add up to make (more or less) one very high-amplitude sound wave, that is, the loud noise we hear. This is a resonance of the total system – both the string and the sounding board acting together.

The same principle applies in marine situations, and the Gulf of Corryvreckan is an example: at one instant there is turbulent water plus forcing conditions brought about by the increasing wind and currents. Suddenly there is an enormous standing wave that creates the overfall or white wall of water. Although this overfall occurs at the mouth of the gulf, it is the product of an oscillation of very long wavelength – theoretically four times the length from the mouth to the most constricted part of the strait. The constriction functions mathematically in the same way as a 'closed end'. The theoretically enclosed body of water is said to exhibit a natural resonance when the largest standing wave occurs.

If the wind gets even stronger, it will increase the size of the standing wave still further. This explains 'the wall of water', sometimes romantically referred to as the 'rim of the Cauldron', that Ronnie Johnson saw at the mouth of the Gulf between the islet of Eilean Mor and Jura and – so he claimed – stretching across to Scarba.

Whirlpools worldwide

Guaranteed whirlpools are found in the eddy stream that occurs when a fast-flowing current interacts with a slow moving or stationary body of water. This kind of whirlpool is sometimes formed in rivers, such as in New York's East River and in the Danube to the east of Belgrade. A small example occurs in England almost at the end of the Severn Estuary just before it narrows and meanders sharply west at Hock Cliff near Frampton-on-Severn.

Most whirlpools in rivers are formed and sustained by a different set of circumstances. A very spectacular whirlpool is found in the Niagara River near the Canada–USA border. It is not associated with one of the falls but occurs a little further downstream where the river takes a very sudden turn to the right. At that point a bowl has been scoured out of glacial drift. This soft material had filled a gorge made by an earlier and different course of the river. The bowl makes the fast-flowing water of the main current spin round and form a whirlpool exactly at the turn of the sharp bend.

Plunge pools are found beneath the falling column of water in powerful waterfalls. They are carved out by the turbulent water. When the eddies break the surface, the swirling foaming water may give the illusion of being part of a whirlpool. Indeed, they are often referred to as whirlpools, but they are not.

There are smaller whirlpools with a vortex in the centre when fast flowing water is deflected around potholes in a river bed. Some of the potholes owe their origin to ice melt at the end of the last Ice Age. Others have been formed more recently by swirling stones and suspended sand grinding away the river's bedrock. The grinding process can produce holes that are perfect hemispheres and the ideal means of rotating a stream of fast moving water.

For the very brave, white-water canoeing or rafting can be another way of getting close to whirlpools in extremely turbulent water. A whole new language has evolved to describe different kinds of turbulent waves and swirling flows. For example, in a 'stopper' the water turns over on itself, rotating with such hydraulic energy that there is no escape. It can seize a canoe and break it up. It then churns around the canoeist who is firmly trapped in the flow until he or she drowns. The only chance is for the canoeist to get through before the canoe is caught up in the rotating water and capsized. This is called 'ploughing through the stopper' and is a risky venture to say the least.

But there is another phenomenon in slow-moving

rivers or small streams and brooks. Often the main part of the stream will flow faster round the outside of a bend and, as it does, a shear develops between it and the slower, sometimes stationary, water flow on the inside of the bend. The vortices look like a series of tiny holes gently floating downstream in the quieter water at a small angle to the main current. Even though each has a lifetime of perhaps a few seconds, they keep coming one after another, continually created at the shear plane. I have seen these in a small stream about 3 metres (10 feet) wide and could have watched them for hours. Although the tiny vortices were only 1 or 2 centimetres (less than an inch) in diameter, I could see the water spinning around inside them.

Strangely enough, there are not many well-documented ocean whirlpools worldwide. With a couple of exceptions, ocean whirlpools are in remote locations that take a considerable effort to reach. Also the words 'whirlpool' and 'maelstrom' – despite their strong literary associations – are invariably applied to an area of strong currents swirling around in all directions rather than to the single terrifying whirlpool with a central vortex of popular imagination.

Perhaps the most famous whirlpool was described by Edgar Allan Poe in his novel *A Descent into the Maelstrom*. To see it would involve a visit to the southern tip of the Lofoten Islands, lying off the coast of northern Norway. With their high mountains rising sheer from the sea, these are some of the most spectacular islands in the world. Between the islands of Lofotodden and Værøy, the fast-moving tidal current is known locally as the Moskstraumen. To get to the maelstrom, that section of the Moskstraumen where the whirlpool is supposed to be, involves a four-hour boat trip from Moskenesøy. Poe's book tells you all you want to know about the drama of such a trip, but he certainly exaggerated the facts. For example, you cannot see any of the Moskstraumen as a whirlpool, only a current of turbulent water trending in a circular direction (clockwise during a rising sea, anticlockwise during a falling sea). Extrapolate the arc you see into a full circle and there is the largest 'whirlpool' in the world – reputedly 6 kilometres (3.7 miles) in diameter.

In the Moskstraumen the eddies define the current shear zone and these can be seen on satellite images. The hydraulic head is a 25-centimetre (10-inch) difference in height between sea water in Vestfjorden, inside the island chain, compared with the open ocean on the outside. It is caused by the transition in ocean shelf width from wide, to the south of Lofotodden, to narrow, to the north of this island. The gradient drives the current to reach its maximum speed of 5 to 6 metres (16 to 20 feet) per second. It flows around the island of Lofotodden and through narrow channels between the islands further east. Topographically, the maelstrom site is above a narrow ridge 50 to 100 metres (160 to 320 feet) down that separates much deeper water on either side. The nearby ocean deeps close to the islands play a part in the channelling and movement of a vast volume of water.

The Moskstraumen is caused by the interaction of an east–west/west–east tidal current with prevailing northward currents. A south-westerly wind will contribute to its strength. Sadly, the Lofoten Islands have lost many men to the sea when their fishing boats have tried to cross this turbulent water in bad weather.

In 1595 Olaus Magnus's map located the position of the whirlpool and he found a ready market for stories, woodcuts and maps of sea monsters. His sea serpent 'revolves in a circle around the doomed vessel' – possibly somewhat reminiscent of the whirlpool. A different Norwegian sea monster, the Kraken (probably a giant squid) emerged later. It lent its name to a masterly work of science fiction explaining the mysterious disappearance of ships: *The Kraken Wakes* by John Wyndham.

Another maelstrom called the Saltstraumen, 33 kilometres (about 20 miles) south of Bodø on mainland

Norway, is easier to get to. The hydraulic pressure powering the currents is very strong. The tide rises by an average of 1 metre over an area of some 200 square kilometres (77 square miles) once every six hours, so the volume flowing through a narrow channel under the Saltstraumen bridge is vast. There is a shallow sill under the bridge some 15 metres (50 feet) down; the bottom then drops to 90 metres (nearly 300 feet) in the fjord. At this point on the surface the current is found to be at its strongest and some reports claim a truly phenomenal water speed of 50 kilometres (30 miles) per hour. Again, there is turbulent water and a lot of swirling eddies. The most powerful whirlpools in the Saltstraumen drop at least a couple of metres (6 feet) in the vortex centre and are up to 10 metres (33 feet) across.

Reliable vortices with rotation speeds of up to 10 knots – a little over 18 kilometres (11 miles) per hour – can be seen in Japan's Naruto Straits. They are caused by a tidal stream between the Pacific Ocean and the Inland Sea of Japan. The difference in height of the tidal current is 1.3 metres (4 feet) over a 1.6-kilometre (1-mile) long strait – not that much different from the Corryvreckan. Depending on the direction of the wind and the state of the tide, probably the largest tidal vortices on Earth occur in these waters. The fastest rotation speed recorded was 13.7 knots – 25 kilometres (15 miles) per hour.

According to Japan's Maritime Safety Guards there have not been any recent fatalities in the Naruto Straits, although each year there are more than ten incidents involving ships of 500 tons or more. Whirlpools are often blamed as the cause of the accidents that happen when the tide is strong. The ships lose control and either collide with each other or go aground in the shallow part of the straits. Sometimes they even get stuck under the Onaruto Suspension Bridge itself. That is 1,629 metres (just over a mile) long and connects Naruto with Awaji Island. It spans the tidal channel and towers over some of the whirlpools. In comparison, over the past eighteen years the Royal National Lifeboat Institution have reported only forty-six incidents in Corryvreckan – but, of course, the maritime traffic there is much smaller.

There is an Italian whirlpool site near Messina, between Sicily and mainland Italy but nearer to the Sicilian coast. Slowly rotating patches of water called *tagli* are tidal 'whirlpools'. If caught up in one, you would gently drift round with no more danger than getting a sunburn from the Mediterranean sun. These *tagli* are brought about by two opposing currents: a south-going surface current meeting with a saltier and therefore deeper north-going undersea current. Introduce the tidal-stream effect caused by the Straits of Messina and it would appear that all the necessary components for these gently rotating whirlpools are present. But because the straits are more than 3 kilometres (2 miles) wide at their narrowest point, they hardly increase the speed of the tidal stream. As the *tagli* are found outside the straits their origins are obviously more complex and it is thought that a small tidal race off a headland near Ganzirra on Sicily might be involved.

On Canada's Atlantic coast, the Bay of Fundy in New Brunswick has the world's largest tidal range – on average 9 metres (30 feet) and at spring tides 13 metres (43 feet). During equinoctial springs the tidal range is 17 metres (55 feet). This means a substantial hydraulic head in the straits between islands in the bay and, therefore, very strong tidal currents too. The area boasts the largest 'whirlpool' in the western hemisphere and the second largest in the world – the 'Old Sow' (so called because it makes a noise like a pig grunting). It is in the narrow straits between three islands – Deer and Campobello in Canada and Moose Island in the USA – and is best seen from the high ground in Deer Island's Point Park. An area of turbulent water with a large swirling eddy, the Old Sow is often surrounded by several smaller eddies known locally as the 'Piglets'.

The really large whirlpool – 15 metres (50 feet) or more

in diameter – is unpredictably occasional, and has a funnel in the centre only when there are exceptional currents and very strong winds. It may be a very short-lived experience when it does occur. There are nineteenth-century accounts of sailing boats that were drawn into the Old Sow and capsized. Today, boat excursions take people out to see it. The Old Sow Whirlpool Survivors' Association issues certificates (for a fee) to those who pass through the whirlpool and survive! In normal weather modern boats are not in danger, partly because they have the extra power to get away from the currents, and partly because the whirlpool is not as strong as it was in the past. In the 1930s a causeway functioning as a tidal-dam was built between Moose Island and mainland USA and as a result the hydraulic pressure of the tidal currents is less than it used to be.

Not far from the causeway, and just a little further upstream from the Old Sow, the water looks like a pot boiling with lots of little whirls. It's known locally as the 'Ebb Tide Boils'. Nearby, small whirlpools can be seen at Cobscook Bay in the narrow channel between Falls Island and Mahar Point in Pembroke, Maine, USA.

On Canada's Pacific coast, opposite north Vancouver Island, are several long narrow inlets. Seymour Inlet contains many locations for those who enjoy the thrill and dangers of sport diving. There is boat transport to the middle of the Nakwakto rapids, said to be the fastest navigable saltwater in the world – extreme tidal currents cause the sea to behave like a river. A tiny fir-covered rocky island is found there, lying in a very fast-flowing tidal stream. It is called Turret Rock – or more appropriately Tremble Island, as it is claimed that it does indeed tremble when the tidal currents reach their maximum, but scientists have been unable to measure it. The island physically splits the flow into two currents which continue either side of it and recombine in its lee. Just in the crux where they meet is a small area of relatively still water, and it is here where a back eddy forms. If the current is fast enough this forms a significant vortex.

Divers can see the animal and plant life that enjoy the tidal extremes – red gooseneck barnacles, for example, and in the back eddy created by the island there are yellow sulphur sponges, parchment tubeworms and anemones.

Another site is further south in British Columbia, near Egmont. The water is greatly constricted in the Skookumchuk Narrows between Sechelt and Jervis inlets and it has a strong tidal current. Boiling tidal rapids and whirlpools can be seen from a viewpoint located high on a rocky bluff in Skookumchuk Narrows Provincial Park. It is one of the easier sites to visit – and spectacular. Diving is only just possible during the two slackest tides of the year. Even then, the water on the surface does not give a true indication of the state of the underwater currents.

Finally, La Rance is a 240 MW power station that operates on tidal energy in the Rance Estuary in Brittany. It boasts constant whirlpools lasting up to three hours each.

There are a few other places in the world where ocean whirlpools and related phenomena have been reported – but a definitive list has yet to be compiled.

Whirlpool research

This takes in vortex theory – essentially physics and fluid dynamics – then things become turbulent and soon lead into mathematics and chaos theory. It all feeds back into how vortices form and how they dissipate. Next there are eddies of increasing size. These can even become huge cyclone-sized features in the ocean, and their importance in the global circulation of ocean water brings in mainstream oceanography. Finally, the animal and plant life that survive in these extreme environments is considered.

To begin with the key question: Where are vortices found? In both water and air, vortices are often invisible – but sometimes they can be seen quite clearly. White trails in the sky are tiny ice crystals that reveal vorticity in the wake of an aeroplane. Leaves or snowflakes show the eddying wind around the buildings of our concrete cities. In the laboratory a jet of coloured dye into water, or salty water into pure water illuminated with a strong backlight, reveals all kinds of eddies.

Vortices are everywhere. For example, they are created by insects and birds as they fly, and fish and cetaceans as they swim through water. Filter feeders create vortices so that particles of food can be lifted up and filtered out of the water. Helical vortices sliding out to the tips of a plane's wings provide 'lift'. Vortices can be experienced in those parts of the world that have waterspouts or tornadoes. On a larger scale, weather patterns in the atmosphere, such as hurricanes, provide dramatic images on satellite pictures. Remote sensing by satellite shows that there are giant eddies in the sea. A constant theme in the research is to 'model' all of these phenomena.

Bjorn Gjevik studies hydromechanics at the University of Oslo in Norway. He has reconstructed the current flow of the Moskstraumen (the Norwegian maelstrom) as a computer model and is currently refining this model in terms of the measurements made at sea. 'The very strong currents produce waves which make it difficult to sail through the Moskstraumen,' he says. 'No measurements have been taken when the currents are at their strongest.' However, in weaker conditions Gjevik has been able to use a current-measuring technique called 'acoustic Doppler'. The ADCP (Acoustic Doppler Current Profiler) will give a series of measurements of the speeds of water flow at any depth between 0 and 300 metres (up to 1,000 feet). Referring the data to a fixed point, such as the bottom of the sea, using echo sounding and confirmed by satellite location using GPS (Global Positioning System), absolute values of the current can be obtained and so can be programmed into the model. The interactive process between computer model and measurement becomes increasingly difficult if the strongest tidal current profiles cannot be obtained.

Similar research is being carried out by David Farmer at the Institute of Ocean Sciences, in Sidney, Canada. He studies the tidal-current phenomena produced in the Quatsino Narrows at the mouth of the Seymour and Belize Inlets in British Columbia. To understand his approach, imagine making a cylinder out of plasticine by rolling it between your hands – one hand moving relative to the other. Similarly, two currents moving – one relative to the other – can promote a circular eddy of swirling water to form between them. The currents do not necessarily have to move in opposite directions for this to happen. For example, imagine that the main flow in a shallow river is momentarily split into two currents by something, such as a rock or a small island. When the currents come together to re-form as the main flow we often see eddies in the dead water in the lee of the rock. This is particularly interesting because it focuses our attention on what happens when two currents with slightly different properties of direction, or speed, or mass transfer (volume of water flow per second), interact and recombine.

A fast-moving flow of water consists of a series of layers. Laminar flow – normal unimpeded flow – means that all the layers move together with the same velocity. But in other conditions, such as when a strong tidal stream passes a point or a headland, a horizontal shear can develop. The layers no longer form a cohesive package and a crescent-like profile of layers of water moving at different velocities can develop. The layers begin to shear. David Farmer is interested in how strong the shear forces are, and the point at which they become so strong that the flow of water becomes unstable and turbulent. He is particularly interested in circumstances that make one side of the shear zone speed up relative to the other. When this occurs the layers get stretched, the shear zone tilts to one side and a whirlpool forms.

By measuring a stream of descending bubbles, very large vertical currents can be detected beneath whirlpools. David Farmer measures the speed at which they descend down to 100 metres (320 feet). This downward current is fast and the upward current surrounding it relatively weak. It is rather like the plasticine story. We can make the plasticine roll by doing much more work with one hand (both pressure and movement) than we appear to do with the other. So, in the whirlpool situation there is a shear between the powerful and focused downward current and the more gentle diffuse upward current that surrounds it. Water starts rolling at various places in the interface (the steeply angled shear plane) and this generates small vortices. The constant stream of small vortices acts like the ball-bearings in a bicycle wheel. If a bike is turned upside-down and the wheel spun round, the ball-bearings reduce the friction and keep the wheel spinning freely. That is why there are often small 'whirlpools' on the edge of the bigger tidal eddies – such as the Moskstraumen and, of course, the Old Sow and her Piglets.

A major contribution to research on tidal whirlpools has been carried out by Tsukasa Nishimura at the Science University of Tokyo. Whirlpools have the unusual ability to aggregate. Smaller ones can join together to form larger eddies. So although they sometimes seem to have disappeared, in fact they have amalgamated. Satellite pictures of the Naruto Straits show the mixing of nutrients taking place as the whirlpools coalesce into large eddies over a kilometre across. Tsukasa Nishimura has measured this inverse cascade process.

In the Naruto Straits, whirlpools with the same sense of rotation revolve around each other and amalgamate into larger whirlpools. Pairs of whirlpools with opposite rotations form vortex-dipoles. These are self-propelled and move in the water for some distance. They, too, grow and get larger and larger by the inverse cascade process.

So, the first feature needed for a whirlpool to form is a very strong current created by a hydraulic gradient that has caused a flow which is literally 'downhill'. The hydraulic gradient (referred to more generally as a 'pressure gradient') can be brought about by a change of sea level forcing the current, local thermal and salinity variations or winds.

The second requirement is the presence of very strong shears in the flowing water. These arise from flow separation of the current caused by an obstacle, such as a small island in mid-current, a headland near an inshore current or irregularities in sea-bed topography.

A third factor which will make whirlpools more likely is that the energy in the system is increased by other eddies, forcing agents or an opposing current.

Once the rotation has occurred, it needs to be sustained. With vortex formation, two-dimensional movement becomes three-dimensional. Here, the effect of gravity and the low-pressure gradient in the whirlpool's core is balanced by the inertia of the rotating water.

The mathematics of whirlpools is fascinating. At this point it might be a good idea to make a cup of coffee, beat into it some energy by stirring it around vigorously and stare into the vortex you have created. Add cream and this will give you another example to study. As an experiment: drop a tiny piece of paper anywhere in the spinning liquid – the paper will eventually make its way into the centre of the vortex. Every aspect of the behaviour of that piece of paper (or 'parcel' of water in the vortex) is such that the geometry of its trajectory, its direction of movement, its velocity, are all governed by a mathematical property called a 'critical point'. This determines the behaviour of every particle in the spinning water. There are functions in mathematics that behave in the same way and 'whirlpool theory' is a mathematical approach to understanding them.

Chaos theory is a branch of mathematics that seeks to understand why ordered systems become chaotic and vice versa. Start to turn off a tap, and at a critical point an ordered flow of water turns into a highly complex chaos of disordered vortices. How then can we predict the behaviour of a chaotic system and its critical point? Sergei Nazarenko is a mathematician working on weather prediction at the University of Arizona and is affiliated with the University of Warwick. He thinks that although some features of vortices can be approached from chaos theory, beyond that the application to fluid dynamics is overestimated in the chaos science literature – where many different views have been expressed. He agrees with David Dritschel of the University of St Andrews that weather is determined by large-scale vortices (cyclones and anticyclones), the behaviour of which is largely predictable, and not by small-scale turbulence – which is not. However, the 'averaged' effect of turbulence is believed to be predictable – despite the unpredictability of any isolated element – otherwise no mathematical treatment of turbulence would make sense.

Sergei Nazarenko is also interested in energy transfer when small eddies aggregate into larger ones – and how three-dimensional turbulent energy transforms into two-dimensional horizontal and two-dimensional vertical components in whirlpools. He suggests that whirlpools may start off with powerful horizontal vorticity and then transform their energy downward to give the vertical flows we associate with them.

Moving from maths to oceanography, giant eddies shape large-scale ocean circulation patterns. These can have an impact on climate and biology and their occurrence has been described as the ocean weather, analogous to 'the atmospheric weather'.

Walter Munk, Emeritus Research Professor in Geophysics at the Scripps Oceanographic Institute, San Diego, points out that the circulation of warm and cold ocean currents has been known for a long time. The first detailed chart showing ocean currents was made by E. W. Happel in 1675.

Interestingly, Happel shows several whirlpools – including one off the northern coast of Norway. This whirlpool also appears in Athanasius Kircher's map of 1678. Kircher speculated that the maelstrom was one of the places were the ocean drained into an abyss which was believed to exist deep in the Earth's interior!

Three hundred years later, modern maps of the oceans show a pattern of cold water flowing away from both poles and, at the same time, warm water flowing towards them from the equatorial region. There are also circulation cells (circular flows of water called *gyres*) around the Earth's major ocean basins, together with currents linking them. Northern Hemisphere gyres move towards the right or clockwise. Southern Hemisphere gyres move in the opposite direction. This deflection is brought about by the rotating Earth and is often referred to as the Coriolis effect. Eddies must be bigger than 15 kilometres (9.3 miles) in diameter for the Coriolis effect to play a role so, contrary to popular wisdom, which way water happens to drain out of a bath does not demonstrate the direction of the Earth's rotation.

While the general circulation pattern described above is important, it does not take into account the daily changes that affect the global pattern of ocean currents. The day-to-day variations in current speed, temperature and salinity are caused by the state of the eddies in the ocean at any given time. The global 'eddy field' comes about because even small changes in physical parameters can cause the ocean currents to be sheared vertically or horizontally. Eddies develop at the interfaces where these shears occur and they grow in size. Complicated mathematics and computer modelling are required to work out the effect of the eddies on the permanent currents – which together make up the ocean weather.

A computer model by K. J. Richards and W. J. Gould of the Southampton Oceanography Centre gives an impression of the effects of eddies on an ocean current. The model starts with a jet, or fast-moving current, in the middle of a long linear channel: 500 kilometres (310 miles) wide, 1,000 kilometres (620 miles) long and 2,000 metres (6,500 feet) deep. The velocity profile shows a tight distribution around a maximum. Add a random disturbance with a wide range of horizontal effects and, day-by-day, the perturbation of the jet increases. After sixty days or so it is no longer a jet but an unstable flow of water moving at increased speed in every direction in a chaotic eddy field. The effect spreads out across the entire width of the channel.

The size of ocean eddies range from 10 to 200 kilometres (6 to 125 miles) in diameter and appear in every guise from the horizontal on the surface to the vertical within the water column. Such is their variation in orientation, size and form that they are called the 'eddy zoo'! Some of the 'animals' in the eddy zoo can be seen by remote sensing from a satellite. For example, images of the eddies can be produced from infra-red radiation giving sea surface temperature, or GPS giving sea surface height, or visible light measurement giving ocean colour – related to the presence (or absence) of certain species of phytoplankton.

Particularly interesting in the eddy zoo are Gulf Stream rings. These eddies take only a few days to form by 'pinching off' the meanders from this highly energetic and sheared flow. Some eddies have a cold core surrounded by a ring of warm water; others are the opposite way round, and both can be up to 100 kilometres (62 miles) across. In contrast, and in the quieter conditions of a mid-ocean for example, eddies may take three months to form.

Essentially, eddies are 'mixers' and 'spreaders' of water, that is, warm and cold, high-salt and low-salt, surface and subsurface. As a consequence, they affect the distribution of plankton and therefore the marine food chain. They affect the large ocean circulating currents – the gyres. They are involved in sediment transport and the transfer of gases between the atmosphere and the ocean. But the most important thing that eddies do is to deal with the excess energy given to the oceans by the constant rotation of the Earth. Eddy–eddy interactions can produce intense vortices. Vortices can break up when they meet other vortices (they are said to 'commit suicide'). The net result is that eddies 'homogenize or balance out vorticity', and as a result dissipate energy that is eventually absorbed in the viscosity of the sea water. Without eddies our oceans would be very tempestuous indeed. But there are still many unanswered questions about eddies and this topic is right at the research frontier.

An allied research field looks at the very large anomalous currents created by the ocean's massive internal waves. In 1976 Al Osborne, a physicist at the University of Turin, discovered a new type of internal wave – a 'soliton' (named after its mathematical properties). Solitons have been responsible for the loss of at least one submarine and they have caused significant difficulties for the oil exploration industry working at depth. Al Osborne's mathematical modelling has suggested that these subsurface waves can cause 'holes' (depressions) to appear suddenly in the surface of the sea.

Mark Inall at Dunstaffnage Marine Laboratory is studying internal tide waves at the ocean shelf edge. These solitary internal waves (SIWs) transport mass and/or energy and produce turbulence at depth which helps stimulate the mixing of warm water above what is called the thermocline with the cool water below. This activity can be spotted at the surface where wind-forcing causes waves to steepen and break – making one patch of sea much rougher than another. Without mixing, the ocean would turn into a stagnant pool of cold salty water within a few thousand years.

Could forcing by internal waves and currents together with winds and surface currents cause giant ocean waves? 'This is a very exciting area of research to be in,' says Al Osborne. 'So little is known about why these monster waves rise up out of the ocean to damage and destroy ocean-going vessels.' The largest wave he has measured was 26 metres (85 feet) high at an oil rig in the North Sea.

Corryvreckan's future

Protect, conserve, enjoy – surely our thoughts about this wild and remote region, the turbulent tidal streams and rocky reefs in the marine area known as the Firth of Lorn. It is being considered as a 'possible Special Area of Conservation' (pSAC). The Gulf of Corryvreckan and the Great Race are within its bounds. But what makes the region so special?

John Baxter manages the SeaMap project for Scottish Natural Heritage. Based at the University of Newcastle-upon-Tyne, SeaMap's field group, led by Jon Davies, a marine biologist, carried out a survey in 1999 of all the sub-littoral habitats and their associated plant and animal life in the Firth of Lorn. 'Sublittoral' refers to marine life from low tide down to about 60 metres (200 feet).

SeaMap's survey was carried out by marine remote sensing using acoustic imaging and the correlation of those images with biological data. The acoustic signature – essentially the absorption of sound frequencies – can be related to different biotopes. A biotope is a consistently recognizable combination of a physical habitat type inhabited by a characteristic set of dominant or conspicuous species. For example, a sublittoral rock face in a particular tidal regime dominated by barnacles would be classed as a different biotope from a similar rock-face habitat dominated by sea anemones. Both would be regarded as part of the same biotope complex, which is defined as a group of biotopes of similar overall character. In contrast, a muddy seabed dominated by burrowing worms would be classified as belonging to a different biotope complex from that of the rock faces.

Forty-nine biotopes and ten biotope complexes were found in the Firth of Lorn and identified from the *Marine*

Biotope Classification for Britain and Ireland. The acoustic signature of each biotope found was checked by remote video recording. The bathymetry (a depth profile of the seabed) was already known. But 'side-scan sonar' (which gives detailed images of undersea topography) proved operationally problematic – it was too slow to be of much use as a mapping technique, though it did produce some impressive images. At some sites 'grab samples' of the seabed were taken to analyse the sediment and to check for biota which were too small to be seen on the video.

Earlier surveys, based on dives by marine ecologists, aided species identification. For example, the 1982 survey for the Nature Conservancy Council reported the results of seventy dives (110 stations) when 156 algae and 384 animal species were identified.

Three stations were dived on the south side of the Gulf of Corryvreckan at places where the bedrock was extremely exposed, semi-exposed and very sheltered. Near the shore the marine community was pretty standard. Further out at about 25 metres (about 80 feet) depth the environment was generally very scoured and only those species that could manage to hold on in the tidal current were found. All that could be seen in any number were *Balanus crenatus,* a barnacle; *Sertularia cupressina,* a hydroid; the bryozoan, *Securiflustra securifrons*; and a colonial ascidian (or sea-squirt), *Synoicum pulmonaria.* Other species were only encountered once or twice in cavities in the rock.

A 1983 survey included the submerged pinnacle at the whirlpool site. Marine biologists who have dived at this site describe the rock surface as being full of hollows scoured out by the action of high currents and sand or rocky debris caught up in the eddies. The sculptural effects are stunning. All the published accounts agree in describing a species-poor community, consisting of hydroid and barnacle turfs on the exposed surfaces, with a few other species such as anemones hanging on in the more sheltered crevices and gullies.

SeaMap's 1999 side-scan sonar images showed very rugged and extremely tide-swept rock reefs at the western entrance to the Gulf. Jon Davies says that massive sponges, *Pachymatisma johnstonia* and *Cliona celata,* were found in the parts of the Corryvreckan area less affected by the tidal currents.

The Gulf of Corryvreckan is one of the most outstanding tide-swept sounds in Europe, of which there are several in the Firth of Lorn. It is too early to say whether any special adaptations to life exist in Corryvreckan's specific circumstances. Certainly there seem to be changes in behaviour. For example, crustaceans may find food more quickly in such currents, but their window of opportunity is shorter. As the slack water starts to pick up strength again, the divers see them scuttle away to find shelter from the irresistible flows. In extremely fast flowing water only the hardiest marine animals survive.

As a whole, the Firth of Lorn supports an exceptional range of habitats and communities. These surveys and conservation initiatives make sure its scientific heritage will be preserved for future generations.

A SENSE OF DISASTER

Karl P. N. Shuker

'I heard a deep rumbling sound that was just a couple of seconds before the shock hit – and the ground was heaving and shaking. I could hear Kathy screaming. I could hear crockery breaking, the chimney ripping through the roof ...'

That was a description of the Loma Prieta earthquake that hit California's San Andreas fault on 17 October 1989. It is fairly typical of the distress and damage caused by moderate earthquakes. Many larger quakes wreak really terrible destruction and claim tens of thousands of lives, yet science, so far, has no accepted method of predicting quakes. There is earthquake folklore from around the world, some of it claiming that there are signs and messages from the Earth and from animals prior to an earthquake. There are even people who claim to be able to sense an oncoming earthquake – but issuing such predictions is seen by most seismologists as a pseudo-science and the preserve of cranks and amateurs. Surely, however, with so many lives lost in earthquakes it is worth exploring the facts behind the folklore and the claims. Is it not just conceivable that a factor leading to better prediction could be lurking in the myths?

What is an earthquake?

The Earth's crust is constantly, but imperceptibly, on the move. Pressures and stresses build up in the rocks, and sometimes so much stress builds up that the rocks cannot take any more and they snap. The result is an earthquake. It is very similar to holding a pencil in both hands, and trying to bend it. Nothing happens at first, despite the pressure, but suddenly it breaks – a brittle fracture. In the ground, a brittle fracture causes faults. If more pressure is applied, those faults can move again and again.

It is a common misconception that earthquakes occur only at the boundaries of the Earth's plates – the vast slabs of lithosphere that make up the Earth's outer rocky layer. It is indeed true that the really big earthquakes are to be expected where plates are moving past each other. So, the Loma Prieta earthquake was caused by the Pacific plate (which carries western California with it) sliding past the American plate. The resultant San Andreas fault, having a horizontal movement, is clearly visible at the surface. The cause of the regular Japanese earthquakes is not visible to the eye. There the cause is the Pacific plate slipping underneath Japan, and going down into the Earth. It does not do so at a constant rate, but in a series of jerks. Every jerk results in an earthquake being felt at the surface.

However, although the vast majority of the world's earthquakes occur at plate boundaries, earthquakes can in fact occur anywhere where there is a fault – even in Britain where the British Geological Survey in Edinburgh record over 300 tremors in Britain every year. Most are not felt, but as recently as 1931 a moderate earthquake was felt throughout Britain when a fault slipped in the North Sea, near the Dogger Bank. The damage caused was only minor, but an earthquake of the same magnitude hit the North African resort of Agadir in 1960 destroying almost all the town, and burying most of its inhabitants. That too was away from the plate margins.

The contrast in the effects on human populations and property between Agadir and the Dogger Bank also illustrates another misconception – that the Richter scale gives a measure of the intensity of an earthquake. It does not: the Richter scale gives an indication of the amount of energy released where the fault actually moves, called the focus. If the focus is very deep, the effects at the epicentre – the point on the surface – immediately above the focus – will not be so bad as for a more shallow earthquake of the same size. For people living at the surface, what they feel will very much depend also on how far away they are from the epicentre and whether they are sitting on solid rock, or on soft unconsolidated ground which will shake very much more. (Agadir was a case in point – where the focus was shallow, and directly underneath a town built on river sediments.)

All of these varied factors make earthquake prediction very difficult for scientists. Every fault is different, every quake is different, and reliable precursors, even if they could be found, might turn out to be present in one part of the world, but not in others. Currently the best science can do is to engineer for the inevitable, in order to make homes and infrastructure more earthquake-resistant, and to prepare for earthquakes by working with planners and engineers, in order to minimize loss and suffering. But will this always be the case, or are these glimpses of apparent prediction from folklore and non-scientists (in the conventional sense) trying to tell us something?

Earthquake sensitives

One of the most remarkable claims aired in recent times is that certain people are able to sense the impending onset of an earthquake. Such people are referred to as earthquake sensitives, and include among their number Ali Rhoden, who organizes an annual central seismic party, at which others professing to possess this mystifying ability socialize at her home in Pear Blossom, California, less than a mile from the infamous San Andreas fault.

But how does such a talent work, and how did it begin? In Rhoden's case, she experiences migraines, and ringing noises in her ears. The intensity of a given migraine or ringing noise seems to be directly proportional to the size of an oncoming earthquake, and the precise location of the migraine or ringing noise indicates the earthquake's geographical location. Fellow earthquake sensitive Terry Loutham experiences a diverse range of warning symptoms, from racing heartbeat and palpitations to adrenalin rushes, dizziness and nausea. And Diane Pope generally suffers severe pains in the back of her head prior to a volcanic eruption or a very deep earthquake.

Rhoden had been experiencing her own symptoms for a long time, but it was only during the past few years that she recognized that they were occurring prior to earthquake activity, and she began to correlate such activity with the intensity of the earthquake and location of the symptoms she was suffering.

For instance, she has found that when she develops headaches in the centre of her forehead, this presages an earthquake in southern California, whereas over her right eyebrow indicates the Kurils and Kamchatka in Siberia, and further up is a warning for the Aleutian Islands. Sounds in her right ear foretell earthquake activity in the general area of Indonesia, the Philippines and the South Pacific. Low-frequency drones in her left ear augur seismic upheaval in the desert region of California's Yucca Valley and Palm Springs, whereas high-frequency sounds in this ear can denote the Wyoming region of the USA. Even her knees are apparently responsive, with pain in her right knee alerting her for Japan and China around the 35°N latitude, and her left knee detecting for the southern USA. Her other principal warning sites are her right hip, responding to Baja California in Mexico, and her right shoulder, monitoring the South Pacific–Indonesia–Philippines area again.

The intensity of Rhoden's migraines, ear sounds and other effects normally denotes the size of the impending earthquakes but, if it is a local quake, its effects upon her can be intense even if the earthquake is only relatively small, often inciting bouts of extreme nausea and profound physical sickness. Indeed, she has become so attuned to the specific levels of nausea, pain and the pitch of the ringing sounds in her ears that she claims to be able to judge accurately from these the size of the impending earthquake as measured on the Richter scale. For example, experiencing pain in her head sufficient to cause a bad migraine 'means a 6.0, or if it puts me into bed then it's going to be a 7.0. If it doesn't give me a two-hour headache or a really bad migraine headache, then it will probably be less than a 6.0, and I usually don't get a migraine for anything less than a 5.0.'

Inevitably, there are many people who do not, or cannot, believe Rhoden's claims. However, she has prepared extensive documentation and charts to support her earthquake predictions (which are even available on the Internet), and claims a success rate of 83.4 per cent.

'I have two forms where I register them [her predictions]. I work as a special contributor on a computer programme called Prodigy on the seismology bulletin board, and I have my own topic called LACQ Watch, and I record down different predictions I have. Then I come back with what has occurred and then I scale it out, as each prediction will have three qualifiers. There'll be the size, the location, and the time. If it meets all three requirements in my prediction scale, then I give that 100 per cent. If it only meets two, well then it's 66.6 per cent, or whatever. Usually what I do is I'll put out a vast prediction of, say, ten different areas or five different areas on one post, and then I take all that information from there, and whatever works out it generally comes out to 83.4 per cent accurate. You know, I do have a couple of misses here and there.'

Rhoden also uses a few instruments, such as a seismograph and compasses, which are of particular benefit in assisting her to gauge local activity. 'Sometimes I am afraid that I may be over-expecting a size because I do get very sick for local activity, and so if I look at my equipment and my equipment says it's not going to be as big as my symptoms say it's going to be, I know that I'm over-experiencing it myself, and tone down my prediction.' She also closely monitors the behaviour of her pets, particularly her birds, which become very excited, squawking and even falling off their perches, prior to earthquake activity. Similarly, her dogs howl, and her wire-haired fox terrier digs into the ground fifteen minutes or so before a quake begins, whereas her goldfish will jump up out of its bowl prior to any quake activity on the nearby San Andreas fault.

Bearing in mind how ill Rhoden becomes prior to quake activity, especially local activity, one cannot help but wonder why she chooses to live so near the San Andreas fault, of all places. However, she has found that she experiences these symptoms wherever she is, so it makes little difference where she lives. Her father lives in Reno, Nevada, and she was very ill at his house prior to a plus 8-magnitude quake in the Kurils. On another occasion, she visited Hawaii and anticipated having a very enjoyable vacation there. Instead, her visit took place just prior to a quake, and once again she became very ill there. Moreover, while she was in Hawaii she predicted quite a few quakes in southern California. In one instance, she even telephoned her sister-in-law in California, telling her that there was going to be an earthquake in Newhall. Sure enough, there was, rating around 6.5 on the Richter scale.

But what value is there in being an earthquake sensitive? In Rhoden's opinion: 'The only value is what you get in your heart. There is no monetary value. It's certainly no fun being sick all the time, but to help people out, that's what it all comes down to. There are a lot of people that are scared of earthquakes, and if you can just alleviate that scaredness from them a little bit, they're happy. It makes me feel good – I like to help people.'

Earthquake prediction and western science

Quake prophecies aired by Rhoden and other earthquake sensitives attract a great deal of attention from the general public, not merely because they echo ancient folk beliefs but also because they fill a vacuum left by science. Certainly, many mainstream western seismologists still tend to look upon accurate earthquake prediction as a distant dream.

Perhaps the most significant problem faced by seismologists who do entertain hopes of making accurate predictions one day is the sheer size of the timescale involved with regard to major quake activity. As succinctly expressed by Allan Lindh, a seismologist with the United States Geological Survey: 'Great earthquakes come hundreds of years apart, you don't know where they're going to be. Since you don't know how to predict them, you don't know where to go to look for them. So how are you going to make measurements of them, how can you try to predict something that you can't predict? And the answer is: you've got to guess, you've got to be lucky, but most of all, you've got to sort of stick with it, you've got to take this as a problem that you're really going to commit to and work on and then you've got to make observations over a very long period of time.'

Needless to say, any project that requires that level of commitment also requires an appreciable financial input, but in the West there is presently little sign that earthquake prediction is likely to receive this in the near future. It is certainly not a current priority at the United States Geological Survey, as it claims only 3 per cent of the survey's Californian budget. And most of this is spent on assessing long-term earthquake risk – primarily by determining how much stress is accumulating along known faults – rather than on short-term predictions.

Having said that, two short-term quake predictions were announced in California during the late 1980s by the Survey, via State-released public warnings, but both were disappointing. The first one, which was the first ever public warning of a quake in the San Francisco Bay area, was announced following a modest-sized earthquake (magnitude 5 on the Richter scale) in June 1988, on the San Andreas fault 130 kilometres (80 miles) south of San Francisco. As major earthquakes in California often occur after a series of smaller quakes or foreshocks, which indicate the release of stress on the fault, Lindh informed the State that there was a slight chance of a 6.5 earthquake within the next five days – but nothing happened. Another foreshock occurred in August 1989 in this same area, and a second public warning of a possible major quake occurring within the next few days was issued – but once again nothing happened ... until the afternoon of 17 October, that is.

Six weeks after the second warning had been released, an earthquake of magnitude 7 struck, killing sixty-seven people and causing 7 billion dollars' worth of damage. Its epicentre was under a mountain called Loma Prieta on the San Andreas fault, nearly 100 kilometres (60 miles) from San Francisco, but the most extensive devastation occurred in the Bay area. It had not been forecast by either of the earlier predictions though, as Lindh pointed out, it was in the general area covered by them and hence had not come entirely out of the blue. Ten years earlier, working on long-term earthquake forecasts, Lindh and colleagues at Columbia University had identified the segment of the San Andreas fault where the Loma Prieta quake happened as being the most likely place in northern California for a big quake to occur – if one was going to occur. Even so, it was hardly an example of accurate, short-term quake prediction.

Some scientists consider the whole subject of earthquake prediction, even within mainstream scientific research, to be an outright folly. Certainly, Dr Robert Geller, a leading geophysicist based at the University of Tokyo, is highly sceptical about the prospect of success being achieved in this field: 'Now why is prediction so difficult?

Well, the reason is, first of all, an earthquake happens very deep inside the Earth. The Earth is very complicated, heterogeneous, we don't know the physical law governing the way earthquakes happen. How is the stress built up inside the Earth? How much energy is available to be released? We also don't know how the fault slips. In view of all of those difficulties the question is not, why can earthquakes not be predicted? The obvious question is, why does anyone seriously even think earthquake prediction is worth discussing at the present time?'

Lindh, conversely, remains optimistic that accuracy in short-term predictions will indeed improve, thanks to the ever-increasing sophistication of seismological technology, coupled with the ever-expanding wealth of data being recorded. In particular, he is hopeful that future recognition of reliable precursors – natural warnings of impending quakes – will play a major part in enhancing the veracity of short-term predictions.

'Almost everything that we know today has been said to be not possible at some time in the past. We have lots of signals coming out of the Earth; it's not like the Earth sits there dumb and quiet. We record lots of stuff. We don't know how to translate that into good estimates yet of the future behaviour of faults. But it's not that there's no signal there. So it's sort of like a translation problem. It's sort of like trying to translate the Egyptian hieroglyphs before you have the Rosetta Stone. One day you can't understand anything, the next day you have the Rosetta Stone and you can understand everything! How can anyone possibly know if that kind of a fundamental breakthrough in understanding will or won't occur? That's up to the future ... The trick is, our responsibility is, I think, to make sure we're collecting the signals with as high fidelity as possible. It's our responsibility to record what the Earth is saying and to try to translate it ... But to say that no one will ever be able to predict earthquakes ... how can one possibly know that? ... In this century [twentieth], the world has been completely turned over. Almost everything that was thought not to be possible a hundred years ago is possible today.'

During the past decade, precursors have indeed been recorded that may yet revolutionize earthquake prediction – and some are quite unexpected. Take, for instance, the geyser at Calistoga, in northern California, which has been watched for over twenty-five years by its owner Olga Kolbek, who lives nearby. Normally, this geyser displays a regular interval of eruption, occurring every forty minutes or so, when water flowing deep underground meets hot rock, becomes superheated, and is forced upwards under pressure. Kolbek's observations, however, have revealed that before an earthquake, the geyser's wholly predictable pattern of eruption is dramatically disrupted.

She was first made aware of this on 1 August 1975 when, while sitting in a picnic area nearby, she became startled when the geyser failed to erupt for two and a half hours. Later that day, however, she learned from the radio news that a 5.9 earthquake had occurred 160 kilometres (100 miles) further north, at Orville. This remarkable coincidence spurred her interest, and since 1980 the geyser has been continuously monitored by computer, recording the time of every single one of its eruptions. Of particular note among this vast collection of data is that just prior to the Loma Prieta earthquake of October 1989, the interval recorded between successive eruptions of the Calistoga geyser nearly doubled. Other, comparable examples on record confirm that although the reason for it has yet to be fully ascertained, the geyser's pattern of eruptions is indeed affected by earthquakes. But what about other precursors?

Recording the music of the Earth

The Earth is a noisy place, not just on the surface but also deep below, due to the occurrence of many different kinds of electrical activity in its crust and interior, altering the planet's magnetic field, and detectable to those with the correct equipment. Could some of these electrical signals be earthquake precursors? One remarkable event has provided some tantalizing evidence.

Professor Antony Fraser-Smith is a radioscientist at Stanford University, California, and for more than twenty years he has been conducting research for the US Navy into ultra-low frequency (ULF) radio waves, in the region of 0.01–10 Herz. These electromagnetic radiation signals, which are far below the minimum frequency audible to the human ear, travel enormous distances around the world. This is why the US Navy is interested in them, because they can be used for communicating with submarines. They penetrate deeply into the sea and the Earth – and, of particular note, they also emerge out of the Earth.

Before October 1989, Fraser-Smith was interested only in these signals, not in earthquakes – but then came the mighty Loma Prieta quake, which was attended by a wholly unexpected and most exciting revelation for him and his colleagues. It just so happened that one of Fraser-Smith's monitors recording these ULF radio signals was sited at the home of the sister of one of his research assistants, electrical engineer Paul McGill, at Coralitos, which is less than 8 kilometres (5 miles) from Loma Prieta. McGill and sister Kathy Mathew had been regularly monitoring the signals at Coralitos for about two years, and were used to the normal daily pattern of fluctuation recorded here – an area specially selected for such monitoring on account of its secluded location, shielded from other potential sources of electromagnetic radiation by a dense redwood forest.

One day in autumn 1989, however, when Mathew checked the monitor, she discovered that a most unusual signal had been recorded, of a kind that they had never seen before. Suspecting a malfunction, Mathew contacted Stanford University and informed them that their equipment was not performing correctly. Even though she had not seen this particular kind of signal before, she was not greatly concerned by it, because fluctuations in the equipment's pumps and other devices had produced odd signals before, and they had always gone away. So they simply waited to see what would happen this time.

Twelve days later, they found out – the Loma Prieta earthquake struck. It severely damaged their house, and completely shut down all of their electrical equipment, including the radio signal monitor. It took a week to restore contact with the monitor's computer, and when Fraser-Smith, McGill and colleague Iman Bernadi examined its recordings, they were very surprised indeed by what they found. The odd signal noticed by Mathew twelve days earlier had oscillated for a long time before dying away. And three hours before the quake had occurred, the monitor had recorded a further huge increase of ULF signal activity – but this time it was so immense that the monitor's computer had put out error messages stating that the signals had exceeded its range of operation!

Three hours afterwards, the earthquake had occurred. At that point, the house's electricity had been cut off, preventing any further signals from being recorded. Of considerable interest, however, was that when the electricity had been restored a week later, the monitor had recommenced recording an abnormally high activity of ULF signals. True, aftershock activity was taking place, which would have been responsible for some signals, but these abnormal ones were occurring even during periods when there was no ground-shake.

As McGill noted: 'We'd never heard of or seen signals like this before, and it would be as if you had a radio in your house turned on for two years and you heard nothing but static, and then one day you start to hear music. I mean, that's going to get your attention.'

Similarly, Fraser-Smith's response to these extraordinary pre- and post-quake recordings was a mixture of amazement, excitement and bewilderment: 'It was the one time in my life when I have been absolutely stunned by something taking place in my measurements. It was really one of the big events of my life, which actually made it all worthwhile. If nothing else ever happens again after that, I would be quite happy, but it was very exciting, and, as I say, there's lots of people looking for electromagnetic fields from earthquakes now, and I think as far as society goes it'll really pay off very nicely in the long run.'

But was there really any correlation between this exceptionally high activity of ULF radio signals and the occurrence of the quake, or could it simply have been coincidence? Even today, Fraser-Smith is still not absolutely certain that the mystifying phenomenon of these signals is directly linked with earthquakes, and claims that it is have elicited a degree of controversy in scientific circles. Dr Robert Geller is not convinced that there is a relationship between the two: 'If we see it before one earthquake, and we continue the observations and we do not see the same phenomenon before other earthquakes, then we have, at least statistically, an overwhelming likelihood that it was just some sort of random coincidence.'

However, Fraser-Smith has studied them for a very long time, '... and seeing an occurrence of most unusual signals that gradually get bigger and bigger prior to an earthquake, and reach their greatest amplitude just before an earthquake, that to me is a very good indication that they must tie in with it'. Moreover, as he also points out, although no identical signals have been seen since, there have not been any other earthquakes monitored by his kind of equipment either, so there has been no opportunity to continue observations and compare them against his original recordings. As to whether an earthquake actually could generate electrical and magnetic signals, Fraser-Smith and his colleagues have come up with several very reasonable mechanisms by which such a process should occur – but *does* it occur?

As emphasized by Fraser-Smith, the answer may lie with the specific geological make-up in California. In many locations elsewhere with faults, the faults slip underneath one another, that is, involving an element of vertical movement. California's San Andreas fault, conversely, is what is known as a strike slip fault, in which the ground is moving horizontally, that is, sideways, with a difference of movement on either side. What this means is that, as the Earth moves, it changes the pressure on the rocks inside that fault zone and it pushes around fluid present inside the Earth. And once fluid begins to be moved around, this results in changes in the way that electricity can flow through it, which in turn can lead to changes in magnetic fields. Most researchers nowadays accept that as the pressures build up prior to an earthquake, it may well crush the ground a little and squeeze water out of it, as well as moving water around inside the Earth. This is precisely the kind of mechanism that can ultimately generate electrical and magnetic fields. And if any ULF radio signals are engendered in this way, they can certainly penetrate many hundreds of miles out of the Earth.

So could they act as quake precursors? Fraser-Smith sees no reason why not: 'Obviously if an earthquake does produce a clearly defined magnetic or electrical signal, it can be measured on the Earth's surface and it can be used ultimately for a prediction, and I believe that will prove to be the case with earthquakes. We have very little data as yet because very few people have actually managed to be next to an earthquake and make good measurements. You can't just go out and sit down at a place and wait for an earthquake to occur.'

Having said that, this is basically what he has been doing ever since the Loma Prieta incident, using monitors stationed at several different Californian sites that seem likely epicentres of big earthquake activity in the future, in the hope of obtaining more of these elusive yet fascinating signals: 'I might have years to catch an earthquake, but I really do feel that there should be electric and magnetic fields from earthquakes, and I think that if we can get enough of them we will be able to predict earthquakes without any trouble whatsoever.'

Sceptics might suggest that the aberrant signals recorded at the time of the Loma Prieta earthquake could have been generated by something other than the quake – some human agency, perhaps? Fraser-Smith discounts this, however, because unless highly sophisticated equipment is used, such as a mass transit system (a huge antenna, hundreds of miles long), such signals can only be generated over very short distances, with ordinary objects like pumps or electric fencing. Thus, if anyone did generate the Loma Prieta signals they would need to have been at Coralitos itself. Even the encompassing redwood forest there was checked, to see whether there was anything within the range of the monitor's antennae that may have been responsible for the signals, but nothing was found.

In Fraser-Smith's view, the way forward for mainstream scientific earthquake prediction is to adopt an interdisciplinary approach. He freely admits that during his own research he has benefited enormously from talking to seismology colleagues in his university's physics department, as well as to geological experts on waterflow. Progress is unlikely to be made by concentrating upon one facet in isolation. On the contrary, Fraser-Smith feels that electromagnetic research, for instance, may well assist in elucidating hitherto opaque aspects of seismology, and vice-versa.

But ever likely to be the driving force for all of his investigations are those enigmatic signals recorded back in that fateful month of October 1989 at Coralitos. Because of their ultra-low frequencies, they are normally inaudible to humans but, if recorded on to magnetic tape and played back at 200 times their normal speed, they can be detected by our ears. Yet even then, they are still bizarre, for instead of resembling the more typical whistling noises associated with radio waves, these signals sound more like a chorus of croaking frogs!

A strange song, for sure, is this subterranean cacophony. Nevertheless, as eloquently expressed by Fraser-Smith: 'It is like listening to the Earth. You have heard of the music of the spheres. This might be the music of the Earth. It is not a very harmonic noise at all, but it is listening to what comes out of the Earth, and it is very fascinating for that reason.'

Applying Chinese philosophy

Whereas earthquake prediction and the investigation of quake precursors in the West is still in its infancy as an accepted scientific discipline, there is one country – unfamiliar to many western seismologists – that already has an entire system of earthquake prediction based on the routine monitoring of precursors. The country in question is China, whose scientists believe that their unique fusion of ancient philosophy and modern science explains how they have been able to issue accurate predictions that have saved thousands of lives.

One of the most celebrated Chinese seismologists associated with earthquake prediction is Professor Chen Li De of the Sinian Seismological Bureau (SSB). He is well known for good reason. Thanks to his accurate prediction of a huge quake in June 1995 at Menglian, in Yunnan, south-western China, which resulted in the area being swiftly evacuated, a great many lives were saved.

In China, there are hundreds of observation stations that keep a continuous record of potential earthquake precursors – engaging over 10,000 researchers and powerful computer technology to monitor, analyse and interpret more than a hundred of these putative indicators of future quake activity, which range from foreshocks to bird behaviour. Based upon this extensive data, no fewer than four different kinds of earthquake prediction are attempted here – long-term, medium-term, short-term and imminent.

The most important of these is the imminent category, as this can save lives if acted upon, but by definition it is also the most difficult. Nevertheless, there are a number of precursors that have been found to be beneficial when attempting imminent earthquake predictions. These include foreshocks, changes in the Earth's magnetic field, water level (for example in wells), crustal deformation and the level of radon gas in groundwater. This latter precursor is particularly telling. When rocks in the Earth's crust are under pressure (a process occurring prior to an earthquake), radon gas seeps into groundwater – so a sudden increase in groundwater's radon level has become recognized in China as a reliable earthquake precursor.

Indeed, this was a major indicator of the massive earthquake that would devastate Menglian in 1995. Another one was the occurrence of a dramatic electromagnetic signal (comparable with that recorded by Fraser-Smith's equipment before the Loma Prieta quake), which was recorded nearly 300 kilometres (186 miles) north of Menglian at the same time as the elevated levels of groundwater radon. Within the next ten days, two medium-sized earthquakes occurred not many miles from Menglian, but across the Chinese border, in Myanmar (formerly Burma).

These earthquakes worried Chen Li De a great deal. The first had been a 5.5 magnitude quake, followed by a 6.2 not long afterwards, but Chen knew that the 6.2 quake could not have been a main quake with the 5.5 as a foreshock. He had learned from experience with previous earthquakes in this part of China that a foreshock of 5.5 is normally followed by a main quake of around 7.0, not a mere 6.2. Therefore, both of these quakes had to be foreshocks, which meant that there was a major quake still to come – and soon.

Consequently, after the two foreshocks had occurred, Chen publicly issued what would have seemed to western seismologists to be an amazingly precise prediction – a major earthquake would strike Menglian within the next three days. Assisted by the government in Yunnan province, he also ensured that everyone in the area was evacuated into earthquake shelters, forcibly in some cases where opposition to the evacuation was encountered. But his actions paid off – before the three days had passed, an earthquake greater in magnitude than the Loma Prieta quake did indeed hit Menglian, destroying homes and schools that would have contained large numbers of people if his warning had not been heeded. Even so, eleven people still died, but this was only a minuscule proportion of the number who would have been killed had the evacuation not taken place.

So what is the secret of his success in predicting the earthquake? Chen firmly believes that the answer to that, and to earthquake prediction in general as practised in China, lies in the Sinian approach to scientific thinking, which differs markedly from that of American and other western scientists. According to Chen: 'American scientists start from the place where the earthquake happens, that is, the epicentre, and predict what kind of precursors might happen and when. And then they observe the real thing. If those precursors don't happen before the earthquake, then they say, "This earthquake couldn't have been predicted." Because they didn't see those things in this area. But the Chinese think the science of the Earth is the science of observation. We do not know how earthquakes come about. We observe the facts, then track down how the earthquake comes about. We start with observations, and then track backwards to how the earthquake might happen. This is a difference in scientific thinking.'

However, this is not the only difference between the western and Sinian approach to earthquake investigation as defined by Chen: 'The second thing is we have a difference in our philosophy and culture, in how we think. American scientists believe those precursors near the epicentre could be the signs of the earthquake. We don't think so. In our culture, we think the Sun, the Earth, and the atmosphere are one. It is called *di qiu zheng ti guan* – whole Earth philosophy. This is a very old philosophy. In our traditional culture, we think nature and mankind are one. We think everything under the Sun is interconnected. Let's say, there's an earthquake here. Something strange might be going on 300 kilometres [180 miles] away. We will think, this strange phenomenon might have something to do with the earthquake.'

Chen likens this attitude to the basic principle in another traditional Chinese practice – acupuncture, in which an ailment in one part of the body is treated by inserting pins into another body region, often some distance away from the afflicted part.

Even so, Chen is no stranger to failure as well as to success with regard to earthquake predictions. In cases of failure, which he freely admits are presently in the majority 'because earthquake prediction has not proved itself yet', most take the form of wrong predictions, or unpredicted earthquakes, or incorrect estimates of magnitude and intensity. Also, it is often difficult to know whether it would be best to make, or not to make, a prediction in a given instance. If, for example, the prediction involves a busy urban area but only a small earthquake, the disturbance to economic productivity that evacuation of the area or other precautions would involve might cause a greater loss than the quake itself – always assuming, of course, that it does actually strike.

Nevertheless, Chen remains convinced that although earthquake prediction is still not sufficiently precise to give accurate predictions all the time, in certain regions, such as Menglian, where there has been considerable study, he and his colleagues can predict some kinds of quake very successfully. And he feels that even greater success will be achieved in the not-too-distant future: 'I predict that in the twenty-first century – around the mid-2050s – we will be able to predict most earthquakes around magnitude 7.0.' Moreover, Chen considers that his present level of success might not be restricted to the prediction of quakes in China, but could also be achieved by him elsewhere in the world: 'I think if I went there and did some research and studied the geology of that area and the earthquake cycles of that area, after I had done a lot of research on those things and got data from observation sites, I think probably I could predict some earthquakes to a certain extent.'

However, he confesses that it would be difficult – much more difficult, for example, than predicting quakes in Yunnan, the region of China that he has studied most extensively. As with comparisons featuring acupuncture, Chen notes that this situation has readily perceived parallels with the medical field: 'If a doctor always deals with one patient, he knows that patient very well. He knows his body, his disease, and the reasons why the disease is caused very well. So he can give the right medicine to him. It's the same with us. After studying this area very well, I can give it the right medicine.'

Due to his success at Menglian in 1995, Chen is hailed there as a hero by its grateful inhabitants. As a local government official in Menglian remarked after the earthquake, 'The people were very happy because we did a very good job. They think the government is great! They think those earthquake experts are like gods!'

Tragically, however, even gods, it seems, can be fallible – as demonstrated by the harrowing history of an earthquake prediction made in 1976 concerning the industrial city of Tangshan, in north-eastern China.

Tragedy at Tangshan

In 1976 Professor Huang Xiang Ning, a leading earthquake-prediction scientist, was in charge of the Sanho earthquake team, whose job was to set crustal stress stations, collect data, and give annual and imminent predictions. Huang's specific task was to lead the national short-term and imminent prediction team, as well as the short-term and imminent crustal stress research. The equipment that they used for testing crustal stress had actually been invented by a Swedish scientist who used it for testing stress in mine shafts, but once Chinese scientists learned about it and how to use it, they improved the basic design, adapting it for use in crustal stress testing.

This is an important task because, according to the theory of the late Li Shu Gang, a prominent Chinese seismologist, earthquakes happen when the stress upon rock increases until the rock can no longer sustain it, finally breaking apart. According to this theory, therefore, if the changes in crustal stress can be mapped, earthquakes can be predicted. Li Shu Gang also believed that the Earth is changing all the time, and that this should be studied too. This is known as spatial or space prediction – predicting in which area an earthquake might be. From this, the locations of the crustal stress stations can be arranged, in order to observe the structural changes, and thus predict the quake itself.

From 1966 to 1970, Huang conducted earthquake research in Tangshan, Luan County, Qian An and Qing Long, after being sent to Hebei province by Li Shu Gang, who considered that this general region may well be vulnerable to quake activity. After the team worked there for five to six years, its findings led to the establishment of crustal stress stations and also fault-line observation stations at Tangshan, Luan County and Changli. Indeed, at Tangshan there were eventually no fewer than eleven different observation sites, which was more than enough to suggest to the team that this locality was at risk from quakes.

Moreover, two of its stations recorded some unique changes in Tangshan's crustal stress. One station was at the Douhe Reservoir, the other at the Zhaogezhuang Mine. In

November 1975, very frequent changes in the crustal stress were recorded at both stations – a situation never previously experienced by the team. They also observed abnormal changes in the crustal stress in outlying provinces, such as Beijing, Shandong, Shanxi, Liaoning and more than twenty other stations. Accordingly, in early 1975 the team submitted a report to the SSB in which they predicted that in the area of Laoting County in Hebei Province, Jinzhou and Aohanqi (which is in Inner Mongolia), there would be an earthquake of 6.0 or so.

The centre of the triangle created by plotting Laoting County, Jinzhou and Aohanqi is roughly 200 kilometres (125 miles) from Tangshan. On 14 July 1976 they submitted another report to the SSB, but now containing an imminent prediction – of an earthquake estimated at around 5.0, expected between 20 July and 8 August, either west of Beijing or in the triangle, centred upon Tangshan. This was based on some extraordinary data that had been obtained from the crustal stress stations at Douhe and Zhaogezhuang, and which was so high that it had overshot the paper on which it had been recorded. In addition, a sudden, very sizeable change in crustal stress had lately been recorded in the observation stations at Beijing and Dalian.

Needless to say, Huang fervently hoped that the government would pay attention to his team's report and warn the people in the vulnerable region about the possibility of a quake occurring there, especially as he had no power to issue such warnings himself. Tragically, however, officialdom took no notice of his prediction – with one notable exception. Recently appointed as administrator of the tiny earthquake bureau of Qinglong County, just over 95 kilometres (60 miles) from Tangshan City, a diligent young civil servant called Wang Chun Qing learned of Huang's prediction at a meeting in Tangshan. As a result, he lost no time in putting into action the county's earthquake protection programme, even though the prediction had claimed that the earthquake would be fairly small, no more than 5.0 or so.

In the middle of the very hot summer night of 27 July, however, Tangshan, totally unprepared as it was for any quake activity, would soon learn differently – and at a terrible cost. Far from being a modest-sized quake, the earthquake that abruptly began that nightmarish night was a seismic monster. Rating a devastating magnitude of 7.8, it mercilessly reduced the once bustling, thriving industrial city of Tangshan to a pile of rubble. One of the most destructive earthquakes of the twentieth century, it killed a quarter of a million people. Only in Qinglong was there relief: although over 180,000 buildings had been destroyed, this apocalyptic quake had claimed only a single life, thanks to Wang Chun Qing and his earthquake protection programme.

Even today, the indescribable pain felt by Professor Huang because of his team's underestimation of the Tangshan earthquake's magnitude has not left him, and he is easily moved to tears by the memory. But this tragedy did at least spur China into increasing its monitoring of every possible precursor – thereby nurturing and shaping today's Sinian researches into earthquake prediction. Indeed, assisted by enhanced technology, during the mid-1990s Huang and his team made a very successful prediction about an earthquake in Xinjiang. And thanks to the cooperation of the local government, they did reduce the loss of lives that would otherwise have resulted there.

Sadly, however, China's utilization of crustal stress to predict earthquakes remains a little-known technique outside this vast country. Indeed, according to Huang, until a major UN conference took place here a few years ago the outside world did not even know about this line of research: 'After we have been studying this theory for thirty years, foreigners have just learned about it ... [but] they don't understand it at all. That's why we think earthquakes are predictable and foreigners do not.'

Earthquake precursors in Japan

The Kobe earthquake of 17 January 1995 was the most damaging to strike Japan for seventy-two years. Over 6,000 people died, and whole sections of the old quarter collapsed without warning – just as there was no public warning before the quake. Yet the Japanese government has made its Earthquake Prediction Programme a priority. It has already cost billions of dollars, making it the world's most expensive programme of this type, and currently employs hundreds of scientists, with satellite technology monitoring every movement of the Earth's crust. Moreover, investigation of the Kobe disaster revealed that certain precursors, such as increased radon levels in well water, had been present, and that the city's inhabitants had noted other odd occurrences that may constitute additional portents of quake activity.

They included reports of flowers and vegetables that danced agitatedly, animals exhibiting atypical behaviour, clocks and radios that mysteriously stopped, and electrical equipment acting in a thoroughly bizarre manner. One of the numerous eyewitnesses to such events as these was Hatsumi Hirayama, whose mother's house was destroyed in the quake.

'About ten days before the Kobe earthquake, while having an evening piano lesson, I looked at the clock – and the hand suddenly dropped down. There were a number of other things. For example, the air-conditioner worked of its own accord, without the use of a remote-control switch. The television remote control had stopped working some time around New Year. Finally, on the day just before the earthquake, the Moon looked very pink that evening. "That's strange," I said, and went out with my mother into the garden over there, and looked at it for a long time. After that, we returned to the house and switched on the television with the thought of watching the news before going to bed, but the TV channels kept switching every so often of their own accord. There were things like that.'

These and other curious accounts came to the attention

of Professor Motoji Ikeya, a physicist at the University of Osaka. He was well aware that Japanese folklore and superstition have a long tradition of odd events occurring prior to earthquakes, but after hearing the post-Kobe testimony, he began to wonder whether at least some of it might have a basis in reality, especially as all of these stories, new and old, seemed to share a common connection – electromagnetism. Could electromagnetic changes in the Earth be occurring prior to a quake be responsible? Well worth noting here is that, prior to the Kobe quake, scientists at Kyoto University had picked up a strange electromagnetic signal like the one recorded by Antony Fraser-Smith's equipment just before the Loma Prieta quake. Consequently, Ikeya decided to see if he could reproduce in the laboratory, with the aid of electromagnetism, some of the odd events said to have occurred prior to the Kobe earthquake and also featuring in traditional Japanese legends. His results were certainly thought-provoking.

Some Kobe inhabitants claimed that the leaves of lettuces rattled and shook, and flowers such as begonias danced, just before the earthquake struck. So Ikeya used a Van de Graaff generator to create an electrical charge, and then exposed some lettuces and begonias to it. Sure enough, when the plants became charged he found that he could reproduce the odd movements described by the Kobe quake eyewitnesses. Similarly, legends affirm that the mimosa plant will bow before an earthquake; once again, when Ikeya exposed this plant to electrostatic charge, it closed its leaves and folded its stem, causing it to bow down.

There is an old Japanese proverb which says that the candle flame in front of the altar before a Buddhist shrine will bend prior to an earthquake. Ikeya tested this by applying an electric field to a candle flame and, because the flame consists of positive and negative ions, it did indeed bend. Another oft-quoted phenomenon on file is earthquake-associated lightning, though few mainstream scientists recognize it. Ikeya, however, can accept the reality of earthquake lightning, because such phenomena are caused by a high-intensity electric field, whose own presence can in turn be explained by realizing that rock fractured underground produces an electric field. Earthquake clouds were another phenomenon: one amateur Japanese seismologist predicted an earthquake by observing an unusual cloud beforehand, but meteorologists denied the existence of any earthquake-associated clouds. When Ikeya and colleagues applied an electric field to supercooled moisture, however, clouds were indeed created. Consequently, he believes that earthquake clouds can be explained by an intense electric field at the epicentre producing ionization, leading in turn to condensation and cloud formation.

Some of Ikeya's most intriguing findings, however, concern pre-quake animal behaviour. According to Japanese fables, earthquakes are caused by the movements of a gigantic subterranean catfish, which in Ikeya's opinion may have been based upon sightings of catfish thrashing violently in ponds just before the onset of a quake.

Another ancient legend tells of a warrior who went to a river to catch an eel, but could not find one, and noticed that a catfish there was twisting and thrashing violently, seemingly in turmoil, so he caught that instead. He then returned home at once, concerned that an earthquake might be coming, and not long afterwards a quake did indeed occur.

These are just stories, but Ikeya was curious to discover whether they might have any factual origin, and experimented to see if he could reproduce their described fish behaviour in the laboratory. He discovered that only very small electric fields, no more than 4–5 volts per metre, were sufficient to agitate catfish into violent movement, yet when he placed his own finger into their tank he was unable to feel anything. Clearly, therefore, catfishes are exceedingly sensitive to electric fields, adding further support to Ikeya's belief that they do respond to electric fields generated in the Earth prior to an earthquake. So too, moreover, must eels, bearing in mind that his experiments with these revealed that they are even more sensitive than catfish to electric fields. Consequently, his interpretation of the warrior legend, in which the eels were absent from the river, is that they must have sensed the impending quake even more emphatically than the catfish and had therefore moved elsewhere, in the hope of avoiding it.

Ikeya's experiments with electromagnetism also duplicated the weird pre-quake behaviour exhibited by the electrical equipment of Kobe's inhabitants. He was even able to reproduce the abnormal, rapid hand movements reported with quartz clocks, using electric fields in laboratory experiments. Tellingly, no such behaviour with mechanical, non-electrical clocks had been reported in Kobe prior to the quake – only with quartz versions.

Based upon his experimental findings, Ikeya looks favourably upon the monitoring of animal behaviour as a significant aid to earthquake prediction, and considers his work in this controversial field to be both useful and important: 'I do not say that if one observes these phenomena, an earthquake will definitely come, because thunder [for example] can cause similar phenomena. But when there are such phenomena all over the place, there is a high probability that we may have an earthquake. And if we know that unusual animal behaviour is not superstition but a scientific result, then I think that people will watch their pet animals and will be careful, which could save lives, especially of people living in the country where there are no scientific studies going on.'

Biological magnetism

In 1975, biologists studying sediment-inhabiting bacteria at Woods Hole, Massachusetts, made a remarkable discovery. Unlike any previously monitored life forms, these tiny organisms' direction of movement was determined by magnetism. Throughout their short life, rarely lasting more than an hour or so, they burrow down into the sediment, attracted not by the Earth's gravity but instead – as confirmed by laboratory experimentation – using its magnetic field. And when their micro-anatomy was observed, it was found that these bacteria possessed chain-like structures of crystals composed of the mineral magnetite (magnetosomes). Also termed lodestone, but chemically a type of iron oxide, this substance was subsequently discovered in many other life forms too, including various insects, fishes, amphibians, reptiles, birds and mammals. It is the only magnetic compound that animals can synthesize biochemically.

Forming the basis of an additional sense – magnetoreception – the presence of internal crystals of magnetite acting as magnetoreceptors is believed to assist such animals during migration, enabling them to use the Earth's magnetic field as a directional guide, by conveying specialized information concerning this magnetic field to the brain via the nervous system. Is it possible that this newly discovered sense also serves a second, equally significant role – as an early-warning system for impending earthquake activity?

Professor Joe Kirschvink, a leading expert in the field of biomagnetics, based at the California Institute of Technology (Caltech), has long been interested in such a possibility, leading on from his early, key discovery that bees were sensitive to magnetic fields. As an undergraduate student, he carried out a research project with Professor Heinz A. Lowenstam, the scientist responsible for revealing in 1962 that primitive molluscs known as chitons possessed teeth composed of magnetite – the first known example of this mineral being synthesized by living things. Until then, it had been assumed that magnetite was created only deep in the Earth, at high temperatures and pressures.

It had been this project that had first brought the subject of magnetite in organisms to Kirschvink's attention, and later, while working as a graduate at Princeton in 1977, he learned that a biologist there had successfully replicated a German experiment showing that honeybees could detect the Earth's magnetic field. The experiment had revealed that the manner in which bees danced upon a honeycomb when deprived of normal gravitational or illumination cues depended upon the direction of the prevailing magnetic field.

Recalling the earlier chiton studies, Kirschvink realized that this clearly indicated that the honeybees contained magnetite somewhere in their bodies, which was acting as a magnetic compass. Finally obtaining permission to examine some of the bees utilized in the geomagnetic experiments, he took them to a laboratory containing highly sensitive magnetometers, and confirmed that these insects did indeed contain magnetite. Subsequent experiments pinpointed a dorsal region in the bee abdomen's anterior portion, and also examined the extent to which the bees were able to detect geomagnetic fields.

This elicited an extremely interesting result – particularly with respect to the ostensibly unrelated subject of earthquake prediction. It turned out that honeybees are exceedingly sensitive to very strong electromagnetic fields within a range of frequencies spanning 0.1–10 Herz – a range that just so happens to contain the frequencies of those strange ULF signals recorded by Fraser-Smith's equipment prior to the Loma Prieta earthquake.

This intriguing coincidence is not lost on Kirschvink: 'Of course, nobody knows the source of the magnetic signals that Fraser-Smith and his group at Stanford were able to measure before the Loma Prieta earthquake. However, if you look at the power levels of the various band widths of the Stanford recordings, it seems pretty clear that honeybees should have been able to detect those signals. So what we have is actually a kind of unique thing – a possible precursor to an earthquake which ought to have been detectable by an animal. Whether it means anything, we don't know, but that's a unique observation ...[and] can lead to predictions and things that are testable.'

On closer reflection, however, the acute sensitivity of honeybees to this particular frequency range relative to electromagnetic fields is not so surprising after all. For as Kirschvink points out, sources of electromagnetic fields with a frequency reaching 50 to 60 Herz are almost exclusively man-made ones; prior to humanity's appearance on the planet, the frequencies of such fields were normally below that level. This explains why bees, which evolved long before humans, are sensitive only to ULF fields, and not to those higher-frequency ones generated by man-made devices and hence of only recent origin. Indeed, these latter fields are as invisible to bees as UV light (visible to bees) is to humans.

Even more exciting is the prospect that while this magnetoreception sense has been evolving down through the ages in bees and other animals, one external factor that may have been influencing its evolution is earthquake activity. As acknowledged by Kirschvink: 'The way evolution works is kind of interesting. You can select for something very strongly over a short period of time, but a weak selection pressure acting over long periods of geological time is just as effective. If these electromagnetic precursors happen with some regularity prior to great earthquakes or large earthquakes, and if an earthquake could cause some fraction of the population to die or lose fitness, there's every reason to think that an avoidance mechanism might evolve. And in fact, it might have evolved aeons ago.'

Animals that would be particularly vulnerable to the destructive activity of earthquakes include burrowing species, whose subterranean dwellings were at risk from collapse during seismic disturbance, killing anything inside them; nesting birds, whose eggs would be thrown out of trees during quake-engendered tremors; and honeybees, whose hives could be obliterated if the trees or cavities containing them are themselves destroyed by the quake. Consequently, it would greatly aid survival if such creatures as these were able to detect quake precursors and thereby take action to avoid the oncoming quake by moving elsewhere.

Such detection mechanisms may not be confined solely to magnetoreception either. For example: any burrowing animals that can sense changes in humidity within their underground homes, such as certain burrowing insects possessing specialized humidity sensors known as hygroreceptors, are also equipped to sense an impending quake. This is because groundwater moves up and down before earthquakes, and hydraulic changes such as these can alter the humidity level within underground burrows.

Certainly, there does appear to be more to the alleged ability of animals to predict earthquakes than mere coincidence or hearsay. A few years ago, a quake was successfully predicted in Japan after scientists had noted abnormal behaviour displayed by animals at Tokyo Zoo. Moreover, there are countless reports on file of localities (including some in Kobe) normally inhabited by rats and mice that suddenly became unaccountably free of their vermin – only for a quake to strike that locality a few days later. Studies of rodents' magnetic sense, featuring species such as the African mole rats, have confirmed that these mammals are indeed magnetoreceptive, and burrowing rodents display a particularly good magnetic sense. Presumably, therefore, this would be of benefit in detecting quake precursors of a geomagnetic nature and could explain the pre-quake exodus reports of rats and mice.

And in the USA, charities are no longer surprised by increases in reports of lost pets not only *after* an earthquake (understandable, as quakes would certainly terrify animals, causing them to flee from their homes), but also several days *before* one – indicating that the pets may have somehow sensed its impending arrival.

Of particular note is an observation recorded on a certain momentous day in 1989 by Nick Corini, a champion pigeon racer from California. On that day, he noticed that about an hour before he released them for some exercise, his pigeons seemed very agitated, and the hens were off their eggs. Nevertheless, he did release them, and they all flew away, but they did not come back. Later that day, the Loma Prieta earthquake struck – and it was two to three days later before some of his pigeons finally returned. It is now known that pigeons possess a small black magnetite-containing structure situated between the brain and the skull, which is believed to serve as a compass, enabling the birds to use geomagnetic clues to find their way back home. Why, therefore, following the quake, did Corini's birds take time to return? Could it be that they sensed its impending arrival by detecting geomagnetic signals acting as precursors, but that these signals temporarily confused them, so that they could only find their way home again once the signals had died down after the quake?

It has long been known that racing pigeons become very disoriented on magnetically stormy days induced by sunspots, which seem to conflict with the normal geomagnetic signals used by the birds as cues when homing. Perhaps a similar scenario is played out before, during and for a time after, an earthquake. If only pigeon racers had a good magnetic observatory with them, or at least a simple flux gate, to record magnetic anomalies when their birds are disoriented, this may yield some enlightening data.

Similarly, the quite frequent, large-scale stranding of some species of deep-water whale when venturing near certain unfamiliar coastal beaches has been shown to correlate significantly with the presence in these areas of geomagnetic disturbances. Some species are even guided to certain feeding grounds by slight fluctuations in the Earth's magnetic field. Once again, these whales are now known to possess magnetite, found in parts of their cerebral cortex.

There are also some aquatic animals reputedly able to predict earthquakes, such as various fishes, that may be utilizing electrosensitivity to achieve this. The examples we have already seen, catfishes and eels, have highly developed electroreceptors, which should be more than adequate to detect the electric currents that create electromagnetic fields, especially as current is much more important than voltage in the generation of these fields. This is clearly another area of animal physiology for consideration and investigation with regard to earthquake prediction.

As for magnetoreception, if animals can indeed predict earthquakes using such a mechanism, this in turn provides an important insight into the geomagnetic nature of earthquakes themselves. Namely, rather than being erratic and, as some scientists believe, wholly unpredictable, these seismic phenomena must produce magnetic anomalies that are repetitive, sufficiently at least for there to be a pattern that can be recognized by magnetoreceptive species. As yet, however, no such pattern has been detected by scientists.

Yet pattern recognition is a well-known phenomenon in animal behaviour, so perhaps, as suggested by Kirschvink: '... maybe there are patterns of earthquake precursors stored in the genome of animals. And by dissecting the behavioural genetics associated to that, we might get some idea of what those patterns could be.'

Even within the field of animal-associated quake prediction, let alone quake prediction across the entire spectrum of putative precursors, it is increasingly evident that an interdisciplinary approach is vital – drawing upon animal physiology, behaviour, even genetics, together with seismology and geology – if the multifaceted key to this highly complex enigma is ever to be uncovered.

But perhaps the most thought-provoking aspect of biological magnetoreception was Kirschvink's revelation back in 1981 that even humans contain magnetite. Moreover, scientists at Manchester University revealed that humans possess a concentration of magnetite in the sphenoid-ethmoid sinus complex – that is, the bones at the back of the skull forming the nasal cavity. It has also been found in the brain – but what is its function?

Unlike magnetite in birds, fishes, bacteria and so on, crystals of this mineral found in humans are not lined up in chain structures. Instead, the cells simply seem to have several thousand of these tiny particles inside their perimeter. In Kirschvink's view, it is unlikely, therefore, that these particular magnetite-containing cells are being utilized for magnetoreception, though he does not rule out the possibility that there are other cells that accomplish this. As he also emphasizes: 'The morphology of the individual particles, however, still bears the fingerprint of their role of something for magnetism; but the problem is, if you really wanted to make a cell magnetic you would line all these crystals up and produce a nice magnetic field in the cell. That's not what these things are doing.' All of which makes it unclear at present just what function they are serving: 'So there's maybe some reason for a cell needing a strong magnetic field but not lining these crystals up. It may be a chemical reason of some sort, but again we are nowhere close to understanding or being able to think of what type of reaction it might be.'

Even so, it is very tempting to speculate whether the presence of magnetite in humans may help to explain in some way the experiences reported by alleged earthquake sensitives. Kirschvink himself admits that some such testimony is compelling, though he is highly sceptical of stories in which sensitives claim to be able to predict earthquakes across continents (for example, a sensitive in the USA predicting a quake in China).

As he points out, these claims 'at face value do not meet any tests for reliability. On the other hand, if a honeybee or a rodent might be able to predict earthquakes, there's no reason why humans might not have some vestige of that sense, of that ability. And some of them may be able to do it. But that requires rather rigorous testing. It's difficult since great earthquakes are rare [and] seismologists are few in number.'

It is a frustrating yet overriding paradox that the single biggest hindrance to the advancement of scientific studies into earthquake prediction must surely be that scientists cannot predict earthquakes. Hence it is not possible to position all the necessary equipment and researchers at a given site to reveal and record whatever precursors may occur prior to the quake itself. Being in the right place at the right time is of paramount importance in this field, but we have little more than luck for guidance at present. Perhaps this is why some western mainstream seismologists are deeply sceptical of what might be learned from listening to the Earth. Yet the effects of a major quake in a populated zone that has not been warned of its imminence can be so devastating that it is surely time to begin investigating seriously all aspects of earthquake prediction, including people who claim to have 'a sense of disaster'.

After all, as so succinctly expressed by Allan Lindh of the US Geological Survey: 'I think the hardest thing for a scientist or a human being to remember is that what we know is this big, and what we don't know is THIS BIG.' He moves his hands from a couple of centimetres apart to as far apart as he can reach. 'So to rule out earthquake prediction, which today clearly lies in the unknown, is just silly. The things that we can't imagine today will be discovered in the next 100 years. I hope one of them is how to predict earthquakes.'

TIMELINE

Era	Period	Years ago	Event
Quaternary		3,700	Last mammoth
		1,100	Younger Dryas
		14,000	End of last main cold stage in Britain
		35,000	Modern humans take over from Neanderthals
	2 million Tertiary		Start of Ice Age in Britain
Tertiary		58 million	flood basalts in Scotland
Mesozoic	*65 million* Cretaceous	65 million	K-T extinction. Deccan traps
	144 million Jurassic		
	208 million Triassic		
			First dinosaurs
		210 million	End –Triassic extinction. American and West African traps
Palaeozoic	*250 million* Permian	250 million	End – Permian extinction. Siberian traps
		270 million	First therapsids
	286 million Carboniferous		
		300 million	First mammal-like reptiles

Era	Period	Date	Event
Paleozoic	*286 million* Carboniferous		
		348 million	Origin of reptiles
	360 million Devonian		
		367 million	Late Devonian extinction
		370 million	First land vertebrate
	408 million Silurian		
		430 million	First vascular plants on dry land
	438 million Ordovician		
		450 million	End – Ordovician Extinction
	505 million Cambrian		
		520 million	First vertebrate. Five mass extinctions
Precambrian	*550 million* Proterozoic		
		600 million	First multi-celled animals
		3000 million	Oldest rocks in Britain
		4000 million	Probable origin of life
		4500 million	Origin of the Earth and Solar System

EQUINOX: THE BRAIN

The Introduction

Everything you experience, you experience because of your brain. All your thoughts and dreams; your perception of yourself and the world around you; that indescribable sense of awareness; your memories – all these are created by the wet, interconnected grey and white lumps inside your skull.

The scientific study of the brain provides us with insight into the very nature of ourselves, and attempts to solve one of the most important puzzles of science and philosophy: just how does the brain work? This scientific endeavour has many different branches – some of the main areas are investigated in this book. Each chapter, based on subjects covered by Channel 4's *Equinox* series, provides different pieces of the puzzle, producing a clearer picture of the brain. I have been able to extend the scope of the television programmes, and to present the science, history, philosophy and politics of these very human stories in greater detail.

Of the various branches of neuroscience (the all-encompassing term for the scientific study of the brain), neurophysiology is the attempt to work out how the structures of the brain and nervous system actually function: how they communicate with each other to endow us with the powers of thought, language, memory and perception. Neurology is very similar, but focuses more on the diagnosis and treatment of brain damage or disorder, such as epilepsy and the effects of strokes. Both neurology and neurophysiology have been organized, sophisticated areas of study for more than a hundred years. Relative newcomers to the field are neurochemistry – the study of the brain's chemical environment – and behavioural genetics, which aims to discover whether any aspects of our behaviour are inherited. Psychology – the study of mind and behaviour – is also part of the study of the brain, by virtue of the fact that as far as we know the mind is produced by the brain. Psychology has made great strides in fathoming human behaviour, but there is still a gulf between an understanding of the brain that produces the mind and the behaviour that the mind produces.

The human brain is incredibly complicated – the most complex object known. However, its basic structure can be described fairly easily. It has distinct areas, each of which is connected to many others. The most obvious feature is the cerebrum, the familiar convoluted grey structure that obscures the rest of the brain. It consists of two cerebral hemispheres – although the brain is not quite spherical – which are connected by a thick bridge of nerve fibres deep inside. The wrinkled surface layer of the cerebrum is the cortex, consisting of thousands of millions of interconnected neurones. The neurone is the fundamental unit of the brain. Neurones produce or conduct electrical impulses that are the basis of sensation, memory, thought and motor signals that make muscles work to produce movement. There are other types of cell present, but they only give support and nourishment to these cellular workhorses. Neurones are like other cells in many ways: they have a nucleus and a membrane, for example. However, they differ in the way they function. A neurone has long fibres, called axons, coming from its cell body. Emanating from an axon or from the cell body itself are other, smaller fibres called dendrites. Neurones communicate with each other: electrical signals pass along the axon and the dendrites, and the brain is constantly buzzing with these signals. The matter that makes up the cortex is grey, distinct from the white matter that makes up the bulk of the cerebral hemispheres it surrounds.

Beneath the all-encompassing cerebrum are other, smaller, parts of the brain, also composed of neurones. In the middle of the brain there is a cluster of smaller structures: the basal ganglia, the thalamus and hypothalamus, and the limbic system. Unlike the cerebrum, we share these structures with all other vertebrates, and this suggests they are involved in some kind of automatic processing. In fact these structures seem to be involved in the production of emotional responses, necessary for automatic reactions to fear, for example. Tucked underneath the back of the cerebrum is a cauliflower-shaped 'little brain' – the cerebellum. This sits at the top of the brainstem, which connects the brain to the top of the spinal cord.

The brain has not always been held as the centre of human thought. Most of the ancient Greek anatomists and natural philosophers, for example, believed that thinking and feeling took place in the heart or the lungs. They saw the brain more as the 'seat of the soul'. However, as early as the sixth century BC the Greek philosopher Pythagoras suggested that the brain was the organ of the mind. The Greek physician Herophilus of Chalcedon realized that the brain is the centre of the nervous system, and several other Greek thinkers began to recognize the importance of the brain in perception, thought and movement. But the brain was to remain a total mystery for many hundreds of years. During the second century AD the influential Graeco-Roman anatomist Galen of Pergamum made many important advances in the study of the nervous system. He carried out ground-breaking experiments – mainly on animals – in an attempt to understand how the human body works. For example, in one of many experiments on live animals he demonstrated that the brain controls the voice, by tying off one of the nerves that connects the brain to the larynx.

Galen also correctly surmised the existence of sensory and motor nerve fibres. Despite his many useful contributions to medical science, and to anatomy and physiology in particular, many of Galen's theories were misguided or simply wrong, but they survived more or less unchallenged for 1,500 years.

During the late Middle Ages, the ideas of the ancient Greeks were modified or extended. Many people believed, for example, that the three main ventricles of the brain (those spaces filled with cerebrospinal fluid) were each responsible for a different aspect of brain function: sensation, reason and memory. During the sixteenth and seventeenth centuries, the long-standing ideas of Galen and his contemporaries, as well as the misguided modifications of the Middle Ages, were challenged by the eager endeavours of Renaissance scientists. Anatomists and physiologists, mainly in Italy, began a complete overhaul of the understanding of the brain. One by one, the various parts of the brain were identified – although, of course, knowledge of their functions was very limited. The nature of the grey and white matter was also uncovered: in the 1660s the Italian anatomist Marcello Malpighi became the first person ever to see, under a microscope, the neurones that make up the cerebrum. And at the end of the eighteenth century another Italian anatomist, Luigi Galvani, demonstrated that nerve impulses are electrical. In a series of classic experiments, he made the muscles of frogs' legs twitch by stimulating them with static electricity. His observations led him to postulate the existence of 'animal electricity'. This was the beginning of the modern understanding of how neurones work – but only in the past eighty years or so have neuroscientists been able to understand the details of how nerve impulses are generated.

Today, science is well on the way to explaining how the brain functions. To date, it has worked out how the different structures and regions within the brain are interconnected.

It has shown which parts of the brain are important in various tasks, such as remembering, feeling sad, reading, using grammar or looking at something. And it has uncovered the biological and biochemical processes that underlie all the electrical signals that make up our thoughts, movements, memories and emotions. Despite these impressive advances, though, neuroscience is still in its infancy. For example, we do not yet understand in detail how the electrical signals in the nervous system actually create thoughts, memories and emotions. The pathways of perception and the production of movement are fairly well understood and in great detail, but little is known about how movement and perception actually relate to our conscious awareness. Just how our brains make us aware at all is a mystery that is not likely to be worked out for some time, if ever. Consciousness may turn out to be one of those fundamental, elusive phenomena – like the ultimate origin of space and time – that science may never fully explain.

This does not mean that the journey towards understanding it is a waste of time. At each step along the way, new wonders and mysteries become apparent. Besides, the study of the brain is not restricted to finding answers to ultimate questions. It is carried out as much with medical goals in mind as with scientific or philosophical ones, and each advance in neurophysiology brings new hope to those suffering from disorders of the brain, whether caused by physical damage or disease.

Of the various branches of neuroscience, neurophysiology, neurology and behavioural genetics are all in the spotlight in 'Mind Readers', which looks at the origin of social aspects of behaviour, such as the ability to empathize. Neurophysiologists have begun to work out which parts of the brain contribute to our social awareness, but neurologists are involved, too: social behaviours can be impaired in certain mental disorders, including autism. People with autism find it difficult to relate to other people, and to understand why people behave as they do. The chapter ends by considering the genetic factors involved in our social behaviour. 'Natural-born Genius' considers the nature of human intelligence, and involves psychology and behavioural genetics. The study of the nature of intelligence, and attempts to measure it, are tasks of the psychologists, but it has long been wondered whether a person's intelligence is determined by his or her genetic makeup. As will become clear, philosophy and politics are never far from the debate about intelligence. 'Phantom Brains' explores a neurological phenomenon that was once no more than a medical curiosity – but that now helps to give new insights into how the brain produces a body image, a sense of self. That phenomenon is the vivid sensation felt by an amputee that the lost limb is still present. New theories to explain this 'phantom' sensation are challenging some long-held beliefs about the brain. The subject of 'Living Dangerously' is the origin of thrill-seeking behaviour. What makes people want to take risks, and why do some take more risks than others? Many people today believe that mind and behaviour are produced by electrical signals and chemical reactions in their brain, and it turns out that the chemical environment of the brain may be related to a person's desire for risk. This situation suggests a kind of biochemical fate – as if we have no control over our behaviour. This is a recurrent theme in neuroscience: an example of the nature versus nurture debate.

Another recurrent theme of neuroscience is the use of new developments in technology, which help to achieve breakthroughs at an ever increasing speed. For example, new brain-imaging techniques are helping neurologists in their diagnoses and helping neurophysiologists to test their theories about how the brain works. These imaging techniques include MRI (magnetic resonance imaging) and PET (positron emission tomography). They are non-invasive –

there is no need to open up a person's head to study the living brain – and produce clear visual representations of brain activity. Scanning techniques rely heavily on computing power, and computers are important in neuroscience in other ways. All scientists have benefited from the information age: they can collect and analyse data more quickly and efficiently, and can communicate their results effectively using computers. But computers are useful to neuroscience in other ways. They are being used to model the way individual brain cells – neurones – work together to achieve memory and learning. Truly intelligent computer systems, which can think for themselves, may be the ultimate result of such research but, for now, computer models of how the brain works are helping to test the latest theories.

The brain is so vital to us that we can easily take its amazing capabilities for granted. 'Thin Air' demonstrates what happens to the brain when it begins to run low on energy, as we follow a scientific expedition to the summit of Mount Everest. 'Lies and Delusions' looks at some of the most bizarre symptoms investigated by neurologists, including claims made by patients whose brains have been damaged in such a way that they think their parents are impostors. It is specific damage to the outer surface of the brain – the cortex that surrounds the walnut-shaped cerebral hemisphere of the brain – that causes these strange symptoms. Study of the effects of damage to the cortex is helping to reveal the way the brain works. Finally, a brief Afterword looks ahead to the fascinating future of developments in neuroscience.

MIND READERS

...in search of the social brain...

Most human beings are mind readers: we can often tell what people are feeling without asking them. You could call this amazing ability a sixth sense, but as far as we know this is no mystical quality – not a kind of telepathy, for example. Instead, mind reading seems to be a facet of the normal functioning of the human brain: we can see beyond people's eyes, and engage with the minds that lie behind them. Neuroscientists believe that various parts of the brain work together to enable us to sense what other people are feeling – to empathize with them. These brain components, working together in this way, are sometimes referred to as the 'social brain'. The varied tools of modern neuroscience are helping scientists to find out what those components are, and to begin to understand how the social brain works.

Can you mind read?

To give you an example of the mind-reading capabilities of the social brain, imagine watching the following scene. A man standing at a bus stop is approached from behind by a woman who thinks she recognizes him. The woman taps the man on the shoulder but, as he turns around, she realizes that she does not know him after all, and rushes away looking embarrassed. Most people watching this scene would easily work out what is going on, by observing the woman's and the man's reactions. We are able to 'get inside' the minds of the two people involved. As far as we know, this ability to empathize sets humans apart from all other animals.

An important part of this skill is the way we interpret facial expressions. When shown a photograph of an expressive face, most people can recognize the emotion being expressed at the moment the photograph was taken. Even if the photograph shows only a person's eyes, it is normally possible to identify what they are feeling: eyes can reveal joy, anxiety, sadness or disinterest, for example. Simon Baron-Cohen, at the University of Cambridge, is determined to find out how the social brain works, and how it relates to disorders such as autism as well as to differences between men and women. He and his team have put together a set of pictures of facial expressions for use in his tests and, as he says, 'Sometimes when an emotional expression changes ... the whole face changes. But often, changes in emotion are much more subtle, and you may not see a big change in the mouth, but you may see a very subtle change in what goes on around the eyes. We appear, as mind readers, to be very skilled in picking up these minute changes.'

Some people refer to eyes as the 'windows to the soul'; reading another person's eyes plays a major role in our ability to empathize. Other primates – chimpanzees, for example – seem to communicate using facial expressions. However, it is unlikely that the interpretation of faces is as sophisticated in chimpanzees as in humans. And unlike all other social animals – from ants to elephants and including chimpanzees – we engage each other in meaningful conversation and, perhaps more crucially, understand that others have their own views on the world and their own feelings. This helps us to act in a manner appropriate to a particular situation.

When you think about it, social interaction is sensitive and extremely complex. Our brains are constantly bombarded by sensory information – from our eyes, ears and nose, from touch and taste, from pain receptors – and are somehow able to act on this information. What actually goes on in the brain during a social interaction – for example, when we meet someone for the first time? First, the brain must evaluate what is going on. To do this, we analyse the behaviour of others: their facial expressions, their body language, their tone of voice. This kind of response involves our emotions, which are, at least in part, automatic functions of the brain. It has been shown that our body responds to emotional stimuli even when we are unaware of what the stimuli are: the heart rate quickens or we sweat in fearful situations, for example. This kind of reflex action is particularly important in frightening or novel situations. Once the brain has generated this kind of response, it elicits an action: it must interpret, or experience, the emotion. Based on this experience, the brain must then produce an output in the form of appropriate behaviour. This third stage involves our intelligence and our memories. The appropriate behaviour in a particular situation may be to hold back – it is not always acceptable to say the first thing that comes into our heads, for example. People who are unable to hold back are perceived as impulsive and insensitive – antisocial – rather than tactful and intuitive, with an active social brain.

So the social brain probably consists of a few distinct functions – perhaps even three distinct structures within the brain. If we are to unravel the mysteries of the social brain, we must first locate parts of the brain that are involved in our emotions, then parts of the brain that make us aware of what is going on, and then parts that generate the appropriate action.

One clue to the whereabouts of the elements of the social brain might be the fact that some people are better at mind reading than others – they have more effective social brains. By comparing the brains of different people and relating them to psychological observation and evaluation, significant differences might show up. These differences might reveal the source of our social behaviour.

Sexual stereotypes?

What kinds of people are more sociable than others? There seems to be a perception that women are generally more intuitive and sensitive than men, for example. This would involve a very active social brain: a vivid ability to mind read. How does this relate to stereotypes, such as the idea that women spend their time discussing each other's feelings, and take caring jobs such as nursing or teaching? The fact that there actually are more women than men in caring professions must be at least partly due to the very existence of these stereotypes: the result of social expectations and history. But is there an element of truth behind the stereotypes? Are women really better mind readers? Psychological tests indicate that, generally, they are. When, for example, women are shown photographs of faces, they are generally better than men at identifying the emotion being expressed. Of course, these tests might simply be illustrating the very stereotypes they attempt to look beyond. For by analysing the behaviour of adult women, researchers are recording the behaviour of brains that have been developing over many years, all the time surrounded by a world in which there is pressure to 'fit in'. You could ask: 'Do women tend to be more intuitive just because that's what women do?' The obvious approach to overcome this test bias is to study the behaviour of very young babies, who are as yet unaffected by society's role modelling. There does seem to be a large body of evidence that apparently supports the idea that females naturally have different approaches to social situations from men.

There have been many books in recent years that have analysed the differences between men and women. This is partly a reaction to the way in which gender roles have shifted so dramatically over the past hundred years or so, and partly a result of interesting new discoveries and ideas in physiology, biochemistry and psychology. In *Brain Sex: The Real Difference Between Men and Women*, authors Anne Moir and David Jessel lay bare what seem like innate differences between men and women. They report some amazing and consistent findings. For example, girls are considerably more sensitive to touch and to sound at just a few hours old. Girl babies generally respond better to emotional speech, and are more easily calmed down by a soothing voice. At a very early age, girl babies are more interested in communicating with other people than boys, who tend to be more interested in inanimate objects. These differences in behaviour between babies of the opposite sex may be the purest signs of differences in the social brain, before social conditioning can have any effect. But there is evidence that the differences in the social brains of girls and boys do underlie the behaviour of adults, too. Another book about differences between the sexes, *Sex on the Brain: The Biological Differences Between Men and Women* by Deborah Blum, tells us an amazing tale of gender confusion in Dominica in the West Indies. Some families there pass on a gene that makes young boys appear to be female – even down to the genitals – until they are in their early teens, at the onset of puberty. Then, what looks like a large clitoris swiftly becomes a penis and, where once there was nothing, two normal, healthy, sperm-producing testicles drop down. These children are raised as girls because, until the changes appear, they look like girls. As soon as it is realized that they are male, the boys apparently begin to make the switch to 'typical male behaviour' without too much fuss.

So it seems that there is a biological basis to differences between male and female behaviours: the two sexes do seem to be fundamentally different from each other. Perhaps women really are generally more sensitive than men. Does that mean that men are unsociable, unable to mind read? There are certainly countless stories of men being insensitive or uncaring; unable to read people's minds. Again, this is certainly not true of every man, but most people would more readily associate lack of sensitivity with men than with women. And the male brain's perceived greater interest in physical objects than in people also seems to persist into adulthood – again, in general. Meet Chris: he is what some people might call a typical 'nerd'. He works with computers, and has always had an interest in electronics and all things technical. As a boy, he spent far more time with his chemistry set than with people. More recently, he has reorganized the family's compact disc collection according to the dates of birth of the composers, and he regularly joins in competitions in which the sole aim is to contact as many people as possible on his citizen band radio over a twenty-four-hour period. More importantly, Chris has never really had many friends, and finds it hard to function in social situations. His wife Gisele says that he is not at all good at gauging when she is upset. Often, when he comes through the door after a day at work, he will not think to say hello to her, perhaps assuming that having heard the door slam she will know he has returned.

'These things – you just tend to accept them and think they are eccentricities,' says Gisele. 'And yet, thinking back, it's quite strange to lie down and go to sleep at social occasions ... I have learned that I have to tell him when I'm upset, and precisely why.'

If women really are better at reading social cues, how might we explain the difference? Human behaviour is extremely complex, and is no doubt shaped by a huge number of factors, many of them probably far too subtle ever to be fully understood. But where behaviour differs in a general sense between the sexes, can science identify the underlying causes of that difference? If women are born mind readers and men are born nerds, then we should find some physiological differences between men and women – but where might we find this evidence? Our mind-reading capability comes from our brain, so perhaps a man's brain is different from a woman's. Physiological difference between males and females is called sex dimorphism: is there any sex dimorphism in the brain that might account for the different social behaviours of men and women?

Sex and sensitivity

As we grow, the connections between neurones in our brains strengthen or weaken, and new ones are created, as we lay down memories and learn from our experiences. So the brains of any two people differ, because everyone has different experiences. This difference at neurone level does affect our behaviour, and could account for some of the differences between the 'typical' behaviours of men and women. This will be particularly true where male and female roles are well defined, and developing children are expected to follow them. In this case, women who have broadly similar experiences could develop similar nerve pathways in the brain. But these neural pathways are too small to be detected, and will differ from person to person and between different cultures. Are there any sex dimorphisms that occur consistently enough to explain the perceived ability of women to have a more finely tuned social brain than men? Neuroscientists do not have enough evidence to give a definite answer to this, although during the past twenty years or so more and more sex dimorphisms have been reported. One of the best-known examples concerns the corpus callosum, a bundle of nerve fibres that connects the two hemispheres of the brain. In 1982 cell biologist Christine De Lacoste-Utamsing and anthropologist Ralph Holloway found that certain regions of the corpus callosum were significantly larger in women than in men. Their result has been disputed in several more recent studies. In some experiments, involving rats, it has even been found that the corpus callosum grows larger under the influence of male hormones.

The corpus callosum, like most other structures of the brain, is composed of neurones that carry electrical signals. But the brain is alive with chemical as well as electrical messages. Among them are hormones. An orchestra of hormones, largely produced in or directed by the brain, is carried by the blood, and can have effect quickly, efficiently and body-wide. Many hormones are released by the pituitary gland, including those that affect growth, the conservation of water by the body and the amount of sugar our cells 'burn' to obtain energy. Hormones help to regulate our bodies, but they can also affect our minds. They can influence our mood, inducing anxiety or producing pleasurable feelings. Might hormones have a role in the social brain?

Sex hormones account for the physical differences between men and women. The main group of female sex hormones, oestrogens, help to cause the growth of breasts during puberty, as well as more structural differences such as a wider pelvis and shorter vocal cords. Similarly, the main male sex hormone, testosterone, causes development of the male sex organs, and causes men to grow more body hair than women. But can they account for sex dimorphism in the social brain, and for the differing behaviour of men and women? Some studies, mainly in animals, have shown that oestrogens make a female more sexually active during her most fertile times, while testosterone has long been known to affect behaviour – increasing 'macho' actions, violent tendencies and sex drive in men.

In 1959 neuroscientist Charles Barraclough found startling evidence of the roles of sex hormones, while carrying out experiments on rats at the University of California, Los Angeles. He injected female rats with testosterone while they were still in their mothers' wombs, and found that the rats grew up permanently sterile, unable to ovulate. More interestingly, they adopted behaviours typical of male rats. In fact, female rats injected with testosterone were found to mount other, unaltered females. This kind of observation led to theories about another facet of behaviour: sexual orientation. Some studies in the 1970s found levels of testosterone in gay men were lower than in heterosexual men. These studies are now largely discredited – one of them included bisexual men, who were found to have higher levels of testosterone than either heterosexual or gay men. Nevertheless, hormones might have an important role in the social brain.

Also during the 1970s, several studies showed sex dimorphism in certain structures in the brains of rats. In particular, parts of the hypothalamus were larger or more densely populated with neurones in females than in males, while other parts were smaller or less dense. One of these sites was actually named the sexually dimorphic nucleus. The hypothalamus is the controller of many hormones, including those released by the pituitary gland. Among these pituitary hormones are those that trigger the release of sex hormones from the testes or the ovaries. The hypothalamus, right at the centre of the head, has long been associated with many different human behaviours. It has a mass of only about ten grams (a third of an ounce), but it has been shown to have major roles in regulating body temperature, controlling thirst and appetite, and influencing blood pressure, sexual behaviour, aggression, fear and sleep. And sex dimorphism in the hypothalamus has been found in human brains, just as in rats' brains.

It seems that there are several differences between men's and women's brains. Might differences in sex hormones, the hypothalamus or the corpus callosum account for women's supposed superior mind-reading capabilities? Although our hormones do affect our behaviour, it is unlikely that their differing levels can be responsible for the sex dimorphism of the social brain. The secretion of hormones by the body changes with temperature, over time and with age – it is not constant enough to explain why women are consistently better at mind reading. Studies of the corpus callosum have not shown consistent differences between men and women. And as yet, reading social cues is one of the few tasks with which the hypothalamus has not been linked: it is associated with control over body maintenance, rather than the subtleties of mind reading.

Comparing the anatomy of the brains of men and women is daunting, and has so far proved inconclusive when it comes to the social brain. Is there another way that we might be able to unravel its workings? Can we pin down which parts of the brain are involved in basic mind-reading tasks, such as interpreting a facial expression? Modern brain-scanning techniques, along with some scientific detective work, have begun to do just this, and neuroscientists are beginning to understand how the social brain works. As we shall see, the amygdalae and the orbitofrontal cortex have been identified as two of the main parts of the social brain. This is a fascinating quest in itself, but it may also benefit the study of autism. People with autism live in a frustrating, closed world. Their social skills are generally impaired, and they seem unable or unwilling to take other people's feelings into account. Indeed, Chris the 'nerd' has recently been diagnosed as having a mild form of autism, called Asperger's syndrome. This diagnosis, along with the better understanding of how the autistic brain works, has helped Chris and his wife Gisele come to terms with his lack of social awareness that has plagued their relationship and even threatened to destroy it. Are there any links between autism and typical male behaviour, including the behaviour of the social brain? Might we be able to obtain clues as to the functioning of the normal social brain by studying the brains of people with autism? The answer might just be yes.

All alone

The word 'autism' – meaning 'aloneness' – was first used in 1912 by Swiss psychologist Eugen Bleuler to refer to the inner world of schizophrenics. It was chosen as the name for the disorder that we know as autism by the scientist who first identified the condition in 1943 – American Leo Kanner at the Johns Hopkins Children's Psychiatric Clinic in Baltimore, USA. Kanner identified a set of characteristics that he had observed in a group of his patients, including the characteristic aloneness and a desire for routine. A year later – but independently of Kanner – a Swiss physician, Hans Asperger, described the same set of symptoms, and also used the word 'autism'. Both these scientists believed that the medical condition they had described was present from birth. Their contemporaries considered the symptoms of autism to be the result of bad parenting, since the disorder does not become evident until the child is three or four years old. As we shall see, this is when the development of the social brain in children is becoming apparent.

Most neuroscientists estimate that some form of autism is found in one in every five or six hundred people. This means that in the UK more than a hundred thousand people have autism; in the USA almost six hundred thousand. It is difficult to obtain an accurate estimate, because not everyone has exactly the same set of symptoms. In fact the medical community now considers autism to be one of a range of related conditions, known collectively as PDDs (pervasive developmental disorders). These are 'spectrum disorders', meaning that their symptoms can be anything from moderate to severe. The most severe cases can be somewhat distressing: such people are frustratingly locked away inside their own minds; their language is seriously impaired; they have much lower than average general intelligence; and they often make repetitive physical motions such as rocking to and fro. People whose symptoms lie towards the moderate end of the spectrum are often diagnosed as having Asperger's syndrome. People with this syndrome, like those with more severe autistic symptoms, are less able than most people to empathize and to be sociable. But in contrast, their language is normally more or less unaffected, and their intelligence as measured by standard IQ (intelligence quotient) is often above average. Such people can lead a normal life, to the extent that they may not be diagnosed as having the syndrome until they are adults – that is what happened to Chris.

In the past ten years, a good deal of research has been carried out into the causes of autism. As well as the behavioural aspects of the disorder, it has been found that people with autism often have other symptoms in common. These include a dietary problem called 'leaky gut' and elevated levels of a neurotransmitter called serotonin. In leaky gut, protein molecules that would normally pass straight through the digestive system are absorbed into the blood through the intestinal wall. Some of these proteins, such as casein, found in milk, can break down in the blood to form substances that act like opium. Certain behavioural aspects of autism – including perhaps the characteristic 'aloneness' – might be connected with the effects of these substances on the brain. The same aspects might also be attributed to the higher than normal levels of serotonin. People with schizophrenia, a disorder that shares some features with autism, have a similar imbalance of serotonin. Leaky gut and imbalances of neurotransmitter might be to blame for some of the effects of autism, though what causes them is another question. Because the disorder is a syndrome – meaning that sufferers have some or all of a collection of symptoms – it probably has a number of different causes. A small proportion of the cases of autism can be shown to be caused by rubella (German measles) during pregnancy. Genetics may ultimately be the cause of most cases, however. Families in which one member has autism are more likely to see occurrence in subsequent children than other families. Siblings are significantly more likely both to have autism if they are twins than if they are not. As autism has many symptoms, you would expect that it is caused by abnormalities in several genes. In 1997, the first of these genes was identified. Researchers at the University of Chicago investigated eighty-six children with autism and found that all of them had an abnormal version of a gene that is responsible for the transportation of serotonin around the body.

So much for the possible causes of autism. What are people with autism like? How do they experience the world? Perhaps the most famous person with autism was a fictional one: Raymond Babbitt, played by Dustin Hoffman in the 1989 film *Rain Man*. In Hoffman's sensitive, Oscar-winning portrayal of Raymond, he illustrates some of the characteristics that are common to many people with autism. Raymond is not interested in other people, or in how they are feeling. Like many autistic people, he has exceptional arithmetical skills. Although people with autism often have lower than average general intelligence, it is not uncommon for them to have amazing skills of computation – mental arithmetic – or incredible memories for facts and figures. Many autistic people are able to carry out different but equally remarkable computational tasks: playing a piece of music note for note after hearing it just once, or producing accurate drawings of a scene they saw just once, for example. In *Rain Man*, Raymond's hustling brother Charlie cynically takes advantage of his arithmetical ability; but Raymond is unaware of being used. This, too, is a common characteristic of people with autism: because they are unable to understand that other people can feel differently about things – they are unable to empathize – they easily fall prey to the most straightforward pranks or confidence tricks.

People with autism live in a closed world. Alison Gopnik, at the University of California, Berkeley, gives an insight into their perceptions: 'What they're seeing is a bag of skin ... stuffed into clothes, draped over pieces of furniture, with these two little dots on top that are moving back and forth, and this hole underneath that opens and closes and has noises coming out. You can imagine how terrifying, how unpredictable, how confusing it would be to make sense of what these bags of skin are doing.' Confusion and anxiety lead to antisocial or solitary behaviour. While most other schoolchildren are busy playing games with each other, sharing experiences, children with autism are spending most of their time with their own thoughts, investigating not the people but the objects around them.

Children with autism mostly attend special schools, where teachers can provide them with the most appropriately stimulating, and understanding, environment. The most able children can be taught what to look out for in order to interpret what other people might be feeling, to help them survive in a confusing world. Alec, Josh and Robert are such children. They sit around a classroom table, looking at photographs in which other children are expressing obvious emotions. But the boys are bewildered by what they see. They are shown a photograph of a girl who is clearly looking very upset because she has fallen off her bicycle. When asked what the boys think the girl in the picture is feeling, they try to analyse what has happened, and to work out logically, rather than emotionally, how that might make the girl feel. So Josh says – after a long pause – 'I think the girl is unhappy, because she has fallen off her bike.' Josh was not able to get an immediate sense of the sadness of the girl by looking at the expression on her face.

Learning about human behaviour in this way – as if it was in a textbook – is like observing an alien species. The title of Oliver Sacks' book, *An Anthropologist on Mars*, aptly sums up how adults with autism often describe growing up. The title chapter of that book describes the life of Temple Grandin, a professor of animal science at Colorado State University. Grandin does not feel love or compassion: she finds it almost impossible to empathize. Like many people with autism, her overriding emotion is fear. A large proportion of the abattoirs in the USA are built according to designs worked out by Grandin. She can understand the behaviour of cows better than she can understand human behaviour. She interacts quite naturally with them – explaining that fear is their main emotion, too. Her ranch and abattoir designs are therefore among the most humane. Grandin compares herself to Mr Spock in *Star Trek*, and refers to non-autistic people as 'emotibots', who function according to feelings rather than reason alone. As she grew up she realized that, unlike her own, other people's thoughts and actions were related to their feelings. Unlike people who do not have autism – who have insight into social behaviour and are able to engage with other people's minds – Grandin has had to work out how other people form and retain relationships, by observing but not participating. She calls other people's normal social interactions ISPs (interesting sociological phenomena). She describes her mind as a network of supercomputers, filled with experiences and observations stored over many years. In any situation, she downloads the relevant data from these computers so that she can select appropriate behaviour based on analysis of previous similar situations. This takes a bit of time, so Grandin admits that she does not work particularly well under pressure.

The brains of non-autistic people work in this way to a certain extent, but they also rely heavily on automatic responses, based perhaps on emotional reactions to what someone says or their facial expression. To help her to choose the right response, Grandin has concocted metaphors, such as visualizing relationships between people as glass doors. If you push too hard, a glass door will break; in the same way, you must often hold back and be sensitive to other people's needs if you want to get on with them. Using these metaphors, together with her mind's supercomputer filled with her observations of other people, she is able to lead a fairly normal life in what she sees as an alien world. Grandin has little sense of why other people may get upset about things to do with feelings – to her, thoughts are much more substantial and important. When the archive of the university's library was flooded, and a large number of unique, original books were ruined, she found the loss of recorded knowledge in those books painful in the same way that other people experience the loss of a friend or relative. This is not to say that people with autism are necessarily uncaring or unkind: they just need a reason to care, rather than simply feeling it.

Development in mind

The supercomputer in Temple Grandin's brain works hard to replace the automatic mind-reading ability of most people. Children who show normal development begin to acquire this ability when they are around eighteen months old. It is worth comparing the development of social skills in 'normal' children with the development of children with autism. Since the 1920s, social psychologists have carried out tests – in observation rooms with one-way mirrors, for example – to explore and document the development of social skills. It turns out that from about eighteen months old, most babies indulge in what has been called shared attention – pointing to things just to share the experience of seeing it. They spend lots of time gazing at the faces of their parents or other adults, focusing mostly on the eyes. When something unusual happens, and a child is unsure whether it is safe or dangerous, he or she refers to the adult, vying for their attention and communicating, without words: 'Look at this; is this OK?'

Simon Baron-Cohen has carried out research in this area. What he has found is that children who do not instigate social referencing or shared attention are likely to develop the symptoms of autism, normally evident around the age of three or four. Such children fix their attention more frequently on inanimate objects than on the faces of adult carers. Other studies have shown that baby boys spend less time than baby girls looking at the eyes of a carer. Another important tool in the developing social brain is pretending. From the age of about two, most children play rich, imaginative and generally participatory role-playing games, in which they might be a dragon one minute and a magician the next. Again, children with autism rarely if ever engage in such role-playing games, and girls tend to play vivid pretend games more than boys tend to do.

Also at about two or three years old, children begin to understand how other people can have feelings that are different from their own – the beginning of empathy. This is illustrated by an experiment carried out by Alison Gopnik. She presents a young child with two bowls: one with tasty crackers and one with not-so-tasty broccoli. Gopnik pretends to taste each, but demonstrates to the child that to her the crackers taste horrible and the broccoli is delicious. This she does with exaggerated gestures and facial expressions, and simple language such as 'yuk' and 'yum'. She then asks the child to choose one of the foods to give to her. Children younger than two years do not seem to take on board Gopnik's obvious feelings about each food, and nearly always offer the crackers, because they prefer them themselves. Children of two years and above do tend to offer Gopnik the broccoli, even though this is counter-intuitive since they would prefer the crackers. At this stage, it seems, children become able to understand that other people can have different views about the world from their own. A bit later on, at age five or six, children become able to understand that the same object can be interpreted in different ways. Gopnik uses ambiguous figures, such as a line drawing that could represent the head of a rabbit or the head of a duck. Children older than about six experience a switching of their interpretation of the drawing between 'duck' and 'rabbit'. Younger children – and people with autism – do not experience this switching.

One test of a child's increasing ability to 'get inside' other people's minds is the Sally-Anne Test. Originally developed by two Austrian developmental psychologists, Heinz Wimmer and Josef Perner, it was adapted by Uta Frith and her colleagues at King's College, London, for their study of the effects of autism. The test involves a scenario that is played out in front of a child, using dolls or puppets to represent the main characters. Sally has a basket, in which she keeps a marble, and Anne has a box. While Sally is out playing, Anne steals Sally's marble and puts it into her box. When asked where Sally will look for her marble when she returns, children above four years old almost always give the correct answer: in the basket where she left it, since she would be unaware that Anne had stolen it. Children younger than four tend to answer that Sally would look in Anne's box – the marble's real location – because they are unable to put themselves in Sally's position. Frith found that most children with autism – in a sample with a mental age of nine – gave the wrong answer too.

The general intelligence of people with autism – particularly those with severe symptoms – is normally lower than average. Could the fact that people with autism are not good at interpreting social cues – that they are not skilled mind readers – simply be a consequence of this lower intelligence? Two pieces of evidence suggest not: first, people with autism who do have normal intelligence, such as those with Asperger's syndrome, still have problems socializing and understanding other people's behaviour. Secondly, people with another disorder – Williams' syndrome – who also have lower than average intelligence, do have normal social skills.

Dr Helen Tager-Flusberg, a psychologist at the University of Massachusetts, is an expert in the study of children with developmental disorders, particularly problems they have in acquiring language. She has studied children with autism and those with Williams' syndrome, and has shown that children with Williams' syndrome are interested in other children, and they are very aware of when other people are happy or sad. They communicate freely and express sympathy – both happiness or sadness – about other children's feelings. They use rich and imaginative language, and take part in pretend games. Tager-Flusberg says that you see none of these behaviours in children with autism. When you give children with Williams' syndrome tasks that require them to understand what people are feeling, they do extremely well. Maude is six years old, and has Williams' syndrome. When shown a photograph of a girl being scared by a dog, she immediately gets involved with the emotions expressed in the photograph. She says that the girl is scared, because the dog is like a monster. Compare this with the way that the able children with autism had to be taught what to look out for to work out when someone looked happy. One of Tager-Flusberg's colleagues carries out a version of the Sally-Anne Test with Maude's friend Becky, another six-year-old girl with Williams' syndrome. Children with autism do not do well at this kind of test, but Becky does just fine. The experimenter shows her a crayon box, which Becky assumes contains crayons. Becky opens the box, to find that there is a sticking plaster inside. When asked what her mother would expect to be in the box if she walked in, Becky immediately answers 'crayons'. Children with autism would be likely to answer 'a sticking plaster'. Because children with Williams' syndrome are so able to carry out tasks such as these, while children with autism are particularly bad at them, Tager-Flusberg, like many psychologists, sees Williams' syndrome as the opposite of autism. The difference is in their mind-reading capabilities, their social brain.

So the deficiencies in the social brain of people with autism cannot simply be a result of lower intelligence. Instead, could they be due to damage in specific areas of the brains of these people? If so, might these areas be key parts of the social brain? In recent years, various studies have been conducted on the brains of people with autism. They have helped neuropsychologists to improve their understanding of the disorder, and have indeed provided vital clues in our quest to find the social brain.

Looking inside

Abnormalities have been found in the brains of people with autism. Some of these are found in parts of the cerebellum (the 'little brain' at the top of the brain stem) known to affect language, general cognition and attention. This might well account for some of the symptoms of autism. More relevant to our discussion of the social brain, some studies have implicated a different brain structure – the amygdalae – in the symptoms of autism. There are two amygdalae, one on either side of the brain stem. You can picture them as almond-shaped knots of nerve cells just behind your nasal cavity. These dense clumps of neurones form part of the limbic system, which is associated with emotional behaviour. The amygdalae receive signals from the eyes, nose and ears, and seem to carry out some sort of low-level, automatic computation on these signals – important in the functioning of the social brain. They also receive signals from the cortex, which carries out high-level information processing and makes associations with memories of past events. Neuroscientists have uncovered increasingly convincing evidence of the role of the amygdalae in the social brain. One way of accumulating such evidence is to use brain-scanning techniques such as MRI (magnetic resonance imaging) to investigate the activity of the amygdalae during various tasks. In 1996 Nancy Etcoff and Hans Breiter at the Massachusetts Institute of Technology did just that. The amygdalae of their experimental subjects 'lit up' on their MRI monitor screens when they showed them a series of pictures of people with fearful or threatening expressions.

The amygdalae have long been suspected to be responsible for the recognition of emotion in other people. Just how do such ideas about brain regions come about? How did neuroscientists have any idea which parts of the brain did what before the invention of MRI and other scanning techniques? There are two main ways: indirectly, by observing consistently abnormal behaviours in people or animals with damage to specific areas of their brains; and directly, by manipulating or stimulating a brain while a person or animal is conscious and recording their response or reported feelings. The amygdalae are common to many animals, including reptiles, birds and all mammals. Removal of the amygdalae in animals has been shown to reduce the animals' avoidance of strange objects, suggesting that the amygdalae are involved when we respond to fear. In several studies, mostly on rats throughout the 1980s and 1990s, Professor Joseph LeDoux of the Center for Neural Science at New York University reinforced the idea that the amygdalae are involved in an automatic response to fear. He showed that our bodies – through the amygdalae – respond to fear before we are aware of it.

Physical evidence that the amygdalae play a part in autism came in 1994 from a famous study carried out by Margaret Bauman at Harvard Medical School and Thomas Kemper at Boston University. They showed that the amygdalae of an autistic brain were smaller and more densely packed than in the normal brain. The actual neurones of the amygdalae were smaller than normal, too. In 1998 a team led by Wendy Kates of the Johns Hopkins School of Medicine in Baltimore examined the amygdalae of seven-year-old identical twin boys using an MRI scanner. One of the twins showed much more severe symptoms of autism than the other. Kates found several differences between the brains of the two boys. The hippocampus, also involved in emotional response, and the cerebellum and the caudate nucleus, involved in shifting attention from one thing to another, were all much smaller in the autistic twin's brain. So too were the amygdalae, and they were also significantly less active in the autistic twin's brain than in his brother's.

Simon Baron-Cohen and his team at the University of Cambridge have carried out similar studies. Using MRI, Baron-Cohen compared the activity of the amygdalae in people with autism to non-autistic people. Each person involved in the study was shown a series of photographs of expressive faces while their brain was scanned. Baron-Cohen's results were conclusive: people with autism do not use their amygdalae when evaluating these photographs, while others do. Again we can see that our amygdalae must be heavily involved in our social behaviour – people with autism, whose social behaviour is impaired, have deficiencies to their amygdalae. Interestingly, and perhaps not surprisingly, the brains of people with Williams' syndrome are smaller than average, but their amygdalae are the same size as in a normal brain.

Thinking it over

Just behind your forehead, above your eye sockets, lie other parts of the brain that have long been suspected to contribute to social behaviour: the orbito-frontal cortex. These sophisticated masses of neurones seem to be responsible for high-level processing of social behaviour. Our modern understanding of their function is based on evidence from case histories spanning the last 150 years. The most famous case is that of Phineas Gage, whose left orbito-frontal cortex suffered extensive damage as a result of a horrific accident in 1848. The accident happened as Gage was using an iron bar to push, or 'tamp', explosives into a hole at a construction site in Vermont. The powder exploded, firing the tamping iron out of the hole at great speed. The bar – about one metre long and six centimetres wide – shot straight through Gage's head and landed about thirty metres away. Remarkably, he survived the accident and lived for another ten years. You might imagine that such an incident would affect your behaviour: perhaps it would make you grateful to be alive. The accident did alter Gage's behaviour dramatically, but the changes were quite specific: he became irritable, bad-tempered, even rude. The accident had impaired his social brain.

Scientists at the University of Iowa have used computers to study Gage's skull which, together with the bar, has been preserved at the Harvard University Medical School Library. Using three-dimensional visualization techniques, together with knowledge of the anatomy of the brain, they have been able to work out exactly which parts of Gage's brain the iron bar destroyed. The bar passed through his chin and behind his left eye, destroying his left orbito-frontal cortex. This single case is not enough to locate the social brain – or any part of it – in the orbito-frontal region, but there have been hundreds of similar, well-documented cases. And changes in personality similar to those experienced by Phineas Gage are still occurring today – in road accident victims with severe head injury, for example. Professor Damasio, one of the team who worked on Gage's skull, says that some of his patients who have suffered severe head trauma have shown such personality changes. High-speed motorcycle crashes are perhaps the most common cause of head injury. Even though modern crash helmets are effective in dampening the force of an impact, the brain may still smash against the inside of the skull. This causes little damage to the rear of the brain, since the inside surface of the skull there is smooth. At the front, however, the inside of the skull has jagged bony ridges. When the brain hits these areas, damage is often sustained to the orbito-frontal cortex.

More evidence of the role of the orbito-frontal cortex is gleaned from a surgical procedure called lobotomy. This operation involves cutting the fibres that lead to the frontal cortex. One form of the operation was carried out by hammering an ice-pick into the brain through the orbit of the eye. Lobotomy was pioneered in 1935 by a Portuguese doctor, Antonio Egas Moniz, who called it 'leukotomy'. Between 1939 and 1950, 18,000 lobotomies were performed in the USA alone. In 1949 Egas Moniz won the Nobel Prize for medicine for his contribution, but the procedure fell out of favour during the 1950s as people began to realize that it seemed to be causing as many problems as it solved. The lives of many patients were effectively ruined after lobotomy; Egas Moniz himself was shot in the spine by one of his former patients. Often, lobotomy changed people's personalities in undesirable ways: many became unaware of other people's feelings, uninhibited, unable to conduct themselves in a civilized manner.

A little more evidence that the amygdalae and the orbito-frontal regions are major parts of the social brain is provided by the bizarre case of one of Nancy Etcoff's patients, referred to as 'LH'. In an accident, he had suffered severe damage to areas of his cortex associated with facial recognition. He cannot recognize his wife's face – she wears a ribbon that helps him identify her – or his children's faces, or even photographs of himself. He describes an experience: 'Several years ago, when I was attending a conference ... I had to go to the lavatory. On the way back, I came around the corner and saw someone who I thought was a former supervisor of mine and greeted him. But there was no response. I looked again and found that I was looking at a mirrored wall, and therefore at myself.'

LH refers to himself as 'the stranger in the mirror'. But while he cannot identify faces, he can still recognize the emotions expressed by them: in other words, his social brain is intact and he can act accordingly. Neither his orbito-frontal regions nor his amygdalae were damaged in the accident. So his amygdalae continued to carry out the initial, automatic processing of signals from the senses, evaluating whether a face is sad, happy or fearful, for example, and sending the verdict to the orbito-frontal cortex. It is probably in the orbito-frontal cortex that the decision is taken on what to do with the information supplied by the amygdalae.

We have seen how this model of the social brain relates to people with autism, but can we relate it to the social behaviour of men? There is no evidence as yet in normal men of the damage to the amygdalae found in autism. Nor is there any evidence of any abnormality in the orbito-frontal cortex of men. In fact, no one has yet pinned down exactly how men's and women's social brain differs. However, one possible clue is provided through work carried out by a team led by David Skuse of the Institute of Child Health in London.

Crucial chromosomes

Skuse's study involved a rare genetic disorder called Turner's syndrome. Previous psychological studies have shown that some girls with Turner's syndrome seem to have impaired social skills. This research could point the way towards an explanation of why men are more likely to lack the social sensitivity that women seem to have. This work is highly speculative, but it could be the first step in finding the genes for social behaviour, and for understanding disorders such as autism.

It is our DNA (deoxyribonucleic acid) that determines whether we are male or female, and so consistent physical differences between men and women will be due ultimately to differences in DNA between the two sexes. The DNA is packaged as separate bundles called chromosomes, which occur in pairs, and a complete set of chromosomes is found in most cells of the body. A gene is a specific section of the DNA that makes up a chromosome. Each gene carries a coded instruction on how to build a particular type of protein. Our bodies are made mainly of proteins, so an individual's DNA in its entirety (the human genome) is like a complete build-a-human instruction manual. Under certain conditions, the various chromosomes are visible and identifiable by shape under an ordinary microscope, and their paired nature is clearly apparent. Genes on a particular chromosome correspond to the genes on the other member of the chromosome pair – although there may be many different versions of the same gene. For example, there is a gene that determines blood group, which occurs at a particular position on a particular chromosome. Each person has two versions of this gene – which may be the same or different – one on each of the two corresponding chromosomes. The blood group gene has several different forms, and the blood group of an individual will depend upon the combination of the two versions of that gene. One member of each chromosome pair comes from a person's mother and one from the father. So a person with two copies of the A version of the gene will be blood group A, while a person with one A and one B will be type AB.

A normal human being has twenty-three pairs of chromosomes, including two sex chromosomes, which may be type X or type Y. A male has an X and a Y, while a female has two Xs. Turner's syndrome – the disorder involved in Skuse's study – is caused by an irregularity in these sex chromosomes. The scientific shorthand for the chromosome complement of a normal male is 46,XY – twenty-three pairs of chromosomes, with an X and a Y for the sex chromosomes. For a female it is 46,XX. In Turner's syndrome, an individual has only one sex chromosome, an X; here, the chromosome complement is 45,X. In terms of their DNA, babies born with Turner's syndrome are neither boys nor girls. However, they grow up as girls since they do not have a Y chromosome. It is the presence of the Y chromosome that leads to the production of testosterone, which alters the usual female form of the human body, producing male body characteristics. Girls with Turner's syndrome have much in common with girls: they share the same physical characteristics (although development of their reproductive organs is impaired). However, in that they have just one

X chromosome, girls with Turner's syndrome (45,X) have something in common with boys (46,XY), too. Might they show similar traits of behaviour to boys? Can a study of the behaviour of girls with Turner's syndrome tell us anything about male-related behaviour? Skuse and his colleagues think that it can.

To what extent our DNA – our genes – determines our behaviour is an issue that has been investigated and hotly debated for as long as genes have been known to exist. This area of study, once called sociobiology, is now more often called evolutionary psychology. It is modern science's version of the age-old nature–nurture debate. Skuse and his team studied a total of eighty females, between the ages of six and twenty-five, with Turner's syndrome, administering psychological evaluations as well as carrying out genetic investigations. The psychological tests included an interview, a questionnaire and careful observation of the behaviour of the experimental subjects. The girls in the study who had social problems also reacted unfavourably to changes in routine – a behaviour commonly found in people with autism, and more often found in men than in women. Kylie, one of the girls involved in Skuse's research, was referred to him after she had shown signs of impaired social behaviour – some of the behaviours that are found in people with autism, and in typical males. If just a small part of the breakfast routine was different from normal, Kylie would react aggressively, and would be terribly confused. Her mother says that it is as if Kylie has to work through a list, that everything must be ordered. She also shows the signs of impaired social behaviour that Skuse was focusing on in his study. She has few friends, and often upsets or offends people without realizing she has done so – until she has thought it through logically, or had it explained to her in detail.

Skuse had a theory that could explain the differences between girls with Turner's syndrome who have social problems and those who do not. There are two possible sources of the single X chromosome found in a girl with Turner's syndrome: her mother or her father. All boys inherit their single X chromosome from their mother. Skuse wondered whether the females with what he refers to as 'impaired social cognition' had inherited their single X chromosome from their mother – just as boys do. Genetic tests on the sample of eighty females showed that fifty-five of them had inherited the maternal X chromosome, while the remaining twenty-five had the paternal X chromosome. What did the personality tests reveal about the social cognition of the two groups?

The results were as Skuse had suspected. He sums up: 'When we looked at those girls who had just one X chromosome, and divided them into two groups – those in whom that single X came from the mother and those in whom the single X came from the father – overwhelmingly, the social problems were in the girls whose single X came from the mother.' One illustration of this was that educational 'statementing' (identifying special needs) was three times as common in those with the maternal X chromosome than those with the paternal one.

How can we explain these findings? Skuse appeals to a relatively recent finding in genetics – that some genes are switched off on a particular member of a chromosome pair. For example, a gene on chromosome number 18 might always be inactive on the edition of that chromosome inherited from, say, the mother. This process is called imprinting: an imprinted gene is one that is deactivated. With most, non-imprinted, genes the two copies – one on each of two corresponding chromosomes – are both active. We have considered the example of the two genes that determine blood group: if a baby's mother has blood group A and the father has blood group B, then the baby will be group AB. If the gene for blood group were an imprinted one, and it was always the gene inherited from the father that was deactivated, that particular baby would have blood type A. This deactivation of a gene according to its parental origin has been found in at least twenty human genes. The process of imprinting is thought to occur when small molecules literally attach to the DNA, blocking its ability to manufacture the protein for which it holds the instructions. How does the idea of imprinted genes relate to the findings of David Skuse's research?

Skuse proposes that there might be genes found on the X chromosome that are responsible for social cognition. He thinks that the versions of these genes on the maternal chromosome are switched off – they are imprinted genes. Usually girls possess the paternal chromosome, on which the genes would remain active. Girls with Turner's syndrome who have only the paternal chromosome also benefit from those genes. However, girls with Turner's syndrome who possess only the maternal chromosome will have only switched-off copies of the genes. The maternal X chromosome is the only X chromosome that all normal males possess, so they too have only switched-off copies of the genes. Might this explain why normal males (46,XY, with X from the mother and Y from the father) and some girls with Turner's syndrome (45,X, with X from the mother) are at much greater risk of having impaired social behaviour than normal females (46,XX, with X from both the mother and father) and the other girls with Turner's syndrome (45,X, with X from the father)?

Another finding of the research was that the incidence of autism in the girls with Turner's syndrome was much higher than normal in the group with the maternal X chromosome than for normal girls or for those with only the paternal X chromosome. Again, this is true of boys, too: the occurrence of autism in males is as much as ten times as frequent as in females. Skuse's work does seem to suggest a link between our genes and our social behaviour. The social brain is almost certainly not determined by just one gene, or even many genes residing only on the X chromosome. If it did, then impairment of the social brain would be found in all males and in all of the girls with Turner's syndrome whose X chromosome came from the mother. Perhaps the social brain depends upon the collaboration of many genes, on several different chromosomes. Maybe some important ones are brought into action only when the genes on the paternal X chromosome are active – in females. Whatever genes actually are at the root of the functioning of the social brain, the situation leaves men with a much higher chance of developing autism. What possible evolutionary advantage could that bring, for it to have remained as part of our DNA?

Simon Baron-Cohen thinks he knows what the advantage of having a Y chromosome might be. He has carried out extensive psychological tests on people with autism or Asperger's syndrome and on their parents. He has carried out the same tests on a control group. The tests require very different mental skills: one involved guessing people's emotions from photographs of their eyes, the other was a visual–spatial task involving analysis of shapes. Baron-Cohen found that people with autistic-type disorders did worse than women at the first of these tasks, and so did normal men. They fared much better with the second task, however, as did normal men. The parents and grandparents on both sides of the family also did worse than the average. About five per cent of the fathers of non-autistic children are employed in some kind of engineering, compared with about twelve per cent of the fathers of children with autism. This could be explained by a bias in the brains of people with autism towards abstract, physical things and away from people, an idea that Baron-Cohen's tests seem to bear out. This would also explain why people with autism are not given to engaging with people socially. Compared with other babies, those who grow up to be autistic spend far less time looking at their parents' eyes. Similarly, boy babies spend far less time than girl babies making eye contact. Perhaps children with autism are just at one extreme of a continuum of male behaviour.

Of course, this perceived preference of males towards physical, technical things might be as much a result of socialization as the perceived superior mind-reading abilities of women. We have to exercise care in interpreting the results of studies like that carried out by David Skuse and Simon Baron-Cohen. Time and time again in the history of science, there have been examples of how the same facts can be interpreted in different ways, according to prevailing theories, religious beliefs or, of course, simply lack of enough information. For example, before the discovery of oxygen and its role in burning, most of the evidence seemed to support the prevailing idea at the time that burning involved the release of a hypothetical substance called phlogiston. Because there was no direct proof of the existence of phlogiston – only evidence that supported the idea that it existed – the theory held sway. In the same way, perhaps the conclusions of neuropsychologists will be overturned in the future, by new discoveries in genetics, physiology or psychology, or by new perspectives on existing knowledge. This does not mean that science is wrong or not worthwhile – the scientific method is a way of treading carefully along the road towards truth. But it is important to remember that our knowledge and understanding of the world are continually shifting. The results of the fascinating studies that attempt to find a link between biology and behaviour may well be interpreted in a very different way in the future. But it is also worth remembering that it was science that highlighted the fact that children with autism are not simply badly behaved because of bad parenting. And for now, current, rapidly developing theories are helping us to find the next step in the journey to find the social brain.

Social behaviour requires intelligence – a different type of intelligence from that used to take standard intelligence tests, which are the subject of the next chapter of this book.

There do seem to be fundamental differences in social intelligence between men and women, which probably do ultimately have a genetic component. Finding out how and why DNA can cause these differences might challenge some of our beliefs and our political and social values. But the real value of the quest for the social brain might be a better understanding of people with autism. Meanwhile, those lucky enough to have a fully functioning social brain can appreciate the amazing and quite automatic skills that allow human beings to be such remarkable mind readers.

NATURAL-BORN GENIUS

...the search for the nature of intelligence...

You are more intelligent than a dog, an elephant and a chimpanzee. That may be obvious, but have you ever stopped to think how amazing your intelligence is? It is easy to take for granted the incredible things that human intelligence allows us to do: read, write, explore, understand, hope, talk, make music, represent things, wonder, solve problems, work out patterns in things or the relationships between things. Parrots can speak, dogs can explore, nightingales can make music and dolphins can solve certain problems. But humans are the only animals that do all of these things, and do them well. Our minds are rich, creative, sensitive, and have given us the power to gain an understanding of the Universe around us.

It is our genes, made up of molecules of DNA, that determine our physiological characteristics and define our species. In your DNA is a complete blueprint for how to build a human body – and that includes your brain. So if there are fundamental physiological differences between humans and other animals, their origin is in DNA. Does that hold true for differences between any two human beings? It certainly does for qualities such as sex, eye colour and height. But what about behaviour and intelligence? Are some people born more intelligent than others, as a result of differences in DNA? This question is intensely controversial for many reasons, not least in terms of the lack of a precise definition of intelligence. Despite this controversy,

American behavioural geneticist Robert Plomin thinks he has found the first evidence that it is our genes that determine how intelligent we are.

Gene genius

Plomin is based at the SGDPRC (Social, Genetic and Developmental Psychiatry Research Centre), part of the Institute of Psychiatry, in London. He is one of the major figures in research into the nature of intelligence. His latest results were the culmination of six years' work, in which he followed his stern belief that there must be a genetic link to intelligence. He used the tools and procedures of molecular genetics – the study of the science of the inheritance of characteristics at the level of DNA molecules. Plomin's research was well timed: recent advances in the field of molecular genetics have made it possible actually to identify the location and functions of thousands of specific genes in plants, and in humans and other animals. In particular, a massive international effort to map out all the seventy thousand or so human genes – the Human Genome Project – is well under way. The initiative for this massive quest came from the US National Institutes of Health and the Department of Energy; in the next couple of years, the mapping process should be complete.

The human genome is the entire DNA of a human being: the twenty-three complementary pairs of chromosomes. The new genome is fixed when an egg is fertilized, defining the new individual with a complete set of forty-six chromosomes. The chromosomes present in that fertilized egg are copied to most of the cells that make up the new, growing individual. If we all have the same set of chromosomes, how can we all be different? And how can anyone assert that our DNA accounts for differences in eye colour, let alone something as subtle as intelligence? This is where genes come into the story: genes have a number of different forms. And so we are physiologically different because we have different versions of genes.

Each gene along the length of the DNA has a very specific function: it holds the instructions to produce a single protein that is used in the body. Some proteins are the building blocks of the body, including the brain; others are enzymes that regulate various functions of the body. Different versions of the same gene produce different versions of the same protein, which behave differently within the body. In some cases, this can lead to congenital diseases – for example, the inherited disease sickle-cell anaemia is the result of a particular form of the gene for the protein haemoglobin, which carries oxygen around the blood. Proteins are also important in the brain: they are the basis of many hormones and neurotransmitters, as well as being involved in the actual building of neurones. So two people with different versions of certain genes will have differences in the brain – but can those differences account for inborn differences in intelligence? Indeed, are we born with different amounts of intelligence?

Robert Plomin moved to London from Pennsylvania in 1994, the year the SGDPRC opened. The author of over two hundred scientific papers, he has a long and impressive history in the field of behavioural genetics. As long ago as 1978 he claimed to have identified a genetic basis for many behavioural aspects of laboratory animals, including learning, sexual activity, alcohol preference and aggression. In studies of twins involving humans, he also discovered evidence of a genetic component in aggression, emotional sensitivity and sociability. Unlike height or eye colour, these qualities are not easy to measure or even to define consistently and unambiguously, making any supposed genetic link difficult to prove. Intelligence suffers from the same problem, and this is one of the main reasons that his work is controversial. Nevertheless, Plomin and his colleagues set about trying to find a link, using standard intelligence tests.

Plomin considered this latest project his most challenging yet, both in scientific and political terms. It certainly has its detractors, both within the field of molecular genetics and beyond. Dr David King, editor of *GenEthics News*, a major force against Plomin's work, can see how a positive result from Plomin's work might have terrible consequences. He claims that the media would treat the results simplistically: the translation of such a finding into popular culture would be simply that those who do not have the gene for intelligence are inferior. But the over-simplification of scientific ideas could lead to more than just prejudice against individuals. Genes are the basis of race – what would it mean if it was found that certain races or ethnic groups were more likely to have the supposed 'right genes' for intelligence? David King is Jewish, and is very aware that the beliefs of the Nazis during the twentieth century were based on just this sort of premise: that breeding is everything, that we are determined purely by our genes.

As well as being challenging scientifically and politically, the possibility of a genetic basis to our intelligence is important philosophically: Plomin's work could change the way we think about human beings. It is fair to say that many academics in the field have been unwilling to believe that a genetic component exists, asserting instead that at birth we are like a blank sheet, a piece of putty to be shaped by our experiences as we develop. The question to what extent our intelligence is determined by our genes is a modern manifestation of the age-old nature versus nurture debate. Many people think that nurture builds on nature – that we are all born with about the same potential intelligence, the development of which is governed by the environment in which we grow up and the things we experience. Many scientists – even in the field of molecular genetics – saw the work as a waste of time. Plomin has always been undeterred by criticism: he sees his research as lying beyond the scope of nature–nurture debates. He thinks of it as part of a scientific quest, not a political one, to discover how the world works, and believes that the social and political consequences of any scientific discovery are the business of philosophers and politicians. This sort of argument is often used to defend research into other areas of scientific exploration, such as nuclear physics. Plomin maintains that his work really is pure science: in strict scientific terms, the funding for his project was simply to be used to search for genes that influence general intelligence.

The overall approach of Plomin's experiment was straightforward enough. He analysed specific regions of the DNA of two groups of people: one whose members had average general intelligence and one whose members had super-high intelligence. The people in the second group were each 'one in a million', as Plomin puts it. But Plomin was not actually looking for genes that create geniuses. He was searching for the first of many genes that contribute to an individual's general intelligence – the idea that underpins the work is that intelligence is a multiple-gene trait. In other words, Plomin believes that there are many genes that contribute to intelligence, at many different sites along the DNA on our chromosomes. Each gene can have many different forms. Plomin believes that some forms of each gene contribute positively, while others contribute negatively to our general intelligence.

An analogy with playing cards is useful: imagine a game where you are dealt one card from each of the four suits. In terms of this analogy the individuals with super-high intelligence in Plomin's research were each dealt a hand with four aces. A player with an average hand might have all 8s or, say, two aces and two 2s. Now imagine that there are in fact tens of thousands of different suits, and the pack consists of millions of cards. Each person is dealt a unique hand, with thousands of cards, again with one card from each suit. Suppose that only some of the many suits are important in this particular game. Then the people with the 'best' hands will have aces in most of those important suits. If you choose one of the suits important in the game, and ask each winning player to give you the card they have from that suit, most of the cards you collect will be aces. If you do the same with people who did not do so well in the game, there will be far fewer aces. If you repeat this for a suit that is not important in the game, then the likelihood is that you will have two sets of cards that you cannot tell apart. Fitting this analogy to Plomin's research, each suit is a human gene, and the face values of the cards in a particular suit are the different versions of a particular gene. So all Plomin had to do was look at the versions of particular genes in his two groups of people and determine whether a particular version occurred more frequently in the people with super-high IQ scores. With tens of thousands of genes, this is quite a task, but Plomin's previous studies had led him to a particular part of chromosome 6, which is well mapped out thanks to the Human Genome Project.

Clever Definition

Before we can begin to investigate whether or not intelligence is determined by genes, we really need to be clear about the definition of intelligence. Even here, we find controversy and disagreement. Just what is intelligence, and how can we measure it? The kind of intelligence that intelligence tests attempt to measure is called 'general intelligence'. It is supposed to be a catch-all term for cognitive ability, which is largely based on reasoning and remembering. These are vital to thought processes such as problem solving, comprehension and the use of language and mathematics, which are involved in the tasks used in standard intelligence tests. This description of intelligence may sound straightforward, but an exact definition of intelligence is almost impossible to pin down, and a reliable way to measure it is therefore just as elusive. In academic circles, the accepted meaning of the word has shifted significantly and frequently over the past hundred-odd years, and so have the theory, content and approach of intelligence tests.

The modern history of the study of intelligence begins in Victorian Britain, with Francis Galton. In 1869 Galton published his book *Hereditary Genius*, in which he set out his idea that intelligence is inborn and that some people are born with more of it than others. In that sense, Robert Plomin's research can be traced back to Galton. British naturalist Charles Darwin, who was Galton's half-cousin, was impressed by his ideas about the hereditary nature of intelligence. Darwin had previously believed that the differences in cognitive ability between any two individuals was the result of hard work alone – that people were born with almost equal potential. He was not the only one to be influenced by Galton's ideas, and Galton's methods for measuring intelligence soon became popular. In 1884 Galton set up a laboratory in South Kensington, London, where people were tested on a variety of different psychological and physiological bases, including the size of their heads and whether they could tell which of two weights was heavier. The results of these tests, Galton claimed, would give an indication of a person's general intelligence.

Galton's approach was part of an attempt to quantify everything in the natural world – biometrics – which is still a major part of scientific study. One part of Galton's biometric approach that has survived, in some academic circles at least, is the idea that intelligence is related to the speed of mental processes. Some modern intelligence theorists use reaction times or decision times as a measure of the speed with which the brain processes information, and therefore as an indication of general intelligence. The most popular test of this kind involves a box with eight lights, each with a button to press next to it. There is another button, on which you rest your hand. One of the lights illuminates, at random, and you have to react by pressing the button adjacent to that light as quickly as you can. The electronics inside the box measures how long you took to move your finger and how long it took you to move to the correct button. The test is repeated many times, so that averages of these measurements can be calculated. When compared with intelligence test scores, these measurements do correlate fairly well, and there may be some truth in them. And there are other scientists today who are continuing in Galton's tradition of measuring physical and physiological quantities to determine the nature of intelligence.

At the University of Edinburgh a differential psychologist, Ian Deary, is carrying out research into how the brain processes information, and how it relates to intelligence. He measures cranial capacity, just as Galton did, but he also uses brain scans to work out which parts of the brain are involved in the processes that give us our intelligence. This kind of research is an attempt to bridge the gap between the physiology of the brain – the hardware – and the elusive quality that we call intelligence – the software. In Deary's studies, and several others, some connection has been found between people's cranial capacities (or rather brain sizes) and the most popular measure of general intelligence, their IQs (intelligence quotients). This idea may seem crude – an elephant has a much larger brain than a human being, for example. Of course, comparing the size of the brain with body weight, humans fare better than elephants. A man's brain is slightly larger on average than a woman's, but a man's body weight is also greater on average. The value of studies into brain size and intelligence is, as yet, unknown.

Just as the roots of Plomin's work can be found in Galton's ideas, so can some of the arguments against it: Galton was the person who coined the term 'eugenics' for the idea that the human race could be bettered by 'selective breeding'. Galton was aware – or perhaps too aware – of the importance of biology and 'pedigree' in determining human traits, and proposed that careful breeding, between distinguished men and well-to-do women, would lead to a world filled with geniuses, for the good of everyone. Incidentally, this underlying concept of eugenics pre-dates Galton and the science of genetics by 2,000 years: the ancient Greek philosopher Plato put forward the idea in his great work, *The Republic*. And eugenics lived on after Galton died: it was the basis of mass sterilization programmes and propaganda in several countries that urged young people to choose someone 'fit to marry', to avoid 'pollution of the blood'.

One of Galton's claims – that a person's intelligence measured using his methods should be a predictor of his or her school or college grades – was shown to be woefully inaccurate, and Galton quickly fell out of favour. By the time he died in 1911, a new approach to intelligence – and to intelligence testing – had emerged. That new approach was pioneered by Alfred Binet, who actually died in the same year as Galton, but whose ideas dominated much of the twentieth century. Alfred Binet was a French psychologist, and impressed by Galton's attempts to use standard tests to measure differences between people. The French minister for education at the time was concerned that children with behavioural problems were receiving less teaching than their peers, because the teachers did not want to teach them. He was convinced that many of these children would therefore not reach their full potential, and ordered the creation of psychological tests to determine children's general intelligence. Binet was given the task of originating these tests, and he decided to break with the tradition of Galton and his followers. Instead of measuring what he saw as unimportant quantities, such as cranial capacity and the strength of a person's grip, Binet focused on abstract but more relevant mental processes such as judgement, comprehension, reasoning and memory. These are the same kinds of skills measured on most intelligence tests today.

But perhaps Binet's greatest contribution was the idea that we can obtain a useful measure of a person's intelligence by comparing their scores on standard tests with the average. Binet faced a fundamental problem when constructing intelligence tests for children: mental ability develops gradually over many years. The average five-year-old would not be able to solve a problem that the average ten-year-old could only just solve, for example. By giving many children of the same age a set of standard questions of varying difficulty, Binet could figure out what level of problem, say, an average seven-year-old could solve. Doing this for every chronological age would bring to light the development of the average child's mind. That is interesting in itself, but the master stroke was the idea that, by comparing a child's results with the standardized set of results, these tests could be used to compare a child's mental abilities with the average. Specifically, by comparing a child's results with the average scores for each age, you could ascertain a 'mental age' for that child. By comparing the mental age with the child's actual, chronological age, you could work out whether a child was above or below average for his or her age.

Binet's ideas were taken up by the German psychologist William Stern, who used simple mathematics to work out the 'intelligence quotient', or IQ. A simple sum is all that is needed: divide a child's mental age by his or her chronological age and multiply by 100. If a child's development exactly matches the average child for his or her age, then mental age and chronological age are the same and the calculation works out as 100 – the average IQ. If a six-year-old child is measured to have a mental age of nine according to Binet's tests, then his or her IQ is nine divided by six multiplied by 100: 150. Development of mental capability slows down dramatically as we leave childhood, so the process of estimating mental age becomes redundant in adults. But the concept of IQ – as a mathematical comparison with the average – was extended to adults. And eventually, the definition of IQ itself has shifted away from the idea of mental age. IQ is now defined using a different statistical method. Despite Binet's important contributions to the measurement of general intellectual ability, he did not really believe in the idea of general intelligence. He saw intelligence as something that pervaded all areas of mental activity, and this is why he used tests that examined skills across a wide range of mental abilities. But he believed that you could never pin down what intelligence was. To him, it was almost meaningless to say that one person was more intelligent than another.

While Binet was busy constructing his tests for French schoolchildren, a British psychologist, Charles Spearman, was setting out his ideas on what general intelligence might actually mean. Spearman's approach appealed to statistics: he was looking for correlation between test scores in several different types of ability. In other words, was a person who was good at language equally good at mathematics or remembering facts? He found that the correlation was good enough to define general intelligence, which he abbreviated to 'g'. The concept of testing people's general mental abilities was given a significant boost by the recruitment demands of World War I. The British army used psychological testing, based on Spearman's idea of g, to help them match recruits to the right jobs. The concept of general intelligence grew in popularity and, in the USA in particular, the idea of intelligence testing really took off. However, during the 1920s, rising immigration into America uncovered an inherent problem with intelligence tests: cultural bias. The tests included verbal reasoning tasks in English and visual problem solving tasks involving pictures of objects such as gramophones and tennis courts. Many of the immigrants did not speak English; some had come from places that did not have gramophones or tennis courts. The tests were clearly culturally specific and not at all relevant to these people.

Another problem associated with intelligence testing – and not unrelated to the new wave of immigration into the

USA – was the resurgence of eugenics, the idea pioneered by Francis Galton. Apart from Hitler's Germany, nowhere was the principle of eugenics applied on a larger scale than in the USA. In more than half American states during the 1920s, eugenics laws were passed that made it unlawful for certain people to marry unless they agreed to sterilization. It was appealing to some, and indeed became common, to apply labels to people according to their mental age or IQ, as defined according to Binet and Spearman. So, for example, people with IQs of over 130 were 'gifted', while those with IQs below average were labelled with varying degrees of 'retardation'. It is the latter of these two labels that was seized upon by the eugenicists of the 1920s. 'Mental retardation' was close to the term 'feeblemindedness', a label beloved of the originators of eugenics. Feeblemindedness was assumed by eugenicists to be a precursor of criminal behaviour. And when intelligence tests indicated that an adult had a mental age of twelve or below, eugenicists normally defined that person as feebleminded. Intelligence tests, then, despite their obvious cultural bias, provided some people with a seemingly objective way of showing that immigrants from Italy, Greece and countries of Eastern Europe were 'inferior'. And these were the people who were most at risk of being affected by eugenics laws. In addition, US citizens who were not immigrants but were 'mentally retarded' through, say, some congenital disorder, were often subject to the same laws. It is worth pointing out that Hitler was not in favour of intelligence testing. His eugenic theories – and eugenics in general – were based on misconceived ideas of race that went deeper than differences in intelligence.

So the concept of intelligence testing, which Binet had developed in response to a desire to provide relevant education for every child according to his or her abilities, became associated with categorizing people. School-based standardized tests became popular in many countries, including

Britain. Between the world wars, the eleven-plus examination was introduced in British schools, in order to select children who would go to grammar schools, a requirement for university entrance. After World War II, a 'tripartite system' was introduced: children went to one of three types of school, according to their ability. In addition to grammar schools, there were technical schools, which taught vocational courses, and secondary modern schools, which taught a general curriculum to a much lower level than in grammar schools.

The eleven-plus became the subject of criticism, as it seemed to determine a child's future based on a test carried out at eleven years old. During the 1950s and 1960s its use began to decline as university entrance requirements became more flexible and people began to object to the fact that standardized tests labelled children and prejudiced their educational futures. The anti-testing mood prevails to this day, as many educationists feel that such labels act as self-fulfilling prophecies. David King of *GenEthics News* says that labelling and ranking at an early age sends a signal to children perceived to have low ability along the lines of: 'Because you have a low IQ, you will be a cleaner, refuse collector or a labourer.' He sees this as a waste of people's potential – not because these jobs are unimportant, but because labelling a child at an early age narrows his or her scope in the future. King worries that if a more objective, genetic test for intelligence does result from Robert Plomin's investigations, this situation would become worse. Perhaps babies would be tested for IQ at birth ... or even before?

As we have seen, the prevailing attitude of many thinkers on intelligence – that the many facets of intelligence are all manifestations of a single 'general intelligence' – had its roots in the work of Charles Spearman. There have always been those who disagree with that view, including Binet himself. During the 1980s American psychologist Howard Gardner suggested a radical approach, based on the idea that intelligence is the ability to adapt to new situations. Gardner came up with the idea of 'multiple intelligences' that widened the scope of the debate, and undermined the labelling that was the result of standardized tests of general intelligence. According to Gardner, intelligence includes linguistic, logical–mathematical, spatial, musical, bodily kinaesthetic, interpersonal and intrapersonal skills. In other words, people can be intelligent in a host of ways that help them to adapt to new situations. So an athlete who does badly on tests of intellectual ability is intelligent in a bodily kinaesthetic way. The concept of intelligence can even be extended to the social domain: people can be emotionally intelligent, for example. This view is echoed by Stephen Ceci, a psychologist at Cornell University and an expert on intelligence. He rejects the notion of IQ tests, insisting that many people with only average or below average IQs think in very complex ways. He claims that IQ measures 'school smarts' – that being a good carpenter or a good partner, for example, also requires intelligence but of a different kind. He also says that many high IQ people 'couldn't find their way out of a wet paper bag'. And yet, notwithstanding the arguments about labelling and self-fulfilling prophecy, a person's IQ does seem to correlate with their performance and their career. The average IQ of lawyers is about 128, while that of labourers is about 96. How much this correlation is due to self-fulfilling prophecy and how much to the existence of general intelligence is disputed.

Despite the approach to intelligence of people like Gardner and Ceci, and all the problems associated with IQ tests, testing remains an essential part of the education system in most countries, particularly the USA. The ETS (Educational Testing Service of America) is a huge organization. Dr Ernest Anastasia, executive vice-president of the ETS, says that the organization employs about two and a half thousand people, including a 200-strong research division. The ETS was given a national charter in 1947, and now conducts nine million tests each year. These tests, like the eleven-plus used in the UK, are an essential part of the path to a university education. For ease of marking and to avoid subjectivity, they are based on multiple-choice questions. There is a variety of tests on offer, costing between 20 and 100 dollars each, and they are marked at a rate of 8,000 per hour by an optical scanner. Many critics say that multiple-choice questions are too crude to capture something as complex as intelligence and there have been several very public attacks on the ETS. In 1980 consumer champion Ralph Nader wrote *Reign of ETS: The Corporation That Makes Up Minds*, which stimulated an outcry against the organization. Five years later, another educationist, David Owen, wrote *None of the Above*, another polemic against the ETS. This book led to the formation of an organization called FairTest, which took legal action against the ETS. In response, the ETS made changes to its methods, addressing potential test bias. But even addressing test bias does not take away the possibility of cheating, or undermining in other ways the principles of fair testing that the ETS attempts to uphold. For example, certain private organizations run courses that teach students how to increase their scores on these kind of tests. The tests are, of course, supposed to give an objective score, not just a rating on how well the students do in this sort of test. So, can an organization such as the ETS really provide information about students that could be accurate and useful?

The supporters of educational testing are very aware of the pitfalls of their testing methods, and they design tests as fairly as they can. With that in mind, many people see educational testing in the way Binet had originally devised it: as a way of providing teaching that is appropriate to each individual. This can open up new possibilities for those who have good natural ability but who may grow up in an environment that lacks intellectual stimulation. Robert Plomin himself sees the value of tests that attempt to measure natural intellectual abilities. He grew up in a large family in which educational expectations were low. It was as a direct result of educational testing that he was encouraged to attend an academic school, and therefore find his way into academia. So, for Plomin, natural-born ability is important, and tests that attempt to highlight it are certainly worth while.

Let it flow

The idea that intellectual ability is both innate and learned was an important part of the work of British psychologist Raymond Cattell. During the 1960s and 1970s, Cattell managed to find a compromise between IQ-based theories of intelligence, whose roots were in the work of Galton, and 'environmental' theories, claiming that environmental factors muddied the water too much, leaving intelligence tests measuring nothing but learned abilities and therefore inherently unfair. He agreed with Spearman's idea of g (general intelligence), but he separated it into two parts. Cattell simply formalized the existing idea that people are born with varying natural mental abilities, but their intelligence is also shaped by environmental factors, such as their schooling and their upbringing. He defined these two parts as 'fluid ability' – inborn general intelligence – and 'crystallized ability' – the result of learning. The two are clearly related: children of differing natural ability will learn different things from the same situations. This relation works in the opposite sense, too: children who grow up surrounded by books and stimulating educational experiences have a much greater chance of developing a wide range of verbal and numerical skills than those of similar ability who are starved of such experiences. These children will therefore do better on tests of crystallized ability than children of similar 'fluid' ability who do not have the same exposure to intellectual input. Tests that purport to measure fluid ability but which use questions based on verbal and numerical abilities are therefore inherently biased. Where differences in educational opportunity are related to factors such as race or economic status, the bias becomes social or cultural. This is the basis of much of the objection that has been raised to the kind of tests administered by the Educational Testing Service.

While the idea of crystallized ability might highlight bias in supposed tests of general intelligence, it can also support both the idea of g and the possibility of measuring it. In separating innate from learned abilities, Cattell presupposed that people differ in their innate ability. Assuming this is true gives credence to studies like Plomin's that attempt to investigate biological – inborn – factors in intelligence. And although tests of general ability cannot be totally free of bias, the people who design the tests are aware of the difference between fluid and crystallized ability. They make the tests as fair as they can, trying to avoid too many questions that rely on specific knowledge. Still, many popular IQ tests include general knowledge questions such as this genuine example: 'Which area of scientific study was Copernicus famous for?' (The possible answers were: 1 Biology; 2 Astronomy; 3 Chemistry; 4 Genetics.) Questions that completely eliminate this kind of specific knowledge would be abstract and perhaps difficult to understand. The closest that intelligence tests seem to come to this is the completion of series of shapes or numbers by recognizing patterns. Of course, this sort of task does improve with practice as well as with age and mental ability. However, the measurement of intelligence, along with other mental behaviours and capacities, is considered by some as an increasingly finely tuned process. Psychometric testing, as this field of endeavour is called, is sophisticated, and is becoming more and more widely used, well beyond the school environment.

In recent years, the British Army has rejected traditional interview and school examination results in favour of IQ and other state-of-the-art psychometric tests. Major Chris Allander, an army recruitment officer, reminds us that there is no army training in the school curriculum, so all new recruits must start from scratch. It is for this reason that natural cognitive abilities, and not school results, are important. The British Army is proud of the fact that this approach can offer people opportunities that they would not have otherwise had. There are many examples of recruits who have done far better in the army than their school qualifications would have allowed them to. It is worth pointing out that the questions used in the psychometric tests still rely on verbal and numerical skills, which do not necessarily reflect innate ability. To design a test that really did measure fluid ability, we would perhaps need to revert to determining reaction times or other quantities that may indicate the speed of mental processing. That explains why some researchers have tried to do just that, but the value of such research is not yet fully understood.

There are, of course, those who think that there is really no such thing as natural ability, or at least that it plays only a very minor role. One such person is Michael Howe, a psychologist at Exeter University. Howe thinks that innate ability is the wrong way to look at intelligence: that parental support is more important than natural ability, and is the key to the successful development of mental abilities. In studies of musically gifted children, he has found that parental support was a key factor: one or both parents of nearly all musically gifted children play musical instruments to a high level; music fills the house as they grow up. Michael Howe claims that experience and practice are more important than genes when it comes to intelligence. Plomin counters such arguments: when you study family members and do not take into account the genetic link, it is plausible to say that parental influence is important. But he maintains that it is also important to look at genetics: parents provide genes as well as support and experience.

Looking at the evidence

There is plenty of indirect evidence to support Plomin's view of innate mental ability. Meet Sandra Scarr, educational psychologist and expert on the influence of upbringing on children's intellectual abilities. She has visited countless schools, studying what influence the school surroundings have on children's learning and on their IQs. She is always impressed at schools that provide adequate resources and a modern curriculum, and thinks it is essential to children's development to offer as wide a range of stimulating experiences as possible. Some of the schools she visits are impoverished, with limited resources and large class sizes. At the other end of the spectrum, she sees schools that have stimulating surroundings, excellent resources and small class sizes that mean teachers can afford much more time per pupil. However, when it comes to the question of whether these factors influence IQ, she answers with a resounding 'no'.

Scarr has carried out four separate, extensive and long-term studies that have investigated possible environmental effects on IQ. In one study, she located people who had been adopted in the 1950s when they were just two months old. Scarr measured the IQs of the adoptees when they were eighteen years old. If the effects of family and schooling were important in determining IQ, then such influences would show up after the first eighteen years of life. What she found was remarkable: the IQs of the people she tested bore no relation to the IQs of the families in which they were brought up. The results could just as well have been randomly arranged. Scarr is convinced that people's genes are paramount in determining cognitive as well as other types of ability. Time and time again, while watching young children at play or engaged in educational activity, she has observed how some children naturally learn more than others. If our genes really are important in determining our intelligence, then there should be similarity between the IQs of biologically related family members. And yet different children with the same parents often have very different IQs. It is possible to explain this anomaly by returning to the analogy involving playing cards, and remembering that intelligence is assumed to be determined by a large number of genes. When a new life is conceived, two packs of cards – the genes of the mother and the father – are shuffled together. There are many different 'hands' that can be dealt from these same two packs, so it is still possible to have one child with better natural intelligence than his or her brother or sister.

But that does not really help to convince us either way – what we need to do is to compare the IQs of two individuals with exactly the same genes. Such people do exist: identical twins. However, most identical twins grow up in the same family and attend the same school, so they not only share genetic factors but environmental ones as well. Several large-scale studies of twins 'reared apart' have indicated that genes contribute at least seventy per cent to intelligence. It has been found that the similarity of identical twins' IQs seems to increase with age, suggesting that the genetic component of general intelligence is increasingly important with age, too. In one study, sixty-eight-year-old identical twins Caroline and Margaret Chang, separated at birth, had IQ test scores that were as similar to each other as the same person tested twice.

Well over one hundred studies of IQ in various contexts – twins, birth families and adoptive families – have been carried out. The most useful in helping search for a genetic basis to intelligence are the studies that involve identical twins reared apart. However, such people are rare, so most of these are of a fairly small scale. Perhaps the best-known twin studies were reported by British psychologist Cyril Burt during the 1950s. Burt claimed to have proved that intelligence is an inherited trait, based on the results of two studies of identical and non-identical twins who had been reared apart. In both studies, he arrived at a figure of 77 per cent 'heritability' – in other words, intelligence is 77 per cent genes and 23 per cent environment. In 1975, four years after his death, Burt's results were called into question. It is now largely believed that he falsified some of his data, and made up the results to match his beliefs about intelligence.

One of the largest studies ever carried out, the Minnesota Study of Twins Reared Apart, was begun in 1979 and reported its findings in 1990. The study involved more than one hundred sets of twins or triplets who had been reared apart, and put them through more than fifty hours of medical and psychological testing. The results were very similar to Cyril Burt's findings, and the Minnesota Study was certainly not faked. This does not mean that Burt's work was necessarily genuine after all, but it does suggest that genetic influence really is significant. And several other large-scale twin studies have come to the same conclusion: that inheritance – genetics – determines between 60 and 75 per cent of the intelligence of an individual.

There is generally little difference in IQ between identical twins. How does IQ vary in the population as a whole?

Scaling the heights

Intelligence – as measured in tests for g and measured as IQ (intelligence quotient) – seems to be distributed across the population in a similar way to other human attributes, such as height. A bar graph showing the heights of, say, 1,000 men at age twenty would demonstrate the distribution of heights in the population. Most men are not extremely tall or extremely short, but have heights within a range centred on the average. The bars of the graph reflect the number of men who have a particular height. Near the middle of the horizontal axis, around the average height, the bars would be tall, and near either end they would be shorter. A similar graph for 1,000 twenty-year-old women would have the same shape, but would be shifted slightly to the left since women are shorter on average than men. The profile of the graph, tracing out the tops of the bars, would look like the outline of a hill. This is called a Gaussian distribution, after the German mathematician Carl Friedrich Gauss who worked out the mathematics of this kind of curve. It is more commonly known as 'the bell curve'.

It has been found that the variation in IQ across a large, random sample of people follows the bell curve. In fact, half the people in any large, random sample will have an IQ within ten points of the average value of 100. The fraction of the population in which people have an IQ of over 130 is about 2.5 per cent. The same goes for people with an IQ of less than 70. In other words, a few people have a very low IQ, most people have an average IQ, and a few people have a very high IQ.

The fact that the distribution of IQ across any population follows the bell curve is support for the idea that intelligence is determined by many genes. To see why this is so, first consider the case of a characteristic that depends upon only one gene. A good example is the height of pea plants – in particular, the pea plants that were the subject of the first ever scientific study in genetics. During the 1860s an Austrian monk, Gregor Mendel, began a remarkable project. In an attempt to discover the rules behind the inheritance of physical characteristics, he manually fertilized over ten thousand pea plants and noted down the distributions of certain characteristics over successive generations of the plants. When he cross-bred tall plants with plants of a dwarf variety, he found that all the plants in the next generation were tall. There is a single gene that determines whether a pea plant is tall or a dwarf. A gene is a section of DNA along a chromosome found in the plants' cells, and you will remember that chromosomes occur in pairs. One member of each chromosome pair comes from each parent. So the gene that determines whether a pea plant is tall or dwarf exists twice in each plant, one copy from each parent. That gene exists in two distinct forms, which can be called T (for tall) and t (for dwarf). So any particular plant can be TT or Tt or tt, depending on which two versions of the gene it possesses. Only plants with two dwarf versions of the height gene (tt) grow up as dwarfs. Where both versions exist in the same plant, the T version 'wins' and the plant grows tall – the T version of the gene is said to be 'dominant'. This explains why the first generation of cross-bred plants were all tall: coming from parents whose height genes were TT and tt, they had to be Tt, since they received one gene from each of their parents.

What happens if you breed these first-generation plants with each other? There are four possible genetic outcomes from crossing a Tt with another Tt: TT, Tt, tT and tt. In a large sample of plants, you would expect that only one quarter of pea plants in the second generation would be dwarfs (tt). This is exactly what Mendel found. It is important to note that however many generations down the line you go, you will still only have tall and short pea plants, since the only possible combinations of T and t are TT, Tt (which is the same as tT because T is dominant) and tt. You never find pea plants with intermediate height.

The IQ of a population is not distributed in this way – only high or low – and so it is clearly not determined by a single gene. If it were, then perhaps one quarter of the population would have very low IQs (gg), while three quarters would have very high IQs (GG, Gg or gG). Could general intelligence be determined by two genes? In that case, the number of possible combinations grows from two to nine. To see how that can be so, imagine that there are two genes, A and B, which determine a particular characteristic. Assuming that each one can have two forms (A and a, B and b), then the following combinations are possible: AABB, AABb, AAbb, AaBB, AaBb, Aabb, aaBB, aaBb and aabb. With three genes, the number of different combinations is twenty-seven.

Human skin colour is thought to depend upon four genes, of which there are eighty-one different combinations. Each genetic combination produces a different concentration of melanin in the skin, and therefore a different skin colour, from pale to very dark. There is a slight natural variation in colour caused by factors such as nutrition, and so the various possible colourings merge together. People are not either black or white: the variation in skin colour in an integrated population would appear continuous. A graph of skin colour in truly integrated populations has a shape similar to the bell curve described above. With five or more genes controlling a characteristic, the variation fits the bell curve almost perfectly. Characteristics that are determined by many genes are called polygenic traits. Many examples of polygenic traits have been found, in plants, in humans and in other animals, from the colour of wheat grains to the milk yield of a large herd of cows. In each case, a graph of the variation matches the bell curve. And this is why the fact that a graph of IQ across a population follows the bell curve lends support to the idea that general intelligence is a polygenic trait. Plomin was aiming to find one of the many genes that, he believes, contribute to general intelligence.

Big news

In 1994 statistical studies of intelligence were the subject of a hugely controversial book: *The Bell Curve – Intelligence and Class Structure in American Life* by Charles Murray and Richard Herrnstein. Weighing in at a hefty 552 pages, with a further 280 pages of appendices, notes and a bibliography, it collated and interpreted the results of a large number of studies of IQ, and claimed to 'reveal the dramatic transformation that is currently in progress in American society – a process that has created a new kind of class structure led by a cognitive elite'. One of the most important studies quoted throughout the book is the National

Longitudinal Survey of Labor Market of Youth (abbreviated to NLSY). The NLSY was set up to collect information on family background, economic status and educational achievements of 12,686 Americans, all aged between fourteen and twenty-two in 1979. The survey did not originally include intelligence tests, but various psychometric tests – including IQ tests – were included from 1980 because the American Department of Defense wanted to 'renormalize' its enlistment exams. In other words, they needed the IQs of a large number of people who would be representative of the American population, so that they could adjust the scores of potential recruits based on the national average. So the IQ tests involved in the NLSY were the ones used in army enlistment. They were based on verbal and non-verbal reasoning tasks, but throughout *The Bell Curve* only one measurement – IQ as a supposed indication of general intelligence – is used consistently in the data.

Many of the findings of studies into IQ test scores presented in *The Bell Curve* were already well known – though not all undisputed – by researchers into intelligence. For example, IQ test scores seem to be rising slowly but steadily, perhaps reflecting improvements in education or health during the past century. This is why the Department of Defense wanted to renormalize their enlistment tests, which had been based on test results from enlistment during World War II. The authors divide people into five 'classes', from 'very dull' to 'very bright', based on their IQ scores. Many different aspects of American life are analysed in terms of these classes. The authors of *The Bell Curve* show that, independent of other factors, low IQ seems to increase the risk of being below the poverty line; of being on welfare; of having illegitimate children; of getting a divorce; and of being involved in crime. The evidence presented to back up these claims is impressive: there is a correlation between low IQ and each of these situations. But, as critics of the book point out, correlation is not the same as causality. In the same way, ancient Egyptians found a correlation between the appearance of certain star constellations in the night sky and the flooding of the Nile delta. These shifting constellations are not the cause of the floods, however. A correlation between low IQ and poverty might be explained in all sorts of ways, including the fact that a child growing up in poverty is not usually surrounded by an intellectually stimulating environment. Murray and Herrnstein briefly consider this idea, but conclude that 'low intelligence is a stronger precursor of poverty than low socio-economic background', and 'the traditional socio-economic analysis of the causes of poverty is inadequate and that intelligence clearly plays a role'.

The most controversial section of the book deals with IQ and race. It has apparently been shown in many reputable studies that the average IQ of people of Chinese or Japanese origin – whether in the USA or not – is higher than that of white Americans. The section of the white American population found to have the highest average IQ was made up of those people who described themselves as being of European Jewish origin. The most controversial data presented concerns black Americans (referred to in the book as African–Americans or simply 'blacks', depending on the context). Studies have consistently shown that black Americans have lower average IQs than white Americans – by as much as fifteen IQ points. This puts the average IQ of black Americans at 85.

Many questions spring to mind about this finding. For example, 'Are the differences in black and white scores attributable to cultural bias?' and 'Are the differences in black and white scores attributable to differences in socio-economic status?' and 'How do African–Americans compare with blacks in Africa on cognitive tests?' The last question is perhaps the most important one, since it would be easy for people to make the jump from 'black Americans equals low average IQ' to 'black equals low average IQ'. If the average IQ of black Americans really is lower than that of white Americans, is this because of genetic differences caused by ethnic origin – in other words, because of ancestry? Or is it because of educational disadvantages due to racist social policies? Or is it because of inherent test bias? The tests administered in the NLSY relied in part on verbal reasoning that depended on cultural reference, but several studies using tests that were supposedly culturally unbiased have also come to the general conclusion that black people's IQs are lower on average than white people's IQs. Whether these tasks were truly unbiased and objective is not certain. The tests, while supposedly culturally unbiased, still assumed a specific definition of intelligence, and still assumed that general intelligence is a meaningful concept. And even if black people in America really do have lower average IQ than white people, this could be due to environmental factors: black people in America are more likely to live in poverty than white people – and therefore less likely to grow up in an intellectually stimulating environment. And it has been found that the average IQ of black Americans is higher than that measured in black Africans. America is far richer per head than any African nation: the numbers of televisions, books and schools per thousand people are much lower in Africa than in America. Could this have something to do with the lower IQs measured? And remember – a lower IQ does not necessarily mean that you are inherently less intelligent: just that you do less well in IQ tests. In any case, you can see how this subject is a veritable minefield.

To their credit, the authors of *The Bell Curve* did embark on detailed discussions of many of the possible reasons behind their findings. But their conclusions leave a strange taste in the mouth that seems more than just having to swallow a difficult truth. At times, the language they use goes beyond pure science and into political rhetoric. Here are a couple of examples: 'The United States already has policies that inadvertently social-engineer who has babies, and it is encouraging the wrong women'; 'An immigrant population with low cognitive ability will – again, on the average – have trouble not only in finding good work but have trouble in school, at home and with the law'. Many critics objected to the book's tone and its conclusions. Charles Murray was described in an article in the *New York Times Magazine* as 'the most dangerous conservative in America'. In each of the social situations analysed in terms of IQ – such as poverty, divorce, unemployment and crime – the authors of *The Bell Curve* do consider the possible effects of environment on IQ scores. In each case, however, they remind us that most research – twin studies, for example – suggests that between 40 and 80 per cent of our intelligence is the result of our genes. The link between genes and intelligence is an important one to establish or to disprove.

We have seen that there is disagreement about whether IQ is a measure of general intelligence, and even whether the term 'general intelligence' really means anything. However, whatever it is that IQ does measure, we have seen that there is plenty of evidence that suggests that it is dependent upon genes. There are twin and adoption studies, and there is the fact that the bell curve observed with IQ is also found in many proven polygenic traits. All this evidence is important, but indirect. Robert Plomin was searching for direct evidence. He was comparing certain genes in the DNA of two distinct groups: one with super-high IQs and the other with average IQs. The group with average IQ would appear at the centre of the bell curve (at its highest point), while the second group would be out at the extreme right (near the very lowest point there). It is not too hard to find people with average intelligence, since they are the most common in the population. But how do you go about finding people with IQs of 160 or more? Where do you find one in a million? Plomin went to Iowa.

Bright sparks

Iowa State University holds an annual summer school for talented and gifted children selected, from the Midwest of the USA, for their incredible intellectual capabilities. The parents of the children come from a range of professions, and there is financial assistance for people whose parents cannot afford the fees. The summer school was founded by Camilla Benbow, an expert on gifted children, who sees it as a voyage for the mind in search of 'the least known' but 'most wondrous' knowledge, and who was also involved in the research with Plomin. During the three-week course, the teenagers study one topic in great depth. The school offers welcome intellectual challenges to people who find their normal school work easy. It also offers a chance for such people to meet others like them. This, it seems, is just as important to the children attending, since being labelled 'extremely bright' or 'gifted' can put pressure on a child.

In 1997 the course was on the subject of genetics – very fitting. One of the teachers, Jay Staker, began his preparation six months before the summer school began, and was confident of being able to deliver his part of the course. But on the first morning he was shocked by the pace at which he had delivered the material. At the end of the three weeks, he had taught the equivalent of a one-year undergraduate course in genetics.

Each of these extraordinarily intelligent children was the 'one in a million' Plomin was looking for: their genes would be studied as part of his research project. A blood sample was taken from each of them, and also – to act as an experimental control – from children with average IQ in a similar age range. A little of each of the samples was sent to two laboratories: one in Hershey, Pennsylvania, and the other in Cardiff, Wales. Both sets of samples were cultured, to produce more cells for further study, and frozen. Each new cell made in culture in the laboratory has a complete copy of the DNA of the individual from which the sample was taken.

At Hershey, in the Department of Microbiology in the College of Medicine, Pennsylvania State University, Mike and Karen Chorney, together with Nicole Seese, worked on the samples. These three researchers were aware that traditional geneticists avoided this kind of study, for two main reasons. First, as we have seen, the ethical and political issues surrounding the whole idea of a genetic link to intelligence had always steered researchers away from such a direct investigation of genes and intelligence. Second, the chance of any positive result – the discovery of a genetic link to intelligence – was perceived as minuscule. For Plomin, it is the truth that is important, not necessarily success. This research was not his first effort to find a genetic link to inheritance. Previous studies had failed to find that link and, although this was disappointing in terms of not proving his theories, Plomin's overriding emotion was relief at the avoidance of further controversy. But this latest research showed great promise for finding irrefutable evidence of that elusive first link.

The researchers at Hershey spun the blood at high speed in a centrifuge, to separate the red and white blood cells. Red blood cells are among the very few types of cells in the human body that have no nucleus, and therefore do not contain the genome. The white blood cells, on the other hand, do contain a nucleus, and the team extracted the DNA from these cells. Similar procedures were carried out in Cardiff, at the Department of Psychological Medicine at the University of Wales School of Medicine, by Mike Owens, Peter McGuffin and Johanna Daniels. Peter McGuffin has since moved to join Robert Plomin at the Institute of Psychiatry.

Both teams – in Hershey and in Cardiff – were looking for certain genetic markers. A marker is a specific section of DNA used in studies of inheritance. Proteins called restriction enzymes are used to cut the DNA, at known locations, into short pieces called restriction fragments. Markers are genes that lie near to the points at which the DNA is cut. The sizes of the fragments hold the key to which versions of genes are present in the fragments. To sort the DNA fragments, they are placed in a gel under the influence of an electrical voltage. The voltage draws them through the gel; long, heavy fragments move more slowly than short, light ones. Once this has been done for a fixed amount of time, the fragments of different lengths have been separated. By attaching a fluorescent or X-ray molecule to the DNA fragments, the positions of the fragments in the gel can be visualized. Then, the lengths of the different fragments can be worked out, along with the order of genes along the DNA. This whole process is painstaking and time-consuming, but the procedures are standard in modern genetics research. Restriction fragment length polymorphism, as this process is called, has found many applications, including DNA fingerprinting used in forensic science and paternity testing. It is also the basis of the Human Genome Project. The teams in Cardiff and Hershey were both using the technique to study genes on chromosome 6, which has been well mapped in the Human Genome Project. Specifically, the teams studied thirty-seven gene sites, looking for particular versions in the two groups. The team in Cardiff was looking at the short arm of chromosome 6, while the team in Hershey was searching the long arm. (A chromosome has a constriction along its length, called a centromere, that divides it into a 'long arm' and a 'short arm'.)

The data from Cardiff were the first to come in. Plomin and his colleague, Thalia Eley, loaded them into a computer, using statistical software to analyse the results. They began with tables showing whether a particular version of a certain gene occurred in the two groups. The results of a previous study had indicated that this gene was present more often in the group with high intelligence than in the average, control group. In this latest study, Plomin decided to double the size of the control group, to check the results. The computer quickly worked out whether a gene occurred more frequently in the high intelligence group. The results were a little disappointing: although the incidence of the gene was higher in the children with super-high intelligence, the difference was not significant enough to prove a link. And so Plomin and Eley had to wait for the next set of results, from Hershey, which looked at the long arm of chromosome 6.

When the results came in from Hershey, they were puzzling. A gene called IGF2R (insulin-like growth factor 2 receptor), was the one that Plomin and Eley were most interested in. About forty-six per cent of one group had a particular version of the gene on at least one of their two chromosome 6s, compared with 23 per cent in the other group. The results were significant, but the opposite of what Plomin had expected: it was the group with average IQs that seemed to possess the sought-for version of the gene more frequently than the group with super-high IQs. Plomin and Eley were confused by the results, and decided to re-evaluate the data the following day.

Thalia Eley came to work early the next morning, and switched on her computer, ready to begin a review of the data from the samples analysed in Hershey. She soon found that the samples had been labelled incorrectly in the computer. The set of results for the high-IQ group had been labelled as the results for the average IQ group, and vice versa. So, the results had shown a significant correlation between a particular version of the gene IGF2R and IQ after all. These results confirmed earlier suspicions about IGF2R, and provided the first hard scientific evidence of a link between IQ and our genes. Plomin estimates that this gene might cause a 2 per cent variation in IQ – about four IQ points. This may sound small, but IGF2R is probably one of many genes that influence IQ. It is not known how IGF2R, or the protein that it manufactures, might affect the brain.

There is another uncertainty: the researchers cannot even be sure that it was IGF2R that was varying between the samples. There are other, as yet unmapped, genes near to IGF2R, which would have been on the same fragments of DNA used in the restriction mapping explained above. It could be that one of these genes, and not IGF2R, is involved in intelligence. The important point is that a gene has been found that has something to do with IQ. Plomin says that people can choose to reject twin or adoption studies, but it is harder to argue with a piece of DNA.

Since the study was published in 1998, another interesting piece of evidence has arisen – this time in mice. In September 1999 a team led by Joe Tsien at Princeton University genetically engineered mice to be more intelligent. The team used a tiny glass needle to inject a gene into a fertilized mouse egg, and transferred the egg to the mother's womb. The gene the researchers inserted into the mouse genome is responsible for making a protein called NR2B. As the fertilized egg divided, to form the growing mouse foetus, the gene for NR2B was copied, along with the mouse genome, into the newly forming cells. Because it had extra copies of the gene, the mouse made more of the protein than a normal mouse. NR2B is vital in the brain, as part of a structure in neurones. Another chemical, called NDMA, which has long been associated with memory and learning, locks into particular sites on brain cells. NR2B is a vital part of those sites.

The mice with the extra gene were better at standard tasks that require intelligence, such as recognizing pieces of Lego they had seen before and learning the location of hidden platforms underwater. It seems as if the more NR2B you have, the more acute your memory. (The strain of intelligent mice was called Doogie, after *Doogie Howser, MD,* an American television series in which a boy graduates at the age of ten from Princeton University, where the research was carried out.) In the long term, this sort of research might lead to the possibility of prenatal screening of embryos to determine the likely intelligence of the baby that develops, or genetically engineering humans to be more intelligent. These possibilities are at the centre of ethical issues surrounding the work of Plomin and other molecular geneticists. If you are familiar with Aldous Huxley's futuristic fable *Brave New World*, you will remember that the population consisted of people cloned in batches, each batch having a different level of intelligence, and therefore being assigned specific roles in society. This may be a far cry from research into mouse genes, but many thinkers believe that research carried out into the genetics of intelligence could set us on a slippery slope towards something like Huxley's vision.

Sandra Scarr says that some intellectuals are fearful of the kind of research Plomin and his team are carrying out. She says that some academics propose that it should not be done or that the results should be suppressed. Scarr thinks this is unwise and patronizing to the public and the scientific community. It is better to have the information, she says: ignorance is not bliss. Camilla Benbow agrees: 'We should never be scared of knowledge.' While this may be true, what is the actual value of carrying out investigations on the influence of genes on intelligence? Benbow says that if people differ in their cognitive abilities, then knowing about the source of the differences will enable educators and politicians to respond better to them. Whether politicians and educators would respond in the best way – by providing the most nurturing and stimulating environment for every developing child according to his or her intelligence – is of course uncertain. And it is the definition and measurement of intelligence, rather than the underlying causes, that are important in providing the best possible education. Nevertheless, definite answers to the questions of whether genes determine our intelligence and how the genetic determination of intelligence works could bring new insights into our origins as a species. For example, how did intelligence develop from other animals, through evolution, and how does human intelligence differ from the intelligence of other animals? Also, discovering the link between the hardware – the physiology of the brain – and the software – intelligence – could help to treat or prevent certain brain disorders. The researchers who discovered the gene that increased the intelligence of mice have suggested that it may one day lead to drugs that could treat Alzheimer's disease or stroke.

If intelligence – and other human behaviours – is explained by biology, then philosophers may need to re-evaluate our concepts of free will and our spiritual beliefs. Plomin is more pragmatic: he says that the knowledge that intelligence is largely determined by our genes, and is only slightly influenced by our experiences, might stop over-anxious parents from flashing vocabulary cards in front of their newborn children's faces to improve their intelligence.

PHANTOM BRAINS

...the search for ghostly remains...

Driving around the S-bends on the A57 road to Blackpool, a car went out of control and left the road, hitting some trees and ending up at the bottom of a ditch. As the car stopped moving, everything went quiet. Both people in the car were alive, but in shock and had not escaped the accident unscathed. Jackie immediately realized that she had lost her left hand. An ambulance soon arrived and took Jackie and her husband to a hospital in Sheffield. Halfway there, a police car drew alongside the ambulance, and stopped it. A police officer who had recovered the hand gave it to the paramedics in the ambulance. Doctors in Sheffield worked hard to reconnect the hand, and for ten days Jackie had two hands again. But infection had set in, and began to take over; the doctors decided that the hand would have to be removed again.

This sort of experience must be traumatic and disturbing enough, but, after a few days without a left hand, Jackie began to feel something even more alarming, and for a time almost as traumatic. She felt as if she still had her missing hand. She did not like to mention this to anyone, because everyone – including Jackie – could see that there was no hand there, just a healing stump. Jackie began to doubt her sanity, because at times the hand felt as real as the one she had before the accident.

This sensation is referred to as a phantom hand. Until recently, phantom hands, arms, feet and legs – sometimes referred to as 'stump hallucinations' – were a minor scientific curiosity, but, in the last few years, scientists have begun to listen carefully to what patients say about their phantom experiences. Numerous experiments and observations in monkeys as well as in humans have provided compelling evidence that challenges existing ideas about the brain. Radical new theories have been suggested to explain the strange sensations experienced by phantom-limb patients, including the idea that the brain adjusts its internal map of the body by physically rewiring neurones in the cortex. There is an increasing number of neuroscientists actively searching for patients with phantoms, and searching too for answers.

Seeing ghosts

Digital artist Alexa Wright is a phantom hunter of a different kind. She helped to bring to light just how real a phantom can feel, working with patients such as Jackie to produce detailed pictures of what their phantoms felt like. Using digital photographs of the patients as they really were, she worked on a computer with photo-manipulation software to add in the phantoms as patients described them to her. Wright discussed the images with the patients as she was creating them, to ensure that what she captured on the computer screen – and afterwards in print – was as close as possible to a visualization of their phantom limb. In Jackie's case, the phantom that Wright visualized on the screen was about the same size as a real hand, but a bit flatter and the wrist thin and stick-like. When Jackie saw the finished image for the first time, she wept: 'When I first saw Alexa's image, I was really upset, not because I was angry about it, it was just that someone had got that image for other people to see what I feel like.'

There is tremendous variety in the images Wright has created. Phantom-limb patients often describe their arm as floating, unattached; that they can feel just a thumb attached to their stump; or that the phantom limb is stuck in a cramped and unnatural position. But the bizarre nature of these phantoms does not stop there. Patients can experience phantoms in other parts of their bodies. Patients suffering from extreme appendicitis sometimes still feel the pain after the appendix has been removed. About a third of all women who undergo mastectomy experience phantom breasts, including tingling in the phantom nipple. Some men who have their penis amputated to remove cancers have experienced phantom erections; and women have even felt phantom menstrual pains after hysterectomy.

Phantom limbs were first reported in the sixteenth century by French surgeon Ambrose Paré, who described sensations felt in absent limbs. Admiral Lord Nelson also experienced vivid phantom pain after losing his arm in an attack on Tenerife in 1797. Nelson is reported to have said that the phantom sensation gave him direct evidence of the existence of the soul. The phenomenon was first documented in detail during the American Civil War in the 1860s. American neurologist Silus Weir Mitchell wrote about the symptoms he observed in an injured soldier in a hospital in Pennsylvania. Mitchell wrote the account anonymously, and in the form of a short story, 'The Case of George Dedlow', in a popular magazine. Perhaps he did this because he was worried about what the reaction would be from the scientific community if he published them in an academic journal. The story told of how the young soldier woke up after surgeons had amputated both his legs, and asked for relief of a terrible cramp in his calf muscles. When the bedclothes were pulled back, he realized that he had no legs, let alone calf muscles.

Since the earliest accounts of phantom limbs, another phenomenon has been reported in most patients. The phantom limb can be stimulated by touching other parts of the body. Examining Jackie, neuropsychologist Peter Halligan gently touches the stump at the end of her left arm with a pen. Jackie can feel the pen touching the stump, as you would expect. But when she looks away or closes her eyes, she feels the pen touching her phantom hand. She attempts to indicate the point where she can feel the pen on her left hand, using her intact right hand. It is a point beyond the stump, in thin air. Halligan is one of the new breed of phantom hunters. Based at the Rivermead Rehabilitation Centre, Oxford, he scours the whole of Britain looking for patients who report phantom-limb symptoms. Jackie explains that inside her mind the phantom is as real as an actual hand. She was relieved to meet and talk with Halligan and Wright: at last she felt that what she was feeling was not unusual, that she was not losing her mind. In fact, some kind of phantom sensation is felt at some time by up to eighty per cent of amputees. Halligan says that he, like other neuroscientists, had tended to ignore the phenomenon of phantom limbs, treating it as a minor curiosity. 'We have neglected something that was actually telling us something about how brain processing was changing,' he says.

The sensation that a missing part of the body still exists, and the fact that this 'sensory ghost' can be stimulated by touching other parts of the body, is scientifically curious. But the story goes beyond science: into medicine. As well as just feeling that the lost limb exists, patients report feeling extreme pain in their phantom. Strong painkillers, or in some cases anticonvulsants that are normally used to treat epilepsy, do sometimes reduce the sensation, as they would do in a real limb. These phantoms exist in the mind, like hallucinations, but how and why should the brain produce such vivid and disturbing feelings? And how is it possible for normal painkilling drugs to reduce pain in part of the body that no longer exists?

When people lose someone close to them, they sometimes go through a period of denial, taking time to accept the loss; some even claim that they can still strongly feel the presence of the dead person. Could this psychological effect explain the origin of phantom limbs? Some researchers think there is a link here, while others – probably the majority – believe that the phantom-limb sensation is too vivid and too common to be explained in this way: the imaginings of a mind that cannot bear to accept the loss of part of the body cannot account for phantom sensation.

To examine the possible sources of the phantom-limb sensation, we need to know a little about what happens when something touches a real limb – one that still exists. Sensations of pressure, change in temperature and pain begin as electrical signals in nerve endings. There are two types of nerve, or neurone: sensory neurones that carry signals from around your body to your brain, and motor neurones that carry signals from your brain to your muscles. Some sensory neurones in the skin have endings that produce a signal when pressure is exerted on them, while others are sensitive to hot or cold. So when something touches the back of your hand, signals from nerve endings in your skin pass along nerve fibres that lead up the length of your arm and meet your spine at a point between your shoulders. Nerve fibres from around the body are like tributaries of a great river – the spinal cord – that flows into the brain. The adult spinal cord is normally between 40 and 50 centimetres (16–20 inches) long, and runs between your head and a point about level with your navel. However, unlike a river, the nerve pathways are not continuous. Instead, a fibre that originates in your hand terminates at the spine, passing its signals on to a different fibre in the spinal cord. So instead of water flowing into a river, you can perhaps imagine a person carrying a bucket of water: at the spinal cord, that person passes the bucket to someone else, who continues the journey. The signal passes along the second nerve fibre, up into the head, where the fibre fans out, connecting to one or more of a number of important structures.

One of these structures is the thalamus, an egg-shaped ball of neurones that acts as a relay station for nerve signals. There are two thalami, next to each other at the top of the brain stem, near the centre of the brain. Connections in the thalami transmit the nerve signals to the most sophisticated part of the brain, the cortex. Once inside the brain, nerve signals from around the body – including signals from eyes, ears, nose and mouth – are interpreted, and we have a sense of the world around us.

Mapping the body

Signals from a particular part of the body always end up in the same part of the thalamus. For this reason, there is a kind of map of the entire body within the thalamus. And in the same way, the signals from a given point in the thalamus go to a particular part of the cortex – there is a map of the body on the cortex, too. This was brought to light by remarkable work carried out in the 1950s by Canadian neurosurgeon Wilder Penfield. Working with patients who suffered from epilepsy, Penfield exposed their brains and subjected the outer cortex to low-level, localized electrical stimulation. The patients were conscious during these procedures, in order for Penfield to record their feelings as well as their actions. The point of Penfield's work was to locate the focus of epileptic seizure, but he also wanted to discover just how the brain senses the world.

He found that stimulating some areas of the cortex would produce twitches in muscles – always the same muscle for the same point on the cortex. Other areas of the cortex would bring forth vivid memories of events in the patients. Most important in our quest to discover the cause of phantom limbs, Penfield found that stimulation of certain areas in the cortex produced the sensations that seemed to come from some part of the body. By careful and systematic experimentation, he produced a detailed map of the cortex, which corresponds to sensory input from the whole body. There are actually two maps, one for each side of your body, which occur in thin strips that extend from the top of your brain down each hemisphere to points just behind your ears.

The map is distorted in two ways. In some areas of the body, nerve endings are more densely packed than in others – this means, for example, that the area of your cortex devoted to receiving sensations from your lips is larger than that receiving sensations from your upper arm. Also, parts of your body that are adjacent to each other do not necessarily appear next to each other on your cortical map. For example, your genitals are represented at the top of the map, adjacent to the representation of your feet. Similarly, your fingers are represented next to your face. Despite these distortions, the map of your body on your cortex can be represented as a little person drawn out on the surface of the brain. It has become known as Penfield's homunculus. The homunculus is clearly involved somehow in the perception of our bodies, and must also be involved in the sensation of phantom limbs. The map of the body found in the thalamus – the gateway from the spinal cord to the cortex – is also likely to be involved. But the mystery of phantom limbs – including phantom-limb pain – is far from solved.

A phenomenon related to phantom limbs is 'referred pain', where pain in internal organs is felt in (referred to) other parts of the body. So, for example, a pain originating in the heart can often be felt as pain in the wall of the chest or in the shoulders. The reason that the pain from the heart is referred to these areas in particular seems to be due to the fact that sensory neurones from the two regions enter the spinal cord at the same point. The signals originating from these two regions also arrive in adjacent areas of the cortex. The heart makes up only a tiny part of Penfield's homunculus, which explains why you are not actually very aware of your heart. The skin and muscles in the shoulder and chest wall form a larger proportion of the homunculus. It is thought that nerve signals originating in the heart 'spill over' into the area of the cortex normally involved in sensing the shoulders and chest wall. This idea is important to some of the theories that are now emerging to explain phantom limbs.

Staff and patients at the Douglas Bader Institute in Roehampton, Surrey, are familiar with phantom limbs and phantom-limb pain. The Institute specializes in making prosthetic limbs and other hardware for amputees. One of the patients, Rod, had to have the lower half of his left leg amputated after a paragliding accident. After the operation, he was relieved to have lost his leg, as up to then he had experienced intense pain. 'I was glad to lose it because the pain in the old leg was too intense, so when they took it off, my initial reaction was "great",' he recalls. 'And then the shock hit me about a week later, I suppose. The feeling of the phantom is just like permanent pins and needles. The actual pain is like a stabbing with a blunt knife.'

Phantom-limb pain can sometimes be so excruciating that patients ask doctors to try whatever they can to relieve it. Perhaps the most obvious explanation for phantom-limb pain is that nerves that were severed when the limb was amputated begin to grow back, becoming more sensitive. The severed nerve fibres that once served the amputated limb do often grow into clumps called neuromas, just inside the stump. Stimulation of the stump might produce sensations that seem to come from the lost limb because, travelling from the neuromas along the original nerve pathway, they will end up in that part of Penfield's homunculus which corresponds to the original limb. Perhaps the neuromas produce random bursts of activity, explaining why patients can feel their phantoms even when their stumps are not being touched.

There are several reasons why the explanations of phantom limbs based on stump neuromas are thought to be false. Firstly, phantom-limb pain can even be felt in limbs that are not lost. Conrad has three arms: two real and one phantom. This strange situation was the result of a motorbike accident, in which nerve fibres that connected nerve endings in Conrad's left arm to his spinal cord were ripped away. His left arm is physically still attached, however. Conrad feels intense pain in his phantom, but feels nothing from the real arm, which now dangles, paralysed. The phantom arose within a month of the accident, and Conrad describes the phantom pain in the early days: 'like your hand was being crushed and wrenched at the same time – like someone was literally trying to pull your hand out of its socket while a sixteen-ton truck is parked on it at the same time ... The pain is really in my phantom hand, not in my real hand, which is strange. You look at it, and you think, "My left hand hurts," but my left hand does nothing, it's the left hand in my head that hurts.'

As with many phantom-limb patients, pain increases in the phantom when Conrad is stressed, or if he thinks about the arm. The pain of Conrad's phantom cannot be due to a neuroma or new nerve endings in his stump: he has no stump. The case of Conrad's painful phantom, co-existing with his real arm, challenges the idea that the growth of neuromas can explain pain in a phantom in another way, too. Conrad's phantom arm is stuck in a painful, cramped position that relates directly to his accident. It is common to find that phantom limbs reflect a memory of the accident that created them, or to the pain that necessitated the removal of the real limb. In Conrad's case, the phantom hand is stuck tight as if it is gripping the handlebar of a motorcycle, and it feels as though the phantom fingers are digging into the phantom palm. Alexa Wright found this phenomenon in many of the patients with whom she worked, and has been able to visualize it in her pictures. A man who had lost his hand as a result of a firework accident, for example, felt pain emanating from the hand, as if there was an imprint of the explosion within the phantom. He asked Wright to add a red area at the centre of his hand, because he could still feel the pain of the explosion there. Some phantom-limb patients feel a watch on their wrist, a bunion on their toe or a ring on one of their fingers. These phantom memories cannot adequately be explained by the growth of neuromas or new nerve endings.

Another reason why phantom-limb pain cannot be explained as pain referred to the missing limb from new nerve endings in the stump comes from the experience of neurosurgeons, who have carried out operations to cut the nerve fibres a little further back from a stump. This isolates the newly formed neuromas from the brain, and would remove phantom sensations if they were indeed the result of neuromas. But such operations rarely have any effect on the phantom pain. Even cutting the nerves as far back as their roots in the spinal cord – an operation called a rhizotomy – normally has no effect. One theory for the cause of phantoms was that neurones in the spinal cord, starved of their normal stimulation, would produce spontaneous bursts of electrical activity. These bursts were shown to happen in certain cases, but there is plenty of evidence that it is definitely not the cause of phantom limbs. For example, phantoms experienced by patients with a complete break of the spinal cord right near the top of the spine cannot be explained by the bursting activity of neurones in the spinal cord below the break. Perhaps neuronal bursts do give rise to phantoms, but higher up – inside the head itself. One theory involved the thalamus: could bursting in the neurones of the thalamus be responsible for phantoms? Again, the answer was no. Neurosurgeons burned holes in the thalamus, with little or no effect. It is as if surgeons have chased the phantom all the way up into the brain, to the homunculus that lies on the surface of the cortex. But even removal of certain areas of the cortex has not always been successful in exorcizing the ghost of a lost limb.

Painkillers may bring some temporary relief from phantom pain. In some cases, acupuncture has also been found to ease the pain, again temporarily. A technique called TENS (transcutaneous electrical nerve stimulation) also has some benefit. In this procedure, electric voltages applied at the surface of the skin around the stump stimulate pain receptors inside the skin. Painkillers, acupuncture and TENS are effective with normal pain (in real limbs, for example). Are there any other treatments that have been effective in reducing or eliminating phantom pain in particular?

Phantom pain comes in several varieties: burning, crushing, cramps, shooting and stabbing. Phantom-limb patients normally describe one or two of these, which tend to remain the same for as long as the phantom exists. So a patient who feels only burning pain will continue to feel burning pain, and not crushing or shooting pain. This may be due to the imprinted memory of whatever led to the amputation of the real limb, as described above. However, there is evidence that each type of pain has a different cause. It has been found, for example, that burning pain is associated with a decreased blood flow in the stump, caused by a reduced temperature there. Muscles in a stump are often observed to tense up a few seconds before cramping pain, and remain tense for as long as there is pain – even though the cramp is felt in the phantom limb. More evidence for these mechanisms behind phantom pain comes from the fact that treatments to address the supposed causes do often have some effect. So ensuring good blood flow tends to reduce burning pain, while treatments that reduce muscle tone tend to reduce cramps. For burning pain, a type of biofeedback sometimes works. Patients using this technique are taught to control the temperature of their stump, at first by listening to a sound whose pitch rises and falls with temperature and then unaided.

Another, rather surprising, remedy for phantom pain involves wrapping the stump in a material, invented in

Germany during the 1960s, and now sold under the name Farabloc. The fabric is made of linen interwoven with fine steel wires. According to the manufacturer, the metal mesh shields stump neuromas from electromagnetic fields outside the body. Metal meshes like the one in Farabloc do stop electromagnetic fields, but the manufacturer's claim – that phantom limbs are caused by electromagnetic fields acting on the neuromas – does sound unlikely. However, in one strict scientific study, involving Farabloc and a similar material to act as a placebo, two-thirds of the patients involved did report some pain relief. And many users of the material do highly recommend it.

One final intriguing feature of phantom limbs is that they often shrink gradually, so that, for example, what begins as a phantom arm emanating from a stump at the shoulder slowly becomes smaller, ending up as just a finger or thumb.

Pain in the net

Ronald Melzack, professor of psychology at McGill University, Montreal, has come up with a theory to explain phantom limbs and the pain associated with them. Melzack is a world expert in phantom limbs, and has been studying them for over forty years. In the early 1990s he put forward an explanation of phantom sensations that involved a network of neurones within the brain. Because the brain is largely composed of extremely complex interconnections of neurones, modern interpretations of many brain functions – including intelligence, memory and consciousness – depend on the idea of neural networks, or 'neural nets'. Melzack proposed that a sophisticated and extensive neural network, which he called the neuromatrix, produces a body image within our brains. He says that the reason why we can feel phantom sensation 'is because the neural network that makes us feel our arm as we normally feel it is the same neural network that makes us feel the phantom when we don't have the arm. The arm is gone, but the representation of that arm in the brain is still there.'

This idea incorporates the sensory map – Penfield's homunculus described above – but goes beyond it. Melzack points out that the brain does not simply react passively to sensory information, but actively creates a sense of the body, which sensory inputs from nerve endings, via the thalamus, modulate over time. This includes an input from our emotions and a sense of 'self'. These three parts – sensations, emotion and a sense of self – together create something he calls a neurosignature. So Melzack's neuromatrix – the hardware that creates the neurosignature – consists of three distinct parts working together. First, there is a sensory circuit involving the thalamus and sensory parts of the cortex. Second, he includes an emotional circuit involving the limbic system, which has long been associated with emotions. The third circuit of Melzack's neuromatrix involves the parietal lobes, one on either side of the brain. Patients who have damage to their parietal lobes seem to lose something of their concept of self. Such patients sometimes deny that parts of their bodies belong to them. The neurosignature that Melzack suggests is produced by these three circuits is a complex pattern of impulses from the neurones involved, and gives a stable impression of the complete body. Melzack uses an analogy with classical music: the neurosignature is like the theme of a piece of music, that can be played on different instruments in many different ways, but which lends stability and identity to the piece.

Melzack suggests that, because the neuromatrix is hard-wired, it must initially be determined by our genes, but that it is updated over time to incorporate changes such as chronic pain or even a wristwatch. These changes, once incorporated into the neuromatrix, and of course then into the neurosignature, would remain after the loss of a limb, explaining why amputees can still feel pains – or wristwatches – that they had before the amputation. Support for the genetic component of the neuromatrix comes from the fact that children born with missing limbs often report phantoms. These phantoms cannot be explained by a sense of body plan that is only learned, since they have no experience of a real limb on which to draw. About twenty per cent of children born with body parts missing report phantom sensations, normally starting at about the age of five, although Melzack has heard children as young as two describe phantom limbs. Perhaps children younger than this can feel phantoms, too, but do not have adequate language to describe them or enough experience of other people to realize that they are unusual.

One of the most compelling parts of Melzack's theory is his explanation of phantom-limb pain. Melzack is particularly interested in the sensation of pain – the least well understood of the human senses. He has made several important contributions to the study of pain including co-founding, with his British colleague Patrick Wall, the 'gate-control' theory. According to this theory, pain pathways up the spinal cord can be modulated, or even blocked, by another signal entering the spinal cord, from neurones whose nerve endings are near to the ones signalling pain, or even by signals from the brain. This would explain why applying pressure, heat or electrical stimulation can help to relieve many types of pain, and why pain can often be overcome by psychological means – 'mind over matter'. Melzack believes that the neuromatrix in a person with functioning limbs would be involved in sending out signals to make the limbs move, and would also collect feedback via the senses. If a limb is missing or paralysed, the neuromatrix would send out more frequent or stronger messages, which Melzack suggests would be sensed as pain. Whatever the cause of phantom pain, Melzack thinks that the mechanisms behind it are probably the same as those behind ordinary pain, which – despite his gate-control theory – are still not well understood.

What possible explanation can there be for all the bizarre effects of phantom limbs? If one can be found, it must account for intense, imprinted pain in the phantom, which may or may not be reduced by painkillers, acupuncture or TENS. It must also account for the fact that phantom pain can be produced by physical pressure on a stump, while it can also be felt after the spinal cord has been severed (a process that would normally eliminate sensation). It must explain how a phantom limb can wither away over a long period, but rarely vanishes altogether. Finally, it must explain how the non-existent limb can be stimulated by touching different parts of a patient's body. We have seen that Penfield's homunculus plays an important role in our ability to sense the world around us, and may be at the centre of the phantom-limb phenomenon. Vital evidence in favour of this possibility came from a controversial study carried out, not on humans, but on monkeys.

Monkey business

The existence of phantom limbs seems to challenge an old assumption of neuroscience – the idea that sensations in the brain are produced only by external factors: heat and cold, pressure, light, sound. It seems that vivid sensations are created within the brain even without external stimulus. During the 1990s another cherished assumption of neuroscience was overturned, after the neural pathways in the brains of monkeys were studied. The research involved the world's most famous macaque monkeys – unwitting media superstars at the centre of an animal welfare row.

In 1981 police confiscated seventeen monkeys from a laboratory in Silver Spring, Maryland, near Washington DC. The police were called in after a volunteer working at the laboratory reported that the monkeys were being treated cruelly. The volunteer was Alex Pacheco, who the previous year had co-founded an animal rights organization called PETA (People for the Ethical Treatment of Animals), which aimed to 'establish and protect the rights of animals'. Pacheco had taken the job, at the Institute for Behavioral Research, with the intention of gathering evidence that would highlight the plight of laboratory animals. He was working for Dr Edward Taub, who together with his colleague Professor Tim Pons was carrying out research that involved cutting nerves that led from the monkeys' arms at the point at which they entered the spinal cord – dorsal rhizotomy. The researchers wanted to see whether there was any way to rehabilitate the arms.

Pacheco collected video evidence of the hardships endured by the monkeys, showing them with paralysed limbs and unbandaged wounds, living in squalor. Some of the monkeys had chewed off their fingers; their excrement filled the cages, and there was serious cockroach infestation. Pacheco took his evidence to a judge, who issued a search warrant. The police seized the monkeys, arrested Taub and collected documentary evidence from the laboratory. The case aroused widespread public interest, and people sympathetic to the monkeys' plight organized candlelit vigils. After a long trial, Edward Taub was eventually found guilty of cruelty to animals, under the Animal Welfare Act. However, the conviction was overturned in an appeal court, on a legal technicality. The monkeys were held in captivity, but were not involved in research, for several years after they were taken from Taub's laboratory. When one of the monkeys was nearing the end of its life, in 1990, it was released back to Tim Pons (who had been cleared of all charges). Pons, and several other researchers, then followed up research that had been carried out elsewhere some years earlier.

In 1983 Dr Michael Merzenich at the University of California, San Diego, had carried out an experiment to investigate the effect on a monkey's brain of amputating its finger. Merzenich monitored the electrical activity in the monkey's cortex when he touched parts of its body.

Conventional wisdom would have predicted that the area of the cortex that corresponds to the missing finger would receive no signals, and that he would detect no signals there. However, Merzenich did record electrical activity in that area – when he touched the remaining fingers, adjacent to where the amputated one had been. So a small region of the monkey's cortex, within Penfield's homunculus, which previously received nerve signals from the finger, was now receiving signals from nerves in the adjacent fingers. The area was small – just over a millimetre square – so Merzenich did not think that new neurones had grown in the cortex to receive the signals. Instead, he reasoned that the signals were arriving at the branches of existing neurones, in the neighbouring areas of the cortex, which receive them from the unaffected fingers.

Now, in 1990, Pons and his colleagues carried out further investigations. They anaesthetized the returned Silver Spring monkey and placed electrodes on to its brain. The map on the cortex of the monkey whose finger Merzenich had amputated had undergone a small change. In Pons's version of the experiment, the neural damage was more extensive: a major nerve, from the monkey's arm, had been cut at the spinal cord. The timescale was greater, too: the nerve had actually been cut eleven years previously. So you would expect Pons and the other researchers to find reorganization on the cortex, just as Merzenich had done, but on a much larger scale. The researchers touched various parts of the monkey's anaesthetized body, and monitored the electrical activity on the cortex. To their surprise, it was touching the monkey's cheek that stimulated the area on the cortex that had once received signals from its arm. This area was about nine millimetres wide – much larger than the area affected in Merzenich's experiment. The results were repeated in tests on another seven of the monkeys. In each case, signals from the monkey's face were received across the entire area. So it seemed unlikely, if not impossible, that existing neurones were at work here. The branches of neurones in the brain do not extend far enough to explain it that way – Merzenich's explanation could not be correct.

What seems to have happened is that the neurones that once received signals from the monkey's arm atrophied, and neurones in neighbouring areas of the cortex spread to fill the space. A good way to think about what is happening when one area of the cortex map invades another is to imagine a flower bed planted with two different varieties of flower. To begin with, both varieties are healthy and they occupy separate but adjoining parts of the bed. If one variety dies off, the neighbouring plants – of the other variety – begin to spread across to the emptying patch of ground. After Pons cut the nerves that led from the monkey's arm to its brain, the area of Penfield's map that corresponded to the arm received no signals. On Penfield's map, the cheek is adjacent to the arm. So nerve signals from the cheek were ending up in the part of Penfield's map perceived as the hand as well as where they should be, in the part that corresponds to the cheek.

Neurophysiologists are uncertain to this day what causes this effect, known as cortical reorganization, but its effect on neuroscience has been dramatic. The discoveries show that the brain is flexible and adaptable, not fixed and static: it is not just the software that is changing as we go through life – the hardware rewires itself, too. Since the 1960s neuroscientists have believed that the brain is able to undergo large-scale wiring and rewiring only before birth and in the first few years of life. This explains, for example, why infants pick up languages better in their early years (and why their native tongue remains much more dominant throughout their lives).

The classic experiment that led to the view that the brain is not able to rewire itself was carried out by David Hubel and Torsten Wiesel in 1963. Hubel and Wiesel covered a new-born kitten's eye and left it covered until the kitten had developed into a mature cat. While nerve signals from the nerve endings of the body go to the area of the cortex mapped out by Wilder Penfield, signals from the eyes travel along the optic nerve to a different area – the visual cortex. Hubel and Wiesel found that the part of the cortex that would normally have received the signals from the cat's covered eye became wired to the good eye. The mapping remained the same after the cat's eye was uncovered, and the cat remained blind in one eye for the rest of its life. More importantly, when one eye of an already mature cat was covered for a significant period of time, the visual cortex underwent no changes. This is why Hubel and Wiesel concluded that the cortex is hard-wired early in life. This idea was not challenged until the experiments on monkeys carried out by Pons and his colleagues. The ability of the cortex to rewire, or at least to reorganize, which those experiments demonstrated is called 'plasticity', and it has created quite a stir in many areas of neuroscience, including the study of phantom limbs.

Primate suspect

One person who was fascinated by the consequences of plasticity and its connection with phantom limbs was Professor Vilayanur Ramachandran, professor of neuroscience at the Center for Brain Cognition at the University of California, San Diego. He was amazed to read about Tim Pons's discovery that parts of the monkey's cortex that used to receive signals from the arm was receiving them from the cheek. 'When I looked at this experiment, I nearly fell off my chair,' he says. Ramachandran wondered whether the monkey would have actually felt sensations in his nonexistent, phantom hand when its cheek was stroked. It seemed natural to suggest this, since if nerve signals from the cheek were being received in the area of the cortex that normally corresponds to the hand, the monkey might feel as if its hand was being stroked. More importantly,

Ramachandran wondered whether the cortex reorganizes in the human brain in the same way as in the monkeys' brains. He suspected that it would. Humans and monkeys are quite close relatives – the anatomy and physiology of their brains are very similar, as they are both primates. Ramachandran could not really justify carrying out surgery on humans – as had been carried out on the monkeys – in order to test his theory. But he hit upon a simple and entirely non-invasive way of finding out whether there could be a connection between the brain's plasticity and phantom sensations. He found patients who had lost an arm, who had reported feeling a phantom limb. He figured that if he touched the amputees' cheeks, they might feel a sensation in their phantoms. This would prove or disprove his idea.

One of Ramachandran's patients was Derek, who had replied to a classified advert he saw in a local newspaper that called for volunteers to take part in the study. The advert was a request for amputees, and offered ten dollars per hour to those who took part. Derek, who had damaged his left arm badly in an accident six years before, wasted no time getting in touch. While he was sitting in a chair in Ramachandran's office, the professor lightly touched Derek's forehead with a pen and asked him where he had felt the pen touch. Derek replied, 'On my forehead.' When Ramachandran touched Derek's chest, Derek felt the pen touch only there. The same went for the top of his shoulder, but, when Ramachandran touched Derek's cheek, Derek could feel it touching his phantom hand as well as his cheek. Ramachandran found that the matching of touch on the face to sensation in the phantom hand was consistent – touching one part of the cheek would always trigger sensation in the middle finger, for example. Ramachandran could map out the phantom hand on Derek's cheek.

Other patients also reported feeling Ramachandran touching their cheek and their phantom hand simultaneously when he touched only their cheek, particularly when they had their eyes closed. However, Ramachandran did discover that he could also stimulate the phantom by touching the patients' arms near the shoulder. That area of the body is also represented on Penfield's map right next to the hand, on the other side from the cheek area. Again, particular locations on the shoulder always stimulated the same points on the phantom hand. So it seemed that the neurones from all around the redundant area of cortex were moving into the area that once sensed Derek's hand. Ramachandran proposed that the reorganization of the cortex was at the root of the phantom phenomenon – he calls his idea the theory of cortical remapping.

Ramachandran wondered next whether what was true of the sense of touch might also be true of the sense of whether something is hot or cold. The nerve endings that give us our sense of touch are different from those that allow us to sense heat. And the nerve pathways along which these signals travel arrive at different parts of the brain. There is a separate map of the body, in a different part of the cortex from Penfield's homunculus, which is like a temperature map of the body. Ramachandran thought that if his theory of cortical remapping could happen in Penfield's homunculus, it might happen in the temperature map too. So he dipped a cotton bud into cold water and touched Derek's face again. Derek did feel a sensation of cold in his phantom hand, and felt cold water trickling across his phantom fingers as the cold water trickled down his cheek.

To gain further evidence to support his theory, Ramachandran and some of his colleagues carried out a study using a brain-scanning technique called MEG (magnetoencephalography). This technique relies on the simple fact that electric currents produce magnetic fields. This effect is put to use in an ordinary loudspeaker: a coil of wire attached to a cone sits next to a permanent magnet. The wire carries a signal from the hi-fi amplifier – the signal is an electric current that flows backwards and forwards. The current produces a magnetic field that changes direction as the current does. This magnetic field interacts with the one produced by the permanent magnet, and the coil moves backwards and forwards making a sound. In a similar way, there are electric currents in the neurones that make up the cortex, and each one produces a tiny magnetic field. By measuring the strength of the magnetic field above particular regions of the cortex, neuroscientists using MEG can monitor which parts of the cortex are active – without opening up the skull or exposing the head to powerful X-rays. It is not possible to 'listen in' to the signals in the brain and, if it were, we would not be able to interpret what a person was thinking. However, MEG can be used to check Penfield's map, by touching various parts of the body and observing where activity is stimulated in the cortex. When Ramachandran and his colleagues used the technique to examine phantom limbs, they found that the cortex really had been remapped. On the MEG display screen, the images were clear: when the scientists touched the upper arm or the cheek of a patient who had lost a hand, the region of the cortex previously associated with the hand, as well as those associated with the cheek, became active.

Ramachandran predicted that cortical remapping would happen when other parts of the body were lost. What would happen if you lost one of your feet, for example? The feet are represented on the cortex adjacent to the area representing the genitals. So were there any people who had lost one of their feet and now felt the foot when their genitals were stimulated – during sex? Amazingly, the answer was yes. Ramachandran published his ideas in the *Proceedings of the National Academy of Science* in the USA, in 1993. Soon he was receiving phone calls from phantom-limb patients with tales to tell. And some of them had indeed felt sensations in their phantom feet while they were having sex. Some of these sensations were sexual in nature.

Ramachandran tells the story in his best-selling book, *Phantoms in the Brain* (1998). Apparently, a colleague of his suggested a different title, based on the title of another best-selling book on neuroscience, Oliver Sacks's *The Man Who Mistook His Wife for a Hat* – *The Man Who Mistook His Foot for a Penis*!

There are those who think that Ramachandran's ideas are too simplistic, but the idea of brain plasticity, at least, has become established. Other researchers have repeated his findings, and the plasticity of the brain in cats, monkeys and in human phantom-limb patients is now well documented. In some cases, reorganization of the cortex has been observed within hours of an amputation. The mechanism by which the cortical remapping takes place is still not known. There are two main theories to suggest why touching an amputee's cheek should stimulate part of the cortex normally devoted to the missing hand. First, nerves from the cheek might always have been connected to that area of the cortex. The signals entering that area from the cheek would have been inhibited before the amputation, but connections between existing neurones could strengthen when there is no input from the hand, after the amputation. This would explain why reorganization of the cortex can happen so quickly, and why some people first feel their phantoms within hours of an amputation.

The second idea is that new neurones actually grow as the existing ones die, and make new connections with other established neurones. A recent study of the brains of people who have died has brought new evidence that neurones can still grow even in older people. The patients in this well-publicized study were all suffering from terminal cancer. A chemical that is commonly used to follow the growth of cancerous cells, called Brdu (bromodeoxyuridine), was administered to the patients, to see if it might show any signs of neurone growth – molecules of this chemical attach to the DNA in dividing cells. After the patients died, their brains were examined. The researchers found Brdu in the brain – in particular in the hypothalamus – showing that new neurones had been growing there. This contrasts with the idea previously accepted, that every day from age thirty or so around 100,000 neurones die. The idea of brain plasticity has created a wave of new ideas about the brain and the suggestion of potential treatments for a range of disorders. This will be explored further in 'Lies and Delusions'.

Derek's cheek is subject to small but constant sensory input – breezes, his pulse through the blood vessels in his face, a smile. This might explain the very existence of his phantom. Some neurologists are very cautious about this explanation of phantom limbs. However, there is supporting evidence: reorganization of the cortex is definitely related to phantom pain. In 1997 Professor Niels Birbaumer and his team at the University of Tübingen temporarily eliminated phantom pain in amputees, using anaesthetics in the patients' stumps. The patients' brains were scanned before and after the anaesthetics took effect. When the patients could feel the pain, the researchers observed remapping of the cortex. But the remapping was found to be very transient: under anaesthetic it disappeared, and the cortex went back to the way it was. Another general finding of this study points to a connection between remapping and phantoms – the more extensive the remapping, the more pain is felt.

While he was working with phantom-limb patients, Professor Ramachandran was very keen to find a way to relieve phantom pain. Derek, who had participated in Ramachandran's previous studies, explained to him how, in the first two years after he lost his arm, his phantom had given him constant and terrible pain. After that, the pain had lessened a little, but when he met Ramachandran, six years after his accident, Derek still felt his phantom hand immovably cramped in an awkward position, and still painful. Ramachandran came up with another experiment that is remarkably low-tech compared with most of the techniques of modern neuroscience. He uses what he calls his mirror box, or the 'virtual reality' machine, to present phantom patients with the illusion that they have two undamaged arms.

The mirror box is constructed simply: it is made of an ordinary cardboard box with one end removed. The box sits on a table, in front of the patient, with its open end facing him or her. An ordinary mirror stands vertically down the middle, dividing the box into two chambers along its length. If the patient places one arm into the box, into one of the chambers, he or she sees a reflection of the arm, which looks as if it is in the other side of the box. The fact that our hands are symmetrical – our thumbs point towards each other when we put them face down – is important in creating the illusion that the patient has two hands. Derek was one of the first patients to try out the box. He recalls: 'The moment I put my hand in there, and saw that throbbing left phantom hand in the mirror, moving like my other hand – instantly, I felt the phantom hand move; I felt the pain withdraw a little bit; I felt the different fingers; I felt the palm; I could never move it before. I made a fist, slowly the hand started to move ... I was floored – I get teary just thinking about it.'

Ramachandran explains that after Derek had used the mirror box for about two weeks for around ten minutes every day, the phantom disappeared almost completely – all he had were the phantom fingers hanging from the shoulder. 'In a sense, we have amputated his phantom limb using a mirror,' he says. Ramachandran's explanation for this phenomenon is that perhaps the visual feedback somehow 'jump-starts' the brain so that the phantom starts moving again. In some cases, including Derek's, this can relieve the cramp in the phantom limb and ease the pain.

Ronald Melzack has tried Ramachandran's mirror box with a number of patients, and has only had one success: a patient whose phantom hand opened for the first time as a result of the procedure, and even in this case the patient's pain was not relieved. He is not convinced of the efficacy of Ramachandran's virtual-reality approach, and proposes instead that what has happened in the successful cases is due to the trust in the doctor and in medicine. In other words, Melzack thinks that there is a kind of placebo effect at work. Ramachandran admits that this could explain the effect he has witnessed with the mirror box.

Back in Oxford, Peter Halligan decided to test Ramachandran's theory and to give the mirror box a try himself. Neil is one of Halligan's patients, who has suffered extreme phantom pain since his nerves were torn in a motorbike accident. As Neil places his arms into the box, Halligan asks him whether he finds that the phantom hand opens up as he looks at the reflection of his good arm. Neil says it does: 'It seems to release the pressure of my phantom ... [reduces the pain] just like that.' The pain went from its normal seven out of ten down to essentially zero as soon as the arm was in the right position in the mirror box, and the pain came back again as soon as Neil removed his hand or closed his eyes. Whether or not the mirror box actually works, Ramachandran's theory of cortical remapping does seem to be backed up by evidence from other researchers using brain-scanning techniques, as well as by direct observation of patients. The whole idea has stimulated interest in trying out a range of new techniques to overcome phantom pain.

Opening up

Neurosurgeon Mr Tipu Aziz works at the Radcliffe Infirmary, Oxford, trying a radical and controversial approach to tackling phantom-limb pain. Wheeled into the operating theatre is Elke, ready for surgery on her exposed brain – while she is conscious. Twenty years ago, Elke was knocked down in a car accident that left her with excruciating pain in her phantom arm. Over the years doctors have tried a large number of treatments, including acupuncture and chemical and surgical nerve blockade – but all to no avail. This is why she has agreed to undergo this incredible procedure. There is no way of knowing what the long-term results might be – it is a measure of how desperate she is for relief from her phantom pain that she is willing to go through the operation.

There are no pain or touch receptors in the brain, so it is only necessary to inject local anaesthetic into the scalp. A week before the operation, under general anaesthetic, Elke's scalp was peeled back and a section of her skull 15 centimetres across was cut through in preparation for her operation. Today, Aziz staples back the scalp, and this time lifts the loosened section of skull away to reveal the brain. 'I'm just opening your skull – it might be a little uncomfortable,' he says. Once the brain is exposed, Aziz uses a small metal electrode to deliver tiny electric currents to the cortex. He needs to stimulate the cortex, just as Wilder Penfield did, in order to locate which part represents Elke's phantom arm. As the electrode becomes active, Elke reports what she feels.

At first she feels nothing, and Aziz increases the current. Suddenly, Elke's face and shoulder go into spasm – the current has caused her to have a small fit. The operation continues, with a lower current, and when Aziz locates the relevant part of the cortex he stitches the electrode in place, on to the membrane that surrounds the brain. He then attaches a wire to the electrode. The wire was previously installed underneath the skin on Elke's face, and it runs down inside her neck and to a power pack in her chest wall. The hope is that electrical pulses delivered by the power pack, arriving at the relevant spot on the cortex, will counteract the phantom-limb pain. Once everything has healed, Elke will go back to hospital and the surgeons will adjust the current from the power pack, in an attempt to change the sensation of pain to something more tolerable.

It is too soon to know whether the sort of operation that Elke endured will have the desired effect, but similar procedures are being tried elsewhere. These procedures do not only give a little hope to some patients who suffer from phantom pain, but can help neuroscientists to find out more about brain plasticity, and phantom-limb sensation in particular. In 1997 an important study that also involved 'open head' surgery was carried out by a team led by Karen Davis at the Toronto Hospital, Canada. In this case, the manipulation of the brain included subcortical structures – parts of the brain that are deeper than the cortex. The particular structure that Davis wished to investigate was the thalamus, which as we saw earlier is the relay station for nerve signals coming from the head and body. Signals relayed by the thalamus end up in the cortex, some of them contributing to Penfield's cortex map. You will remember that a sensory map of the body exists in the thalamus as well as on the cortex. A different research project, published in 1996, had shown that this map can also undergo reorganization after amputation. The role of the thalamus in the remapping could easily be forgotten, because surgery normally only exposes the cortex, and magnetoencephalography – used to investigate the remapping non-invasively – looks only at activity near to the surface, in the cortex. The electrode used in Elke's operation was more than a centimetre long and about half as wide. Davis's team used microstimulation, with much smaller electrodes, so that they could monitor and activate tiny regions of the thalamus at a time. This is necessary since the thalamus, while larger than many of the brain's structures, is still only a little more than two centimetres wide and three centimetres long.

Davis found that parts of the thalamus normally devoted to receiving nerve signals from a limb now received signals from the nerve endings in the stump. These areas of the thalamus probably still pass on the signals to the area of the cortex that represents the limb. This seems to suggest that stimulation of the stump would produce the phantom sensation – something that is quite common, though not universal, in amputees. The researchers also found that the phantom sensation could be produced by stimulating those same areas of the thalamus, when the stump was not being touched. Interestingly, the study also involved amputees who had never experienced phantoms. Stimulating the area of the thalamus corresponding to the missing limb in these patients produced no phantom sensation.

So it seems that when our brains stop receiving signals from a particular part of the body, due to amputation or to the severing of nerves, the sensory maps in our thalami and in our cortex reorganize. This probably happens through a combination of the growth of new neurones and the strengthening and weakening of the connections between existing neurones. This remapping might explain the occurrence of phantom limbs – and why touching parts of the body other than the stump, as well as in the stump itself, can produce sensations in the phantom limb. However, the cortical remapping theory alone seems to be an incomplete and inconsistent picture of how phantom limbs come to be. Not all amputees experience phantoms, for example. And in many cases, removal of the relevant part of the cortex or the thalamus does little or nothing to remove the phantom or to relieve phantom pain. Sometimes painkillers work effectively to reduce phantom pain, sometimes not. And sometimes other approaches to phantom-pain relief work in some patients, but not all. And how does cortical remapping relate to Ramachandran's mirror box – which also works in only some of the cases?

Perhaps phantom limbs can be caused by a number of different factors, some or all of which are present in a particular patient. The discovery of cortical and thalamic remapping, the plasticity of the brain, has reawakened the fascination with the way in which our brains are able to create our body image. Ramachandran says that due to plasticity, the body image is far from hard-wired, as Melzack suggests in his theory of the neuromatrix. Instead, he says it is a 'temporary construct' that the brain has created for passing our genes on to the next generation. In a sense, the whole body is a phantom that has been constructed in the brain.

Curiouser and curiouser

This fascination with the brain's creation of body image extends beyond the study of phantom limbs. Other well-documented but unexplained neurological curiosities are receiving a fresh interest since the discovery of remapping in the brain. An interesting example is a rare condition called AIWS (Alice in Wonderland syndrome), in which the sufferer's body image is severely distorted. This bewildering phenomenon is related to several different diseases. For example, an episode of AIWS normally precedes an attack of one type of migraine. There is a link, too, with glandular fever (mononucleosis). Just as Alice in the book by Lewis Carroll keeps changing size, sufferers of AIWS may feel their necks becoming longer like a giraffe's, or their whole body becoming wider. Sometimes they may feel that one half of their body is much larger than the other.

One sufferer, Lesley, explains what she commonly feels during an episode: 'I get a sensation that my hands have enlarged. They become almost as though you've put some rubber gloves on you've filled them with water and the water is expanding. And they do feel odd, the hands themselves: not only do they feel tingly, but they're huge. I feel like I could reach out and touch something a long way away. And at the same time my body is shrinking. I'm left with a feeling of only my hands – that I've gone ... that I'm reducing right down to a tiny walnut.' On other occasions, she feels herself growing taller. She feels as though she is walking on spongy, springy platform soles, and that she is looking down on everything.

AIWS was first linked to migraine in 1952, and a further link, to glandular fever, was established in 1977. But a convincing explanation for the condition has always evaded neuroscience. Brain plasticity might just be able to supply an answer. One idea is that before a migraine attack, the blood supply to the brain is disturbed, and different parts of the brain receive less blood than normal. Evidence in support of this comes from the existence in the brain of the neurotransmitter serotonin, which acts as a vasoconstrictor – a chemical that causes blood vessels to constrict, reducing blood flow. Perhaps when Lesley feels her body shrinking relative to her hand, the blood supply to the part of the cortex representing her hand remains normal, while blood supply to the rest of the cortex is diminished. If the blood supply in a particular part of the brain is less than needed, the neurones in that part will be less active, and a short-term version of the phantom-limb phenomenon could occur: some parts of the cortex 'take over' others. If this is true, then our body image really is just a fragile and temporary construct, and is just as plastic as the brain that makes it.

Ronald Melzack thinks that the new theories about phantom limbs and the new approaches to treating phantom-limb pain are overly simple: 'Simple-minded approaches – like "Let's stick an electrode here and zap it or burn it out, or electrically stimulate it, or give it magnetic waves" – these are fine; give them a try, let's see what happens. But I think the real answer is going to be hard slugging down the complicated roles of endocrinology, immunology, genetics and the like.'

And Melzack is probably right. For a long time, the relationship between the wiring in our brains and our perception of our bodies and the world around us have been a question as much for philosophy as for neuroscience. And it will almost certainly continue to be so for some time to come. However, the discovery of plasticity in the brain has opened up major new avenues of thought. Some of these might one day provide a more complete understanding of some of the most bizarre effects that neuroscientists and their patients have come across – including Alice in Wonderland syndrome, and perhaps phantom limbs.

LIVING DANGEROUSLY

...the search for the thrill seekers...

It is just before dawn on a calm December morning. Elliot and John set out in a car, in pursuit of danger. They compare the excitement of what they are going to do today with that of having sex for the first time. The two thrill seekers arrive at their destination – a 200-metre-tall radio mast – while the sky is beginning to lighten, and prepare themselves. They make the slow climb up to the top of the mast, take a look out across the surrounding countryside, and jump out into the cool, calm air. As they fall, they release parachutes that will slow their descent and save their lives ... they hope. Without their parachutes, they would hit the ground at more than a hundred and fifty kilometres per hour (100 miles per hour) after just five seconds. Elliot says: 'I don't want to die, but, if that moment comes, I would deal with it. I ask myself, "Why am I doing this? Am I doing this for me?"'

Elliot and John are BASE jumping. The acronym stands for Building, Antenna, Span and Earth, and refers to the types of objects that BASE jumpers leap off. BASE jumping is illegal in most countries, although special dispensations are often made – for example, there is a 'Bridge Day' annually in the USA. There are many thousands of people involved in other dangerous sports, including high-speed downhill in-line skating, sky surfing and even canoeing down steep mountainsides. Why do these people risk their lives participating in these sports?

The world is too safe

One possible answer to the question of why people engage in extreme sports or put themselves in other dangerous situations is that the world has become too safe for them. Perhaps we humans need risk in our lives and thrive on it, and some people feel that they are not in enough danger. No one really knows how the brain goes about assessing risk, but there is evidence to support the idea that we subconsciously keep track of it. Professor Gerald Wilde of Queen's University in Ontario, Canada, has found plenty of such evidence. Wilde is an expert on risk, and in a recent study the Canadian government asked him to investigate the behaviour of drivers at railway crossings.

There are thousands of unattended level crossings, without barriers or warning lights, in Ontario. Only a handful of trains pass through them every week. The experience of Wayne, one resident of Smith Falls, Ontario, illustrates how dangerous these level crossings can be if a train arrives at just the wrong moment. For fifteen years, Wayne had travelled more or less daily over a level crossing near his home without incident. One day in 1996 he approached as normal, slowed a little as he came close to the railway lines, but kept moving. He glanced to his right as he was crossing, and saw a train hurtling towards him. The train hit his car, and turned it around. Wayne was lucky to escape with his life: 200 people have been killed crossing Ontario's level crossings during the past ten years.

Professor Wilde observed the behaviour of motorists at a level crossing over a five-week period, studying a total of 517 cars from the back of a van. As with many of the crossings, there were trees along the railway track, which obscured the motorists' view. Taking this into account, Wilde worked out at what speed motorists should approach the crossing to remain safe if a train should come. He measured the speed of each vehicle as it approached the crossing. To his surprise, he found that about seventy-five per cent of the motorists would have been hit if a train had been approaching. The fact that more people are not hit at these level crossings indicates that drivers have an awareness of the fact that trains are rare. The obvious solution, which would perhaps reduce the number of people who were not safe, was to remove the trees that were obscuring the motorists' view. Wilde and his team did just that, and Wilde recalculated the safe speed of approach given the improved view. When he observed the behaviour of drivers who now had a better view, he made a startling discovery. You would expect that the motorists, having a better view of the railway, would naturally be more safe. In fact, the effect of the improved view was that the motorists increased their speed – the level of risk stayed the same. The same effect can be seen on ordinary roads: the advent of wide, straight motorways, and safety technology such as air bags and seat belts, does not affect the level of risk of having an accident per hour spent on the road. Instead, we naturally adjust our driving to account for the risks we perceive. So, on a winding country lane in a car with no seat belts, people tend to drive more slowly and cautiously than on a motorway in a car fitted with safety equipment. It seems that this principle applies to our behaviour in general, not just to our driving.

In his book *Target Risk* (1994), Wilde documents many other examples of this phenomenon, in which people naturally and quite unconsciously maintain a certain level of risk in their lives. He calls this effect the 'risk homeostasis theory', using the term 'homeostasis' in a similar way to how French physician Claude Bernard defined it during the nineteenth century. Bernard was referring to the natural and unconscious ability of systems within the human body to maintain body temperature and chemical balances. Wilde thinks that something very similar is going on when we assess the risks of an action or activity. He defines 'target risk' as an individual's 'accepted' or 'preferred' level of risk, and explains that we seem to be happy only within a narrow window either side of this level. If the risks become too great, we will change our behaviour and take precautions to avoid undesirable outcomes. However, we also seem to adjust our behaviour if the risks become too small. He uses his theory to argue that providing technological safety measures such as seat belts, anti-lock brakes and air bags is not the most effective way to reduce the occurrence of accidents on the road. He presents plenty of convincing evidence, in addition to the study described above, that indicates that people do indeed have an awareness of how much danger they might be in, in any given situation such as driving a car.

The principle of 'risk homeostasis' sometimes stands even through dramatic changes in our way of life, such as when the motor car was invented in the late nineteenth century. Before the invention of cars, most traffic fatalities were associated with horses or horse-drawn vehicles. As car use rose during the early 1900s, so did the rate of traffic accidents associated with cars. At the same time, the rate of accidents associated with travel using horses and bicycles declined, so that per head of population, per hour of travel, the accident rate remained about the same, fluctuating slightly above and below a constant level. Wilde likens this to a thermostat, which automatically allows the temperature to drop if it is too high, and causes it to rise if it is too low. He also presents evidence that the introduction of seat-belt laws does not affect only the wearing of seat belts. The rate of accidents has been shown to rise after the introduction of such laws, again perhaps indicating that people increase their speed when they are wearing seat belts. And although people in the front seat are less likely to die in an accident if they are wearing a seat belt, the increase in the number of accidents means that the death rate remains about the same. So, if safety measures such as seat belts do not actually reduce the risks associated with driving, what is the point of introducing them?

Wilde is sure he knows the answer: 'These things are designed to improve performance, not to increase safety,' he says. 'That's why people like them.' He proposes that a person's level of target risk is determined by four factors: 'the expected benefits of comparatively risky behaviour', 'the expected costs of comparatively risky behaviour', 'the expected benefits of comparatively safe behaviour' and 'the expected costs of comparatively safe behaviour'. In the last category, he includes factors such as the discomfort and inconvenience of wearing seat belts, while in the first he includes factors such as the fact that 'making a risky manoeuvre fights boredom' – the influence of thrill-seeking behaviour.

The science of risk brings fascinating insights into human behaviour. The level of risk in an activity is directly related to the mathematical probability of the unwanted happening. Probabilities are usually given as a number between zero and one. Zero probability means that an event has no chance of happening, while a probability of one means that it definitely will. In some cases, it is easy to calculate the probability of certain events occurring. When throwing dice, for example, it has been found that the various sides of the dice land in a completely random way. This means that the probability of scoring a six on any particular die is 0.167 (one in six), since there are six faces each with an equal chance of landing face up.

We are surrounded by dangers of different types and, in many of these cases, risk cannot be calculated so easily. Risk analysts cannot rely on the mathematics of random events, like rolling dice, to calculate risks in complex, real-life situations. In these situations, risks can only be estimated using figures collected in large-scale studies or surveys across a large population and over a significant amount of time. So figures for the risk of accident per hour behind the wheel, for example, are calculated using data about the number of people driving, and the numbers of accidents that occur. Using this sort of analysis, the probability (risk) throughout a person's life of dying from a heart attack is about 0.25 (one in four), while the probability of dying as a result of being hit by lightning is about 0.00000001 (one in ten million).

Many of us are unaware of some of the very ordinary everyday risks that surround us. From the moment we wake up, being alive is dangerous. Every year in the UK, twenty people are electrocuted by their bedside lamp or alarm clock. Another twenty die as they fall getting out of bed. Thirty people die from drowning in their morning bath, and sixty are seriously injured putting on their socks. And a shocking 600 people each year die from falling down stairs – nearly two people per day. In the USA, around 6,000 people manage to injure themselves with their bedclothes. If we tried to make the world perfectly safe, we would have to remove stairs, beds and baths. Gerald Wilde puts it simply: 'Zero risk is not an option.' We can install as many safety measures as we like, and try to reduce the risks we can control (the risk of heart attack, for example), but we can never do away with risk altogether. And if we did, the world would cease to be interesting and enjoyable.

Imagining that the world can be or should be totally safe can not only make it boring – it can be dangerous. One evening in 1996 Verity and her husband were watching a *World in Action* programme about the risks to women who take contraceptive pills. The brand that Verity was taking was featured. The programme did say that the risks to women's health of taking this pill were within government limits, but it also showed the newly discovered and nasty side effects of the pill on some of the people who had suffered them. In particular, it showed women who had suffered blood clots as a result of taking the pills. The blood clots had been transported in blood vessels to the brain, where they blocked the blood supply, causing convulsions and sometimes death. Verity threw her pills into the bathroom bin the following morning, and two weeks later found out that she was pregnant. Thirty-two weeks into the pregnancy, she went for a routine check-up and was diagnosed with pre-eclampsia. This dangerous condition is one of the major risks involved in pregnancy, and is the result of a build-up of toxins in the blood. It causes high blood pressure and swelling of the hands and face; if not treated quickly and effectively, it can be fatal.

Verity was treated by consultant obstetrician Richard Johanson, who says that her kidneys were not functioning properly as result of the condition. She had an emergency caesarean section and, even after this operation, her condition did not improve for a long time: her heart could not cope, and there was a build-up of fluid in the lungs. It was a 'near miss', as Johanson puts it, and 'in times past or in other places present, she might have died'. He adds: 'Pregnancy is more of a risk than taking the pill. If half a million women who were taking Verity's pill stopped taking it, maybe one life would be saved. If half a million women became pregnant, between twenty-five and thirty would die.' There are many good arguments for and against the contraceptive pill. But in terms of risk, since the pill reduces the number of pregnancies, it reduces the risk of complications of pregnancy, such as pre-eclampsia. This applies to other methods of contraception, too, of course; but the point is that every action has a risk.

Television news programmes and the newspapers bring us horror stories every day, painting a picture of our world as a place filled with danger. We see a world in which aeroplane crashes, mass murder, rape and other violent crimes seem to be rife, and there is danger around every corner. And we see a world filled with incurable diseases, and in which the food we eat or give to our children can affect our health. In 1982, for example, there were reports of a new disease, which was spread by a virus and seemed to affect only gay men. Initially called 'gay-related immune deficiency' or GRID, it was renamed as AIDS (acquired immune deficiency syndrome) when it was realized that it did not affect only gay men – intravenous drug users, heterosexuals and even babies were also contracting this fatal disease. By 1994 AIDS was the leading cause of death among Americans aged between twenty-five and forty-four, and there was a worrying increase in the number of reported cases worldwide. AIDS was an epidemic and, due to some effective campaigning, no one was unaware of its dangers.

In 1986, this time in the UK, another health issue began to hit the headlines: it was in this year that the first case of BSE (bovine spongiform encephalopathy) was reported. BSE is sometimes called 'mad cow disease' because it affects cows' nervous systems, causing the animals to exhibit disturbing repetitive behaviours, and eventually to degenerate completely. In 1992 people began to wonder whether BSE might affect people who ate beef from cows infected with the disease. In 1993 two farmers with BSE-infected herds died of CJD (Creutzfeldt–Jakob disease), the human form of the disease. Also in 1993, the first case of CJD that proved to be the result of eating infected beef caused a media storm. A worldwide ban on British beef soon followed, and hundreds of thousands of cows were destroyed.

More recently, the public has been made aware of toxic compounds called phthalates. Investigations of these compounds have brought to light evidence that they seem to survive for longer than was thought without breaking down into harmless substances. Furthermore, they have been found in food packaging, and in foods. A study of the effects of phthalates in rats' and in human fertilization showed that fertility was reduced by an intake of some of these compounds. There was definite evidence of a reduction in size of rats' testes, for example. A further study of phthalates found them in powdered baby milk. Add to these risks the horror stories brought to us by television and the newspapers, and the aeroplane crashes that claim hundreds of lives and it might seem that the world is more dangerous than ever, partly due to the modern technological lifestyle we now live.

With so many dangers in the world, why would people want to take part in dangerous sports such as BASE jumping? Perhaps the risks from AIDS, beef, baby milk, violent crime and aeroplane crashes are not as high as we might be led to believe. Even at the height of the AIDS epidemic, the risk of contracting AIDS, for those not in a high-risk category such as intravenous drug users, was less than the risk of dying from falling out of bed. In the UK, there were between thirty-three and eighty-one cases of CJD, and around fifteen deaths each year between 1990 and 1998. This is one quarter of the number of people who die from drowning in their baths. Current estimates of the risks of eating beef on the bone suggest that one person might die as a result of eating such beef in the next ten years. In the same period, 6,000 people will have died falling down stairs. The tests that showed how phthalates can affect fertility involved concentrations of the compounds far higher than those found in foods and packaging. Aeroplane crashes, while horrific for those involved, are rare. Air travel generally carries considerably less risk than driving – the risk of dying in an aeroplane crash over a period of one year is about 0.000002, or one in 500,000. Before the invention of motor vehicles such as cars and trains – despite the accidents they undoubtedly cause – travelling great distances was very hazardous. And for every person murdered today, it is thought that ten were murdered in the Middle Ages. The murder rate has been halved in the past two hundred years. Diphtheria and polio, once common life-threatening diseases, have been all but eradicated in most countries. For every death from an infectious disease in the twenty-first century, there were probably at least a hundred in the Middle Ages. So, for many of us, the world seems to be becoming safer, not more dangerous. Of course, people in poorer, developing countries are generally at greater risk from disease and natural disasters than people in rich, developed countries. For example, of the 33 million or so people estimated to be living with AIDS at the end of the twentieth century, the large majority were living in poorer countries, where people have far less access to new medicines, education and health-care workers.

When the media report to us the terrible consequences of eating certain foods or of travelling on aeroplanes, they tend to highlight the bad cases – safe air travel is no news. In fact, it is easy to lose sight of the benefits of technology when you are exposed only to the 'bad news'. The concept of 'benefit' is an important part of our assessment of risk. Misunderstanding the benefits of an action can be as hazardous as miscalculating the dangers. We can make mistakes when we exaggerate or underestimate the danger, or if we the ignore or magnify the benefits.

In November 1995 poisons expert Professor John Henry of St Mary's Hospital in London was contacted about the case of a girl, Leah Betts, who had become ill at a birthday party after taking an Ecstasy tablet. Leah died, and Professor Henry correctly diagnosed that the cause of death was an excessive intake of water. Ecstasy, or MDMA (3,4-methylenedioxymethamphetamine) is a hallucinogenic drug that gives people who take it feelings of euphoria. Extreme thirst is one of the side effects of the drug and, surprisingly, too much water in certain circumstances can kill you. The media interest in the case of Leah Betts was intense. Henry says that the media used him to put the 'anti-drugs' side of the Ecstasy story. And yet, he maintains that alcohol is a far more dangerous drug than Ecstasy. 'If you ask people, "Which is more dangerous, alcohol or Ecstasy?" people will immediately say "Ecstasy" – and yet we all know that alcohol presents a far greater risk,' he says. People who take the drug report significant recreational effects. That is not the same as saying that it is not dangerous. There are certainly risks associated with using the drug, perhaps some we do not know about. However, the point is that compared with alcohol, which in one way or another claims around a hundred lives every day, Ecstasy kills only a tiny number of people – no more than one person per month. In fact, since 1985, there have been fewer than a hundred Ecstasy-related deaths, while estimates of deaths related to either long-term or short-term abuse of alcohol number around thirty thousand each year.

Perhaps part of the reason we underestimate or forget the risks associated with alcohol is that they are not reported in the media. Soon after the Leah Betts case, another young person, Mark Dogget, was out for the night at a birthday party, in his local pub. It was Mark's wish to drink one whisky for every year of his life – twenty-four. He managed this, and drank a little more on top. Outside the pub, Mark vomited and fell over. He choked on his vomit and died from lack of oxygen to his brain. That evening, Mark's father Geoff received a phone call – 'The one phone call that no parent ever wants to receive' – that his child had died. Like Leah Betts, Mark Dogget died as a result of irresponsible drug use at a party. Leah Betts's story dominated the newspapers and television news for many days, while Mark's story was reported as a small item in only one national newspaper. Professor Henry says that in his hospital cases of Ecstasy toxicity are rare, but 40 per cent of emergency cases are directly related to alcohol.

We tend to ignore this huge cost of alcohol – we perceive the benefits, but not the dangers. Geoff Dogget is bemused: 'There have been situations where people are concerned about eggs, there was the BSE scare; and it's plastered all over the daily newspapers and on the television news – but one thing they don't do is mention alcohol.' Professor Henry calls this a 'national scandal'.

It seems that our concept of risk is a little distorted. Life is safer than it has ever been, and people seem obsessed with ever smaller risks. Perhaps an explanation of this is that we need risk in our lives. Gerald Wilde says there should be 'risk in your life like salt in your soup: not too much, not too little'. We have seen that we attempt to keep risk at our own acceptable or desired level. The fact that the world is actually safer than ever might explain why people go out of their way to put themselves at great risk, by participating in extreme sports. It may also explain our fascination with 'bad news': no one wants to hear how safe the world is. Similarly, we seem to have a hunger for danger in the films and television programmes we watch. The average American teenager has seen thousands of violent deaths on television or at the cinema by the age of fifteen, while most people never see one at first hand in their entire lives. Is this our way of getting our fix of the thrills of danger? If this is the case, we are still left with the question of why some people take more risks than others. The world may be more safe than ever before, and people may strive to achieve a constant level of risk, but the level of risk that is acceptable to an individual seems to differ from person to person. Just as different people like different amounts of salt in their soup, different people seek different levels of risk. However, while the taste of soup may be a matter of preference, risk-taking seems to be programmed into us, as a result of our genes or our brain biochemistry.

Seeking the thrill seekers

If we are to discover what makes people take risks, we must be able to identify risk takers. This will not only help us to 'formalize' their behaviour, but it may allow us to extend the definition of people who seek thrills and take risks, perhaps to include those who take risks without participating in dangerous sports. Only once we have a way of identifying who the risk takers are will we be able to investigate their brain and body chemistry, and their genes, and find out what gives them their drive for danger. So our quest to understand the behaviour of thrill seekers begins with an attempt to understand and categorize the personalities of these people.

Can psychologists offer us a profile of these people? Personality testing is an inexact science, despite more than a century of thought and experiment – human personality is elusive. Part of this elusiveness comes from the fact that everyone is unique, and that people change over time. People identified by one personality test as being in the same personality category might behave in very different ways from each other in some situations, and indeed might be in different categories according to a different personality test. In fact, since we are all unique, we could each be put into a different category. Personality tests can also be tautologous – in other words, they may tell you nothing that you do not already know. Organizing people into groups according to, say, the colour of their shoes, would enable you to say nothing more about each group than what colour shoes they are wearing. And so, in a personality test, you might find out nothing more than the fact that people who engage in extreme sports enjoy, well, taking part in extreme sports! The moment you make conclusions about people's personalities that go beyond the questions you ask them, those conclusions are ultimately unsupported, and are pure conjecture.

Having said this, personality tests do have their uses, and have formed the basis of scientific research into behaviour. Standard personality tests are based on theories of personality – theories about what factors influence people's feelings, thoughts and behaviour. There are many such factors, including, no doubt, genes, biochemistry of the brain, culture, childhood experiences and pure chance. It is the first two of these that we shall be examining in detail in this chapter. Is there something in the genes and the brains of thrill seekers that is different from other people's genes and brains?

The study of personality has come a long way since the ancient Greeks first attempted to explain what influences the way we think, feel and behave. Like modern neuropsychologists, the ancient Greek philosophers believed that our personalities are determined by our body chemistry. However, their ideas were a little less sophisticated than modern ones. The accepted wisdom of the ancient Greek philosophers was that personality is a direct consequence of the four humours – fluids that were thought to make up our bodies. These humours are blood, black bile, yellow bile and phlegm. Many Greek thinkers believed that blood made people enthusiastic, black bile made people melancholy, yellow bile made people angry and phlegm made people apathetic.

The history of the modern approach to personality begins with Austrian psychologist Sigmund Freud. It was Freud who set the study of personality on scientific grounds. He realized that there are hidden factors, such as innate drives and early childhood experiences, that are important in shaping our thoughts, feelings and behaviour. He believed that these can be teased out and analysed by using such techniques as free association and the interpretation of dreams. The psychoanalytic approach was extended by Swiss psychologist Carl Jung, who included a spiritual and a cultural dimension as well as attempting to classify behaviours according to four functions of the mind: sensing, feeling, thinking and intuition. He proposed that people could approach a situation in an 'introverted' or an 'extroverted' fashion, depending on whether they drew their inspiration from inside or outside themselves.

Another way of looking at personality is the behaviourist approach, pioneered by Russian psychologist Ivan Pavlov and American psychologists John Watson and Burrhus (B.F.) Skinner. According to the behaviourists, the hidden inner workings of the consciousness are not important. Instead, they believed that our behaviours, and therefore our personalities, are learned through our experiences in a manner that is understandable, objectively testable and measurable. This was an attempt to give psychology the same firm footing as other sciences. The theories of behaviourism arose through experiments, largely on animals, in which behaviours were learned, in a predictable way, through carefully controlled stimuli. Pavlov's dogs, which he taught to salivate at the sound of a dinner bell, is a famous example, but perhaps more interesting are Skinner's pigeons, which he taught to play table tennis.

Elements of both the psychoanalytic theories of Freud and Jung, and the behaviourist theories of Skinner and Watson, survive in modern approaches to personality. Modern psychology attempts to be as objective as possible, while acknowledging the fact that the human mind is extremely complex and elusive. Advances in neuroscience have begun to bridge the gap between biochemistry and behaviour, but there is still a great deal that remains unexplained. Without a complete and consistent understanding of personality or behaviour, how can we begin to categorize risk takers? If we cannot, we will be unable to compare their brain biochemistry and their genes with those of other people. Modern personality tests normally include one or both of the following components: a personality inventory and a projective test. A personality inventory is a set of questions about a person, answered by the person himself or herself, in an interview or a written questionnaire. A projective test is one with no definite answers – instead, the person being tested is asked to interpret a story or a picture. The rationale is that the person will 'project' their personality through their interpretation. The most famous example of a projective test is the Rorschach inkblot test, which involves ten abstract pictures that look like stains made by ink soaked into blotting paper.

There is a number of different approaches to testing for and categorizing people by personality type. One widely used test is the Myers-Briggs Type Indicator (MBTI). The categories defined by this test are based on a personality inventory, and have their roots in the work of Carl Jung. The questions that make up the test are based on a person's preferences in four aspects of living: how we gather information, make decisions, organize our lives and how we find 'energy' day to day. Each category has two options, each denoted by a letter. You might prefer to find energy inside (I) or outside (E) yourself; you might prefer to gather information using only your senses (S) or using your intuition (N); you might prefer to make decisions based on thought (T) or feelings (F) you might prefer to organize your life using judgement (J) or perception (P). There are sixteen combinations using these four dimensions, including ENTJ and ISFP, for example. People categorized as ENTJ are generally self-confident, with good verbal communication skills, and make good leaders, while ISFP people have a good aesthetic sense, are trusting and sensitive, and may make good artists or teachers. Since the 1970s the MBTI has been used by many organizations to help build effective teams, by identifying who has the correct personality to be a leader, who might provide the best creative input, who might be the best communicators, and so on. According to the MBTI, risk takers fall into the categories ISTP and ESTP.

Another categorization of personality – one that has become part of the language of self-help books and magazines – divides people into three types: A, B and C. Type A people are intense, driven and often uptight. Type B people are more laid-back and easy-going. Type C people are in between A and B: not too highly strung or too laid-back. Again, a personality inventory is the basis for determining whether a person is type A, B or C. In recent years, psychologists have added another – type D. Such people apparently are insecure, socially inhibited and have a negative outlook. In which of these categories do we find the risk takers? Not in any of them: in 1996 American psychologist Frank

Farley added another – type T. According to Farley, the type T personality seeks thrills, risks and arousal. Studies based on this classification have shown that people who are type T satisfy their needs in their job – as firefighters or paramedics, for example – or in other activities, such as skydiving and, yes, BASE jumping. Apparently, not all thrill seekers fulfil their desires in physical activities: many take risks in art, business or politics. Perhaps rather more destructively, some type T people also have a greater propensity towards taking drugs or drinking excessive amounts of alcohol than the general population. If Farley is correct, then, T-type people are predisposed to risk-taking behaviour, which can bring positive or negative results.

Farley drew, in part, on the work of psychologist Marvin Zuckerman, who has carried out extensive studies of personality and behaviour. Zuckerman devised a test whose aim is to find out just how risk taking an individual is. The test results are assessed along four dimensions: thrill and adventure seeking, inhibition, susceptibility to boredom and general experience seeking. Like Farley's categorization of thrill seeking, this four-dimensional approach takes into account the expression of risk-taking behaviours in art, music, sex and in social situations, as well as a propensity to take part in dangerous sports. In his book *Behavioural Expressions and Biosocial Bases of Sensation-seeking* (1994), he suggests a strong genetic component to thrill seeking. Evidence of a genetic link comes from studies of twins, which suggest that thrill seeking is a trait which is as much as sixty-nine per cent inherited. He also looks at some of the features that thrill seekers seem to have in common: they are less likely than others to suffer from stress-related disorders, for example. He investigates what biochemical mechanisms within the brain might make thrill seekers behave as they do. One biochemical that seems to be important in thrill-seeking behaviour is an enzyme known as MAO B (monoamine oxidase type B).

Another biochemical associated with thrill seeking is the neurotransmitter dopamine.

Taking a chance on MAO

The story of the role of MAO B in risk-taking behaviour begins more than twenty years ago on the prairies of mid-western Canada when a biochemist, Dr Peter Yu, embarked on an unusual experiment. At the time he was carrying out his research, MAO had already been linked to behaviour: the level of the enzyme was known to be related to mood. In particular, high levels of MAO were found in people suffering from depression. Yu decided to study the levels of MAO in the blood of the most violent criminals he could find. Near the University of Saskatoon in Saskatchewan where he works is a penitentiary for the criminally insane, and Yu tested the concentration of MAO in the blood of some of the most violent, psychopathic inmates. 'I was hoping to be able to find some biological markers for mental disorders,' he says, 'with the reason that, first, we can understand the mechanism of the disease, and, second, maybe we may be able to provide a better strategy for treatment.'

There are two forms, or isomers, of the enzyme MAO – A and B. Both isomers are found in abundance in the brain. In the rest of the body, MAO B is found in blood platelets, while MAO A is largely locked away inside organs such as the liver. Because MAO B is found in the blood, the concentration of that form of the enzyme in the body can be worked out using a simple blood test. Yu's research involved taking blood samples from the psychopaths, spinning them in a centrifuge to separate the blood cells from the platelet-containing plasma, and determining the concentration of MAO B present using radioactive molecules that bind to the enzyme. Yu took blood samples of thirty-five prisoners, and found that the levels of MAO B in their blood were much lower than in the general population. The more violent the offences they had committed, the lower the level of MAO B – as much as one-third lower in the most violent prisoners.

Yu's results have been confirmed in several more recent studies, including one in Sweden in 1997. That study found that twenty-one out of twenty-seven juvenile offenders had levels of MAO B significantly lower than the average. At the time Yu carried out his study, he was unsure how to interpret his result. 'We were very excited by these findings ... but we really didn't quite understand what that meant to us,' he explains.

Soon after Peter Yu had carried out his research, 11,000 kilometres away in Madrid clinical psychiatrist Jose Carrasco wondered whether the reduced levels of MAO B in violent criminals were related to an increased desire for risk or sensation seeking. He wondered whether the behaviour of some of his mentally ill patients might also be partly due to the same sensation seeking desire. To check the link between sensation seeking and MAO, Carrasco chose to look at the levels of MAO B in people without mental disorders or violent criminal tendencies, but who exhibited obvious risk-taking behaviour. He knew just such a group of people: matadors. Bullfighting is extremely dangerous – every year in Spain around a hundred and fifty matadors are seriously injured, and deaths are not uncommon. And yet the matadors enjoy what they do – they relish danger. One member of Carrasco's study, Fernando Gallindo, has been injured many times, including being gored in his neck and his left lung. And yet he says: 'When you are in front of the bull, and you see his eyes, and you put the bull near you – you are the centre of the world at this moment. I like risk, because behind the risk is pleasure.'

Carrasco studied seventeen bullfighters altogether. He subjected them to two tests. First, he identified how risk taking they were, according to standard personality tests including the Rorschach inkblot test, and then he tested their blood for MAO B. All the bullfighters in Carrasco's sample had lower-than-average concentrations of MAO B. Carrasco found that the bullfighters who had come out as the most risk taking on the psychological tests had the lowest concentrations of MAO B in their blood.

Back in Canada, Peter Yu decided to take the investigation into MAO B a step further. He wondered whether people who go out of their way to avoid risk might have higher than normal levels of the enzyme. At the University of Saskatoon, Nurse Judy Hawkes helps people with various phobias. Penny-Jane, one of her patients, has a range of phobias, including extreme agoraphobia, an irrational fear of open spaces, and ochlophobia, fear of crowds. In fact, her anxiety is so pervasive that she has been diagnosed with panic disorder, in which breathlessness, faintness and palpitations are common and may happen suddenly and unpredictably even in everyday situations. She feels at ease only in her home or in her car. She has not visited the local supermarket for more than six years. For her, everyday life is full of unacceptable risks. She says that she is terrified of going into shops and that she feels 'very trapped, like an escape feeling, like I want to get out; as if there was a fire – I feel that same feeling'.

Under the care of Nurse Judy Hawkes, Penny-Jane practises overcoming her fears. At the supermarket, Penny-Jane ventures inside while Hawkes remains outside in the car park. The two speak to each other on mobile phones, so that Hawkes can monitor Penny-Jane's progress and reassure her if she becomes anxious. After just a few minutes inside the shop Penny-Jane, breathing heavily and feeling numb, can take it no more, and asks the nurse to come and rescue her. What would the levels of MAO B be like in the blood of people like Penny-Jane?

Peter Yu carried out the same test on Penny-Jane – and another twenty-eight people with a similar panic disorder – as he had done on the violent inmates of the nearby penitentiary. As he predicted, the MAO B concentration in the blood of these people was significantly higher than average. Phobias have always been assumed to originate in traumatic events during early childhood – and no doubt many specific ones do. However, here was a biochemical basis for general fearful anxiety, seemingly caused by an excess of the same biochemical as, when in lower concentrations, seemed to be responsible for the risk-taking behaviour of the violent criminals and the matadors. Moreover, since the studies of Yu and Carrasco, other studies have shown that men generally have lower levels of MAO B, which correlates with the fact that there are more risk-taking men, involved in dangerous sports or violent crime, than women. It has also been found that the levels of MAO B generally increase with age, perhaps explaining why people seem to become more cautious and less risk taking as they grow older. There seems to be a link between MAO and risk-taking behaviour. To explain what neuroscientists think that link is, we need to understand the function of MAO in the brain.

Chemicals in the brain

The two forms of MAO (monoamine oxidase) are part of a complex cocktail of biochemicals in our brains. Their job is to break down neurotransmitters – chemicals that play a vital role in the transmission of nerve signals between neurones. A neurone is surrounded by a thin membrane, and is made up of the cell body, a fibre called an axon that carries 'output' signals, and thinner fibres called dendrites, which carry 'input' signals. A neurone is subject to the inputs of hundreds or thousands of dendrites together – each one the signal from a different neurone. It is as if each one is registering a vote – the result of all the votes is either the neurone 'firing' or remaining silent. When a neurone does fire, the signal passes along the length of the axon, and neurotransmitters are released through the cell membrane at the many ends of the axon. The neurotransmitters burst forth from the end of the axon, and attach themselves to the end of a dendrite of another neurone sitting very close by. As they do, they may cause holes to open in the membrane of the second neurone. Positively charged sodium or calcium ions flood in through these holes, initiating another nerve impulse in the second neurone. The gap between one of the ends of an axon and the end of one of the dendrites of another neurone is called a synapse.

It may seem unnecessarily complicated to have this indirect transmission of nerve signals involving neurotransmitters across a synaptic gap – why not just have the nerves connected together directly? There are in fact several reasons why it is important that the neurones do not connect directly to each other, but instead form synapses. One of them is that neurotransmitters may be either 'excitory' or 'inhibitory'. In other words, a signal received at some synapses will give a 'no' vote and at others, it will be a 'yes' vote. This gives the whole system a much greater flexibility. Also, the cocktail of neurochemicals in the synaptic gap can change according to mood, and this results in changes in the transmission of nerve impulses across the synapse.

There are many different neurotransmitters, each one concentrated in particular areas of the brain. For example, dopamine (hydroxytyramine) acts as both an inhibitory and an excitory neurotransmitter, and is concentrated in areas of the brain that are concerned with movement. A deficiency of dopamine can lead to Parkinson's disease. Too much dopamine is associated with schizophrenia. Other neurotransmitters include serotonin (5-hydroxytryptamine), adrenaline (also called epinephrine) and noradrenaline (norepinephrine). All of these neurotransmitters have effects in other parts of the body as well as in the brain. For example, adrenaline is produced by the adrenal glands just above the kidneys as well as at the ends of neurones in the brain. It causes the heart rate, blood pressure and blood sugar level to increase, preparing the body for 'fight or flight'. Adrenaline has a profound effect on mood, and thrill seekers crave it. Serotonin, too, has an important effect on our bodies as well as on our brains. It acts as a vasoconstrictor – a chemical that causes blood vessels to contract, reducing blood flow – but it also affects our mood and, indirectly, our perceptions. An excess of serotonin in the brain can produce nausea and migraines, for example. Having too little serotonin is one cause of depression. Post-mortem investigations of suicide victims and samples of the fluid from the spinal cords of depressed patients have revealed a deficit of serotonin, and the injection into a patient of a substance that breaks down serotonin can actually cause depression.

All the neurotransmitters mentioned are examples of a class of compounds called monoamines, and they have very similar molecular structures. In fact, the monoamines are related to each other in another way: adrenaline is made from noradrenaline, which is made from dopamine. Neurotransmitters are very important in the functioning of the brain, and in particular in making us feel good or not so good. Most illicit as well as medicinal drugs directly affect the production or inhibition of one or more neurotransmitters, or their uptake into the neurones either side of the synapses.

Where does MAO and thrill-seeking behaviour fit into this picture of synapses and neurotransmitters? MAO breaks down the monoamine neurotransmitters mentioned above. Specifically, it removes a part of the neurotransmitter molecules, rendering the molecules inactive. This is particularly important at synapses, where MAO 'mops up' after a nerve impulse has passed through. Both versions of MAO – A and B – are found in some of the cells that fill the spaces between neurones in the brain, the so-called glial (meaning 'glue') cells. From there, they can easily clean up the neurotransmitters released by nearby neurones. It seems that if the brain has too much MAO, the neurotransmitters do not work as effectively as they should, and this can lead to depression or lethargy. Since the late 1950s, drugs that inhibit MAO's activity have been used to treat people suffering from depression. The first 'MAO inhibitor' drug was called iproniazid. This drug was originally tested as a treatment for tuberculosis, but its use for that purpose was stopped when it was shown to produce euphoria in the patients who took it. In 1952, researchers found that iproniazid inhibited the action of MAO, and so they proposed that it could be used to treat depression. Its use began in 1958, but was halted after it was found to cause damage to the liver. However, since then, MAO inhibitors have become a major class of antidepressant drugs, and includes isocarboxazid (sold as Marplan). Most MAO inhibitors are generally used as a second line of attack because MAO inhibitors often have undesired side effects, and the exact mechanism by which they work is not very well understood. The first line of attack is normally to administer tricyclic antidepressants, which include imipramine (sold as Janimine and Tofranil). These work in a similar way to MAO inhibitors, by reducing the inactivation of certain neurotransmitters. The drug Prozac (fluoxetine) works in a different way again. It is one of a class of substances called SSRIs (selective serotonin re-uptake inhibitors), that prevents serotonin from being reabsorbed from the synapses, making them more active. Tranquillizers such as diazepam (Valium) work by enhancing the action of inhibitory neurotransmitters, such as gamma-aminobutyric acid. In this way, tranquillizers help to allow neurones to communicate more effectively.

Just as high concentrations of MAO have been associated with depression, many studies, including those carried out by Yu and Carrasco, have shown that low concentrations of the enzyme are associated with thrill-seeking behaviour. This behaviour crops up as a desire for novelty, stimulation or excitement, in many areas of life other than in dangerous sports. Carrasco himself has carried out other studies involving MAO. In 1994 he and his colleagues conducted a study of compulsive, or pathological, gamblers. He used several measurements of personality, including Zuckerman's Sensation Seeking Scale mentioned earlier, and took measurements of MAO B concentration in blood platelets from pathological gamblers and from members of a control group. As you might expect, the risk-taking, sensation-seeking gamblers had lower concentrations of MAO B in their blood than members of the control group.

Many other studies have highlighted a link between alcohol or drug abuse and low concentrations of MAO B, with care being taken to make sure that the drugs themselves had not affected the production of MAO B. In one study, blood taken from the umbilical cords joining newborn babies to their mothers was measured for MAO content. Babies with low MAO levels were found to be more physically active than those with normal levels. Studies have also been conducted in animals – monkeys and rats with low MAO B levels are more active and playful, and tend to sleep a good deal less. All in all, studies have shown, with varying degrees of certainty, that low concentrations of MAO B are associated with all of the following: sensation seeking, impulsiveness and 'monotony avoidance', job instability, poor academic performance, childhood hyperactivity, recreational drug use (particularly alcohol), defiance of punishment, extroversion, and a preference for highly varied sexual experiences.

The most likely explanation of MAO B's apparent importance in our behaviour involves the enzyme's ability to 'mop up' after neurotransmitters. If there is not enough MAO, the neurotransmitters hang around after they have been used, and can cause a kind of over-stimulation of neurones by allowing too many nerve signals to pass through. This may lead to hyperactivity in the brain, and this can pass on to a person's behaviour. Another possibility is that low levels of MAO are the effect and not the cause. The amount of MAO produced may be an indication of how much neurotransmitter is being produced. In this case, individuals who produce only a little neurotransmitter in a given situation will want to over-stimulate themselves, to give their brain a certain level of satisfaction.

T time

MAO is not the whole story in the quest to understand thrill seekers. Looking at the biochemistry of people whom personality tests highlight as 'type T' thrill seekers provides more clues in the quest to understand why the human brain takes risks. Many of these people have levels of the neurotransmitter dopamine that are higher than average. Dopamine is the 'feel good' neurotransmitter. In 1998 researchers at the University of California, Los Angeles, discovered an important link between thrill-seeking behaviour and dopamine receptors, which are vital in the transmission of nerve impulses across certain synapses in the brain. Dopamine receptors are like slots that dopamine molecules fit into on the receiving side of a synapse. In other words, dopamine molecules released by a nerve impulse that arrives at the first neurone burst across the synapse and attach to the second neurone, as described above. The dopamine receptors are very specific – only dopamine molecules can fit into them. When they do slot in, yet another chain of events occurs that results in tiny holes on the receiving neurone opening up, allowing positive ions to flow in. It is these ions that instigate the nerve impulse in the receiving neurone.

Until the early 1990s, three particular types of dopamine receptor had been identified. The fourth, named D4DR, was discovered in 1991, and a few years later it was found that there are several different versions of the gene responsible for manufacturing this receptor. So in 1995 a team of geneticists based at the Sara Herzog Hospital in Jerusalem and the Ben Gurion University in Beersheva set about investigating this gene. They wanted to find out whether people's personalities might be in part determined by which version of the D4DR gene they possess. In particular, did having one version of this gene make people more sensation seeking?

They selected at random 124 people, and subjected them to personality tests as well as genetic tests to determine which version of the D4DR gene they possessed. They found that people with one version of the gene scored 10 per cent higher in their sensation-seeking scores than those who had the other version. This result has been repeated in several studies in several countries, including the one in Los Angeles, and this genetic basis for at least part of sensation-seeking behaviour is well established. It has been discovered that people with one version of the gene have only half the density of the dopamine receptors in their synapses as people with the other. The team in Los Angeles looked at two different dopamine receptor genes, D2DR and D4DR. They carried out genetic and psychological tests on 119 boys aged twelve years, and found that boys who had a particular version of each of the two genes had significantly higher sensation-seeking scores than the other boys. In addition, they found that boys who had only one of the versions in question had slightly higher sensation-seeking scores than the average, but slightly lower than the boys with the relevant versions of both the genes.

The connection between sensation seeking and dopamine is clear. But so is the link between sensation seeking and addictive behaviour such as alcoholism and drug abuse. Further evidence of this second link comes from experiments on rats at the US National Institute of Drug Abuse (NIDA). In 1995, researchers funded by NIDA and working at the University of Kentucky found that the same 'reward pathways' in the rats' brains were involved in the search for novel experiences as are involved in drug-taking. In the first of two experiments, they found that rats' desire for novelty could be dampened down by administering a drug that affected the action of dopamine receptors in the reward pathways, the main one of which is in the limbic system of the brain. In the second, they showed that when rats were engaged in novel activities, such as investigating previously unexplored chambers in a maze, the same reward pathways were stimulated.

The limbic system is involved in many brain functions, including emotion in humans. It includes brain structures such as the hippocampus and the amygdalae. The fact that the major reward pathways, sometimes referred to as the reward cascade, are located in the limbic system was discovered in the 1950s, by American psychologists James Olds and Peter Milner. In their experiments, Olds and Milner connected the limbic systems of living rats to a low-power source of electricity. The rats could stimulate the electrodes, to deliver a burst of electricity to the limbic system, by pressing a lever. The researchers found that the rats pressed the levers repeatedly and continuously, up to five thousand times every hour. They also found that being allowed to press the lever would be a more than adequate reward to encourage rats to solve problems like finding their way around mazes. Later, Olds went on to perform similar experiments in human subjects, who reported pleasurable, even orgasmic feelings when regions of their limbic systems were stimulated. They felt light-headed and their minds were cleared of any negative thoughts they had prior to stimulation. Alcohol, nicotine and cocaine are known to promote the release of dopamine (and another neurotransmitter called acetylcholine) in the reward pathways of the brain, and this explains why they can become addictive: they excite the reward pathways, and reduce the craving for reward in the human brain. Unfortunately for smokers, cocaine users and alcoholics, the brain 'habituates' to the effects of these substances and, after prolonged use, more and more of these substances is required to produce the same effect.

The convincing evidence that increased dopamine activity and reduced levels of MAO are important factors in thrill-seeking behaviours and in drug addiction or dependency could be worrying. It does not mean that all people who engage in extreme sports are at risk of becoming drug addicts. However, it may well be that extreme sports can be a replacement for drug dependency – perhaps the same urge for thrills can be satisfied in either activity. Could BASE jumping and other extreme sports be an effective way of avoiding drug dependency?

In 1995 the National Institute of Drug Addiction created a novel media campaign that was aimed at young sensation seekers. The campaign was based around a booklet and television adverts. One of the scientists involved in the campaign, Dr Lewis Donohow of the University of Kentucky, said that, because sensation seekers have different perceptions of risk, 'ads using scare tactics may actually enhance the attractiveness of drug use for high sensation seekers'. Particularly worrying for children of alcoholics or drug addicts is the fact that a tendency for alcohol or other drug dependency is at least partly inherited. This may lead some alcoholics to blame their inheritance – 'my genes made me do it' – but that is little comfort in the face of some of the social implications of addiction.

Some studies have shown that intervention in the reward pathways in the brain can be effective in helping alcoholics to stop drinking. When a compound called bromocryptine was administered to alcoholics, it was found in many cases to reduce the craving and make it easier for the alcoholics to kick their habit. Bromocryptine is a dopamine agonist, which means that it can stimulate the same pathways as dopamine itself. According to a report by behavioural geneticist Kenneth Blum and his colleagues, there are about 18 million alcoholics in the USA, 28 million children of alcoholics, 25 million people who are addicted to nicotine and about 6 million cocaine addicts. The report suggests that all these behaviours, and various behavioural disorders such as attention deficit disorder (ADD or ADHD), and perhaps thrill seeking, can all be thought of as one condition: 'reward deficiency syndrome'.

Blum and his colleagues foresee effective prevention and treatment of alcoholism and other 'reward deficiencies', as well as an understanding that these behaviours are not entirely the fault of the individual. This, they say, will remove the social stigma of such conditions. However, this sort of idea could be used to restrict people's freedom. In the future it may be possible to screen babies – or even embryos – for the 'genes for alcoholism'. Alternatively, these genes could even be genetically engineered out of the genome. There is no serious suggestion of doing this, and indeed some of the same biochemistry that can lead to addiction can also lead to thrill-seeking behaviour. And thrill-seeking behaviour is important: the world needs people who are full of energy, who are creative and dynamic. These are the sorts of traits that we find in thrill seekers and risk takers. People who engage in extreme sports have often been labelled as 'having a death wish'. Quoted in the magazine *US News*, Marvin Zuckerman says that this cannot be true: 'If these people were real death wishers, they wouldn't bother with safety precautions. The death wish is a myth made up by those who aren't high-sensation seekers who can't understand the rewards.'

There do seem to be biological causes of sensation-seeking behaviour, involving neurotransmitters such as dopamine and adrenaline, and the enzyme MAO. For people who are predisposed by their genes and their biochemistry to be sensation seekers, the world has become worryingly safe. BASE jumper Elliot certainly thinks so: 'Imagine jumping off a housing tower block ... your toes are two inches over the edge of the roof ... you push out, arch your back, and all of a sudden you start accelerating down the side of this building. You see the lights start to speed up as you fall. All of a sudden – crack – the canopy opens. Once you've landed, you look up at this building and there might be three or four hundred people living there ... content to sit watching television ... I wouldn't want to live if that was me.'

THIN AIR

...in search of oxygen...

Climbing Mount Everest is one of the ultimate human challenges. The mountain's peak sits more than eight kilometres (5.5 miles) above sea level, well above most of the cloud tops. It is cold and remote, and the climb is hard and dangerous. The hardships and the dangers are intensified by the effects of a decrease in oxygen in climbers' blood. On top of Mount Everest, a climber takes in about one-third as much oxygen with each breath as at sea level. Oxygen is needed by all of the tissues of our body, but the organ that uses the most is the brain. What happens to the brain when it receives less oxygen than it needs? In 1997 a team of climbers, led by mountaineer and film maker David Breashears, conducted a special expedition to explore and document the effects of breathing 'thin air'. The expedition was sponsored by the public television science series *NOVA*, produced in the United States by the WGBH Science Unit.

On top of the world

David Breashears had already three times reached the summit of Mount Everest before embarking on this scientific mission. In one of those ascents, in 1996, he made a remarkable film of the mountain using an IMAX camera – which produces a huge, high quality, wide-angle picture – so that others might share some of the experience of being on top of the world. At the same time that Breashears was making his IMAX film, eight climbers died in a single day. The tragedy was caused partly by blizzard conditions, but also partly by the number of people who were on the mountain. Sometimes bottlenecks form at certain parts of the main route up to the summit, due to the large numbers of climbers. In 1993 a total of 129 people made it to the top of Mount Everest – a record that is likely to be broken in the next few years. There is ample evidence of the increasing numbers of visitors to this hostile place: as well as Base Camp, there are discarded or abandoned tent poles, tent pegs and oxygen bottles at some of the four camps where climbers stop for a night or two en route to the summit. And, scattered at various points along the route, there are human remains – when climbers die on the mountain, it is usually not practical to recover the bodies and carry them down for a proper burial.

For every six successful summits of Mount Everest, one person dies. Reduced oxygen supply to the brain can cause confusion and loss of co-ordination. The tragedy that killed eight people in 1996 – along with many of the other deaths on Mount Everest – was probably also due to the effects on the brain of breathing the tenuous atmosphere at high altitude. This made Breashears's expedition and experiment all the more important.

Joining David Breashears on this scientific voyage were Ed Viesturs and David Carter. Viesturs is one of the most experienced mountaineers in the world – before joining this expedition, he had already climbed Mount Everest four times, three of them without supplemental oxygen. David Carter was returning to the mountain after an unsuccessful attempt to climb to the summit six years earlier. Another key member of the team was Jangbu Sherpa, who did not take part in the scientific experiments but headed the team of Sherpas supporting the effort. Jangbu had climbed Mount Everest twice before, including in 1996 when he assisted Breashears's film-making expedition. Also part of the team was Pete Athans, who was to operate a digital video camera and also take tests. At the time of writing, Athans is the only non-Sherpa to reach the peak of Mount Everest six times. Also part of the expedition was Dr Howard Donner, the expedition doctor, who was to remain at Base Camp, 3,490 metres (11,400 feet) below the summit.

At a staggering 8,850 metres, the higher of Mount Everest's two peaks is the highest solid ground above sea level on which you can stand anywhere in the world. When asked what the height of Mount Everest is, many people will quote a different value: 29,028 feet (actually 8,848 metres). This figure is familiar because climbers tend to give values of height in feet, and so the figure of '29,028 feet' has stuck in people's minds since it was announced in 1954 by the Indian Bureau of Measurement. However, those people will now have to learn a new figure: in November 1999 the measurement of 29,035 feet (8,850 metres) was officially recognized by the American National Geographic Society. The new, highly accurate measurement was taken in May 1999, by an American team led by Pete Athans using high-tech GPS (Global Positioning System) satellite navigation equipment. A recent geological study has shown that Everest is increasing by as much as three centimetres per year, due to the same forces that have been forming the mountain for the past sixty million years.

The 'new' height of Mount Everest is not the only thing that may come as a surprise. Although the mountain is the highest point on the Earth above sea level, it is not the furthest from the centre of the Earth. This is because our planet is not perfectly round – it is an oblate spheroid, a slightly flattened sphere. Its diameter at the equator is 42 kilometres (26 miles) greater than its diameter from pole to pole. So the peak of a mountain in Ecuador, Chimborazo, which sits only one degree south of the equator, is actually further from the centre of the Earth, although it is more than eight thousand feet (2,440 metres) shorter than Mount

Everest as measured from sea level. And, perhaps even more surprisingly, Mount Everest is not even the tallest landform as measured from its base to its peak. That honour goes to a huge, snow-capped volcano, Mauna Kea, that rises out of the ocean in Hawaii. If you could move it on to dry land and take it to Nepal, Mauna Kea would stand nearly three thousand feet (900 metres) taller than Mount Everest.

Despite these peculiarities about the height of Mount Everest – known as Chomolungma ('goddess mother of the world') in Tibet and Sagarmatha ('goddess of the sky') in Nepal – it remains the aim of many climbers to conquer this mountain. With its peak well within the 'death zone', above about eight thousand metres (26,000 feet), in which climbers become out of breath just breathing, it makes a perfect natural laboratory in which to study the effects of oxygen deprivation on the brain. The brain constitutes only 2.5 per cent of the body's mass, but uses up 20 per cent of the energy. The source of this energy is the process of aerobic respiration, which involves the reaction of glucose (a type of sugar) in the blood with oxygen.

It's a gas

Oxygen – denoted by the chemical symbol 'O' – is the most abundant element in the Earth's crust. This is because it occurs, combined with other elements, in the rocks that make up the crust, such as limestone (calcium carbonate, $CaCO_3$) and quartz (silicon dioxide, SiO_2). It is also the most abundant element (by weight) in the oceans, lakes and rivers: every drop of water is made of countless oxygen and hydrogen atoms joined together in groups of three called water molecules (H_2O). When pure, it is a gas at room temperature – the oxygen atoms join in pairs to form oxygen molecules (O_2). Oxygen becomes a liquid if cooled to below minus 183 degrees C (minus 297 degrees F). The resulting liquid is pale blue. Cool it still further, to minus 218 degrees C (minus 361 F), and it becomes a bright red solid.

Oxygen was discovered in the 1770s, at a time when several eminent scientists, mostly in Europe, were experimenting with the physical and chemical characteristics of gases. In 1772 the Swedish chemist Karl Scheele produced oxygen gas by heating other substances. And in England two years later, Joseph Priestley discovered oxygen in the same way – he noted that a candle would burn remarkably well in the gas. An existing scientific theory postulated that when matter burns it releases a hypothetical substance called phlogiston. Priestley assumed that the reason the candle burned so well in the newly discovered gas was that it had a great affinity for phlogiston. And he guessed that this might be because the gas had no phlogiston, so he named it 'dephlogisticated air'. It was a French chemist, Antoine-Laurent Lavoisier, who realized the role of this gas in burning and in respiration. When a substance burns, the atoms and molecules of which it is made combine with oxygen – this is the real reason that Priestley's candles burned so well in the pure but as yet unnamed gas.

Scientists before Lavoisier had noticed that most substances lose weight as they burn – candles, for example, grow smaller as they burn. They assumed that the reduction in weight was associated with the release of phlogiston. However, some substances – in particular, metals – actually gain weight when they burn, and this posed a problem for the phlogiston theory. Lavoisier was the first person to realize the importance of oxygen in the burning process, and that in some cases – such as a candle – the products of burning were released into the atmosphere, therefore explaining the loss of weight. In other cases, the oxygen remained, chemically combined with the substance, and this explained the increase in weight. If you burn iron wool, for example, the resulting compound, called iron oxide, is heavier than the original iron wool, due to the oxygen that has combined with the iron. In one stroke, then, Lavoisier had overturned the phlogiston theory, and oxygen had been recognized as a very important element. He realized, too, that water was the product of hydrogen gas burning.

He extended his studies of oxygen to bodily processes: he observed, for example, that birds kept in sealed, oxygen-filled cases lived longer than birds in atmospheric air, which contains less oxygen. Unknown to Lavoisier, the oxygen that the birds inhaled was used in the tissues of the birds' bodies, to release energy, in the process of respiration. In fact, respiration and burning have a lot in common. A burning candle releases heat and light energy, as oxygen (O) from the air combines with the atoms of carbon (C) and hydrogen (H) of which the candle is made, to produce carbon dioxide (CO_2) and water (H_2O). In a similar way, oxygen carried by the blood combines with glucose ($C_6H_{12}O_6$) in a kind of burning reaction, releasing energy for use by the muscles and the brain, and producing water and carbon dioxide as waste products.

The oxygen required for respiration comes from the atmosphere, a little of which you take in every time you inhale. The atmosphere is made up of a mixture of many different types of molecules, all dashing around at high speed, bouncing off each other and anything else that gets in their way. Where air molecules bombard a surface, they exert pressure – a balloon stays inflated because the air inside it is at higher pressure than the air outside it, for example. This is because the collisions of billions of air molecules against the inside of the rubber produce greater force than the collisions of molecules on the outside. And although you cannot feel it, the air at sea level presses against you with a force about equal to the weight of a kilogram bag of sugar on each square centimetre. When you breathe in, your chest expands, reducing the pressure of the air inside your lungs; higher pressure air from outside rushes in. The density of the atmosphere – a measure of the mass of air present in a particular volume – decreases with altitude, too. At sea level, one cubic metre of air has a mass of 1.22 kilograms. At the summit of Mount Everest, the density of the air is less than half a kilogram per cubic metre. The reason for this difference in mass is that the air molecules are more spread out at high altitudes. Air at sea level – at the bottom of the atmosphere – is 'squashed' by the weight of air above it. In each cubic centimetre of air molecules at sea level, there are about twenty-five million million million air molecules. At the top of Mount Everest, however, there are 'only' about nine million million million air molecules in each cubic centimetre. About twenty-one per cent of them are oxygen molecules – this is true at any altitude. You take in about three litres (3,000 cubic centimetres) of air with each breath. So, at the summit of Mount Everest, about five thousand million million million oxygen molecules rush into your lungs each time you inhale.

What happens to oxygen molecules once they are inside your lungs? Most of them are expelled again when you exhale, but some of them make it into the alveoli, tiny spaces in your lungs. If you imagine a lung as an upside-down tree, the alveoli are the leaves. And just as leaves are involved in exchanging gases between a tree and the surrounding air, gas exchange takes place in the alveoli. Molecules of oxygen pass from the air through the very thin membrane of each alveolus, and into the blood. Some of the molecules dissolve directly into the blood plasma, but most of them attach to molecules of haemoglobin, a pigment found in red blood cells. Haemoglobin is blue when it carries no oxygen but becomes red as soon as oxygen molecules – four per haemoglobin molecule – are attached. In fact, red blood cells are red only when they are carrying oxygen. At the same time that oxygen is passing into your blood, carbon dioxide – one of the waste products of the reaction of sugars with oxygen – passes from the blood to the air in the alveoli, and out into the atmosphere when you breathe out. The blood, with a fresh supply of oxygen and having rid itself of much of the carbon dioxide, then passes from the lungs to the heart, from where it is pumped around the body. Very little oxygen is actually stored in the body, so if breathing stops, or there is no oxygen in the air, the result is death within a few minutes.

The importance of oxygen in the brain is highlighted by modern medical scanning techniques called functional MRI (magnetic resonance imaging) and PET (positron emission tomography). In one form of PET, a special glucose solution is injected into a patient. Glucose reacts with blood oxygen to release energy required by living cells. The molecules of glucose in the injected solution are radioactive, and decay by releasing particles called positrons. When these particles collide with electrons in the patient's brain, they release gamma radiation (electromagnetic radiation like X-rays), which is detected by sensors positioned around the body. Computer analysis of what the detectors pick up show where glucose is most concentrated, and so which tissues are demanding the most energy, using the most oxygen. PET brain scans can help to identify which areas of the brain are most involved in certain mental tasks. Functional MRI can also visualize which areas of the brain are demanding the most oxygen: it discriminates between haemoglobin molecules with oxygen and those without. The images created by an MRI scanner show areas with high oxygen content, and therefore where oxygen demands are highest. MRI and PET have begun to unravel the mysteries of what part of the brain does what during different kinds of task.

In short supply

As we have seen, each breath at high altitude takes in less oxygen, at a lower pressure, than each breath at sea level. The first effect of breathing air at altitude is that less oxygen reaches the haemoglobin in the blood. The tissues of the body – including those in the brain – are then supplied with less oxygen than they require, a condition known as hypoxia. The lower pressure of air at altitude means that the air presses less hard against the membranes of the alveoli – this is why less oxygen is absorbed into the blood. High-altitude climbers are well aware of some of the milder physiological effects of hypoxia, which together are known as mountain sickness, or altitude sickness. As mountaineers stay at high altitudes for relatively short periods of time, symptoms are normally short-lived. For that reason, mountain sickness is normally referred to as acute rather than chronic. The symptoms of acute mountain sickness include dizziness, severe headaches, lack of appetite, impaired judgement and even hallucinations. When crossing the dangerous crevasses and climbing up the hazardous icefalls of the world's highest mountains, these symptoms can put climbers at great risk. Many of the deaths of high-altitude climbers are a direct result of the loss of judgement associated with hypoxia; the main rationale behind the study carried out by Breashears and his colleagues was to understand the effects of hypoxia on the brain.

This is not the only study to have investigated these effects. Several others have measured various aspects of cognitive performance in experiments carried out at high altitude, while others have investigated the effects of hypoxia in sea-level experiments. It is possible to induce hypoxia at sea level by supplying someone with air at a lower pressure or air that is deficient in oxygen. Some such studies are carried out to assess the physiological and psychological effects of hypoxia on people who commute to high-altitude areas – most astronomers, for example, work in telescopes situated well above the light-obscuring pollution of cities, high up on mountains, and miners in several locations in South America who live near sea level but work in mines at locations as high as 4,500 metres (14,800 feet). Other studies are conducted with air or space travel in mind – many aeroplanes cruise at altitudes greater than the height of Mount Everest. Of course, aeroplanes are pressurized containers, so pilots and passengers do not breathe air from the atmosphere at that height. However, the air inside the cabin of an aeroplane is not pressurized to simulate the atmosphere at sea level. It is not uncommon for the air inside cabins to be at the same pressure as the atmosphere at an altitude of 2,440 metres (8,000 feet). Most people do not have symptoms of altitude sickness or loss of cognitive function at this height – these signs usually appear higher than about three thousand metres (10,000 feet).

It has been shown that mountain sickness and mental dysfunction due to hypoxia affect different people in different ways, so some people might feel slight effects of hypoxia during an aeroplane flight. If those people were pilots – who rely on quick reactions and clarity of mind – then safety could be compromised, so this kind of research is important. Other research – including that carried out by Breashears and his team – is aimed at examining the effects of hypoxia on mountaineers, basically to understand the relationship between mental dysfunction and climbing tragedies. This is not true of all such studies: in 1993, Philip Lieberman, Professor of Cognitive and Linguistic Science at Brown University, Rhode Island, conducted a study of climbers on Mount Everest to test his theory about the evolution of speech. From the fact that the mental dysfunction of climbers at altitude covers a range of abilities, he concluded that speech involves archaic brain mechanisms, involved with muscle control, as well as mechanisms more recent in evolutionary history, such as the elements of language. In his book *Eve Spoke: Human Language and Human Evolution* Lieberman says that the results of the cognitive tests taken by the climbers at altitude 'suggest a linkage between the neural mechanisms implicated in speech motor control and syntax'. There is more to studying the effects of the air we breathe than you might have thought.

It is only in recent years – with a sophisticated understanding of the processes of respiration – that we have been able to study the effects of hypoxia in a scientific way. But humans have been aware of the effects of breathing at high altitude for at least two thousand years. In about AD20 a Chinese general described the routes taken by travellers in the Karakoram mountain range in China, in which is found the popular peak K2. According to the account, the routes were named according to how they made the travellers feel – there was Mount Greater Headache and Fever Hill, for example. The general's report described how people and some animals experienced feverish symptoms when they were climbing over the higher parts of some of these routes. When hot-air ballooning became popular at the end of the eighteenth century, more people began to fly to increasingly higher altitudes. Balloonists' reports of the strange effects of breathing rarefied (low-density) air were sometimes conflicting. Many balloonists reported suffering from headaches and vomiting – the symptoms of acute mountain sickness – but some actually thought they might gain some benefit from breathing rarefied air. One even went as far as to suggest taking chronically sick people up in balloons to high altitudes, as a form of medical treatment. The idea that the fresh air at high altitudes can have a therapeutic effect persists today: for 5,000 US dollars you can buy the 'Altitude Tent', and benefit from increased production of red blood cells and improved lung performance. And indeed, there is an increase in the production of oxygen-carrying red blood cells and an increase in lung performance when a person breathes in less oxygen than they need. The price tag may seem a little steep, but the manufacturers' claim that it can save mountaineers time and money by allowing them to acclimatize before they arrive at Base Camp.

The first person to investigate scientifically the link between the symptoms of acute mountain sickness and oxygen deprivation was German explorer Alexander von Humboldt. In the first few years of the nineteenth century he climbed several high mountains in South America. His assault on Chimborazo in Ecuador, which fell about three hundred metres (1,000 feet) short of the summit, was a high-altitude climbing record for nearly thirty years. All Humboldt's climbs were achieved without oxygen supplies – pure, bottled oxygen was not available until the early 1900s – and Humboldt described the symptoms now diagnosed as acute mountain sickness. It was not until 1937 that the medical community recognized the condition.

The physiological effects of acute mountain sickness – headaches, nausea and fatigue – are no doubt related to the increase in breathing, blood pressure and heart rate and the change in the levels of various enzymes and hormones that are instigated in climbers when their bodies receive less oxygen than normal. The effects of oxygen deprivation on the brain vary from mild mental dysfunction, through permanent damage, to death. Most human cells can survive without a constant supply of oxygen. For example, muscle cells in a sprinter's legs derive energy temporarily using a reaction that does not involve oxygen. In fact, whenever any muscle cell is working it is always deriving its energy without involving oxygen – the oxygen becomes involved after the muscle stops working. Brain cells cannot release energy this way, and die quickly in the absence of oxygen. Permanent brain damage can result from lack of oxygen at birth for just a few minutes, for example. Strokes are caused largely by a lack of oxygen to the brain. A stroke causes massive damage to the brain by interrupting the blood supply there – because a blood clot blocks an artery, for instance. The symptoms exhibited by stroke victims can include lack of co-ordination or even paralysis of an area of the body (or the mind) associated with the part of the brain that is damaged. The lack of oxygen due to the interruption of blood supply is a major cause of these symptoms, as large numbers of neurones may die. While a near-total loss of oxygen supply to the brain can result in permanent or semi-permanent brain damage, what might be the effects of the slight reduction in supply experienced by climbers at altitude? If oxygen supply to the whole brain is reduced, you would expect a general reduction in a person's mental faculties. This would perhaps manifest itself as a general, but temporary, loss of co-ordination, memory, comprehension and language capability – all of which are very important to mountaineers at high altitudes. Measuring and cataloguing these effects will improve safety for climbers – that is the aim of studies such as the one carried out by Breashears and his team.

While breathing less oxygen than you need is harmful, breathing pure oxygen or air at high pressure often has therapeutic effects. Oxygen tents are transportable cabinets in which oxygen-enriched air circulates. Patients suffering from respiratory complaints or heart disease may spend time in these cabinets, which are also controlled for temperature and humidity. Between 30 and 50 per cent of the air inside the tent is oxygen gas – much greater than in atmospheric air. An incubator is very similar, and is used to house babies born prematurely so that they have a greater chance of surviving and avoiding brain damage or other disorders. And another similar idea is the hyperbaric chamber – a large tank filled with air at a pressure higher than in the atmosphere at sea level. The oxygen molecules that rush into the lungs when you breathe such air are pushed through the membranes in the lungs' alveoli more forcefully than in normal air. There are many uses of hyperbaric chambers, including treatment for decompression illness. This dangerous condition occurs when a person who has been in a high-pressure environment – such as deep under the ocean – moves too quickly to a lower-pressure environment. What happens under high pressure is that gases are forced to dissolve into the blood and then into the body's tissues. Moving to lower pressure slowly allows these gases to pass back into the blood and then out into the lungs, causing no problems. If decompression is too quick, however, the gases do not have time to pass into the blood, and instead they form dangerous bubbles in the tissues. The most problematic gas is nitrogen, since it constitutes more than seventy-five per cent of the atmosphere and is not used up by the body's tissues, as oxygen is when it reacts with glucose to produce energy. When a deep-sea diver comes to the surface too quickly, enough nitrogen forms tiny bubbles in various tissues to cause major problems. A diver with decompression illness is often described as having 'the bends', because the bubbles of nitrogen under the skin can cause swelling that means that joints cannot be bent or unbent. Bubbles that form in the brain or the spinal cord can cause serious disorientation and paralysis or loss of co-ordination.

Hyperbaric chambers can also help babies born with heart defects. Such babies can be placed in hyperbaric chambers before an operation, which may give them a better chance of survival and recovery. Hyperbaric chambers that contain pure high-pressure oxygen also have an additional range of uses. Some harmful bacteria, inside or on the surface of the body, cannot live in high concentrations of oxygen, for example, and oxygen is sometimes used as a fast healer. Breathing inside such a chamber, you take in more oxygen than normal, and at higher pressure. The result is that the concentration of oxygen dissolved in the blood plasma rises dramatically – by as much as two thousand per cent. This does not mean that the total amount of oxygen in the blood increases by that amount – most of the oxygen carried by the blood is attached to haemoglobin molecules in the red blood cells. One cannot force more than four molecules of oxygen on to each haemoglobin molecule, and increasing the atmospheric pressure will do nothing to increase the numbers of red blood cells. However, the extra oxygen dissolved in the plasma will be delivered to the various tissues of the body – what might be the benefits of this? Some people use hyperbaric oxygen chambers to treat stroke, coma and traumatic brain injuries. The idea is that the extra oxygen encourages the regeneration of neurones in the damaged brain – though not everyone believes that this approach actually works. Another controversial application of hyperbaric oxygen tanks is to help people lose weight. Here, the idea is that increased oxygen in the blood will increase metabolic rate, to 'burn calories'. Other examples are 'smart pills' and 'smart drinks'. Their manufacturers claim that they enhance mental performance because they contain large amounts of oxygen, or increase oxygen delivery to brain cells. And there are increasingly popular 'oxygen bars', where people socialize and pay to inhale pure oxygen.

Testing time

So much for high-pressure, high-density air, oxygen bars and smart pills. In order to investigate how the effects of too little oxygen relate to altitude and to oxygen concentration in climbers' blood, Breashears and his team were to conduct a number of physiological and mental tests. They would carry out these tests at sea level, at several different elevations on their way up to the summit of Mount Everest, and – if they made it to the top – at the summit itself. The tests at sea level were baseline tests: the results of these would be used as points of reference against which to compare the results of the tests taken at altitude. The baseline tests were carried out at the University of Washington Harbourview Medical Centre in Seattle. There, Dr Brownie Schoene first subjected Breashears, Viesturs and Carter to MRI brain scans, which also allowed the doctor to anticipate some of the effects that being at high altitudes might have on the climbers' brains. First, Schoene carried out a scan on each member of the climbing team while they were resting – simply lying still on the table, inside the scanner's strong magnetic field. Next, he scanned them again while the climbers were over-breathing – hyperventilating. Each climber had to lie down in the brain scanner and purposely hyperventilate for twenty minutes. Breathing at altitude would be laboured and rapid like this, so this second set of scans would show what chemical changes might take place in the brain at altitude. The scans also enabled Schoene to measure the volumes of the climbers' brains – there is some evidence that repeated exposure to high-altitude can actually reduce the size of the brain slightly. Another set of MRI scans would be carried out after the expedition, allowing Schoene and physiologist Peter Hackett to check whether the climbers' brains did shrink as a result of their high-altitude experiences.

After the MRI scans, the climbers underwent other physiological tests, including spirometry – measurement of lung capacity – and measurement of HVR (hypoxic ventilatory response), which is a gauge of how well and how quickly the body adjusts to breathing in less oxygen than normal. The climbers' HVRs were measured by testing the blood for 'oxygen saturation' and by monitoring the heart rate and breathing rate. At sea level, your blood is normally totally saturated with oxygen, meaning that the haemoglobin molecules in your red blood cells are carrying as much oxygen as they can. The oxygen saturation is measured as a percentage and depends upon the concentration of oxygen in the plasma of the blood. The higher the concentration of oxygen in the plasma, the more oxygen is taken up by the haemoglobin in the red blood cells. As you would expect, oxygen saturation falls as people breathe at altitude; this is because the lower atmospheric pressure there means that less oxygen dissolves in the plasma. During the tests for HVR, each climber was made to breathe through a mouthpiece – the air supply was gradually changed to simulate the air they would breathe at high altitude. The climbers' oxygen saturation levels dropped to around seventy per cent, and heart rate rose from around sixty-five to over 100. This is a normal response to breathing air with lower-than-normal pressure and density. Each climber was also tested to gauge the maximum amount of oxygen his body used when working flat out. On a treadmill, the climbers breathed through a mouthpiece that supplied them with normal sea-level air and measured the amount of oxygen their bodies demanded. All the climbers were found to be in excellent condition, ready for the physical challenges that awaited them on the mountain.

Psychometric tests were also conducted at the university, and would be repeated at altitude. These tests would not be able to come to any truly scientific conclusions about what hypoxia can do to the brain: the effects of general fatigue, lack of sleep, diet and climate, for example, would also probably affect the test results. However, the test results would add valuable data to existing findings about the effects of high altitudes on cognitive function. Psychometrics expert Gail Rosenbaum collected together standard tests that would give an indication of how clearly and quickly the climbers were thinking. During the baseline test session, she commented, 'What we will expect to see is that there is a slowing of speech; there'll be a slowing in reaction time; we may see a lot of mis-speaks; they won't say things quite the way they would at sea level; there'll be a lot of slurring and hesitation.'

There were five different types of test, each measuring a slightly different aspect of cognitive function. One test measured the climbers' attention: the climbers were asked to listen to a series of high and low tones, and state how many low tones were present in each case, ignoring the high tones. Another well-known psychometric test, called the Stroop test, measured flexibility in the climbers' thinking by investigating the ability to inhibit one of two responses. In the Stroop test, the person being tested looks at words printed in different colours, and has to say out loud the colour of the ink in each case. Each word is actually the name of a colour, but one that is different from the colour in which the word is printed – so, for example, the person might read the word 'RED' printed in green ink, and be asked what colour the ink is. The person's brain becomes conscious of two conflicting items – the colour of the ink and the word itself – but would be asked to state just one. Success in this task requires the person to choose the desired item and discard the other. 'That'll be fun at altitude,' David Carter joked as he took the Stroop test.

Some PET brain studies have shown that an area of the brain between the two cerebral hemispheres, called the anterior cingulate, demands energy during the Stroop test. If this region does not receive the oxygen it requires when this test is being taken, at high altitudes, perhaps test performance will be impaired. The third type of test comprised sets of simple true-or-false questions. This would provide a way of measuring mental speed. Statements such as 'Roast beef can be eaten' and 'Squirrels are usually sold in pairs' were included – they are very simple, so the time taken to give the answer is then a measure of mental speed, not of the level of cognitive ability. Rosenbaum expected that the climbers would take longer to answer at altitude. The fourth type of test – simple verbal puzzles – was also used to measure mental speed, but also to test verbal reasoning ability. Each puzzle took the form of a statement followed by a question – for example 'Charles beats David in a sprint; which man is faster?' The final type of test administered both at baseline and on the mountain was based on short-term memory and attention. The climbers were asked to listen to and then repeat sentences, such as 'Mike walked around the block three times until he had the nerve to knock on Carol's door'. Helping to conduct the physiological and psychological baseline tests was Dr Howard Donner, who would be going to Nepal with the climbers, as the expedition physician. From Base Camp, he was to keep in regular contact with the climbers, to record their blood saturation levels and heart rates and to conduct the psychometric tests.

To Base Camp

Early in April 1997 the team set off for the Himalayan mountains. The medical and scientific team was to stay at Base Camp, at an altitude of 5,360 metres (17,600 feet). In the early expeditions to climb Mount Everest, climbers had to trek for weeks just to reach Base Camp before they started their assault proper on the mountain. Nowadays, climbers are generally flown by helicopter from Kathmandu to Lukla, at an elevation of 2,440 metres (8,000 feet). The journey from there to Base Camp takes about ten days. Mount Everest is shaped somewhat like a pyramid, with three distinct faces – the north face, the east face and the south-west face. Most Everest expeditions approach the summit via the south-east ridge. There are three ridges that lead to the north face: the north-east ridge, the west ridge and the south-east ridge. David Breashears and his team elected to use the south-east ridge, which is the one most often used because it is the least precarious. Having said that, the ridge is still extremely dangerous, as it falls away sharply 3,000 metres (10,000 feet) on either side. Many climbers have died at this point of the climb. The danger is exacerbated, of course, by the effects of acute mountain sickness – the ridge is only 100 metres (330 feet) below the summit.

On the way up to the south-east ridge climbers generally stop at four camps, usually spending a few nights at each. This is so that they can acclimatize to the increase in altitude on moving from one camp to the next. The journey from Base Camp to Camp I – up through a treacherous icefall – is the most dangerous part of the climb. Camp I is situated on the gradual slope of a long glacier called the Western Cwm. From Camp I Sherpas ferry supplies to the higher camps. Camp II, at the top of the Western Cwm, is perhaps the most important of the camps. From here, climbers ascend one of the faces of a mountain called Lhotse – Mount Everest's next-door neighbour – before crossing the south-east ridge. Camps III and IV are situated at either side of the Lhotse Face. Camp IV – called the South Col – lies at the foot of the south-east ridge, and it is from there that climbers mount their final assault on the summit.

The route taken by most climbers up the mountain today – including Breashears and his team – is the same as that taken by the first people to make it to the summit, New Zealand climber Edmund Hillary and Sherpa Tenzing Norgay. Their historic journey was completed in May 1953, but several expeditions during the 1920s came very close to reaching the summit. Perhaps the most famous of these was the 1924 expedition in which the celebrated British climber George Mallory and his colleague Sandy Irvine tragically lost their lives. Mallory's body was discovered in May 1999 by a team of climbers, on a special recovery mission. The body was found jutting out of the snow below the north-east ridge, just 600 metres (2,000 feet) short of the summit. During that fateful mission, eight climbing assistants, mainly Sherpas, died during an avalanche.

Despite the obvious risks to which they put themselves, Sherpas have been an important part of every expedition on Mount Everest. The Sherpa people live both in Nepal and India, and number about a hundred and twenty thousand in total. About three thousand of these live in the Khumbu Valley in Nepal, which forms much of the approach route to Everest Base Camp. The Sherpa traditionally used yaks to carry goods such as wool for trade, and yaks are used to help carry much of the huge consignment of supplies required by Mount Everest expeditions from Lukla to Base Camp. The tented village that is Base Camp typically houses between three and four hundred people, comprising climbing parties, their guides and Sherpa support teams. Breashears and his team had a tent dedicated to the radio and scientific equipment they needed to record the results of the physiological and psychometric tests that the climbers would be undertaking.

Even at Base Camp, the dramatic physiological effects of being at high altitude could be felt. The expedition climbers – and the medical and scientific team – all experienced headaches and fatigue, and their blood oxygen saturation level was much lower than at sea level, at about seventy-five per cent. Oxygen saturations were measured using a hand-held device called a pulse oxymeter, which fits over a finger and gives an instant reading. It works by shining red and infrared light through the finger and detecting what proportion of each passes through. If only red light were used, then the device would indicate only the amount of oxygen-carrying haemoglobin present. By comparing the amounts of red and infrared light that pass through the finger, a pulse oxymeter is able to calculate the ratio of oxygenated to deoxygenated haemoglobin, and so the percentage of oxygen saturation.

The actual amount of haemoglobin present in the blood changes with altitude: the body increases the rate at which it manufactures haemoglobin-carrying red blood cells as a result of acclimatization to high altitudes. If you were placed from sea level on to the summit of Mount Everest, giving the body no chance to acclimatize, you would quickly die. Because air pressure drops as you climb to altitude, the force pushing oxygen through the membranes of the lungs is reduced. For this reason, less oxygen dissolves in the blood plasma running through the lungs, and this results in less oxygen being taken up by the red blood cells. To offset this, the body steps up its rate of manufacture of red blood cells. There are other, more subtle changes that take place in a body that is acclimatizing, but the increase in red blood cell production is the most important. While the extra red blood cells allow enough oxygen to be carried to where it is needed, they can actually be a problem, too. The extra volume of red blood cells thickens the blood, slowing the flow and increasing the risk of the blood clotting.

The process of acclimatization is initiated by small organs located in the neck, next to the carotid artery that conducts blood to the head. These organs, carotid and aortic bodies, contain clumps of nerve cells that respond to changes in the blood's pH (how acid or alkaline it is) as well as the concentration of dissolved oxygen. Blood's pH is related to the amount of carbon dioxide present – carbon dioxide gas dissolves in water (and therefore blood) to produce a solution that is slightly acidic. The function of the carotid and aortic bodies is not well understood, but they seem to initiate a chain of events that increases the blood's oxygen saturation, which helps to rid the body of carbon dioxide. You can think of these organs as hypoxia sensors. They are connected to a part of the brain called the medulla, one of the roles of which is to control breathing rate. Doctors sometimes administer to certain patients a mixture of oxygen and carbon dioxide – the carbon dioxide causes the medulla to increase breathing rate, so that more of the oxygen is taken in. The carotid bodies are also thought to release a substance that stimulates an increase in levels of a hormone called erythropoietin. It is an increase in this hormone – itself due to the release of yet another hormone by the kidneys – that causes the bone marrow to increase the rate at which it manufactures red blood cells. So when a climber first arrives at high altitude, and becomes hypoxic, his or her body responds first by increasing the breathing rate, and then by increasing the number of red blood cells in the blood. There is also some evidence of a thinning of the membranes of the alveoli in the lungs.

All these changes act to increase the potential of the blood to carry oxygen, to compensate for the reduction in oxygen in the atmosphere. People who live at high altitudes have higher red blood cell counts than those who live at lower altitudes. Strangely, this seems to change over time – almost as if a person's body becomes tired of being acclimatized twenty-four hours of every day. The 'chronic mountain sickness' that ensues is sometimes called Monge's disease, after the Peruvian doctor who identified the condition in the 1920s. It has been studied in a number of research projects, including one carried out on a population of Peruvian people who live in and around the highest city in the world, Cerro de Pasco, which is 4,340 metres (14,200 feet) above sea level. The people who had lived there the longest tended to suffer the most. Yaks and other animals that live at high altitude are born better able to cope with lower atmospheric pressure and density – they are adapted, not acclimatized, to high altitudes.

During the process of acclimatization, symptoms of mild and moderate acute mountain sickness – such as a throbbing headache, nausea and vomiting and debilitating fatigue – are common. These are largely due to the brain actually swelling inside the skull at high altitude. The swelling is caused by plasma fluid leaking out of blood vessels inside the head, and puts pressure on the brain, as well as displacing some of the cerebrospinal fluid – in which the brain is normally bathed – down into the spinal cord. This may explain some of the impairment in cognitive performance as well as the headaches. If the body does not fully acclimatize to each increase in elevation, for example, if it is not given enough time to do so, then a climber may suffer from more debilitating or even life-threatening problems. The most severe problems are caused by fluid leaking from capillaries into certain tissues of the body. When fluid – blood plasma with some red blood cells – leaks into a climber's lungs, the climber will develop a rasping and incessant cough, and the struggle for air will increase with each breath. This condition is known as high-altitude pulmonary oedema. Peter Hackett, the physiologist involved in the expedition, explains that 'the air sacs start to fill up ... the person starts coughing a pink frothy sputum, can't get any air at all; they go into cardiovascular collapse and die'.

Another life-threatening condition is high-altitude cerebral oedema, which is an extreme form of acute mountain sickness. In this distressing illness, where the fluid build-up in the brain has increased to dangerous levels, disorientation and general mental dysfunction can give way to psychotic behaviour and coma. Sometimes people hallucinate. Both cerebral and pulmonary oedema can be fatal if not treated quickly and appropriately. The ideal treatment is rapid descent by a few hundred metres to air with higher pressure and density. In situations where the patient is unable to walk or be transported easily, a clever but simple invention can save his or her life by simulating just such a rapid descent. This invention is the Gamow bag: a sealable, body-sized plastic bag, which acts like a portable emergency hyperbaric oxygen chamber. To simulate lower altitudes, the seriously hypoxic patient is sealed inside the bag, and a foot pump is used to increase the air pressure – and therefore the oxygen pressure – inside the bag. After an hour or two inside a Gamow bag, the patient is usually well enough to walk down to lower altitudes, where he or she can recover.

Move on up

After a few days of acclimatizing at Base Camp, the climbers set off up the Khumbu Icefall – the most treacherous part of the climb. This involved crossing crevasses on metal ladders and climbing steep walls of ice. They were heading for Camp I, which is at an altitude of 5,940 metres (19,500 feet). The icefall is filled with huge chunks of ice, each the size of a house, and these chunks can move without warning, sometimes taking climbers with them. On arrival at Camp I, the climbers measured their pulse and their oxygen saturation levels with the pulse oxymeter, and took the psychometric tests. The saturation levels were as expected for this altitude. David Breashears radioed the team at Base Camp and reported an oxygen saturation of

80 per cent and a pulse of about 78. In fact, as he was radioing and being filmed, he also reported 'test anxiety', which was causing his pulse to race up to 104. Test anxiety may not be the only thing that seemed to affect the results. Perhaps due to fatigue, David Carter gave the wrong answer to a true-or-false question, announcing that the statement 'Lion is a military title' was true. Even taking into account test anxiety and fatigue, the climbers did seem to be thinking more slowly, and their judgement seemed to be impaired.

More tests would have to be carried out, at various altitudes, to collect enough data to be able to come to any useful conclusions about reduction in blood oxygen and mental dysfunction. After a few days at Camp I, the climbers descended to the 'showers, oxygen, warmth and cotton clothing' at Base Camp, where they would recuperate before ascending again. On their way down, they were caught at a bottleneck, where climbers can descend only one at a time. Talking from the top of the Khumbu Icefall, Breashears joked: 'It's amazing – we're on Everest and we're waiting in line.'

The team were to spend more than two months on the mountain, gradually working their way up to the summit, taking time and care to acclimatize and rest between each stage. Their schedule is typical of expeditions to conquer Mount Everest:

UP from Base Camp to Camp I
DOWN from Camp I to Base Camp
UP from Base Camp to Camp I
UP from Camp I to Camp II
DOWN from Camp II to Base Camp
UP from Base Camp to Camp II
UP from Camp II to Camp III
DOWN to Base Camp
UP through Camps I, II, III, to Camp IV
UP to the summit.

The climbers could stop for a few days at Base Camp and at Camps I and II. But they could stay just two nights at Camp III and only overnight at Camp IV, which are in the 'death zone' where humans can be only transient visitors. The reason for this is that where the air is very thin, and oxygen intake is much lower than normal, a climber's body rapidly deteriorates, the muscles waste away and consume themselves just to survive. There is a maximum rate at which a person's heart can beat. This is generally well above the heart rate at rest, so there is scope for an increase in heart rate demanded by physical exertion. However, as we have seen, heart rate at rest increases with altitude, and this leaves less scope for an increase during exertion. Eventually, above about 7,920 metres (26,000 feet), the heart rate at rest becomes very close to the maximum heart rate. Acclimatization can offset this effect only slightly, and climbers without breathing apparatus are in great danger in the death zone. In fact, physician Peter Hackett explains that, in the death zone, 'acclimatization is essentially impossible'. Add to this the fact that, at this altitude, it is impossible for helicopters to reach stranded climbers because the air is so rarefied, and you can see why this is called the death zone. Breashears and his team were carrying cylinders containing compressed oxygen with face masks, for use at and just below the summit. A few people have climbed to the summit of Mount Everest without using supplemental oxygen, including Ed Viesturs on two occasions. The first people to do so were Reinhold Messner and Peter Habeler in 1978.

After their second ascent to Camp I, and the subsequent ascent to Camp II and descent to Base Camp, the climbers made their way up to Camps II and III, and then down to Base Camp again. At Base Camp, a sick Sherpa was being treated by the doctor of a Malaysian expedition. The Sherpa had been found very ill: according to the Malaysian doctor, he was 'ashen grey, breathless' and his oxygen saturation level was a dangerous 20 per cent. He was given pure oxygen, which boosted his oxygen saturation to 80 per cent, and, after spending two hours in a Gamow bag, he was well enough to be flown down to Kathmandu in a helicopter. Another Sherpa, found lying in the middle of the trail below Base Camp, was not so lucky. A Canadian doctor who attended him diagnosed him as suffering from extreme pulmonary oedema – he was 'drowning in his own secretions', and he died soon after he was discovered.

Breashears and his team prepared themselves for the final ascent, which they hoped would take them up to the summit. As they made their way up the Khumbu Icefall, they came across some human remains – the bones of a foot still in a climbing shoe. According to Howard Donner, 'this sort of stuff is spilling out all the time'. The climbers bypassed Camp I, pushed on up the Western Cwm, and stayed a few nights at Camp II, known as Advance Base Camp, at an altitude of 7,040 metres (23,100 feet). The climbers' oxygen saturation levels and pulse rates changed as you would expect with the increase in altitude. The psychometric tests, too, showed that the climbers' brains were taking longer to process information, and were prone to making more mistakes. Over the radio, Ed Viesturs hears the question: 'If Daphne walks twice as fast as Margaret, and they are the only two people in the race, who is most likely to finish last?' Viesturs hesitated a little longer than he would have done at sea level, but he answered correctly and confidently. By this stage of the expedition, David Carter had developed what he described as 'a real violent cough; it comes from deep within ... my main concern is that when I get higher, the cough will get worse, and I'm worried about breaking a rib or vomiting or something'. He had conquered a head cold that had troubled him earlier on in the expedition, but his health was still a real worry. Despite this, he was determined to make it to the summit this time, after his failure to do so in 1991. From Camp II, at the bottom of the

Lhotse Face, the climbers made an arduous ascent up to Camp III, where they would stay for two days and nights before moving on towards Camp IV and the summit.

At Camp III – at an elevation of 7,500 metres (24,600 feet) – the pulse oxymeter readings showed that the high altitudes were putting the climbers under increasing physical stress. David Carter, in poor health, had a pulse of 140 beats per minute and an oxygen saturation level of just 60 per cent. As he was approaching Camp III, David Breashears asked him about how the climb was going. Carter replied, 'It's tiring, it's slow; you're winded, you're dehydrated, you're losing your voice, you're coughing; but the views make it worth it.' The psychometric tests at Camp III indicated that the climbers' brains were definitely beginning to slow down, and their confusion and lack of attention was evident. Ed Viesturs was asked to repeat the following sentence over the radio: 'The video camera captured the bank robber's daring daylight robbery of the First Avenue Bank.' He said, with breathless hesitation: 'The video camera captured the daring bank robber's robbery of the First National Bank.' When repeating similar sentences at lower altitudes, Viesturs had done much better than this, and without hesitation. Similarly, David Breashears was asked to repeat this sentence: 'The action of the brave cyclist kept the small boy from being hit by a ten-tonne truck.' His attempt was: 'The action of the brave cyclist ... um ... helped ... uh ... save the boy ... let's see, I know I have to repeat all I know ... prevent the boy being hit by the ten-tonne truck – oh shit.' The poor performance of the climbers' brains at this altitude – no doubt caused by a combination of hypoxia and fatigue – put them at risk of serious accident or death. Camp III is situated on a 45 degree slope of ice and snow and many climbers have died there as a result of just one poorly placed step.

After a night at Camp III, the climbers pushed on towards Camp IV, called the South Col, situated at the top of the Lhotse Face. The climbers had taken two months to reach this point. Breathing supplemental oxygen through a mouthpiece attached to the nozzle of a canister in their rucksacks, each climber's oxymeter reading increased. But the climbers were still out of breath as they bravely continued with the psychometric tests. Their performance was slower still, and they made more errors than before in repeating sentences. Over the radio from Base Camp, Dr Howard Donner asked David Carter to repeat the following: 'Ed lived by the river for twenty years and only twice before in all those years had it been this high.' Carter replied breathlessly, and with a few errors: 'Ed lived by the river for twenty years and this was the first time it had ever been this high.'

Climbers at this stage of the ascent of Mount Everest, with or without supplemental oxygen, are extremely tired, and it is exhausting just gasping for breath. After just half a night at Camp IV, Breashears and his team set off for the summit. Peter Hackett, expedition physiologist, explained the magnitude of this task: 'When one considers the condition that a climber is in at the South Col on summit day, it's really amazing that they can reach the summit at all. First of all, there hasn't been sleep for usually a couple of nights; there hasn't been enough to eat or drink; even if they've been on oxygen, it's still been very uncomfortable to breathe. The mucous membranes are all dried out – there's always a sore throat, there's always a cough, there's often a headache. It takes a tremendous amount of will to keep going in these conditions.'

Peak performance

The team decided to make their move for the summit in the late evening, and climb during the night. This was so that they would not get caught up with other climbers making the same journey. So at ten o'clock on the evening of 22 May, the team left Camp IV, heading for the summit. The final stages of the climb involved a difficult trek along the treacherous, knife-edge, south-east ridge that rises fewer than a hundred metres (300 feet), to a point just beneath the summit. To get to the summit, the climbers had to conquer one more hurdle – the Hillary Step, a 12-metre (40-foot) face of rock and ice. This they did and, exhausted but jubilant, every member of the team made it to the summit at around six o'clock in the morning. Breashears was the first to pull himself up on to the summit itself, and he erected the Tibetan flag in the snow. The atmospheric pressure at the summit is about thirty-four per cent of the atmospheric pressure at sea level – this is about as high as people can go without supplemental oxygen or being in a pressurized capsule, such as an aeroplane fuselage. The views from this, the highest point on Earth, were incredible.

David Breashears announced on the radio, to Howard Donner at Base Camp, that he was 'on top of the world', and he and Ed Viesturs took their final physiological and psychometric tests of the expedition. Their oxygen saturation levels were between 70 and 80 per cent – higher than they had been for much of the expedition because of the pure oxygen they had been breathing. Donner asked Breashears to repeat the sentence 'Mike walked around the block three times until he had the nerve to knock on Carol's door.' Even on the top of Mount Everest, after the weeks of physical and mental effort that it took to get there, Breashears answered with but one error: he said '... walk on Carol's door'. Ed Viesturs was asked: 'A man who is an engineer came to the store where Alice worked to buy pastries; who bought pastries?' After a long, anguished hesitation, he replied: 'The man ... Jack ... the engineer.' He giggled, clearly confused.

After his tests, Breashears described to Donner what he had seen as he passed the site of the previous year's tragedy, at the Hillary Step. He explained, 'All the bodies that were there last year were covered, but unfortunately we did pass one body right on the fixed rope; it only makes me question my sanity, why I climbed this mountain again, because it is dangerous and cold.' David Carter could not take the psychometric tests at the summit because he had lost his voice. The climbers left the summit and, after five hours of more hard work, made it down to Camp IV. Many of the 150 or so people who have died on Mount Everest have perished on their way down from the summit – the relief of making it to the top causes many climbers to lose concentration.

Back at Camp IV, Carter's condition became worse still. His oxygen saturation was a healthy 93 per cent, with the supplemental oxygen, but he could not breathe at all well and was suffering badly. Ed Viesturs attempted to take him down to much lower altitudes, but because of Carter's condition the two had to stop every few metres so that Carter could catch his breath. In fact, they made it only as far as Camp III. That evening, Viesturs made an emergency call down to Base Camp, shouting, 'David's dying!' Dr Donner was on hand to help him to try to save Carter's life. Carter, already dangerously short of breath, was choking on something that he had coughed up. Viesturs performed the Heimlich manoeuvre several times on him. This involves squeezing the patient hard, from behind, around the abdomen, forcing a sudden burst of air up the windpipe to dislodge any blockage. It saved Carter's life. In the morning, Carter was a lot better, and he was able to make it down the mountain to Base Camp.

All safely back at Base Camp, the climbers and the medical and scientific team were soon on their way back home. Ten days later, the team met back in Seattle, at the University of Washington, to assess the pulse oxymeter measurements and the results of the psychometric tests. The climbers also went back into the MRI scanner, so that any slight damage to the brain or change in brain size during the mission might be detected by comparing the post-expedition scans with those taken before it. As expected, the oxygen saturation levels were generally lower at higher altitudes, apart from when supplemental oxygen was used. It was interesting to note that David Carter's oxygen saturation levels were consistently lower than those of the other climbers – this presumably had something to do with his sickness on the mountain. The results of the psychometric tests also seemed to be dependent on oxygen levels in the climbers' blood, as you would expect. The results of the MRI scans were less clear – more analysis and further tests will need to be done to ascertain whether prolonged reduction in blood oxygen due to altitude leads to any temporary or permanent brain damage. The only damage the MRI scans picked up were, according to Dr Hackett, 'a very mild atrophy in the brain of Ed Viesturs, who was the one that had climbed many times to high altitude without supplemental oxygen. And what we'd like to do is follow him over a longer period of time to see if this is something that might actually progress with his high-altitude career.' On reviewing the psychometric test results, which illustrated the slowing of mental functioning, the expedition's psychometrist, Gail Rosenbaum, told the team: 'The question we ask is, "What happens if you are in an emergency situation – are you able to think quickly and clearly about what you need to do to survive?"'

As more and more people want to scale the heights of the world's great mountains, it becomes ever more important that scientists and mountaineers understand the effects of hypoxia. Eight people died in that single incident in 1996, the year before Breashears and his team carried out the expedition detailed above. And as the 1997 expedition was ending, there was a similar tragedy on the other side of the mountain. Shortly after the expedition was over, in a radio interview on WGBH, the radio station that followed the expedition, David Breashears spoke about the worries he had about the increasing numbers of people climbing Mount Everest. 'And despite what happened last year, not a lot of lessons have been learned. People will continue to come here with great hopes and dreams, and some of them will make it and some of them will die. And that's the nature of climbing on the highest mountain in the world.'

LIES AND DELUSIONS

...searching for the self...

So far, we have looked at some of the remarkable discoveries being made in the quest to understand the brain and how it works. One of the most powerful tools in this quest is the study of what happens when something goes wrong. This last chapter focuses on damage sustained by the cerebrum, the most prominent and the most complex part of the brain. The walnut-shaped cerebrum – which consists of two large hemispheres – takes up most of the volume and most of the mass of the brain. It is thought to be responsible for some of the higher brain functions, so it will come as no surprise that damage to the cerebrum will cause the loss of some of our most remarkable behaviours or abilities. We saw in the first chapter 'Mind Readers' how damage to the orbito-frontal cortex can cause a dramatic shift in a person's personality – changing a calm, thoughtful and sensitive person into a selfish, bad-tempered bore. The cases described in this last chapter explore other changes that can occur in an individual's beliefs, behaviour, physical abilities or perception as a result of damage to the cerebrum. The resulting neurological disorders are fascinating, though sometimes disturbing, and they can be very revealing about how the undamaged cerebrum works. The study of the effects of damage to the cerebrum may also help to realize one of the ambitions of the study of the brain: to work out how your brain creates a sense of your 'self'.

Convoluted story

The role of the cerebrum in producing a sense of self was illustrated in 'Phantom Brains', where we discovered that there is a sensory map of the body on the outer surface of the cerebrum, the cortex. The deep wrinkles of this incredibly complex sheet of brain tissue mean that it can have a surface area almost as large as a pillowcase and still fit around a brain not much larger than a fist. There are many millions of neurones in the cortex, connected by literally thousands of kilometres of fibres. This amazing place is alive with waves of electricity coursing along the fibres that interconnect the neurones. It is here that much of our perception and learning take place, and it seems that it is here that we initiate the conscious actions that make up our behaviour.

Just what kind of disruption a person suffers to his or her mental faculties as a result of damage to the cerebrum depends upon where the damage is sustained, and what type of cortex is damaged. There are three different types of cortex: sensory, motor and association. The last of these generally lies between areas of the other two, and there is a flow of information – from the sensory, through the association, to the motor cortex. The connections between the neurones in all three types of cortex change over time, according to the sensations received and the experiences gained by the person whose brain they inhabit. But it is areas of association cortex that seem to hold the bulk of our memories of past experience, which are used to produce the appropriate action for particular sensory input. The recognition of familiar objects, decisions about what to say, understanding spoken words and producing precise, skilled movements are all carried out in areas of association cortex.

The two hemispheres of the cerebrum are close together but separate. There is a vertical cleft that extends down between the hemispheres, again just like a walnut. As well as the many small furrows shaped by the wrinkles in the cortex, there are two much deeper furrows that conveniently divide each hemisphere into four lobes. These are the frontal lobes found behind the forehead; the parietal lobes, beneath the top of the head; the occipital lobes, right at the back of the cerebrum; and the temporal lobes, situated inside the parts of the skull where the ears are.

Studies of patients with damage to their brains are useful; but such studies are not the only way of gathering an understanding of how the brain works. This is an important point: when part of the brain is damaged, the resulting behaviour of the patient is actually more a reflection of the capabilities of the remaining parts of the brain than of the damaged part itself. Knowledge about the connections between different parts of the brain, its biochemistry and the behaviour of individual neurones is just as useful in the quest to understand this complex organ. However, in the early days of neurophysiology, matching damage in particular areas of the cortex to brain dysfunction was the main method available.

The first person to link a brain function with a particular area of the cortex was a French surgeon, Pierre-Paul Broca, in 1861. Broca studied the strange speech problems of one of his patients, a man known as 'Tan-tan' because that was all he could say. This kind of problem is an example of aphasia, the loss of the ability to use or understand words. In a post-mortem examination of Tan-tan's brain, Broca discovered damage localized to part of the patient's cortex in the frontal lobe of the left cerebral hemisphere. Broca followed up this case by carrying out a further eight post-mortems on patients who had suffered from similar aphasias. The area of the cortex damaged in these patients, towards the front of the brain above the temples, is now called Broca's area. A little more than a decade after Broca made his discovery, the German neurologist Carl Wernicke made a similar one – this time locating the area of the cortex that seemed to be responsible for understanding the meanings of words. Patients suffering from Wernicke's aphasia are able to speak fluently and even construct sentences using correct grammar – but what they say has no meaning, and they have trouble understanding what is said to them. Wernicke's area is further back in the brain, in the left parietal lobe. As you might expect, there are large bundles of nerve fibres connecting Broca's area with Wernicke's area.

Since the days of Broca and Wernicke, thousands more detailed accounts of strange brain disorders, together with examinations of abnormalities in the brain, have continued to provide clues to the functions of most of the regions of the cortex. The range of dysfunctions is almost as diverse – and equally as mind-boggling – as the range of abilities that the normal human brain possesses. Neurology has devised a comprehensive list of names for the various disorders exhibited by patients with damage to the cortex. For example, in addition to aphasia there is alexia (loss of the ability to read); apraxia (loss of the ability to perform complex tasks); agraphia (loss of the ability to write); agnosia (loss of the ability to recognize things or people). Each of these is subdivided according to the specific nature of the loss of ability: for example, auditory agnosia is the inability to recognize sounds, while tactile agnosia is the inability to recognize objects by touch.

The damage causing disorders such as aphasia and agnosia is normally the result of injury or disease. Damage to the cortex of a boxer is a gradual accumulation over many years, and can lead to a general slowing of mental function or specific problems such as those with speech. Road traffic accidents are the most common cause of injury to the cortex. Pedestrians who are knocked down often suffer a deceleration injury, as their head slams against the road. Drivers who are involved in collisions and who are not wearing seat belts may suffer a similar injury as their heads smash against the windscreen. The physical damage produced by a deceleration injury or a violent blow to the head can cause bruising or laceration in the tissue of the cortex, or haemorrhaging inside the skull. A haemorrhage may be caused by blood escaping from arteries, and therefore not reaching the tissues for which it was intended; this can lead to the death of neurones in the cortex. Haemorrhages can also exert physical pressure on the cortex, and this too can cause the death of neurones. Among the afflictions that cause the death of brain tissue are Alzheimer's disease and Creutzfeldt–Jakob Disease (CJD); another is stroke, which is by far the most common. While Alzheimer's disease and CJD generally cause a gradual but widespread reduction in brain function – because they cause slow deterioration across the whole brain – strokes often cause more localized damage. A stroke involves the loss of blood normally due to blockage of one artery supplying the brain, and can lead to sudden and devastating paralysis or loss of memory, or in some cases more specific disorders such as those listed above. Whatever the cause of brain damage, that damage is known as a lesion.

Watching the unseen

One of the strangest things that can happen to a person's sense of the world as a result of a lesion to a specific region of the cerebrum is a condition known as 'blindsight'. Patients with blindsight can correctly 'guess' in which direction an object is moving, or unconsciously grab a moving object, even though they cannot actually see it. Blindsight patients are typically able to put an envelope through a letterbox in front of them without a second thought, even though they are not consciously aware of the letterbox. Most people with this odd affliction are only blind in part of their 'visual field' – typically one half. Such a person looking at the centre of this book would be able to see, say, the right-hand page but not the left.

The term 'blindsight' was coined by psychologist Lawrence Weiskrantz, who pioneered the study of the disorder at Oxford University during the 1970s. At the time, many researchers believed that it was simply due to weakened eyesight. Some neurologists thought that blindsight patients might be making up their remarkable ability. But careful observation of eye movements and monitoring of pupil dilation showed that blindsight patients really are blind in the normal sense of the word.

Professor Colin Blakemore, also at Oxford University, is a neurophysiologist who includes blindsight in the range of brain functions and dysfunctions that he studies. One of the blindsight patients he studies is Graham, who suffered injury to part of his cortex in a road accident more than thirty years ago, when he was eight years old. During careful tests of his condition, Graham sits in front of a desktop computer. The monitor screen is divided into two regions: white on the left of the screen and grey on the right. When Graham gazes at a grey dot just inside the white portion of the screen, the border of his blind field of vision lies at the border between the white and grey areas on the screen. He can see the white side of the screen on the left, just as any sighted person can, but he is blind to the grey side of the screen, on the right. So when Blakemore holds his hand in front of the grey part of the screen, Graham does not see it. But as soon as Blakemore moves his hand up and down, Graham responds, saying, 'You are moving it up and down.' Graham still does not actually see the hand, but part of his brain somehow registers the movement. As Graham continues to gaze at the grey dot on the screen, a series of black squares appears in the right-hand, grey portion of the monitor screen. Graham cannot see them. Each one moves off – up, down or to the right – soon after it appears, and Graham has to guess which way they have moved. He guesses right every time.

Blakemore explains why studies of blindsight are helping to uncover the way vision works: 'If there is one thing that this phenomenon of blindsight teaches us, it is that vision is not entirely seeing; that there can be a disconnection between the capacity to respond to visual information and the act of being visually aware.'

How can we explain the strange phenomenon of blindsight? The current theory involves the fact that sensory input from the eyes travels along two separate pathways in the brain. It seems that one of these pathways allows us to be conscious of what we see, while the other provides our essential but unconscious visual functions, such as the ability to follow moving targets. Input to the eyes is generated when light falls on a layered sheet of cells at the back of the eye – the retina. Cells in one of the layers are sensitive to light, and stimulate the nerve endings of the optic nerve. Each eye collects information from both sides of the visual field – you can still see the whole of this book when you close one eye (assuming you are a sighted person with two eyes). A few centimetres behind the bridge of the nose is a point at which the optic nerve from each eye splits in two. This point is called the optic chiasma; on leaving it, the nerves from the left side of the visual field are separate from those that come from the right side. The bundle of nerve fibres from each side of the visual field now goes along the two separate pathways.

One of the two main pathways leads first to the thalamus, the way station for nearly all sensory information to the brain. One small part of each thalamus – called the lateral geniculate nucleus – is a relay station for visual information from the optic nerves. The information is routed from the thalami to the 'primary visual cortex' in the occipital lobes, at the backs of the two cerebral hemispheres. Just as there is a sensory map of the body – half of it in the cortex of each parietal lobe – the two halves of the visual field are 'mapped out' in the visual cortex. Visual signals from the left side end up at the right occipital lobe and vice versa. The primary visual cortex is connected to many other regions of the cerebrum. Damage to just a tiny part of a person's visual cortex means that that person will be blind in the corresponding part of his or her visual field. It was damage to Graham's left occipital lobe that left him blind in the right half of his visual field.

This first visual pathway – from the eyes through the thalamus to the visual cortex – is more important in humans and most other primates, such as chimpanzees, than in dogs and cats. A dog, a cat, or any other animal less sophisticated than humans has a smaller cerebrum and so this pathway is not as important. This seems to tie in with observations of the behaviour of animals that see their reflections in mirrors. Whether or not an animal recognizes itself in a mirror is the classic test for visual self-awareness. It was devised in 1969 by Professor Gordon Gallup Junior, then at Washington State University. Most animals react as if they are looking at another animal when confronted with their reflection. But humans (from about eighteen months old) and chimpanzees and orang-utans (from adolescence) do recognize themselves. Chimpanzees and orang-utans that have had some experience of mirrors will, for example, notice a mark on the face in the mirror and immediately reach up to their own face to investigate. Reptiles have virtually no cerebrum, and we can assume that they are not really conscious of what they see in the same way. Instead, they rely only on automatic, instinctive reactions to visual information. This information is supplied by the second visual pathway, which developed in ancient evolutionary history: it is common to all mammals, amphibians, birds and reptiles. It passes directly from the optic nerve to a structure at the top of the brain stem called the superior colliculus. In animals with little or no cerebrum, this visual pathway ends in the superior colliculus; in humans and the higher primates, it continues to the parietal lobes. The superior colliculus initiates automatic movements that direct an animal's vision towards an object of interest – a predator or a potential meal, for example. The rapid, unconscious movements that direct your eyes are called saccades, and are initiated by the superior colliculus itself, just as in the reptiles and other animals. But the signals carried along this pathway, which arrive in the parietal lobes, are thought to give a sense of space, directing attention consciously to objects in the visual field.

In Graham's brain, the pathway that leads from the retinas of the eyes to the superior colliculus is intact. For this reason, Graham can react unconsciously to objects moving in the blind part of his visual field. The damage to his visual cortex – the end of the other visual pathway – means that he is not aware of the objects. This explains Graham's blindsight. In reptiles, such as lizards, blindsight is normal because they have no visual cortex. Graham explains that – in half of his visual field at least – he and a lizard are 'distinct cousins'.

Although most of us do not suffer from blindsight, we all benefit from the automatic behaviour of the 'older' pathway involved in this strange condition. For example, if you are driving, your consciousness may be engaged in talking to someone next to you or listening to music – many of the routine aspects of driving are being carried out subconsciously, using the primitive visual pathway. It would seem from the study of blindsight, then, that there is an unconscious aspect of vision. This seems to be true in fully sighted people as well as those with blindsight. People who can see normally do not become aware of these unconscious factors but, in blindsight patients, there is no conscious aspect of vision to mask the unconscious behaviour. This is one of many clues that the cortex is heavily involved in producing consciousness.

One-sided view

In blindsight, the evolutionarily more recent visual pathway – which allows us to be conscious of what we see – is damaged; the older pathway – which initiates automatic, unconscious, responses – is intact. What would happen if this situation were reversed? In other words, what would happen if a person sustained damage to the older pathway, while the newer pathway remained intact? Would we then have the opposite of blindsight? Professor Vilayanur Ramachandran, Director of the Center for Brain Cognition at the University of California, San Diego, (whose research on the phantom-limb phenomenon was discussed in 'Phantom Brains'.) has been very active in recent years in attempting to explain some of the strange syndromes associated with damage to the cortex, including blindsight. He suggests that patients who suffer a stroke focused in the right parietal lobe experience something like the opposite of blindsight.

As described above, areas of the parietal lobes are the end point of the older visual pathway, and the parietal lobes seem to be involved in spatial awareness and in tracking moving objects. It has been shown, for example, that neurones in the parietal lobes become active when an object enters the visual field. So you might expect people who have suffered damage to the older visual pathway, in the parietal lobe, to be visually aware of the world but unable to carry out the unconscious tracking of moving objects. And you would be right: such patients are said to suffer from 'visual neglect'. They may also suffer disruption to their spatial awareness. Asking one visual-neglect patient, Bill, to track his finger as it moves horizontally in front of his face, Ramachandran demonstrates this idea. Bill suffered a stroke in his right parietal lobe, and the left side of his body is largely paralysed. His head and neck are skewed to the right – he describes his upper body as being 'twisted like a pretzel'. Bill's head and eyes remain stationary as Ramachandran's finger moves in front of him, although he can correctly describe the motion and knows that it is the professor's finger. Ramachandran explains that Bill 'can no longer point to something on the left side of the world ... and can no longer orient to what's going on to the left; he's not blind – his visual cortex is still normal'. Another patient suffering from visual neglect is Peggy, who compares her condition to 'like just before you faint: everything disappears'. Her visual neglect becomes clear when she tries to draw: copying a four-pointed star, she draws only three points, leaving out the one on the left. Similarly, drawing daisies, she includes the correct number of petals, but all of them are squashed awkwardly on to the right-hand side of the stalk. 'I've done it on all of them!' she exclaims, looking at her drawings after she has finished.

One way of investigating the extent of visual neglect is to ask patients to use a mirror to bring the missing part of the visual field – left of the patient – into view. You might expect patients suffering from neglect to turn to their left when they see an object of interest in the mirror. This seems reasonable, because these patients have not lost their mind – they are still aware of how things work, including mirrors. Ramachandran holds a mirror on Bill's right and asks him to reach for a set of keys held to Bill's left, but which appear in the mirror to the right. Surprisingly perhaps, Bill attempts to reach into the mirror, however much he tries to concentrate, and despite the fact that he is fully aware that the keys he can see are part of a mirror image. Ramachandran says that the job of working out the relationship between a real object and its mirror image is the sort of process carried out by the parietal lobe, and this explains why Bill simply cannot reach for the real keys. He says that it is as if Bill is saying to himself 'on my planet, left does not exist'. Interestingly, when the same experiment is carried out with the mirror in front of Bill's face, and the keys held behind his head, Bill is able to reach behind him to grasp the keys. In this case, he makes use of the spatial awareness of his undamaged, left parietal lobe to achieve this task.

In some patients with right parietal lobe damage, neglect is so extreme that not only do their brains deny the existence of space to their left, they may even deny the existence of the left side of their own bodies. One patient studied by Professor Ramachandran, a Mrs Sinclair, claimed that she was being shown her husband's hand – 'I've been living with him for twenty-seven years, I know his hand' – when actually it was her own, paralysed, hand that was being held up. An extension of this, and even more bizarre, is a disorder known as anosagnosia, also found in people with right parietal lobe damage. People with anosagnosia lie – even to themselves – about their paralysis or other conditions that result from their brain damage. These other conditions, well documented but not completely understood, include inability to draw, to dress or to construct things. People with anosagnosia do not deny their symptoms simply because they are ashamed of them – other patients with damage to their brains do not behave in this way. Mrs Sinclair showed signs of anosagnosia: during each of Professor Ramachandran's first few visits, she said that she could use the paralysed hand. In one task used to test patients with anosagnosia, they are asked to touch a doctor's nose with a finger of their paralysed left hand. Of course, they cannot direct their paralysed arm towards the doctor's nose. Rather than explaining that their left arm is paralysed, most anosagnosic patients will make up excuses for their arm's inactivity. When Ramachandran asks Bill to attempt the test, Bill explains that he is 'calling up his arm ... explaining to it what to do ... it's tired'. Some patients actually claim that they are touching the doctor's nose with their hand.

Another revealing test of the nature of this strange disorder is asking a patient with anosagnosia to clap. Bill makes the normal motions of clapping with his right hand, fooling himself that the other hand is doing the same. But his good hand ends up hitting the air, and then his chest, as the paralysed arm lies motionless on his lap. Bill seems to believe that he is clapping normally. Anosagnosia was first described nearly a hundred years ago by French neurologist Joseph Babinski, one of the pioneers of experimental neurology. But only now are neurologists and neurophysiologists discovering how the symptoms might relate to damage to the right parietal lobe.

Lesions to either parietal lobe can lead to a number of different symptoms, depending on where the damage is sustained. The sensory map on the cortex is located on the parietal lobe, and damage to that area in either hemisphere leads to a loss of sensation in part of the opposite side of the body. Damage to the motor areas of a person's parietal lobes results in that person being unable to carry out skilled movements – apraxia. When other areas of the parietal lobe are damaged, patients may suffer from agnosia (inability to recognize familiar objects). There are several different types of agnosia, again depending on where in the parietal lobe, and in which parietal lobe, the lesion occurs. Lesions to areas of the left parietal lobe result in tactile agnosia – the loss of ability to recognize objects by touch. Anosagnosia is caused only by damage to the right parietal lobe. The fact that the nature of a neurological disorder depends upon which cerebral hemisphere is damaged has been known since the time of Broca and Wernicke, mentioned above. Both types of aphasia that they described were caused by damage to the left side of the brain.

Which is right?

In 1874 Wernicke published a book in which he described various deficits in cognitive function caused by damage to specific areas of the cortex. As part of his descriptions, he claimed that certain abilities are found in only one of the cerebral hemispheres. Some of the earliest attempts to work out which parts of the cortex are responsible for which brain functions were carried out by German researchers Gustav Fritsch and Eduard Hitzig in the 1860s. Hitzig began this process while working in a military hospital, taking advantage of patients whose skulls had been damaged in battle: he stimulated exposed areas of the cortex using wires connected to a battery. Around 1870 Fritsch and Hitzig found, in experiments on dogs, that electrical stimulation of one cerebral hemisphere produced movement in the opposite side of the body.

Signals to and from the left side of the body are carried by nerve bundles separate from those to and from the right side of the body. Above the top of the spinal cord, in the brainstem, the nerve bundles to and from the rest of the body cross over. Nerves to and from the eyes, ears, mouth and the rest of the head also cross over in this way. This explains why the visual map in the left occipital lobe receives signals from the right side of the visual field, and vice versa, as we saw in the discussion of blindsight, above. Similarly, the two sensory maps of the body – one on each parietal lobe – correspond to the opposite sides of the body. There are similar maps of the body on the motor cortex of the frontal lobes. Stimulating the area on this map corresponding to a finger will cause a person's finger to twitch. As with Fritsch and Hitzig's dog, the map on the cortex of the left frontal lobe corresponds to the right side of the body. Similarly, signals from the left ear arrive in the auditory (sensory) cortex of the right hemisphere.

Startling evidence of the differing abilities of the two cerebral hemispheres was found in experiments carried out by Roger Sperry and his colleagues at the California Institute of Technology, highlighting important differences between the two sides. The experiments involved surgically cutting through the corpus callosum – the dense bridge of nerve fibres that acts as the main pathway for signals between the two hemispheres. This resulted in 'split-brain' patients, whose two isolated cerebral hemispheres were unable to communicate with each other. The experiments were a desperate attempt to cure severe epilepsy. After patients recovered from these operations, they seemed to have all the faculties of people whose cerebral hemispheres have not been separated. However, further study showed that there were some important differences: the two sides of the brain seemed able to act independently, unaware of each other. For example, while the left hand of a 'split-brain' patient was buttoning a shirt, his right hand was busy unbuttoning it. More importantly, careful studies of split-brain patients revealed differences in function between the two hemispheres. So, for example, they found it easier to identify objects held in the right hand – interpreted by the left hemisphere – than in the left; they could copy drawings better with their left hand than their right, even if they were right-handed.

In a more recent split-brain experiment, conducted by Michael Gazzanioa at Cornell University in New York State, a split-brain patient was shown written commands in only one half of the visual field at a time. When the patient was shown the word 'laugh' in the left visual field (linked to the right hemisphere), he laughed but could not explain why. In fact, he made up reasons why he may have laughed, using his left hemisphere, which is more specialized to deal with language. The left hemisphere seems to have observed the laughter, and attempted to invent a story that would account for it.

The results of these and several other studies of split-brain patients have helped to show that the two hemispheres approach tasks differently. The left hemisphere is more concerned with logic, mathematical calculation and language than the right. The right hemisphere is more concerned with recognizing shapes and faces and appreciating music and art; it works holistically, analysing form and other visual-spatial aspects. One of the hemispheres is normally dominant – this explains handedness. If a person's left hemisphere is dominant, that person will be right-handed, and vice versa. Some people suggest that for this reason left-handed people are more likely to be creative, while right-handed people are more likely to think logically and have better language skills. There is some evidence to back up this idea, but it is by no means a hard and fast rule.

The different specialisms of the two cerebral hemispheres might help to explain the self-deceit involved in anosagnosia. Bill, whose anosagnosia made him fail to acknowledge his left-side paralysis, had suffered damage to his right parietal lobe. Anosagnosia is not found in patients with damage to the left parietal lobe. Professor Ramachandran has a theory that might help to explain this one-sided nature of anosagnosia. This involves the concept of the ego, first proposed by Sigmund Freud in the first two decades of the twentieth century.

Pleasing your self

Your senses provide information about the world around you – outside your body – but the information is put together inside your brain. It is there inside your brain that the information is interpreted and, more importantly, experienced. Your brain somehow constructs a feeling of yourself in the world. This sense of self is clearly an important part of consciousness, and is what Freud defined as the 'ego'. Consciousness itself is simply awareness. You can be conscious of a sound, an idea or the presence of a person, for example. Consciousness is elusive – modern neuropsychologists really have little idea how electrical signals in the brain can produce such a rich and ethereal experience. Until a decade or so ago, it was believed that consciousness is produced by the cortex, since that part of the brain is the most complex. The cortex clearly does play an important role in consciousness. Patients with blindsight are unaware of objects in the right visual field because of damage to their visual cortex, but can still react unconsciously to the things they 'see', for example. Research carried out using fMRI (functional magnetic resonance imaging) has confirmed that when blindsight patients react subconsciously to objects moving across their visual field, the superior colliculus – at the top of the brainstem, and not in the cortex – are active. However, most neuroscientists today tend to think that the cortex supplies the information – the thought processes and the ego – for other parts of the brain to be conscious of. While the cortex may not produce consciousness, what is known about it suggests that it is important in the creation of the ego, the self-image. The cortex is the destination for the incoming information, and it is the cortex that produces the 'higher' functioning of the brain. This includes speech, motor skills, memories and the recognition of objects, for example – all essential elements of our sense of self.

Ramachandran's theory of cortical remapping (explained in 'Phantom Brains') provides evidence that the cortex produces a 'body image', which seems like an important part of the self. When a person loses, say, a hand, he or she also loses sensory input from the hand to the relevant part of the sensory map on the cortex. Sensory input from different parts of the body – those that lie adjacent to the hand on the map – arrive at that part of the map previously associated with the hand. This has the strange consequence that a left-arm amputee will feel sensations in his or her non-existent (phantom) hand when touched on the cheek. The cortex seems to hold the blueprint for the body image, and when signals from the cheek arrive at the part of the cortex normally reserved for the hand, other parts of the brain become conscious of the hand. Historically, the first study to indicate that the cerebrum is important in producing conscious actions was carried out by the German physiologist Friedrich Goltz. In 1892 he found that dogs whose cerebrums had been removed lost all their abilities except their reflexes.

So how does the self – the ego – come into the behaviour of anosagnosic patients, who delude themselves about their left-arm paralysis? Ramachandran proposes that the ego might be produced by the left side of the brain, while the right brain includes a 'devil's advocate', which constantly checks to see if the self fits with the information received about the world. In a person whose brain is not damaged, but who has a paralysed arm, the ego may assert that the arm is not paralysed. But the right brain would be there to force the left brain to re-evaluate the ego. If the relevant part of the right brain is damaged, this devil's advocate would no longer function, enabling the left brain to construct the ego unchecked. In other words, once it has constructed the 'self', the brain seems to be prepared to lie to itself if the incoming information does not correspond to the self-image.

There is evidence that the brain does indeed work in this way, making up stories to account for what it perceives. In a famous study conducted at the University of California at San Francisco, brain researcher Benjamin Libet investigated people's awareness of their own will to cause their body to move. Before someone makes an apparently voluntary movement, such as flexing their wrist, a voltage called the readiness potential sweeps across the scalp. You might assume that the readiness potential is a result of a conscious decision to produce the movement. Libet asked his experimental subjects to flex their wrist or fingers whenever they felt they wanted to, and to declare that urge as soon as they were aware of it. He found that the awareness lagged behind the readiness potential by a consistent amount of time – about 0.4 seconds. You might think that this delay is simply the time it takes for a person to communicate the desire. But many other researchers have found similar results, which seem to suggest that consciousness really is a consequence of the unconscious behaviour of our brains, rather than a deciding factor or guiding light in how we should behave.

This calls into question the existence of free will – that ability we all seem to have to make decisions about our actions. As we live our lives day to day, we tend to assume that we make our own decisions about how we behave, or at least that we can do so whenever we want to. If someone asks you whether you want a cup of coffee, you think about it, make a decision and act upon that decision – you seem to have free will. However, some neuropsychologists believe that the brain reacts automatically in any situation, and that the mind works out what went on, making up reasons for behaviour shortly after the fact. According to this view, we are machines – automatons, behaving without conscious control. But at the same time, we are our own social commentators, supplying ourselves with reasons why we behave in the way we do – as if the mind is constantly looking for reasons why. Consciousness still has a role in this picture of behaviour: by monitoring your actions it can work out whether what happened was the best course of action in the circumstances, and learn from it. If this picture of consciousness – as an afterthought – is correct, it may explain why anosagnosic patients are able to lie to themselves. Ramachandran's theory suggests that the accounts of what is happening are produced largely in the left hemisphere and checked mainly by the right.

Ramachandran compares his suggested combination of ego and devil's advocate in his theory with the way that science arrives at its understanding of the world. The current theories at any time define the scientific view of the world. So, for example, physicists of the nineteenth century believed that time ran constantly and lengths were unchanging. Evidence to the contrary came from classic experiments during the 1880s, and from inconsistencies in the theories of electricity and magnetism. Physicists attempted to absorb these inconsistencies into their world view, trying to preserve stability in their theories, which had served them well until then. Albert Einstein's theories of relativity, the first of which he published in 1905, presented a new framework that explained the inconsistencies in the previous theory. Suddenly, the physicists' world view underwent a dramatic change – what philosophers call a paradigm shift. It is important that scientists work in this way: if they abandon their accepted theories whenever an inconsistency crops up, there would be no stability, no basis from which to move on. Ramachandran believes that in the brain, the ego in the left hemisphere is the stable sense of self, like the scientists' world view. The devil's advocate in the right hemisphere picks up on inconsistencies between the ego and the information it receives about the world or the body. It alerts the ego to the inconsistencies, so that it can undergo a paradigm shift.

So, according to Ramachandran, the brains of anosagnosic patients are able to deceive themselves about the state of the body they inhabit because their devil's advocate is no longer present. Freud believed that the ego was able to deceive the mind for the sake of its own stability, in just this way – until pressures on the ego became too much to bear and it would have to change. He suggested that the ego would use humour and self-deceit in defence of its own stability, and Ramachandran says that the kind of denial observed in anosagnosic patients is often tinged with a sense of humour or irony. It is as if the brain knows the truth but attempts to fool itself and other people. And indeed, the fact that the brain knows the truth about its body's condition is brought to light by a remarkable experiment on anosagnosic patients. When researchers trickle cold water into a patient's left ear, the delusion is broken – the patient is able to acknowledge their disability. Denial returns again after a few hours, and the patient cannot even recall being conscious of their paralysis. Cold water in the ear causes the brain to enter into a state something like dream-laden sleep: the eyes flicker rapidly as they do during the REM (rapid eye movement) phase of sleep, in which people dream. It is often in dreams that subconscious information, at other times suppressed, comes to the surface.

According to Ramachandran, the dreamlike state induced by the cold water allows the repressed knowledge of the paralysis to come to the surface.

Deceit is a fascinating aspect of human behaviour. Just why we have the ability to lie to each other and ourselves we may never know, though much thought has been given to it. Many evolutionary biologists consider deceit an important faculty in the animal world: it can be a crucial survival technique, and probably therefore evolved as part of animal behaviour. One such evolutionary biologist is Robert Trivers, one of the major figures in the world of sociobiology – also called evolutionary psychology – which attempts to work out human and animal behaviour in evolutionary terms. In the foreword to one of the classic books on this subject, *The Selfish Gene* by evolutionary biologist Richard Dawkins, Trivers explains why animals might have evolved the power of self-deceit. According to Trivers, animals would have evolved the ability to recognize when another animal was being deceitful; then, like an evolutionary arms race, the deceitful animals may have evolved mechanisms that lead to self-deception. This would 'render some facts and motives unconscious, so as not to betray – by subtle signs of self-knowledge – the deception being practised'.

Trivers believes that the different roles of the two cerebral hemispheres are apparent when a person is being deceitful – that we are still able to detect deceit, despite the proposed evolution of self-deceit that is supposed to mask it. 'If you put a false expression on your face – a forced smile – it will tend to curl up a bit more on the right side of the face, because it is being run by the left hemisphere. The right hemisphere is more involved in a natural expression, so if I could smile warmly at you now, it is supposed to curl up a bit more on the left side, because it is run by the right brain.' He has analysed video footage of President Bill Clinton discussing his alleged sexual relations with a woman who was working at the White House. In the face of this scandal, Clinton lied to the media: 'I did not have sexual relations with that woman.' Trivers says that when Clinton was denying his involvement with the woman, he subconsciously showed the signs of deceit in the imbalance between the two sides of his face. 'Initially, there was a facial imbalance – the right face was into it, the left face said, "Get me the hell out of here".' Trivers notes that the balance was redressed when Clinton eventually went before the media to admit that he had lied; both sides of his face were 'into it'. Ramachandran agrees: 'Your face starts leaking traces of deceit ... the muscles are slightly different when you are lying.'

We have seen that the strange lack of awareness of the left-hand side of the world evident in visual-neglect patients, and the denial of symptoms in anosagnosic patients, are caused by damage to the right parietal lobe. Earlier, we saw that the equally curious symptoms of blindsight patients is caused by damage to the visual cortex in the occipital lobes. Broca's aphasia is one example of the results of damage sustained in the frontal lobe. That leaves the temporal lobes – what might damage to them be able to tell us about their function?

The temporal lobes are tucked in almost underneath the rest of the brain, right behind the ears and the temples. As you might expect, their proximity to the ears means that the temporal lobes are heavily involved in our sense of hearing. Just as there are sensory maps of the body on the parietal lobes, motor maps of the body on the frontal lobes and retinal maps on the occipital lobes, there are auditory maps on the temporal lobes. The cortex here is arranged according to the frequency of a sound, so that two closely pitched notes will stimulate adjacent areas. The temporal lobes are also involved in efforts to understand or recognize sounds. Adjacent to the auditory maps is an area of association cortex. Sometimes, damage to the auditory association cortex results in auditory agnosia. A person affected by this strange dysfunction can hear sounds, but does not recognize them. Also in the temporal lobes, just behind the auditory association areas, lie regions of association cortex involved in visual recognition. And just as auditory agnosia is the result of damage to auditory association cortex, damage to the visual association cortex can cause visual agnosia – the inability to recognize objects by sight.

What are you looking at?

In the 1970s Philip was injured in a serious road traffic accident which put him into a coma for several weeks. Luckily, he regained consciousness and his brain seemed to be functioning normally, despite the force of the collision. It was soon realized that Philip had sustained damage to part of his temporal lobe near to the back of the brain: the visual association cortex. The result of this damage is that Philip has trouble recognizing animals and fruit and vegetables. His neurologist, Dr Rosalind McCarthy, shows him a photograph of King's College Chapel in Cambridge, and asks him if he can identify it. Philip replies, 'That's King's College' – he can recognize buildings. He can find his way around, too, and he picks his daughter up from school every day. When shown a pair of scissors, a pair of glasses and a toy helicopter, Philip names them immediately. However, when McCarthy shows Philip a toy elephant, Philip has no idea what it is. He says, 'It's an endangered species but I can't place it, though. I tell you what gives it away to me are the tusks at the front, and I know what it is but I can't name it. It's annoying me.'

At the zoo, standing next to the giraffe enclosure, he says that the giraffes might just as well be extinct dinosaurs or the Loch Ness monster. In the end, he guesses that the animals he can see must be camels – because he had heard passers-by talking about camels, which are in the enclosure behind him. He eventually discovers that the animals he is looking at are giraffes by reading a sign alongside the enclosure. Philip always looks out for clues like these when identifying objects that he does not immediately recognize. And when a pineapple is held in front of his face, he is equally unable to name it. When the pineapple is handed to him, he correctly states that 'it grows on trees, comes from hot climates, it's juicy, you can't eat the outside' – then 'I've forgotten what it is.' It seems sensible that normal human vision has the ability to draw associations between objects that it sees. This sort of information, which seems to emerge subconsciously and from different, undamaged parts of the brain, is important for survival. Some patients with more extensive temporal lobe damage have trouble even with this kind of subconscious recognition: they put just about anything into their mouths, even razor blades or the wrong end of a lit cigarette. This kind of associated knowledge would have been even more important to prehistoric hunter–gatherers – as would the ability to recall associated knowledge about animals, which could have been predator or prey, for example, or fruit that could be poisonous or nutritious. Philip can function at this subconscious level, but has trouble simply extending his recognition of objects by name.

Philip's agnosia includes the inability to recognize faces, a condition known as prosopagnosia. This is the same condition that 'LH' suffered from, as described in 'Mind Readers'. When shown a photograph of Margaret Thatcher, Philip cannot name her. When shown a photograph of the late Princess Diana, he guesses who it is by looking at her earrings. Marilyn Monroe and Elvis Presley are the subjects of the next photographs, and Philip has no idea who they are. When asked whether he knows who Elvis Presley is, he explains that Presley was a famous American singer and film actor who made his mark in the 1950s, and who is now dead. When he tries to conjure up an image of

Elvis, he sees sequins, but does not know why. When meeting familiar people, he can only recognize them by the sound of their voice or their 'habits, clothes, hairstyles, nose, moustache'. When Ramachandran draws a picture of Donald Duck on a white board, Philip has no idea who it is. He guesses that it might be the late Grace Kelly.

People have long suspected that parts of the temporal lobes are heavily involved in the recognition of faces, but only recently have neurophysiologists begun working out the extent of the role of the temporal lobes in the recognition of other objects. Blakemore and his colleague Tim Andrews at Oxford University are among a number of researchers carrying out imaging studies, using fMRI, aiming to discover which parts of the cortex are involved in visual recognition. Some other researchers are using a more direct method of spying on the brain: by connecting electrodes to the cortex of a living monkey, they can find out at precisely which point the cortex is active when the monkey is engaged in various specific visual tasks.

Philip's 'category-specific' agnosia has affected only very limited aspects of his vision: his ability to recognize animals, fruit and vegetables and faces. This suggests that vision actually consists of a number of different modules, each with a particular job. Until the 1960s it was thought that the visual pathways from the retina led only to the maps on the visual cortex in the occipital lobes. The problem with this idea is that it does not explain our experience of seeing. In fact, the classic idea that the optic nerve leads to the cortex and perceived there is not too far away from an idea by the French philosopher René Descartes in the sixteenth century.

Descartes cut away the retina of a disembodied sheep's eye and replaced it with a translucent screen. He found that the lens in the eye produced an image of his window on the screen, and he correctly surmised that the retina is a simple screen, too. Descartes supposed that the optic nerve carried the image from the retina deep into the brain, where it produced another image. The soul could then view the scene. Many people today think in a similar way – unable to conceive of a way that the tissues of the brain could be responsible for the awareness of actually seeing something. Instead, many believe that there is a soul or spirit that does the seeing – just as Descartes did. However, this does not explain how vision works, because it does not explain how the soul sees. For some people, the soul takes on human form, so that there is a little person inside your head who does the seeing. Professor Ramachandran says that if the image is simply projected inside the brain for a little person to see it, then that little person must have another brain, with another person inside it. He says, 'You end up with an infinite regress of eyes and images and little people without solving the problem of perception.'

The rich experience of vision is many-faceted. Among its many functions are spatial awareness and avoidance of obstacles, reaching and grabbing things, general awareness, communication and the recognition of objects (Philip's problem). The brain needs to interpret colour, form, shape, a sense of space; it needs to search its database of associated memories about an object; it must be able to follow smooth movement of an object across its visual field. As you would expect with so many different functions, there are many different areas of the brain involved in vision.

We have seen that the visual information from the retinas enters the brain along two main pathways: an older one and a newer one. The older pathway – involved in visual reflexes and a sense of space – terminates in the parietal lobes, as discussed above. The new pathway leads through the thalamus to the visual cortex, and is involved in conscious awareness. In a sense, the visual cortex in the occipital lobes is like the hypothetical screen described above – it is a simple map of the visual field. Other areas of the brain interpret what is 'displayed' there, and each area carries out a different, well-defined operation on the information it receives. More than thirty such areas have so far been identified, with specializations for sensing colour, movement, relative distance and depth, the form, movement and orientation of objects. Despite this thirty-fold division of labour, the visual functions of the brain seem to follow two distinct pathways from the visual cortex. The first of these is referred to as the 'where pathway' or the 'how pathway', and is concerned with enabling you to work out the location and distance of objects, and with finding your way around and avoiding obstacles. The 'how' pathway shares some of the functions of the evolutionarily ancient pathway that blindsight patients rely on to invoke unconscious responses to visual information. And, like that ancient visual pathway, it ends up in the parietal lobe. In Philip, the 'how' pathway is clearly undamaged, since he can walk – and even drive – around in the world as well as anyone. However, the 'what' pathway for visual information – involved in visual recognition – is what is damaged in him.

When Philip is handling a piece of fruit, he can call up associations with that fruit, as he did with the pineapple. In this case, his subconscious brain systems know what type of fruit he is handling, but this information cannot be interpreted using his higher-level functions, like language and memory. Colin Blakemore explains that part of the 'what' pathway is involved in condensing the information about the world 'into a succinct description which becomes a memory ... and second, to use that same kind of description as the basis of language ... a word is a wonderful symbol'. Surprising evidence of the role of the temporal lobes in memory is that stimulation of the auditory association cortex in awake patients undergoing brain surgery evokes complex memories of sounds. Visual memories can be evoked by stimulation of the same area, in certain patients. The 'what' pathway – involving the temporal lobes – is involved in recognition of objects, and it is this pathway that is damaged in Philip. More tantalizing evidence about the function of the 'what' pathway comes from another curious syndrome, called Capgras' syndrome.

A case of mistaken identity

After a road accident a few years ago a young man, David, suffered damage to his temporal lobes. He was in a coma for a week but, when he regained consciousness, his behaviour showed no real ill effects. He was not psychotic or emotionally disturbed, but he was suffering from one remarkable delusion: that his mother and father were impostors. He agreed that they looked identical to his real parents, but he simply would not accept that they were who they said they were. David's mother recalls the events of one meal time: 'He probably didn't like the food that day, because he said, "That lady who comes in the morning, she cooks much better than you" ... but the lady was me all the time.' Speaking about his father, David explains, 'He can look like my father but the fact is that it doesn't feel like him because I know that it is not him.' David's disbelief about the identity of his parents is so strong that he does not recognize the family home. One day, he demanded to be taken to his real home. His mother took him out of the apartment building, around the block, and returned to the same apartment, where she left him alone. In the absence of the supposed impostors, David was then satisfied that this really was home. David's symptoms are known by the name Capgras' syndrome or Capgras' delusion.

Capgras' syndrome was first reported during the early years of the twentieth century. Until recently, the textbook explanation of its strange symptoms has been based on one of Sigmund Freud's ideas – the Oedipus complex. The explanation goes something like this: Freud claimed that sexual feelings are felt by a son towards his mother, or a daughter towards her father, in early childhood. These feelings are repressed as a child develops and, in a normal adult, there is no sexual attraction to parents. In certain people, those repressed feelings come to the surface in adulthood, causing unrest in a person's personality. This is the Oedipus complex. According to the Freudian interpretation of Capgras' syndrome, the damage sustained by the patients' brains somehow releases the sexual feelings towards their parent – the Oedipus complex brought on by a nasty knock on the head. The patient is then aware of sexual feelings towards someone who looks just like his or her parent. The brain subconsciously deals with this: it convinces its belief system to assume that this person is not a parent after all.

There are certain reasons why this interpretation does not work, including the fact that the Oedipus complex involves sexual feelings only towards the parent of the opposite sex. David denies the identity of his father as well as his mother. Furthermore, it is not only parents that are the subject of the delusions experienced in Capgras' syndrome. Often it will be a wife or a husband. Ramachandran has even seen a patient who has the same delusion about his dog: 'He'll look at his pet dog and say, "Doctor, this is not Fifi. It looks like Fifi, but it's been replaced by an identical dog."' The Oedipus complex cannot really explain that particular case.

Ramachandran has suggested an alternative explanation of Capgras' syndrome, which appeals to recent advances in understanding the 'what' pathway of the visual system. There are connections between the areas of the temporal lobes involved in recognition – the ones that were damaged in Philip – and the limbic system, which is the emotional centre of the brain. In particular, there are major connections to the amygdalae, whose roles were discussed in 'Mind Readers'. The amygdalae are heavily involved in producing emotional responses, in creating feelings about things. Ramachandran explains: 'When you look at an object, the message goes first to your temporal lobe cortex, where you recognize it ... after you've recognized it, you also need to respond to it emotionally.' This is probably important in our appreciation of great works of art, but it must have developed – back in our evolutionary history – as a mechanism to avoid danger or to find a mate. Seeing a looming predator makes an animal feel differently from how it feels when it sees a mate, and the animal responds accordingly. This seems to be true of humans, too: laboratory tests show that a person's body sweats very slightly when that person sees something that evokes an emotional response. The emotions can be almost imperceptible, but the sweating response can be recorded, as a change in the GSR (galvanic skin response), after a few seconds. The GSR is a measurement of the skin's conductivity – how well it conducts electricity – and is determined by attaching electrodes to a person's palm, where the sweating is most pronounced.

Ramachandran proposes that a specific part of David's 'what' pathway was damaged in his accident – the part that connects the temporal lobe with the amygdalae. When David looks at his mother, his temporal lobe recognizes her and brings forth appropriate memories. But David's temporal lobes do not communicate with his amygdalae, because the connection is broken. So there is recognition of faces, but not the expected feelings that should accompany that recognition. To make sense of the conflict between the visual memory and the lack of emotion, David's brain – perhaps his storytelling left parietal lobe – works out that this person must be an impostor. There is evidence to support this idea. First, strong connections between the temporal lobes and the limbic system do exist – clearly there is communication between the recognition centres of the temporal lobes and the emotional centres of the limbic system. Also, David has no problem accepting his parents' identity when he can only hear them – over the telephone, for example. The Freudian explanation does not seem to fit in with that observation. The connection between the auditory cortex – also in the temporal lobe – and the amygdalae is separate from the one that connects the visual association cortex to the amygdalae.

To test his idea, Ramachandran and his colleagues decided to measure David's GSR while he was shown a series of photographs of people's faces. Some of the photographs showed total strangers, while the others showed his mother and father. Normal people produce a GSR when they look at photographs of their parents – 'Every time you look at your mother, you start sweating,' says Ramachandran. If his theory is correct about the disconnection between the temporal lobe and the amygdalae, then you would expect no response when David sees photographs of his parents. And that is exactly what he found. It seems to show how closely our intellect is linked to basic emotional reactions.

Ramachandran accepts that his theory cannot explain all aspects of Capgras' syndrome. It does not explain why the syndrome is normally restricted to a patient's parents or spouse, for example. Another Capgras' syndrome patient is Oliver, who is delusional only about his wife. 'I thought it was a twin of her,' he says. In fact, Oliver was convinced that his wife was three different people. His delusions led him to believe that one of the people who claimed to be his wife was spying on him to gather information for the other two. Oliver's neurologist is Dr Simon Fleminger of Goldsmiths College, London University. Like Ramachandran, Fleminger believes that the physiological explanation of Capgras' syndrome is not enough. He suggests that a psychological dimension is necessary to explain the syndrome. He says, 'There is a to-and-fro between the information coming from the outside world on the one hand, and our thoughts of what we are about to see on the other ... Our expectations can colour what we perceive.' Oliver's delusion may be caused by his belief that his wife is not to be trusted, which then affects his perception of his wife.

So it seems that the emotional responses of the limbic system feed into the areas of the temporal lobes involved in recognition – how you feel about something can help you to identify it. A dramatic illustration of the relationship between emotions and the temporal lobes is found in epileptic seizures that occur in the temporal lobes.

A religious experience

In Capgras' syndrome, it seems that the connections between the temporal lobes and the amygdalae are disrupted, so that the patient does not feel the appropriate emotional response when looking at a familiar face. Ramachandran believes that a well-known effect in patients who have epileptic seizures in their temporal lobes is the opposite: such people can see emotional significance in almost everything, and often declare themselves to be modern-day prophets or visionaries.

An epileptic seizure is like an avalanche of electrical signals in the brain. Ramachandran describes an epileptic seizure as like an 'electrical storm in the brain; where a group of neurones starts discharging, unco-ordinated from the rest of the brain'. The increase in electrical activity can produce extreme convulsions, during which the whole body begins to shake violently, and a person loses consciousness. Less severe or more localized seizures can cause just one part of the body to shake or can result in a momentary loss of awareness.

The close connection between the temporal lobe and the limbic system – of which the amygdalae are part – explains why seizures localized in the temporal lobes are likely to evoke strong emotional responses. The limbic system is also responsible for emotional response to the sense of smell, and to memory and hearing. During a seizure in the temporal lobe, patients often have visions and hear voices, and sometimes sense noxious smells or tastes. A sense of déjà vu is also fairly common in temporal-lobe seizures. One experience that has been reported by some patients with temporal-lobe epilepsy, since it was first studied a hundred years or so ago, is a religious feeling. This can manifest itself as a visitation by God, or the belief that the patients actually are God. Amid the disrupting of bodily functions – often including a disturbing physical fit – there is a sense of understanding the world, a spiritual high, like being at one with the cosmos.

One patient who suffers from temporal-lobe epilepsy is John, who separates the physical dimension of the seizures from the 'spiritual' dimension. 'The seizures involve my person – that's the seizure I'm experiencing – and my soul, and my spirit,' he says. John actually welcomes the spiritual feelings that he experiences: 'You're fighting with your soul and your spirit afterwards.' Describing a recent seizure, he explains that, 'My attitude was that I was God, and I had heaven and hell in my eyes. You know, I was the grand guy who created heaven and hell.' Sometimes, John has seven or eight seizures in a single day, each one lasting between three and six minutes. During each one, he is in another realm: 'through the gateway and into another reality'. The experiences of temporal-lobe epilepsy are extremely real, and they can be extremely emotional. In tears, John gives an idea of the intensity of his experiences: 'I've been in so much pain that I would rather be shot or whipped to death; and yet I've been in so much joy that I would rather be left alone; take everything away and just let me sit there, and have that much joy ... I feel like I can float, it's the best.' Before John had his first seizure, at the age of seventeen, he was a normal adolescent, and was not religious. In Ancient Greece epilepsy was known as 'the sacred disease'. In some cultures in Asia, India and Australia, people who suffer from epilepsy are thought to be in contact with a transcendent reality, and therefore able to heal or foretell the future. They may become shamans – the word 'shaman' means 'he who knows'. John believes he has special insight: 'I am so right in my own head, I know I could go out there and get people to follow me ... Were all the prophets people who were flopping around on the ground? Is that was this whole message was, this whole time?'

In his book *Phantoms in the Brain* Ramachandran describes a device called a transcranial magnetic stimulator, which is a helmet that can be used to excite any part of the brain. He describes a Canadian neuroscientist, Professor Michael Persinger, who used one of these devices to stimulate his temporal lobes. He was attempting to instil within himself the sort of experience that John has during his temporal lobe seizures, and he succeeded. In his 1987 book *The Neuropsychological Base of God Beliefs* he investigated the origins of religious beliefs. He has also suggested how UFO sightings may be explained by hallucination. In each case, he looked for the source of the psychological experience in terms of the physiological brain. Ramachandran, too, has devised a physiological explanation of why even atheists can have seemingly religious experiences during a temporal-lobe seizure.

According to Ramachandran, the importance of the amygdalae in the creation of emotions and their proximity to the temporal lobes means that seizures in the temporal lobes will produce a welling up of a huge range of emotions. The brain's normal emotional response to the world is important, and is to some extent 'hard-wired': the reactions to aggression, for example, are important for survival. The emotional responses of the limbic system are not all hardwired, however. We develop our own personal emotional reactions, to things that have certain meanings to us. Ramachandran says that we create a 'landscape of emotional salience' in the barrage of different emotions experienced during temporal-lobe epileptic seizures. And repeated seizures can produce permanent changes to the emotional pathways – a process called kindling. Ramachandran proposes that kindling along the pathways between the temporal lobes and the amygdalae is what brings the sensations of glory, so that almost anything can be imbued with emotional significance.

This tendency to assign significance to everyday objects might relate to mystical experience. John's father reports, 'When he has had a seizure he'll want to talk philosophy. He'll want to discuss all the things that are floating around in the stew he's got up here that he's trying to reconstruct.' Religious beliefs are widespread, occurring in every human society, and they can contribute to social stability. Collective worship or shared belief in a supreme being can bring a group of people together, for example. Ramachandran stresses that if the source of religious or spiritual feelings is physiological, this does not undermine or devalue such feelings. He says that they can enrich the patient's life. This poses a dilemma for neurologists: 'What right do we have to treat the patient with medication or with surgery, thereby depriving him of these valuable experiences that often enrich his mental life?' The experiences of patients with temporal-lobe epilepsy may be related to creativity, too. There is evidence that the Dutch artist Vincent van Gogh suffered from temporal-lobe epilepsy, in which case he would have experienced heightened emotional responses to visual images. This may go some way to explaining his emotionally charged, visionary paintings and his delusional mental state.

Opening up the brain

In years gone by, people who claimed that their parents were impostors, believed they were prophets or denied the existence of their medical conditions were usually simply labelled 'mad'. One of the positive things to come from the study of the brain is the potential for understanding strange behaviours like these in terms of physiology. In particular, as we have seen, study of the effects of damage to the most complex part of the brain – the cerebrum – may explain some of the most bizarre things that can happen to the human mind. Ramachandran is excited about the fact that subjects such as creativity, God and religion – once the exclusive province of psychology, theology, philosophy and metaphysics – can now be studied in terms of the physiology of the brain. Consciousness, too – while still elusive to neuroscientists – may one day be understood in terms of signals dashing around the brain, just as memories, emotion and perception are being understood today. As more and more of the incredible behaviours and abilities of the human mind come under study in terms of the brain's hardware, we may be moving closer to the day when the human brain can understand itself.

AFTERWORD
...looking ahead...

Neuroscience is moving forwards ever more rapidly. From a philosophical point of view, it is working towards the same goal as the Human Genome Project, cosmology and the study of evolution: the understanding of the very nature of ourselves. On a more practical level, neuroscience holds out hope for people suffering from conditions such as epilepsy, Parkinson's disease and schizophrenia. Most strategies to overcome or correct diseases or disorders of the nervous system are currently based around drugs or the removal of or deliberate damage to parts of the brain during surgery. The surgical approach to treating diseases or disorders of the brain is often carried out while the patient is awake, since only then can surgeons be sure they are in the desired spot in which they can destroy or remove an area of brain tissue. Although this sort of operation has been carried out for decades, and is growing in success rate and sophistication, it is still an arduous and delicate procedure for patient and surgeon, and does not always work. In the future, when scientific understanding of the brain is far more accomplished, surgery may be more accurate, consistent and free from side effects.

In the long-term future, however, the best treatment of brain damage or deficiency may be the combination of the 'wet' brain and 'dry' silicon-based electronics. Scientists have already managed to connect brain cells to semi-conductors such as silicon, and the development of a 'brain expansion chip' may not be too far into the future. It could perhaps replace areas of brain tissue lost to disease or in accidents. This kind of approach may also one day give sight to some blind people: connecting a light-sensitive electronic chip like that found in a video camera to a patient's optic nerve has already been attempted, with some encouraging results. This is just one of a number of different approaches to helping blind people see again. Neuroscience is changing our relationship with our brains. In the future, paralysis, blindness, deafness, loss of memory, learning disabilities – all of these may be overcome using what will become routine treatment.

Perhaps just as important as the potential neurological treatments is the fact that discovering the pathology – the nature and causes – of brain disorders allows us to understand people with unusual behaviour or strange mental disabilities. Autism, once thought to be the product of bad parenting, is now understood in terms of what goes on in the autistic brain: it is no longer blamed on parents or on the children themselves. People who suffer from schizophrenia or epilepsy are better understood, too. With understanding can come compassion and tolerance.

Understanding how the brain works, how it develops and how it determines behaviour could also have its downside: it might enable people wielding political or economic power to control or manipulate minds, or to discriminate against others. Advertising companies and media-hungry politicians – already often powerful forces in behaviour modification by tapping into our subconscious fears and desires – could launch even more successful offensives on the minds of humans as consumers and voters, armed with sophisticated knowledge about the brain.

The revolution in genetics is also a potential threat to liberty. The modern techniques used by geneticists might enable them to locate particular versions of genes that occur more often in people with certain types of personality than in others. Genes that influence the following behaviours and abilities have all been located: intelligence, depression, alcoholism, thrill seeking and homosexuality (though this is in dispute), as well as a number of congenital brain disorders. In the very near future, scientists will have completed the Human Genome Project – the detailed map of all the genes possessed by human beings. The links between genes and behaviour will be much easier to track down, but perhaps also to change or select. Indeed, one day it may be possible to choose certain behavioural aspects as well as physical characteristics in an unborn child by manipulating the human genome. For example, parents may want an incredibly intelligent baby, or one that is good at languages or musically gifted. For now, this is just over the border from reality, into science fiction. But it may not be too far into the future before it becomes part of reality. Many people are deeply concerned about this possibility, and the spectre of 'designer babies'.

What might the future hold for the study of the brain and behaviour? Perhaps the twenty-first century will bring artificial brain implants, intelligent robots, more consistent and wide-ranging brain surgery, new ways of learning and hyper-intelligent children. However, it may be longer than we think before neuroscience can bring any of these technological marvels, and a radically new view of ourselves: many commentators suggest that neuroscience is at a similar stage of development today as physics and chemistry were around the middle of the nineteenth century. If this is true, then we must recognize how limited our understanding may be, and how some of the currently accepted theories about the brain may be seen as primitive in a hundred years or so. And even then, there will still be many mysteries – just as there still are today with both physics and chemistry.

In fact it is unlikely that we shall ever be able to solve all the mysteries of the brain. The more we discover about the brain, the more we realize there is still to discover. Emerson Pugh, a pioneer researcher into artificial intelligence computer systems, once remarked, 'If the human brain were simple enough to understand, we'd be so simple we wouldn't be able to understand it.'

GLOSSARY

amygdalae	Structures in the brain involved in generating emotional responses, such as fear. They are sometimes considered to be part of the limbic system. There are two amygdalae.
auditory cortex	The part of the cerebrum at which signals from the ear arrive.
axon	A long fibre that is part of a neurone, and along which nerve signals pass within the brain and the nervous system in general.
brainstem	The part of the brain that connects to the spinal cord. Signals to and from the body pass through it – and cross over in it, which is why each cerebral hemisphere is associated with the opposite side of the body.
cerebellum	The 'little brain' at the back of the brain, near the top of the brainstem. It is involved in co-ordination and in keeping the muscles toned.
cerebral hemisphere	One of the two large sections of the cerebrum. The left cerebral hemisphere is more concerned with language and mathematics, the right with spatial awareness and creative ability.

cerebrospinal fluid (CSF)	The liquid, produced in the ventricles, which surrounds the brain and spinal cord. The brain is effectively floating in CSF, and this reduces its weight. CSF also cushions the brain during an impact, protecting it from injury.
cerebrum	The largest part of the brain, associated with its higher processes. The cerebrum consists of two cerebral hemispheres, each divided into four lobes. The outer layer of the cerebrum is the cortex.
chromosome	One of the DNA-containing objects in the nucleus of a cell. Humans have forty-six chromosomes in all, including two chromosomes that determine an individual's sex.
corpus callosum	The large, dense bundle of nerve fibres that connects the two cerebral hemispheres.
cortex	The grey outer layer of the cerebrum. There are three different types of cortex: sensory, motor and association.
dendrite	One of the many slender fibres that emanate from a neurone. Dendrites form synapses with other neurones, so that signals from the other neurones can pass to the dendrite and into the neurone of which it is part.

deoxyribonucleic acid (DNA) Deoxyribonucleic acid, a chemical substance found in chromosomes, which carries information from generation to generation. That information relates to body characteristics, but may also affect behaviour.

hippocampus Part of the limbic system that is involved in generating emotional responses. There are two hippocampi.

hormone A chemical messenger that helps to regulate various bodily functions. Many hormones are produced in the brain, by the hypothalamus and the pituitary gland.

hypothalamus A structure in the brain, literally 'below the thalamus', which produces hormones that regulate body temperature, hunger and thirst, sexual behaviour and aggression. It is closely associated with the limbic system, and most of the hormones the hypothalamus produces directly affect the pituitary gland.

intelligence quotient (IQ) A number derived from psychometric tests that purport to measure general intelligence. It is standardized, so that the average IQ of a large number of people is 100.

limbic system A collection of structures in the brain that lie to either side of the centre, and are involved in generating emotional responses and also in producing memories. Its structures include the hippocampus and the amygdalae.

meninges Three membranes that enclose the brain and the spinal cord. Inflammation of the meninges is called meningitis.

magnetic resonance imaging (MRI) A technique used to produce images of a living brain without having to open the skull. Functional MRI (fMRI) can show which parts of the brain are working the hardest during a particular task.

neurone The fundamental unit of the nervous system. It is composed of a cell body, from which axons and dendrites originate.

neurotransmitter A chemical substance in the brain whose release from an axon facilitates the transmission of a nerve signal across a synapse.

orbito-frontal cortex Part of the frontal lobe of the cerebrum associated with personality, in particular social behaviour.

pituitary gland A small organ, located just behind the nasal passage, which produces a range of important hormones. Most of its functions are controlled by the hypothalamus.

positron emission tomography (PET) A technique used to produce images of a living brain without opening the skull.

psychometrics Tests that aim to measure aspects of behaviour, such as IQ.

receptor A cell or group of cells that responds to a stimulus, such as light, pressure or a change in temperature. The dendrites of a sensory neurone form synapses with receptors, and carry a signal from a stimulated receptor to the spinal cord or to the brain.

synapse A tiny gap between the dendrite of one neurone and the axon of another. Neurotransmitters released at one side stimulate a nerve signal to jump across the synaptic gap.

thalamus Part of the brain that relays sensory information from the head and body to the relevant part of the cortex. There are two thalami.

ventricle A space inside the brain in which cerebrospinal fluid is produced.

visual cortex Part of the cortex on the occipital lobe of each cerebral hemisphere. Signals produced by receptors in the retinas of the eyes arrive at the visual cortex after passing through the thalamus.

EQUINOX: SPACE
The Introduction

Space is where we live, and 'Earth' is our home address. From our beautiful planet, we gaze out into space – in awe of the tantalizing beauty that lurks there and the grand scale of it all. Over the past few decades we have actually begun to venture into space. But the more we look, the more we see there is to understand; and the farther away we explore, the farther we want to go.

Space attracts us for many reasons. Some people are engaged in a quest for understanding, to help make sense of our existence. Some are determined to conquer space, to expand human frontiers. Others see space in terms of its commercial opportunities. All these reasons are investigated in this book, each chapter of which is based on subjects covered by Channel 4's *Equinox* series. Here, I have been able to extend the scope of the television programmes, and to present the science, history, philosophy and politics of these stories in greater detail.

The study of space can be conveniently divided into astronomy and space exploration. Astronomy involves observation, scientific thinking and experimentation, and aims to discover the nature of space and the things it contains. Space exploration involves developing technologies that can escape Earth's gravity and control spacecraft in Earth orbit and beyond. There is often no clear distinction between astronomy and space exploration. Modern astronomy relies on space exploration to enable humans or robot probes to collect information from space. And space exploration requires knowledge about the space environment, as well as knowledge about the nature of space destinations – whether those are the Moon, the planets or, eventually perhaps, the stars.

A major part of astronomy is astrophysics – the study of the physical processes behind the things we see in space. Astrophysics involves knowledge of mathematics, of gravity, of the behaviour of light and other forms of electromagnetic radiation, and an understanding of nuclear reactions, of chemistry and of the nature of space and time. Astronautics is the science and technology of the construction and operation of spacecraft. Much of what is involved in the actual theory and practice of modern astronomy and space exploration is beyond most people's comprehension or interest; but the discoveries of astronomy can have profound effects on everyone. How does it feel to know that we live on an extraordinary but cosmically insignificant planet in orbit around an ordinary star in an ordinary galaxy?

What people see when they gaze into space is a fantastic array of lights – the Sun, the Moon, the planets, stars and galaxies, comets and shooting stars – all shifting around at different speeds, so that the sky is never exactly the same twice. Some of these objects are relatively close to Earth, while others are unbelievably far away. Meteors (shooting stars) are caused by rocks and dust burning up as they enter our atmosphere. The light they produce takes only a tiny fraction of a second to travel the few tens of kilometres from the upper atmosphere to the ground. Light from the Moon takes less than two seconds to reach us. The Sun is about eight 'light minutes' away, while a planet can be between several light minutes and several light hours away, depending on which planet it is and where it is in its orbit. The stars are much farther away – light from the nearest star takes just over four years to reach us. All the stars that people can see with the unaided eye are just a small part of a huge collection: a galaxy, the Milky Way. Ours is a typical galaxy: like an island of stars, 100,000 light years across, in the vast empty ocean of space. The most distant object visible to the unaided eye – as a small, faint, fuzzy patch of light on a dark, clear night – is another of these island communities, called the Andromeda Galaxy. Light from there takes more than two million years to reach Earth. The very farthest objects so far detected in space lie an unbelievable 10,000 million light years away.

Our efforts in the exploration of space have resulted in some spectacular achievements. From Earth's orbit, astronauts have seen our planet from an unprecedented viewpoint, and uncrewed Earth-orbiting observatories such as the Hubble Space Telescope are much better placed for looking into space than their ground-based counterparts. Space exploration is not designed only to give astronauts an exciting ride or to help scientists to look deeper into space: it can bring benefits to many people on Earth, beyond the scientific community. Communications satellites provide international television, telephone and Internet links; there are satellites to help predict the weather, and satellites to help people to navigate; there are even military satellites, secretly doing whatever they do in case the other side does it first. But Earth orbit can be just the departure lounge for travel farther afield. Spacecraft can remain in orbit without using any fuel, and it is relatively easy to blast off from Earth orbit to the Moon, the other planets, and comets and asteroids. Human space pioneers have made it as far as the Moon, but uncrewed robot probes have visited all the planets except for Pluto, the farthest from the Sun. And even the uncrewed space probes have travelled only tiny distances compared to the size of our galaxy. In the foreseeable future, the number of journeys we undertake is likely to increase, rather than the distance of these journeys. Space travel will become more widespread, more routine.

The earliest people probably had no dreams of travelling into space, because they did not understand what and where space was. They were probably too busy surviving to worry about discerning the nature of the Universe, although it is nice to think they at least looked up in wonder. The equivalent of astrophysics for the ancients was workable myths that involved the gods who were in charge of the heavens and Earth. When stable civilizations formed, they began to observe how the Sun, planets and stars shifted across the sky, and learned to construct calendars that would help them to know when to plant their crops. These people used the Sun, Moon and stars as aids to navigation, too. In ancient India and Egypt, astronomers used complicated mathematics to predict when solar and lunar eclipses would happen. The ancient Greeks went further, forming theories that attempted to explain what the Sun, Moon, planets and stars actually are. The prevailing view was that the Earth was at the centre of the Universe, with the Sun, Moon, planets and stars exist on a celestial sphere that rotated around it. Complicated variations of this concept were put forward, which involved spheres within spheres in an attempt to explain the way the planets were seen to move across the sky.

The idea of an Earth-centred Universe was convincing enough until the time of the Renaissance, when the long-standing ideas of the ancient Greeks came under serious scrutiny and re-evaluation. In the sixteenth century, the Polish astronomer Nicolaus Copernicus proposed that the Sun, and not the Earth, was at the centre of the Universe. To many people – particularly leaders of the church, who could not accept that the Earth was not the centre of everything – this idea was not easy to swallow. But several important new discoveries lent support to it. The German astronomer Johannes Kepler worked out mathematically how planets move in their orbits, and his findings corresponded to a Universe like the one Copernicus had envisioned. The Danish astronomer Tycho Brahe discovered a new star in the sky (a supernova), which challenged the idea that the heavens were perfect and unchanging celestial spheres. The Italian scientist Galileo discovered four objects that circle Jupiter in the same way as Copernicus had suggested the planets orbit the Sun. Galileo had also seen surface details on the planets – they were not perfect and unchanging. And perhaps the most convincing evidence of all in favour of Copernicus's view of the Universe came from Isaac Newton's theory of gravitation. Newton realized that gravity could be the mechanism by which Copernicus's Universe could work. Gravity between the Sun, the Moon and the planets explained faithfully their observed movements.

Newton's theory was the beginning of astrophysics proper, because it provided a framework upon which to investigate what the Sun and the planets are: how big they are, how far away and how fast they are moving, for example. And it secured the demise of the Earth-centred views of the Universe.

The telescope led to discoveries of previously unknown planets, and all manner of beautiful fuzzy objects – some of which are now known to be clouds of gas (nebulas) in our own galaxy, while others are now known to be galaxies separate from our own. Using telescopes, astronomers discovered moons around Saturn, Mars and Uranus. The nature of our space neighbourhood, the Solar System, was being uncovered, but the stars and the fuzzy objects remained a mystery. In 1838 the Universe got bigger: the German astronomer Friedrich Bessel made the first measurements of the enormous distances to the stars. Such measurements enabled astronomers to work out the distribution of stars in what they thought was the Universe, though the nature of the fuzzy objects remained a mystery. In 1924 the Universe got bigger again: American astronomer Edwin Hubble worked out the distance to one of the fuzzy objects – the Andromeda Galaxy – and found that it was much farther away than any of the stars. Hubble realized that what astronomers had believed to be the entire Universe was just one of a large number of galaxies.

Many scientific disciplines have contributed to the development of astronomers' ideas. For example, spectroscopy – the analysis of the spectrum of light from stars – revealed what stars are made of, and nuclear physics explained how stars produce their massive energy output. Also, the discovery of infrared, ultraviolet and radio waves opened new windows on the Universe, extending the scope of astronomical observations and leading to the discovery of new types of astronomical object. Photography has been another vital tool for the astronomer: long-exposure photographs – or today, modern electronic detectors – reveal new objects too faint to see otherwise with or without a telescope.

Compared to astronomy, the exploration of space has a very short history. Novels and scientific papers on spaceflight began to appear in earnest around 1900, and rapid development of rockets fit for spaceflight took place from the 1920s. Space was first reached at the end of the 1940s, but a spacecraft first went into orbit around the Earth in 1957. This was the beginning of the space age. As well as launching telecommunications satellites and achieving crewed spaceflight, space scientists and engineers sent probes to the Moon, Mercury, Venus and Mars, and later to Jupiter, Saturn, Uranus and Neptune. These probes – together with orbiting space telescopes – have brought the planets into our homes, presenting for the first time really close-up photographs of them. As far as launching satellites and space probes is concerned, spacegoing technology has become routine.

Initial developments in space technology were rapid. At the time of the first Apollo Moon landings, people had great hopes that space travel would be affordable and routine within their lifetimes. But the early developments were driven more by Cold War politics than by the quest for knowledge about the Universe or the drive towards space exploration. The Americans and the Russians were engaged in a space race. In 'Day Return to Space', we find out what attempts space engineers are making to develop cheaper, reusable spacecraft. Many private companies are becoming impatient with the attempts being made by government space agencies in this direction, and are taking it upon themselves to exploit space commercially. Space is already becoming the next great business opportunity, and tourism is sure to be a part of it – in 'What Shall we do with the Moon?' we discover that hotels are among the money-making opportunities planned for the Moon. There is a cautionary tale in

'Space – the Final Junkyard', which tells of the consequences of irresponsible use of outer space. Orbiting our planet at dangerous speeds are vast clouds of debris from previous space missions, including malfunctioning nuclear-powered satellites. There are many more satellite launches planned – to coincide with and help support the dramatic increase in global telecommunications. Space is becoming congested and dangerous, and we are becoming more dependent on space in our daily lives with every year that passes.

Some of the things upon which we have come to depend may be under threat not only from fast-moving space debris, but also from powerful magnetic storms on the Sun. These sun storms can cause power cuts and perhaps endanger health on Earth, too, and are the subject of 'Sun Storms'. There is a massive worldwide effort to develop our understanding of the Sun, so that we might be able to forecast the 'space weather'.

Astronomy is normally about observing things in the sky – by looking at electromagnetic radiation, such as light, coming from them. But astronomers hunting the most elusive of celestial objects – black holes – have to use some cunning methods in their search. They have recently shown that these mind-boggling objects, once part of science fiction and abstract theory, almost certainly exist. 'Black Holes' tells their story. It may be that the space and time of our Universe was created by the explosion of something like a black hole. That explosion – the Big Bang – happened a long time ago, but just how long ago is a hotly debated topic, and is explored in 'The Rubber Universe'.

Finally, a brief Afterword looks ahead to the fascinating future developments in the study, the exploration and perhaps the exploitation of space.

DAY RETURN TO SPACE

...searching for the space Volkswagen...

The space race of the 1950s and 1960s inspired a generation with the idea that within its lifetime there would be space hotels and bases on the Moon and Mars. Initially, progress in space technology was extremely rapid: the first artificial satellite was launched into orbit in 1957; the first humans made it into space in 1961; and people were standing on the Moon by 1969. Since then, most journeys into space have been undertaken by uncrewed space probes or Earth-orbiting satellites. The space hotels and Moon bases have remained a distant dream, still firmly rooted in science fiction. Will travel into space ever be as routine as long-haul air travel is today? Will *you* ever get the chance to go into space?

Orbital adventures

If a holiday in space were as affordable and available as a cruise on the *QE2* or a month on safari in Africa, would you consider going? Would you like to experience the weightlessness that being in orbit brings? Would you want to see our beautiful blue and white planet beneath you, and a glorious sunrise every hour and a half? Would you want to gaze at the stars from an observation post in a space hotel high above the atmosphere? And if the price was right, would you one day want to go and spend some time on the Moon?

Many people say 'yes' to this kind of question: surveys have shown that between 60 and 70 per cent of people in the USA and Japan would like to take a trip into space. One day, space travel will be a major part of the tourism industry, and will therefore have to be as routine as long-haul aeroplane flights. In the past few decades, many feasibility studies have been carried out into the possibility of space tourism. Most of the recent studies suggest that it would be easy to realize space tourism as a going concern, and they estimate that tourists will have to pay between $10,000 and $100,000 per trip. Despite the perceived commercial value of space tourism, government-run space agencies such as NASA (National Aeronautics and Space Administration) have so far done little to promote the idea. According to a 1998 NASA report, *General Public Space Travel and Tourism*: 'US private sector business revenues in the space information area now approximate $10 billion per year, and are increasing rapidly. Not so in the human spaceflight area. After spending $100s of billions in public funds thereon, and continuing to spend over $5 billion per year, the government is still the only customer for human spaceflight goods and services.'

Government-led space programmes are not making any headway in the space tourism business – which might go the same way as the Internet. In the late 1960s and through the 1970s the Internet was limited to the international scientific community and the US defence agencies. But since the early 1980s access to this growing global information exchange has been possible for anyone with a computer and a modem. Progress in the technology has been astonishingly fast, and now hundreds of millions of people use it to find information, keep in touch quickly and easily, watch videos and listen to the radio. What made this possible, in such a short space of time, was private enterprise. As soon as it became possible for people to make money from this world-wide computer network, people found the best ways to do just that. Technology has made the Internet available, in principle, to everyone, but the development of that technology has been driven by the desire to make money. With private finance, space travel could develop in the same way as the Internet did, quickly becoming accessible to more and more people.

Private organizations – with names such as Spacetopia Incorporated, Zegrahm Space Voyages and Virgin Galactic Airways – have already been formed with space tourism in mind. In March 1997 a German company, Space Tours Gmbh, hosted the first International Symposium on Space Tourism. In attendance were representatives of the aerospace industry, hoteliers and financiers. Several companies presented well-developed plans for tourism in space. That same year, the Japanese construction company Shimusu announced its plans to open a space hotel by the year 2020 – offering anyone with enough money the chance to take a holiday at an altitude of 450 kilometres. And in 1998 a US company called Space Islands announced another well-developed and ingenious plan, for taking holidaymakers up to a hotel 320 kilometres above the Earth's surface. This hotel would be built from discarded Space Shuttle fuel tanks, which at present are jettisoned and burn up as they re-enter Earth's atmosphere. Each of the tanks is 47 metres long, and the proposal is to join twelve of them together to make a huge wheel with a circumference of nearly 600 metres and a diameter of 179 metres. The tanks, on the rim of the wheel, would contain luxury accommodation for up to 350 people. The wheel would slowly rotate, simulating gravitational forces inside the tanks equivalent to one-third that experienced on Earth's surface. For the hub of the wheel there would be a module where people could experience zero gravity. The company hopes to offer tourists a chance to visit their hotel as early as 2005.

This idea may not be pie in the sky: Hilton Hotels, Virgin Airways and British Airways have all expressed an interest in becoming involved. Space Islands estimates that by the year 2010, the cost of a week's holiday at the hotel may be as low as $15,000. With companies selling advance tickets, space tourism is already big business. But there is one thing that is holding it back: a cost-effective vehicle that can launch people into space.

At present, the pinnacle of person-carrying spacecraft technology is NASA's Space Shuttle (more formally known as the Space Transportation System – STS). This remarkable vehicle was tested in the atmosphere in 1981, and made its first orbital flight – in space – in 1983. The Space Shuttle was designed to be reusable and relatively cheap, to make launching into space a routine procedure. And yet each launch costs up to $500 million. NASA's chief administrator, Daniel Goldin, explains the problem: 'We, NASA, are limited in what we can do in space because we spend too much of our budget on launch. Of a budget of $13.8 billion, we're spending close to $5 billion on launch. Think what we could do with that money.' Several thousand members of ground crew are needed for each launch of the Space Shuttle, which can carry up to only seven highly trained astronauts.

To understand why rocket launches are so expensive, it helps to look back over the history of rocket flight.

Countdown to the space age

The first object propelled into space was an American rocket, the V-2-WAC-Corporal combination rocket, which rose to a height of 390 kilometres in 1949. It consisted of a V-2 rocket – of the same design as those launched against several European countries during World War II – with an American wartime rocket attached to the top. The highest altitude reached before 1949 was 85 kilometres, by a German V-2 rocket in 1942. According to space scientists, this is 15 kilometres short of space: they generally define the boundary between Earth and space as 100 kilometres above the Earth's surface. There is no distinct boundary, however – just a gradual decrease in the density of the atmosphere with increasing altitude. At 100 kilometres, the air is so rarefied that no jet engines can function. Jet engines take in air, which contains oxygen that their fuel needs if it is to burn. Rockets, on the other hand, must work in extremely rarefied air or in a vacuum, so they carry both their fuel and oxygen with them.

A rocket accelerates a spacecraft because it produces a huge thrust. The source of this thrust is the expulsion of gases at high speed from the back of the rocket. An inflated party balloon can produce thrust in this way, as air escapes through its neck. As it pushes air backwards, the balloon accelerates forwards. This general principle – that pushing something in one direction will push you in the opposite direction – is a consequence of the laws of motion discovered by Isaac Newton. For a graphic illustration of Newton's Laws, imagine that you are on an ice-skating rink, wearing ice skates, and that next to you is a heavy metal safe standing firm with its legs stuck into the ice. If you push hard on the safe, it stays still while you move off in the opposite direction. The fact that you start moving means that a force is being exerted on you, and it is the safe that is exerting this force. This is an example of a reaction force – an equal and opposite counterpart to a force exerted on an object. The reason you can sit on a chair without falling through it is that the chair pushes upwards on you with a force equal and opposite to your weight, which is pulling you downwards towards the ground. If the chair suddenly disappears, then so does this balancing reaction force, and you fall down to the next lowest object that will provide a reaction force: the ground.

It is not just objects pushing against each other that produce equal and opposite forces in this way. Any pulling force also has its equal and opposite counterpart. When you pull a sled over snow, the sled pulls back on you – by virtue of tension in the rope. As there is less friction between its runners and the snow than between your feet and the snow, the sled moves. If the sled was stuck firm in the snow, the tension in the rope would make you move, however. Objects do not even have to be in contact to exert equal and opposite forces on each other. A paper clip attracts a magnet with as much force as a magnet attracts a paper clip, for example. In the same way, you exert the same force upwards on the Earth as it exerts on you.

In the ice-rink scenario, imagine now that the safe is turned on its side, so that there is very little friction between it and the ice. When you push on it this time, the safe moves away from you. But there is still a reaction force – the safe pushes on you as before – so you start moving, too, in the opposite direction. The mass of the safe is important: a small plastic safe will speed away from you when you push it on the ice rink, while you hardly move. Conversely, a 20-tonne safe would practically stay still, but in this case it would be you moving away at high speed. In fact, if you multiply your mass by the speed you achieve by pushing against the safe, you will get exactly the same answer as if you multiply the mass of the safe by the speed it moves away in the opposite direction.

So much for ice rinks and safes. How does this relate to the party balloon? It seems as though there is nothing pushing the balloon, and yet the balloon shoots forward faster and faster, until all the air has been expelled. The air is expelled because it is pushed by the stretched rubber as the rubber shrinks to regain its normal size. And so, according to Newton's laws, the air exerts a force on the balloon, too. The balloon accelerates until the rubber is no longer stretched. This does not quite explain how a firework rocket works, since there is no stretched rubber to force the air out. In this case, gunpowder explodes inside a rigid cardboard casing. Gunpowder is a solid mixture that includes an oxygen-containing substance. Oxygen is necessary for the other components of the mixture to burn. Because oxygen is part of the gunpowder mixture, and is released as the gunpowder heats up, the gunpowder does not have to take oxygen from the surrounding air and the mixture burns very quickly. As it does so, it produces hot gases. These gases take up more volume than the solid gunpowder, and force their way out at the back of the paper cylinder. The rocket shoots up into the sky.

Rockets have been in existence for about a thousand years, since they were used in firework displays in ancient China. In the thirteenth century, Chinese armies used what they called 'fire arrows' against the invading Mongols. Rocket technology in Europe had a slow development: having arrived in the thirteenth century, rockets were not used effectively in battle until the end of the seventeenth century. All early rockets contained gunpowder in a pointed paper tube. During the eighteenth century, the efficiency of rockets was improved by housing the fuel in a metal tube. In the same century, these rockets were used in warfare in India, against the British Army. The first reliable system designed for military use was developed by a British weapons expert, William Congreve. This was used in 1812, during the Napoleonic wars, when Congreve's rockets rained on several European cities. Congreve's design had an important feature: a long stick that stabilized the rocket in flight. This stability was improved further by a British engineer, William Hale, who designed rockets with angled exhaust pipes that made the rockets spin as they pushed their way along. Many modern rockets still use this kind of spin stability.

The first person to suggest seriously that rockets could be used to launch people into space was a Russian school teacher, Konstantin Tsiolkovsky. In his 1895 book *Dreams of Earth and Sky*, he set out his visionary ideas: for example, he foresaw the possibility of satellites, the use of solar power to provide energy during space travel, and the eventual spread of human civilization across the entire Solar System. In 1898 he worked out many of the principles behind the use of rockets for travel into space and, in 1903, he published a book called *Exploration of Space by Reaction Apparatuses*, in which he wrote: 'Earth is the cradle of the mind, but one cannot live in the cradle for ever.' Tsiolkovsky developed his ideas in many other articles, hoping to inspire others to share his dream. In Russia, he became famous, and was given funds to continue his research into astronautics and aeronautics. But he was not well known outside his own country until long after he died, in 1935.

The first liquid fuel rocket was built by American rocket pioneer Robert Goddard in 1926. Until then, all rockets had solid fuel, often compacted powders as in a firework. Goddard's first liquid fuel rocket climbed to an altitude of 13 metres, in a flight that lasted just two and a half seconds. Liquid fuel rockets have several advantages over solid fuel rockets, particularly for space travel. One of the main advantages is controllability. In a solid rocket, the fuel burns at a set rate. This rate can vary through the rocket's ascent, but only by pre-packing the fuel in a particular arrangement. In a liquid fuel rocket, pumps are used to deliver the fuel to the engine, and therefore control at what rate it burns. By regulating the pumps, you can control the rocket's thrust, even after launch. Like Tsiolkovsky, Goddard was a genuine visionary. As a school physics teacher in Worcester, Massachusetts, he included potential visits to the Moon in the curriculum. He carried out his own research, including proving for the first time that a rocket would work in the vacuum of space – many people believed that a rocket would not work in space because there is no air against which the rocket can push. Initial funding for his work came from the Smithsonian Institution, which in 1920 made public the details of his funding application, and Goddard became mockingly known as 'The Moon Rocket Man'. In 1929 one newspaper featured the headline 'Moon Rocket misses target by 238,799 miles'. Goddard continued his work, undeterred. In 1935 one of his liquid fuel rockets became the first manufactured, self-propelled object to travel faster than the speed of sound.

Also during the 1930s another rocket pioneer – this time in Germany – was pushing back the frontiers of rocket design for the Nazi war effort. He was Werner von Braun, designer of the V-2 rockets that brought terror to Britain, France and Belgium during 1944 and 1945. After the end of the war, von Braun and about a hundred of his colleagues moved to the USA so that they could continue their work on rocketry. The rest of the German rocketry researchers and engineers moved to Russia. Both superpowers had existing rocket programmes, designed to launch intercontinental ballistic missiles with nuclear warheads. They used the expertise and experience of the members of the German team to make their rockets fly farther and higher than ever before. The rocket that reached 390 kilometres in 1949 was a combination of the V-2 and an existing US Army rocket. Seven years after the V-2-WAC-Corporal crossed the boundary into space, another American rocket reached an even higher altitude of 1,090 kilometres.

However, all these objects fell back to Earth again, just like stones thrown into the air. If you could throw a stone with a speed greater than eleven kilometres per second, it would never fall down again. Instead it would continue travelling away for ever – this is not too desirable for a satellite or crewed space vehicle. What you want to do with satellites is to stop them from falling to Earth, but also to keep them from travelling away for ever. If you simply place a spacecraft in space, stationary, it will fall to Earth: it will be attracted back down to the ground again by gravity. The gravitational force on an object decreases gradually with altitude; despite popular belief, there is gravity in space. Even if the satellite was at an altitude of several thousand kilometres, it would – eventually – fall back to Earth. To get around this problem – to prevent falling back to Earth at all – the satellite must orbit the Earth. In this case, it will still be pulled towards the Earth, but it will never move any closer to it.

The gravitational pull between the Earth and an orbiting satellite is just enough to keep the satellite in orbit. If the satellite moves too slowly, it will be pulled back to Earth. If it moves too fast, it will shoot off into space. Out in the depths of space, it will either be pulled back to Earth again or continue moving away for ever. And if it is pulled back to Earth, the satellite will shoot past again only to be pulled back again. It will be moving in a very unusual – or 'eccentric' – orbit. To move in a practical orbit, where its distance from Earth does not vary too much, the satellite has to travel at just the right speed. This speed depends upon the altitude. At an altitude of 300 kilometres, it must be travelling at nearly eight kilometres per second to stay in orbit. It takes about an hour and a half to complete each orbit at this altitude. The orbital speed decreases as it moves farther away from the Earth. At an altitude of 50,000 kilometres it has to travel at only 2.6 kilometres per second. This is still more than four times the top speed of Concorde, and the satellite would be a long way from home. At this distance, it would take about thirty-eight hours to complete each orbit. Most communications satellites orbit above the equator at an altitude of 35,900 kilometres. At this distance, each orbit takes twenty-four hours, and the satellite remains above the same point on Earth – it is geostationary.

Staging a comeback

Any object in orbit around another is a satellite. The Moon is our only natural satellite, but there are now many thousands of artificial ones. Crewed spacecraft generally orbit at altitudes of less than 480 kilometres, in a 'low Earth orbit'. The people-carrying part of the Space Shuttle – called the orbiter – flies to low Earth orbit, stays there for between ten and sixteen days, and then returns to Earth. It was designed to be the first reusable spacecraft. However, not all of it is reusable: to reach the speed necessary to move into orbit, solid rocket boosters are needed in addition to the spacecraft's main engines. At an altitude of 45 kilometres, the solid rocket boosters have used all their fuel and they are jettisoned, recovered and used again up to twenty times. The Shuttle's main engines rely on fuel from a large external tank, which remains attached until orbit, when it too is jettisoned. The empty fuel tank is left to re-enter the atmosphere and burn up (although it could always be salvaged and used to build a space hotel).

So, like all missions that have so far made it to orbit, the Space Shuttle is a multi-stage launch vehicle, and only a small proportion of the mass of the spacecraft at launch makes it back again. This is wasteful and is always going to be costly, as Daniel Goldin of NASA explains: 'Do you think we could build an aircraft to go from New York to London and have the aircraft take off and then a piece drops off and then another piece? That's why it's inefficient.' The cost of carrying people into space will remain prohibitively high until launches become cheaper. And what is needed to make it cheaper is a totally reusable space vehicle: a SSTO (single-stage-to-orbit) spacecraft.

The multi-stage approach to launching spacecraft is a remnant of the early days of space travel, when it did not matter how cheaply you could get into space – just that you got there at all. The launch vehicles looked and worked like missiles, because the American and Russian space programmes both had their origins in the development of warhead-carrying rockets. As well as the development of missiles, several scientific missions were considered. As part of the International Geophysical Year – which strangely ran from July 1957 until December 1958 – the Americans and the Russians both planned to send up satellites that could carry out Earth science experiments from the ultimate vantage point, in orbit. American plans hinged on the Vanguard rocket, which had been put together by a team from the US Navy. During 1956 and 1957, however, attempts to launch the Vanguard ended in disaster. Still, the Americans smugly assumed that they were far ahead of the

Russians in spaceflight technology. But they were wrong: the first manufactured object to go into orbit was Russian.

After several announcements of their intent – which most Americans wrote off as unrealistic – Russian space scientists succeeded in placing an artificial satellite into orbit around the Earth on 4 October 1957. This was Sputnik I – a silver-coloured metal sphere with four long straight aerials and a mass of 83.5 kilograms. The launch of Sputnik I was originally planned to coincide with the hundredth anniversary of the birth of Konstantin Tsiolkovsky, which would have been 19 September 1957; the delay was due to technical problems. Sputnik's aerials transmitted a test signal for ninety-two days, after which the satellite burned up as it re-entered the atmosphere.

Scanning the sky with radio telescopes, to tune into the test signal, was the only way to convince some people that Sputnik was really in orbit. Once proven, the Russian achievement dealt a blow to the national pride of the Americans. Not only was it a matter of pride: the Americans and Russians had been involved in diplomatic rivalry – the Cold War – since shortly after the end of World War II. If Russian rockets could manage to place a satellite into orbit, they could certainly launch a nuclear warhead to strike anywhere on Earth. As Daniel Goldin remarks: 'The objective of the space programme in the Eastern Bloc and the Western Bloc was to show technical superiority ... it was about winning the hearts and minds of the uncommitted countries of this world.' And the Americans were lagging behind.

Things got worse before they got better for the American space programme: in November 1957 Sputnik II, weighing more than half a tonne – nearly six times as heavy as Sputnik I – was placed successfully into orbit. Sputnik II carried a dog, called Laika, around the Earth for 162 days, until the dog and the satellite both burned up as they re-entered the atmosphere. At the time, the American satellite programme was in tatters. The Navy's Vanguard project included a plan to send a small scientific payload into orbit, but the Vanguard rockets were not working as hoped. Fortunately for the USA, Werner von Braun had been developing another launch system, the Jupiter-C, with the US Army. And on 31 January 1958 a Jupiter-C rocket carried a smaller rocket and a payload into space, where the small rocket pushed the payload – a satellite called Explorer I – into orbit. Explorer I was considerably smaller than even the first Sputnik, but did help to make important scientific discoveries, such as the existence of a belt of charged particles encircling the Earth, called the Van Allen Belts. Further American satellites followed, including one launched on top of – at last – a Vanguard rocket in March 1958.

The space race was now in full swing. The next major challenge was to send space probes beyond Earth's orbit to investigate the Moon and the other planets. Again, the Americans lagged behind. They attempted to send three probes to the Moon late in 1958, but the rockets that were to take them into orbit failed on launch. And so the Russian space effort made it to the Moon first, too. In 1959 they sent up two probes: Luna II deliberately crash-landed on the Moon's surface, while Luna III took photographs of the far side of the Moon. As the Moon turns once on its axis in exactly the same time it takes it to orbit, one particular half of its surface is always hidden from Earth, so the Russian achievement was significant. The American space programme did pick up, however, particularly as a result of the formation of NASA in October 1958, in response to the launch of Sputnik I. Until 1958, American efforts to travel into space resided in NACA (National Advisory Committee for Aeronautics) and the disparate plans of the US Army and Navy. NASA provided a way of administering and co-ordinating the American efforts in space.

The next obvious development, after satellites and space probes, was to fire a person into space. Sputnik III had carried a dog, and in 1960 another Russian craft, Sputnik V, carried two dogs – Strelka and Belka – into orbit and safely back again. This was obviously an important break-through: Sputnik V was the first object to be recovered from orbit, and the fact that the dogs had returned alive made this a vital step towards putting humans into space. The first person ever to make it into space was Russian astro-naut Yuri Gagarin, who made slightly more than one orbit of the Earth in April 1961 at a maximum altitude of 300 kilometres, before re-entering the atmosphere to a hero's welcome.

The Americans soon followed, in May of the same year, sending Alan Sheppard into space – though not into orbit. Also in this month the American space programme received an important boost, when President John F. Kennedy announced an effort to redress the balance in the space race: he committed the US space programme to putting people on the Moon, with the Apollo programme. When elected in 1960, Kennedy had made crewed spaceflight a key political goal; now, on 25 May 1961, he proclaimed: 'I believe this nation should commit itself to achieving the goal, before this decade is out, of landing a man on the Moon and returning him safely to Earth. No single space project in this period will be more impressive to mankind, or more important in the long-range exploration of space; and none will be so difficult or expensive to accomplish.'

In February 1962 an American rocket took John Glenn into orbit, as part of the project Mercury, the precursor to the Apollo programme. To achieve the goal of Apollo within the proposed time scale, more than four per cent of the entire US federal budget would have to be committed to the project. The Saturn V rockets that launched the Apollo spacecraft into space were incredibly powerful, but they were also the ultimate in wastefulness. Single-stage launch vehicles would have been much cheaper, particularly if they could come down in one piece, to be reused. But cost was not important, since space programmes in the early years were driven largely by Russian and American political ambitions, with little care paid to the costs involved. If cost had been an issue, development in the space programmes would not have been as rapid. And so it was huge, expensive, missile-shaped rockets that made headway in the space race.

The most obvious way to make a space vehicle that is reusable is to design it like an aeroplane, so that it can land horizontally, on an airfield. The Space Shuttle is a compromise between a reusable spaceplane and a wasteful multi-stage rocket. So, although it is lifted into space by huge rockets and an external fuel tank that are jettisoned during the mission, it does land back on Earth gracefully, just like a glider, to be reused. In 1983, the year of the Space Shuttle's first mission, British aerospace engineer Alan Bond proposed a radical spaceplane, called HOTOL (horizontal take-off and landing). This would have been an elegant spacecraft, with swept-back wings and an engine that would work as a jet when it was in the atmosphere, but as a rocket when in space. Much less oxygen would have to be carried in the fuel tanks, since the engine would take oxygen from the air for much of its ascent – a significant advantage over conventional rockets. However, to achieve orbital velocity from a standing start on a runway, the craft would have to accelerate quickly to hypersonic speeds. The project was developed by British Aerospace and Rolls-Royce, but the technological and financial challenges were too great and the project was shelved in 1989. However, a private company formed by ex-Rolls-Royce engineers is now working on a successor to the HOTOL concept, called SKYLON. Like HOTOL, it is a spaceplane that will have a hybrid engine – one that will breathe air in the atmosphere but will carry oxygen for flight in space.

And so the quest for a truly reusable spacecraft continues. Until the Space Shuttle, all spacecraft that returned from space crash-landed – except one. In 1963 a rocket-powered aeroplane set an altitude record of 101 kilometres, just crossing the boundary into space. This was the X-15, the culmination of a secret US project to develop supersonic aeroplanes. The 'X' of the X-programme stands for 'experimental'. The earliest product of this top-secret technological drive for innovation – the X-1 – became the first aeroplane to travel faster than the speed of sound, in 1947. The X-plane programme was run first by NACA, and then by NASA. It was bold and dangerous, and many pilots lost their lives, but much was learned about the effects of travelling at extremely high speeds. It was not the main aim of the programme to design spaceplanes, but the development of the Space Shuttle programme can be traced back to these heady days. Much was learned, for example, from the X-15's successful glide back to Earth from above the boundary of space. Since Yuri Gagarin's first spaceflight, fewer than 400 people have ventured into space. To increase that figure dramatically, the cost of getting into space will have to be reduced considerably. What is needed is the equivalent of the Volkswagen: a people's vehicle for space travel. What will this spacecraft be like?

Any ideas?

In 1995, in an effort to develop true reusability, NASA invited aerospace companies to bid for a project to build a successor to the Space Shuttle. This was a continuation of the X-programme, though it was far from top secret, and the new vehicle would be called the X-33. The new vehicle would have to carry into space the same weight as the Space Shuttle can – 23 tonnes – but be totally reusable. Furthermore, it would have to be developed within a budget of $1,000 million – only twice the cost of a single Space Shuttle launch. Of the three serious contenders for the X-33 contract, one was the company that built the orbiter of the Space Shuttle – Rockwell International. Rockwell's design was a winged spaceplane, not unlike the Space Shuttle itself. To comply with the requirements of the X-33, Rockwell had developed the idea of the Space Shuttle into a spaceplane that needed no boosters or external fuel tank. The operating costs would be less, too, as the number of people on the ground would be reduced to fewer than a hundred.

Another design – and the one hotly tipped as the favourite to win the contract – was submitted by McDonnell Douglas, a company that also had long-standing links with the aerospace industry. McDonnell Douglas's radical design was based on a spacecraft that they were already testing, in conjunction with the US Air Force, called the Delta Clipper. The DC-X – a one-third-scale prototype of the Delta Clipper – had already undergone successful trials in 1993, in the Army's White Sands Missile Range in New Mexico. There were two radical features to the Delta Clipper that helped make it an attractive proposition as a reusable launch vehicle. First, it landed vertically, using rockets to slow its descent in a delicate hovering motion. This soft landing could be achieved at space ports almost anywhere, without needing a long runway as the Space Shuttle does. Second, fewer than twenty people were needed for each launch – a significant reduction compared with the several thousand people needed for each Space Shuttle launch. Major Shell, McDonnell Douglas's spaceplanes test manager, says, 'With the DC-X, the goal was three people to fly it, in the operations centre, with fewer than fifteen people on what we call touch labour.' In addition, the Delta Clipper would be able to fly again within a week of landing.

The DC-X's first test flight was an incredible success. This eye-catching vehicle – which looked like a 13-metre-tall, upside-down ice-cream cone – moved up into the air, hovered for a few minutes, carried out a horizontal 'translational' movement, by tilting slightly, then eased itself slowly back down to the ground. A few metres above the ground, three sturdy legs emerged from the flat base of the craft, and these supported it as it touched down. This unusual design brought a great deal of interest, both within and beyond the space science community. Although it was designed as a military 'space truck', under contract by the Strategic Defense Initiative Organization, the Delta Clipper's designers had wider goals in mind. The craft would ultimately be able to carry military equipment, commercial satellites or tourists into space. In 1995 NASA took the DC-X under its wing, renaming it the DC-XA – the Delta Clipper Experimental Advanced. This is why most people assumed that the DC-X would be the vehicle chosen by NASA for the X-33 contract. And McDonnell Douglas themselves were confident. Their X-33 programme manager, Paul Klevatt, explains the capabilities of the team that was developing the proposal – the same team that designed the DC-X: 'McDonnell Douglas was able to fly the DC-X in twenty-four months and three days, from a piece of paper that was just a blank proposal ... The learning is there, the processes are in place.'

The third design put forward for the X-33 programme came from the company that assembled the Hubble Space Telescope: Lockheed Martin. Their design for the X-33 – which they dubbed 'VentureStar' – was unusual. It was a single-stage spaceplane, but it had no wings. The shape of the fuselage was called a lifting body: the lift force that would make it fly was created by airflow over the entire body, not over wings as in a conventional aeroplane. The Space Shuttle is a lifting body: although it does have wings, most of its lift force is gained by virtue of the shape of its fuselage. In the VentureStar, there were no wings at all – just small fins for control and stability. The idea for the VentureStar was the brainchild of one person, Dave Urie, who says: 'A lifting body is a thick delta wing; and so we haven't got rid of wings, we've just changed the shape of the wing so that it makes a better hydrogen container.'

Lifting bodies had been around since the 1950s: six of the secret X-planes were wingless lifting bodies, and they were involved in more than 200 test flights between them. These unlikely-looking aeroplanes were carried underneath bomber aircraft, such as the B-52, and released at high altitude. Lockheed Martin's design would have to take off under its own power. And as it was to be a single-stage vehicle, it would have to take all its fuel with it. In fact, the weight of the fuel would have to be around 90 per cent of the total weight at lift-off. In other words, the weight of the spacecraft and its payload must be only 10 per cent of the total. To address this, it was proposed to make the body out of strong but incredibly light new materials and to employ a new type of engine, called an aerospike. Lockheed

Martin's design – with its white body, black nose cone and black and white fins – looked a little like a penguin. It was a plump, graceful, black and white triangle.

In July 1996 NASA's decision on which of the three designs would be taken up as the X-33 was announced: addressing a live television audience, Al Gore, who was then the Vice President, lifted a cover to reveal a model of the winning design. 'This is the craft that can carry America's dreams aloft and launch our nation into a sparkling new century,' he said enthusiastically. To many people's surprise, it was a model of Lockheed Martin's design that was sitting there; Dave Urie's unusual, but slick, penguin-shaped lifting body had won. The team from McDonnell Douglas, whose DC-X design had been tipped to win, were very surprised, and very disappointed. Soon, after four more weeks of successful flight trials, their disappointment turned to disaster. In the DC-X's testing ground in White Sands, the test vehicle toppled as one of its legs failed on landing. As it hit the ground, the whole spacecraft exploded in a huge and devastating orange fireball. This was to spell the end of McDonnell Douglas's DC-X programme.

Dr Hans Mark, an ex-deputy administrator of NASA, believes that the DC-X would never have met the challenges of the X-33 programme: 'The DC-X was a publicity enterprise really, to heighten the tension. Also the DC-X itself had no technology that had anything to do with solving the problem. The engines were thirty-year-old RL10 engines and the aeroshell was made out of a plastic with a low melting point ... I never had much of a use for it.' Those involved in the DC-X project, however, were very upset about the termination of their development programme because of a single failed leg. Major Shell of McDonnell Douglas's Philips Laboratory was one of those for whom this was a bitter blow: 'We put so much of our time, effort and careers into it, we hate to see it end like that.' Daniel Goldin of NASA responded to criticism of the decision: 'They lost; we ran a fair and open competition. This is how democracy works.'

NASA chose Lockheed Martin's design to be the X-33 because it was the one with the most technological challenges to be overcome – the project was the one most likely to drive aeronautics forward. Within weeks of the X-33 announcement, representatives of NASA and Lockheed Martin met at the Skunk Works – Lockheed Martin's once-secret engineering base in the Mojave Desert. Within thirty months, Lockheed Martin would have to build a half-scale prototype for testing in the atmosphere. They would receive $941 million from NASA, and would have to contribute $220 million of their own. Despite these huge sums of money, Lockheed would have their work cut out to meet the deadline. Paul Landry, chief engineer on Lockheed Martin's X-33 programme, spells out the nature and magnitude of the task: 'The difficulty of putting man on the Moon – to me, it was easier to do that than it is to do this programme; here we have a very aggressive schedule, a cost factor that is limiting us on one side, and the technologies that we're having to try to use to get there.'

It was the very asset that won Lockheed Martin the contract – the necessity of developing new technologies – that would make the project extremely challenging. The team would have to design a new type of airframe that would have to be both strong and incredibly light, and would have to withstand the extreme heat of re-entry. To achieve this, the heat shield will be integrated into the airframe. Unlike the Space Shuttle orbiter's heat shield, comprising thousands of glued-on ceramic tiles, the X-33 – and eventually the

VentureStar – will use nickel alloy panels that are attached to the cage-like bracketing system to form the fuselage itself. While the Space Shuttle's tiles are subject to frequent and detailed inspection, re-waterproofing and general maintenance, these panels will need only one waterproofing and can be replaced easily thanks to the bracketing system. Weight is a crucial factor, and is fundamental in every aspect of the design. The fuel – liquid hydrogen and oxygen – will be held in huge, state-of-the-art cryogenic tanks that are extremely light given their size and strength. There are two hydrogen tanks, which take up much of the volume of the craft, and one oxygen tank. The oxygen tank is made of aluminium, and is therefore very light for its size. A further weight saving is achieved by attaching the X-33's landing gear directly to the tanks – to the oxygen tank at the front and the hydrogen tanks at the rear.

Another significant weight-saving aspect of the design is to be found in the engines. The VentureStar's aerospike engines have only about one-quarter of the size and weight of a conventional rocket engine that produces the same thrust. Much of the saving in weight is due to the fact that an aerospike engine does not have a heavy, bell-shaped nozzle like other rocket engines. The familiar bell shape of the rocket engine nozzle is designed to confine the plume of exhaust gases to a concentrated stream. The shape of the exhaust plume does not change as the atmospheric pressure changes with altitude, and so conventional rocket engines work at maximum efficiency only at a particular altitude. In an aerospike engine, the exhaust gas – water that is produced by the reaction of hydrogen and oxygen – is pushed out from the sides of the reaction chamber, along a tapered 'nozzle ramp'. It is as though the nozzle has been turned inside out; the result of this is that the outside of the chamber is open to the atmosphere. And so, as the atmospheric pressure changes while the X-33 climbs higher, the shape of the exhaust plume automatically adjusts to maintain maximum performance.

The simple but fiendishly clever design concept of the aerospike has another advantage over conventional rocket engines. Traditionally, steering a rocket engine has been a complicated task, involving tilting the whole engine assembly on gimbals (pivoting components). In an aerospike engine, there are no gimbals – in fact there are no moving parts at all. The reaction takes place either side of the nozzle ramp, and increasing the flow of fuel to one side means that the engine's thrust is unbalanced, forcing the X-33 to which it is attached up or down. Turning – banking or rolling – is achieved by adjusting the orientation of the fins.

The concept behind the aerospike engine was first developed during the 1960s, and was considered for the Space Shuttle. However, the technology remains untested in spaceflight. In October 1997 aerospike engines underwent their first tests in flight in the atmosphere, as part of a one-tenth-scale model of the X-33 attached to a supersonic jet aeroplane, the SR-71, or 'Blackbird' – a spy-plane built at the Skunk Works, by the then Lockheed Corporation, in the 1960s. The SR-71 is the fastest production-model aeroplane ever built. It can fly at speeds of more than 3,200 kilometres per hour – more than three times the speed of sound – and at altitudes higher than twenty-six kilometres. In December 1999 the X-33's engines were tested successfully at full power for the first time, this time in a ground-based test facility.

Both the VentureStar and its prototype, the X-33, will be piloted robotically, by a sophisticated system of avionics (electronics that control flight). The X-33 is half the size, one-ninth of the weight and one-quarter of the cost of the

VentureStar. Overall, the VentureStar will be considerably shorter than the Space Shuttle orbiter – 39 metres long compared with 56 metres. Despite the fact that the VentureStar will be able to carry the same mass of payload, its total mass at lift-off will be considerably less than the Space Shuttle: 1,190 tonnes compared with 2,045 tonnes, just over half the weight. The overall aim of the X-33 programme is to reduce the cost of putting objects into orbit – from the current figure of $22,000 per kilogram down to $2,200 per kilogram).

And so, soon it will be a wingless, penguin-shaped, reusable spaceplane that you will see instead of the Space Shuttle. VentureStar should be operational by 2004, and will make routine flights into orbit from then on. Even if these aims are achieved, this spacecraft will not be the 'space Volkswagen' that many people are looking for. It is designed to carry pre-prepared payloads – satellites or crew for the International Space Station, for example – which will be loaded into standardized modules that slot into the craft's cargo bay. For all its amazing technological advances, it will not be suitable for carrying fare-paying tourists to space hotels. Another project that is under way – the X-34 – is similarly aimed at cargo-carrying flights. The X-34 is a smaller-scale project than the X-33, and the result will be another robot-piloted, reusable craft. It will be launched in the air, after receiving a piggy-back from an aeroplane, but will not reach orbit as the VentureStar will. The X-34 is really just a test-bed: a way of gauging the performance of certain new technologies. Some people think that the X-33 and even the VentureStar have similar goals. Ex-NASA deputy administrator Hans Mark reckons, 'We are not at the point where the technology is ready to commit to a single-stage-to-orbit vehicle. That's why the X-33 which was picked out of these three concepts is really not a vehicle to go into space; it's an experimental aeroplane ... My guess is that we're looking at a twenty-year timescale before we actually have a single-stage-to-orbit vehicle.'

Too long to wait

For most people who want to see affordable access to space, twenty years is far too long to wait. We have watched in wonder the incredible achievements of the American and Russian space programmes. We have seen the pictures of people weightless in orbit; we have seen pictures of people floating free in space, attached only by cables and tubes to their spacecraft; we have seen pictures of the curvature of Earth and the glorious views of the Earth from space – now many people want to experience these things for themselves. The basic technological challenges that make spaceflight different from flight in the atmosphere have been overcome, so people are naturally wondering why space is still not within their grasp. But all we see are more satellites being sent up to orbit by the telecommunications companies and global television empires, and more, but smaller, space probes. And now the most exciting development in spaceflight technology is a robot-piloted vehicle with no facility for taking ordinary people into space. For some people, the pace of development is far too slow.

In an effort to speed things up, an organization based in St Louis, Missouri, set up the X Prize – a competition aimed at stimulating the design and production of just the kind of reusable launch vehicle that is needed to transport passengers into space. The X Prize was announced in May 1996: the chair and president of the X Prize Foundation, Peter Diamandis, explained that the winning spacecraft must 'be able to carry at least three human adults, go to

100 kilometres altitude, which is above the boundary of space, come back safely, and then launch again within two weeks; it needs to be reusable, low-cost and be able to carry you and me into space.' The rules of the X Prize are well thought out, and ensure that the aim of the competition are upheld. They are:

- The entries must be privately funded – this will ensure that governments cannot buy into the prize, which the organizers believe would reduce the chances of the competitive spirit they hope will encourage space tourism.
- The winning craft must reach 100 kilometres, the recognized boundary of space – any higher and sophisticated heat-shielding would be required, increasing costs.
- The winning craft must be able to carry at least three people – this will immediately open up the possibilities for fare-paying would-be astronauts.
- The same vehicle must be reusable within two weeks – this will mean that only a minimum amount of work can be carried out to prepare the craft for its next launch, which in turn should bring down the costs to the potential fare-paying customers.

The X prize competition is based on the aviation prizes that were held from the very earliest days of flight. One of the first was the £1,000 offered by the *Daily Mail* that led Louis Blériot to become the first person ever to fly across the English Channel, on 25 July 1909. Another prize, also sponsored by the *Daily Mail*, encouraged John Alcock and Arthur Whitten Brown to make the first transatlantic flight, an arduous sixteen-hour journey from Newfoundland to Ireland. The two pilots received £10,000 for their efforts. On 21 May 1927 Charles Lindbergh won $25,000 as he became the first person to fly non-stop between New York and Paris. The amount of money offered in the aviation prizes increased over time and with the magnitude of the task. However, they have one important thing in common: the amount of money spent by people attempting to win the prize far outstripped the actual prize money. In the case of Lindbergh, for example, sixteen other attempts were made to achieve the task, and about $400,000 was spent in total.

So offering prizes is a very shrewd way of attracting large amounts of funding to encourage rapid advances in technology. The effect of the aviation prizes was to increase the safety, efficiency and range of aeroplanes, and this in turn made aeroplane flight within the reach of many more people. The advent of affordable air travel would probably have been considerably delayed had it not been for these initiatives. The organizers of the X Prize hoped that it would do the same for space travel, and are offering $10 million to the first team who can achieve the objectives they have set out. The founders of the X Prize include Byron Lichtenberg – who has flown on two Space Shuttle missions – and Erik Lindbergh, the grandson of Charles Lindbergh. At the time of writing, there are seventeen teams registered for the X Prize. All the entrants have expertise in aircraft or spacecraft design. Twelve of the entrants are from the USA, one is from Argentina and one from Russia. There are also three British entrants, though one of them is based in Stuttgart, Germany.

One of the X Prize entrants is Mitchell Burnside Clapp, who was involved in the DC-X programme. Burnside Clapp is unconvinced about the safety and viability of powered vertical landing, like that used by the DC-X. Instead, he prefers horizontal take-off and landing – 'the way people were supposed to fly'. His company – Pioneer Rocketplane

Incorporated – is designing a spaceplane that will take off from a runway under jet power, and will be given liquid oxygen for the rocket engines in-flight by a tanker aeroplane, through an interconnecting hose. The runway and tanker refuelling operation are existing technologies, making this approach to reaching space routine and achievable. 'We are attempting to adopt an aviation pattern from the beginning,' explains Burnside Clapp. The jet and rocket engines aboard the Pathfinder will be powered by kerosene. The tanker aeroplane will do the hard work of carrying the liquid oxygen to altitude. Transferring the liquid oxygen in the air should also make the launch site safer. At 110 kilometres, the Pathfinder's payload bay doors will open, and the payload can be launched into orbit. The spacecraft will then glide back to Earth in the same way as the Space Shuttle orbiter does, but the jet engines fire again to help control the craft during landing. Because this spaceplane is similar to a conventional aeroplane, and is piloted, it could even travel to a designated landing site to pick up a satellite which it will then go on to launch into orbit. Apart from the transfer of liquid oxygen from a tanker aeroplane, this whole system depends upon tried-and-tested components and techniques.

As with many of the X Prize entrants, Pioneer Rocketplane was not formed simply to compete for the $10 million prize. As their product is composed of existing technologies, they believe that they can offer significant reductions in the cost of launching satellites within a few years. This idea was given support in 1997, when Pioneer Rocketplane was among four companies awarded a contract to develop a 'bantamweight satellite' launcher. Such a satellite has a mass of about 150 kilograms.

Some of the other entries to the X Prize will also rely on conventional aeroplanes to help them on their way into space. For example, the Russian entry, Cosmopolis 21 – designed by a company of the same name – will be a spaceplane that sits upon a powerful carrier aeroplane to an altitude of 20 kilometres. At this height, the carrier aeroplane will turn upwards, into a steep ascent, and the spaceplane's rocket engine will take over, launching the spaceplane above the boundary of space. Another entrant that will use a conventional aeroplane to help it into space is the Eclipse Astroliner, proposed by Kelly Space and Technology Incorporated, based in San Bernadino, California. The Astroliner will be a delta-winged craft that will be towed into the air by a Boeing-747 jumbo jet, rather like the way a glider is launched. The company claims that towing the spaceplane into orbit is safer than attaching it directly to the fuselage of an aeroplane. The Eclipse Astroliner will then be released, and will use a conventional kerosene-powered rocket engine to fly to 182 kilometres, deposit its payload and glide back to Earth.

Another company competing for the X Prize is Scaled Composites, based in the Mojave Desert, near to Lockheed Martin's Skunk Works. They also propose giving its spacecraft a piggy-back on a conventional aeroplane. Their entry is Proteus, a spaceplane that will be taken on a specially designed aeroplane up to 11 kilometres, and from there use rocket power to thrust up to the 100-kilometre point stipulated by the X Prize. The aeroplane part of the design has already been built, tested and exhibited at several air shows. It is made from an ultra-light composite material – the speciality of Scaled Composites, which is normally more concerned with aeroplanes than with spacecraft. The company was founded in 1982 by Burt Rutan, an ex-air force engineer with a great deal of experience in the aerospace industry. On 14 December 1986 Rutan and Jeana Yeager took off from Edwards Air Force Base, also in the Mojave Desert, in a remarkable aeroplane – called Voyager – which Rutan had designed. It carried just 6,000 litres of fuel, but nine days later it arrived back at Edwards Air Force Base, after flying around the world without refuelling. What made this possible was a remarkable composite material designed by Rutan himself. Voyager weighed less (when empty) than a car, despite its 33-metre wingspan. The first stage of Rutan's X Prize entry – the aeroplane – is made from the same material as Voyager, but the second, rocket-powered spaceplane stage is still under wraps. Rutan is determined to make his dream of affordable spaceflight a reality. He is unhappy with the approach taken by NASA: 'They are in the way, instead of helping us, as far as you and I going into space is concerned ... it is as if they don't want us to ever go.'

Most of the entrants to the X Prize are spaceplanes, though not all of them are released by conventional aircraft. The Cosmos Mariner, for example, designed by Dynamica Research, is among those that will take off from a conventional runway, carrying everything it needs to make it into space. The main problem to be overcome by single-stage-to-orbit spacecraft concerns weight: at take-off, a rocket must carry all its fuel and an oxidizer (normally liquid oxygen), in huge quantities, and this greatly increases the weight. The Pathfinder, described above, gets around this problem by taking on oxygen from a tanker aeroplane at altitude. Others, such as Rutan's Proteus system, will carry the spacegoing vehicle up to altitude, so that it does not need to carry so much fuel. The approach taken by Dynamica Research is different: it will use an 'air-breathing' jet engine while in the atmosphere, and then switch to rocket power when the air becomes too rarefied – just like

Alan Bond's HOTOL would have done. The rocket engine will be switched on at a height of about twelve kilometres, as a result of which the Cosmos Mariner will shoot up to above 100 kilometres, to glide back down to Earth again. The same approach is taken by one of the British entrants – Bristol Spaceplanes, based in Bristol – with their spaceplane, called Ascender. Similar again is the X Van, another jet-and-rocket spaceplane, built by Pan-Euro Inc. There is one significant difference with the X Van: the launch will be vertical, like a rocket, and the first phase of the flight – to 10 kilometres – will be near-vertical. The craft will glide to Earth, like the other spaceplanes.

While some entrants to the X Prize intend to use rocket power only, and others intend to use a mixture of jet power and rocket power, there is one who intends to use jet power and no rocket power. He is John Bloomer, of the Discraft Corporation. His proposed spacecraft is called the Space Tourist. Because the jet engines need air to operate, they will be used only below about sixty kilometres. The Space Tourist's powerful pulsed-jet engines will accelerate it to a high enough speed to shoot it above 120 kilometres. The fact that the X Prize rules stipulate only that the craft reaches space does not mean that the winning craft must go into orbit. And so this ballistic approach – one big thrust, then let momentum and gravity do the rest – is acceptable. Indeed, none of the entries is designed to go into orbit. One way or another, they are all designed to travel to extremely high altitude, where gravity will pull them back down again.

So far, we have only heard about spaceplanes. Some other entrants to the X Prize propose using a more tried-and-tested way of getting into space: vertical take-off, missile-shaped rockets. One of the entrants taking a chance on traditional rocketry is Dr Graham Dorrington, based at

Flight Exploration in Stuttgart. His design, Green Arrow, will use kerosene fuel, but with hydrogen peroxide rather than liquid oxygen. All the fuel will be used to produce one powerful thrust, and the spacecraft will coast up to 100 kilometres, from where it will return to have its descent slowed by parachute and its landing softened by gas-filled balloons. Other designs will use rockets like this and various parachutes, balloons or aero-shields, to slow their descent. Thunderbird, put forward by Steven Bennett of the Starchaser Foundation, based in Dukinfield just outside Manchester, is one of them. Thunderbird looks like a conventional, missile-shaped rocket but, like many of the other entries, it will use a combination of jet and rocket power. Bennett comments, 'You don't have to be a rocket engineer to do this. The technology to do what we're proposing to do has been around for the past forty years ... It's just a case of bringing these existing things together in a smart way and bring them together in a unique way to crack the problem.'

The most unusual entry to the X Prize is Roton – a space helicopter being put forward by HMX Incorporated. Roton looks like the DC-X – a huge, upturned cone – but it has one important difference. At lift-off, Roton uses a rocket engine at its base, just like the DC-X. It climbs vertically, burning the kerosene fuel and liquid oxygen. For its descent, Roton will deploy its rotor blades, which spin like the blades of a sycamore seed falling slowly through the air. This will provide Roton with a controlled approach to landing. Burt Rutan, whose company Scaled Composites is producing the material from which Roton is being made, believes that one of the good points about Roton is its safe descent: 'I'm a helicopter pilot, and I know that you can take a helicopter to any altitude and fail the engine, and it's actually a very good way to get down.' The landing will be controlled yet further by firing up the rotor blades, which will have small rocket engines at their tips. In the original form of the design, the powered rotor blades actually provided lift to raise the craft off the launch pad, and to help with the first phase of ascent. The spinning action of the rotor blades would be used to pump fuel at high pressure to the main engines, situated in a ring around the base of the spacecraft, increasing the engine's efficiency dramatically. However, to reduce the financial risks involved in the project, development work on this design has been deferred, and a conventional rocket engine will now be used.

The concept of Roton was devised by Bevin McKinney, who describes it as 'not anything like an existing rocket configuration'. McKinney's colleague Gary Hudson remarks: 'Usually after ten minutes of saying, "Gee, that can't work", they say, "Hmm, actually that helps you here, or here", and by the time they're done they're asking if we need another consultant.' Whether it wins the X Prize or not, Roton may be here to stay: the atmospheric test version of the Roton was 'rolled out' in March 1999, when it hovered its way through a range of tests. A company set up to take on the Roton concept – the Rotary Rocket Company – hopes to conduct orbital tests early in 2001, and to go into commercial service later that year.

In addition to Roton, some of the other vehicles competing for the X Prize will also still be built if they lose the competition, and will probably go into service. In addition to the X Prize, there are several other spaceflight awards on offer – with smaller prizes, but just as much prestige. Examples are the CATS Prize (Cheap Access to Space) and the FINDS Prize (Foundation for the International Non-governmental Development of Space). These will help to ensure that the human race quickens its step in the march to space. In fact, the pace is quickening quite naturally – the prizes are just a catalyst, to encourage the march to become a jog or even a sprint.

As well as the governmental and private rocket-building projects, there is a growing number of private individuals or groups investing time and money into their own space race. Some are simply weekend rocket enthusiasts – hobbyists out to dream of the day when they may leave the Earth. Others are taking part in the private space race, experiencing first hand the thrill of launching something they have made so high that it comes close to the boundary of space. And in the past few years, some have come tantalizingly close. In 1998 a group succeeded in launching their home-built rocket so high that the on-board video camera could capture a stunning view of the Earth. The flight controller was ecstatic: 'Oh my God, the curvature of Earth ... We've got it on video ... I can see the curvature of the ... Look at that, look at that! It's the friggin' Earth, man! That is so cool!' The sense of excitement is clear. One approach taken by the amateurs, which overcomes the problem of increased air resistance at lower altitudes, is to use helium balloons to lift their rockets up into the rarefied atmosphere before launching them. This idea, borrowed from the 1950s, has allowed one team to set an altitude record of 21 kilometres.

There are people who believe that the pace of the human race's march to space should be slowed rather than speeded up. They are not many in number, and their voices are not loud over the excited murmur of the space enthusiasts, both amateur and professional. But there are plenty of reasons why the drive to increase space traffic may be something we need to think about. There is environmental concern over noise and pollution of the upper atmosphere. There is the use of huge amounts of natural resources, and the amounts used will continue to increase as space becomes more and more accessible. And then of course there is the fact that space projects will benefit only the privileged few.

There are other people whose voices are in dissent against the majority: those who think that there is nothing wrong with the human desire to explore space – just with the assumption that it might happen cheaply and very soon. Dr Hans Mark is one of them: 'I've always worried about the term "cheap access to space"; I think we're fooling ourselves. To get into space, you have to achieve an orbital velocity of 17,000 miles per hour, and that's never going to be cheap.' He believes, too, that competitions like the X Prize are irresponsible. He says that the risks involved in air travel are small and well known, but, in space travel, the risks are simply too high. Burt Rutan believes otherwise: that 'if no one dies going after this prize then we are not really going out truly searching for new ideas'. For him, risk is a necessary part of progress: 'If we had shut down things because there's risk being taken, we would still be travelling across country looking up the assholes of donkeys; that's not the future and that's no fun.'

Whatever the concerns of dissenting voices like Hans Mark's, or those who are concerned about space travel for environmental reasons, private enterprise will almost certainly move quickly into a future in which space travel really is relatively affordable and routine. As well as the space hotels or joy-rides into space, there are possibilities that cheap access to space could one day replace long-haul air travel. Michael Wallis is a space enthusiast and entrepreneur. He has a clear view of the future: 'When you're going through space to get somewhere, you can get anywhere on the surface of the Earth in forty-five minutes. It means that London, for example, is thirty-one minutes from California ... You can travel at speeds because you're moving out of the atmosphere.' He also sees great opportunities for transporting cargo, say from Japan or New Zealand: 'Pick up by nine in the morning; delivery to the United States or Canada by 5 pm the previous day – for those occasions when it really did have to be there yesterday.'

Burt Rutan also believes that cheap and reliable access to space will one day replace air travel. 'To sit in an airliner burning fuel continuously at 30,000 feet going to London can't be the best way to go there. It would be a lot more fun – and we're going to find very affordable ways to do this – to make a big-time thrust for a few seconds and then to sit there and enjoy your meal in weightlessness and have a much better view, and then re-enter over London and land. That's got to be a better way to travel.'

WHAT SHALL WE DO WITH THE MOON?

...searching for a future for the human race...

The Moon has been our companion for about 4,500 million years, silently circling our planet in the dark vacuum of space. With its mysterious beauty, the Moon has been the subject of myth and wonderment for millennia, inspiring poets and scientists alike with imaginative ideas. But in the past hundred years or so it has captured people's imaginations in a new way: the advent of space travel has brought new fanciful notions, such as the idea that humans may one day colonize the Moon, setting up permanent bases or even lunar cities. This idea may not actually be fanciful, and the day it is realized may come along sooner than you think. And once the bases have been built, the Moon could be just the first stop-off point for journeys much farther afield.

Bright future

Perhaps the most obvious and magical thing about the Moon is its phases: the fact that it changes its appearance gradually over a period of a month, from a thin crescent to a full bright white disc and back again. The phases are caused by the fact that the Sun illuminates half the Moon. As the Moon orbits Earth, once every month, we see different amounts of the illuminated face. When the Moon is between the Earth and the Sun, we cannot see the illuminated side at all – this stage is called new moon. When the Moon is precisely in line with the Sun – something that does not occur every month because the Moon's orbit is tilted by about five degrees to the Earth's orbit around the Sun – the Moon blocks out the Sun, and we have a solar eclipse. This happens only once every few years, and only at new moon. The day after new moon, when the Moon has moved on in its orbit, a thin crescent of the illuminated face of the Moon is visible from Earth. Just to confuse things, this crescent is sometimes called a new moon.

When the Sun and the Moon are at right angles to each other as seen from Earth, half the Moon's illuminated side is in view. This is a half moon – although this phase of the cycle is called quarter moon, because it occurs when the Moon is one-quarter or three-quarters of the way around its orbit. When the Moon is on the opposite side of the Earth from the Sun, we see its entire illuminated face: the full moon. Again, because of the tilt of the Moon's orbit it is rare for the Earth, Sun and Moon to line up. When they do, we have an eclipse – this time a lunar eclipse, where the shadow cast by the Earth falls on the Moon.

Although ancient astronomers had no idea what the Moon actually was, they were able to observe and then predict its movement across the sky and the way its phases changed. They were also able to predict when lunar and solar eclipses would occur – a task that involves complicated mathematics.

The silent, startling beauty of the partially or fully visible Moon dominates the sky on a clear night. The Moon is the second brightest object in the sky, after the Sun. The main reason for this is its relative proximity to Earth: the Moon is our nearest celestial neighbour. In fact, the distance between the Earth and the Moon is, on average, only 384,400 kilometres – less than ten times around the world. And, unlike travel around the world, in space there is no air and therefore no air resistance. So once you are outside the atmosphere, in orbit around the Earth, it is relatively easy to give yourself the thrust necessary to leave that orbit and head for the Moon. Then the Moon is only about three or four days' journey away. If your thrust was just right, you would move into an orbit around the Moon instead of around the Earth, and from there it would be easy to use your rockets to slow yourself down and descend to the lunar surface. What would you find when you got there?

First, you would have a surface area about four times that of the USA to explore. The Moon's diameter is 3,476 kilometres – a little less than the distance between New York and Los Angeles. Imagine slicing the Moon in half along its equator: one of the resulting hemispheres would cover most of the USA. Although the Moon looks bright, it produces no light of its own and reflects only about seven per cent of the sunlight that hits it. If it were painted white, it would reflect nearly all the incident sunlight, and it would then appear almost as bright as the Sun itself. But the Moon's surface is actually dark grey, so it reflects only that small proportion of sunlight. As most of the sunlight is absorbed, most of its energy is also taken in. This in turn means that the illuminated surface gets very hot. Each point on the Moon is repeatedly in constant sunlight for two weeks and then constant darkness for two weeks. So during the long lunar 'day', the temperature of the surface becomes very hot, reaching 130 degrees C. During the lunar 'nights' the surface reradiates most of the energy into space, as infrared, and its temperature drops to a more than chilly minus 110 degrees C.

The length of the lunar day is a result of the Moon's slow rotation on its axis. The Sun moves very slowly across the lunar sky, taking two weeks to cross from the eastern horizon to the western horizon. The sky remains dark even when the Sun is 'up', since there is no atmosphere on the Moon to scatter the Sun's light and create daylight. The Earth also drifts slowly across the lunar sky when it is in view. It appears nearly four times as large in the sky as the Moon does from Earth.

The gravitational attraction pulling you down on to the Moon's surface would be only about one-sixth of the gravitational force pulling you on to the Earth's surface. This is why astronauts who go there can bounce along, taking giant steps, and can carry heavy objects with ease. And whereas you cannot easily jump more than about thirty centimetres on Earth, you would be able to push yourself several metres off the surface of the Moon, and fall back slowly and gracefully. If there was a smooth surface, like a polished floor, you would have trouble walking along normally – because the traction you need to start walking comes from the friction between the floor and you. That friction depends upon how hard gravity pulls you down on to the floor, and so on the Moon your feet would slip and slide at first.

The lack of atmosphere on the Moon has several fascinating consequences, too. Many of the things we take for granted here on Earth are radically different on the airless Moon. Striking a match would produce a brief burst of flame, as the chemicals in the match tip ignite, but the wood would not burn. A balloon pump would be easy to use, as there would be no air pressure inside it to resist your pushing on the plunger. However, for the same reason, the balloon would not inflate. Anything that relies on aerodynamic surfaces – such as an aeroplane – would not work on the Moon as it does on Earth. No lift force would be generated by the wings, however fast you thrust the aeroplane forward. If you dropped a feather and a hammer simultaneously from the same height, they would reach the ground together, because with no air there is no air resistance. This was demonstrated by Apollo astronauts. Furthermore, sound needs substance through which to travel, and so you could shout as loudly as you like through the visor of your spacesuit, but a friend standing right next to you would hear nothing.

The effects on people are perhaps the most dramatic. If you went to the Moon, you would not be able to survive unless you were in a pressurized container such as a spacesuit, and were supplied with oxygen. If you removed your spacesuit, your eyes would bulge, the air in your lungs would immediately rush out into space, and your lungs would collapse. You would quickly die. One other thing: the Moon offers ideal conditions for observing the stars and planets. On Earth, the best views of the starry night sky are seen on calm, clear, dark nights. Even then, the sea of air through which you have to view the stars is in constant motion – pockets of warm air rise and mix with cold air, for example. When you look at the sky from the Moon, light coming from the stars and the planets does not have to pass through a turbulent atmosphere as it does on Earth. And there would never be any clouds to obscure the view.

Making an impact

Most of the surface of the Moon is covered with regolith, a mixture of rocks and sand, which is formed when large meteoroids hit the Moon. These heavy, fast-moving, interplanetary rocks make craters when they collide with the Moon, shooting thousands of tonnes of material out in all directions. The shockwaves created in a powerful impact heat the surface material, which melts then quickly cools and solidifies, forming tiny glassy globules. The existence of these globules in the regolith was hypothesized before people or space probes made it to the Moon – and it was confirmed by analysing samples of regolith brought back by the Apollo astronauts and by Russian robot probes.

Regolith covers most of the lunar surface because craters have been formed all over the Moon. Some areas do have fewer craters than others, however: these are the 'maria' (plural of 'mare', Latin for 'sea'). There is no water in these seas: they are flat areas of solidified lava that has filled ancient craters. The maria are the dark areas you can see with the naked eye when you look at the Moon in the sky. Maria are less well populated with craters because they are relatively young – most of the Moon's craters are very old. This in turn is because meteoroids struck with much greater frequency in the Moon's younger days, when the Solar System was bustling with the remnants of its own formation. It was at this time – theory has it – that the Moon was formed, from a huge cloud of material thrown into orbit around the Earth. That cloud of material was ejected from Earth when a huge object collided with our planet in the very earliest period of the Solar System – about 4,600 million years ago. Most of the rocks on the Moon, then, date from this period – much older than the rocks found on Earth.

Our planet is a changing place: weathering constantly changes the chemical composition of the Earth's rocks. And new rocks are made, or existing rocks changed, by the movement of the tectonic plates of its crust. For these reasons, there are almost no rocks older than about 3,000 million years left to study on Earth. But the Moon has no tectonic plates, and no weather. Over time, meteorite impacts have changed the physical properties of some of the rocks – producing regolith – but the chemical composition of the rocks has not changed. So the Moon provides a good picture of the chemical composition of the early Solar System.

The craters that form when the Moon is hit by heavy meteoroids are a familiar feature of the lunar surface. The very biggest ones are visible to the naked eye from Earth, because the material that is thrown out by the impact produces large streaks that radiate out in the surrounding areas. When you look at the Moon, one of the most prominent craters you can see is Copernicus, named after the sixteenth-century Polish astronomer Nicolaus Copernicus. At its edges, this crater has terraced walls, which form huge steps that may one day allow people to make their way down the 3.8 kilometres from the rim to the base of the crater, which measures 93 kilometres across. To the right of Copernicus on the face of the full moon is a relatively flat, uncratered area called Mare Tranquillitatis – the Sea of Tranquillity – where the first human being stepped on to the Moon, in 1969. It is also the first place from which rock was collected on the Moon. Before that date, only robotic probes had actually landed on the surface, and none had returned any lunar material.

Throughout the 1960s a series of Russian and American probes flew by, orbited, hit or landed on the Moon. None of these probes carried any people. It was Apollo 8, in December 1968, which first took people to the Moon. The astronauts aboard the Apollo 8 spacecraft made ten lunar orbits before returning home. As became typical of the Apollo programme, the people still down on Earth could monitor the progress of the astronauts in a series of live television broadcasts. During their Christmas Eve broadcast – from lunar orbit – the crew of Apollo 8 read sections of Genesis from the Bible, while showing the people of Earth the beautiful views of their planet and the Moon from space. The pilot of the command module, Jim Lovell, said: 'The vast loneliness is awe-inspiring, and it makes you realize just what you have back there on Earth.' It is no surprise that the Apollo missions inspired many people.

Apollo 9, launched in March 1969, did not go to the Moon. Instead, it went into orbit around Earth, to test the operation of the lunar module, identical to the one that would descend to an orbit close to the lunar surface in the next mission. Apollo 10 lifted off in May 1969, and completed this mission successfully. Two months later, Apollo 11 took Neil Armstrong, Edwin 'Buzz' Aldrin and Michael Collins to the Moon. And so on 20 July 1969 Armstrong became the first person ever to step on to the Moon. It is the fact that only twelve people have done this so far – despite more than thirty years of endeavour – that leaves Moon enthusiasts frustrated and eager to go to the Moon themselves.

Including the regolith they collected, astronauts and space probes have so far brought 382 kilograms of material back with them from the Moon. Some of the minerals available on the Moon could be of economic importance: there are huge amounts of aluminium, titanium, magnesium and iron there, for example, and silicates, from which you can make glass. Pure silicon can be extracted from silicates to make electronic components, which could consist of perfect crystals if they were manufactured in the weightless conditions of a Moon-orbiting factory. The metals could be extracted from the rocks, using the heating effect of sunlight or electrical smelting plants running on solar power. Products of these lunar industries could be used to build space stations or spacecraft far more cheaply than on Earth.

This is because lifting thousands of tonnes of metal off the surface of the Moon requires a lot less thrust than doing the same from the surface of the Earth, and launch – as we have seen in 'Day Return to Space' – is the most costly part of space travel. This is easy to appreciate when you compare the size of the Apollo lunar modules that blasted off from the Moon with the size of the Saturn V rockets used to blast the mission off from Earth in the first place. The difference is in the force of gravity against which the engines were fighting.

There is another valuable resource on the Moon: an isotope of helium, which may be the fuel for nuclear fusion power stations. This could be of importance in the coming century, as the population and its energy demands increase. Fossil fuels are used to generate electricity and to power most forms of transport, but burning them releases pollutants, including millions of tonnes of carbon dioxide, into the atmosphere every year. The use of helium-3 as the basis of fusion, to generate electricity for the power-hungry modern world, would reduce the need for fossil fuels.

Mining and lunar space-vehicle construction may not be the only technological projects to be staged on the Moon. In another effort aimed at reducing our dependence on fossil fuels, Dave Criswell, of Space Systems Operations at the University of Houston, has proposed a way of using solar power collected on the Moon here on Earth. Criswell's idea is to place huge arrays of solar cells on to the Moon's surface, to change sunlight into electrical power. There are no clouds on the Moon to obscure the Sun, and so the cells would provide a reliable source of electrical power for as long as they are in the light. Having solar cells in only one part of the Moon would mean that power would be generated for only two weeks out of every month – when that part is illuminated. So Criswell proposes covering vast tracts of the lunar surface with solar cells, ensuring a constant supply. The electrical energy would be used to produce high-power microwave radiation that would be beamed to Earth, where the energy could be changed back into electricity.

Criswell summarizes the project: 'Solar power is the ultimate resource in the Solar System ... so the challenge is to divert a tiny bit of it efficiently down here to Earth.' But if a massive mining operation or Criswell's lunar–solar power scheme does go ahead, people will need to live on the Moon for extended periods of time. Moon bases will have to be built. And if that happens, Criswell compares what would happen to the Moon as something like the colonization of mining towns in Colorado in the nineteenth century: 'People will find reasons to be there, most of which have nothing to do with the primary activity.'

Despite these thought-provoking possibilities, to nearly all the world the exploration of the Moon seems to have halted. The last human beings to stand on the Moon were Eugene (Gene) Cernan and Harrison Schmitt, during the Apollo 17 mission in December 1972. The Apollo programme was the jewel in the crown of the USA's efforts in space: twelve of its astronauts stood on the Moon, collected samples and even drove around on the surface in a remarkable lunar rover. After the Apollo programme finished, many people assumed that visits to the Moon would continue – NASA themselves had plans for an exciting future. Dr Alan Binder of the Lunar Research Institute in Gilroy, California, recounts this lost opportunity: 'We had a post-Apollo programme planned which would have put a lunar base up by the end of the 1970s, and there would have been fifty to a hundred people living and working on the Moon by the early 1980s; we simply dropped that.' And so, since 1972, only robotic probes have been back to the Moon.

After the flurry of lunar exploration during the 1950s, 1960s and early 1970s, planetary scientists moved on to explore the deeper realms of the Solar System, and probes have now also visited all the other planets except Pluto. You would be forgiven for believing that we have forgotten the Moon in favour of more exciting destinations. But the dreams of building permanent bases on the Moon – so much in currency during the Apollo era – have remained alive, and have recently undergone a renaissance. One reason for this resurgence in interest in the Moon is that the people who were inspired by the Apollo missions are becoming impatient.

Patrick Collins of the Space Development Agency of Japan is one of the impatient ones: 'People went to the Moon thirty years ago; now if you do something thirty years ago, that means it is easy today.' Another person who is very keen for humans to set themselves up on the Moon is aerospace engineer Greg Bennett – Director of the Lunar Resources Company, based in Houston, Texas. He says, 'If you were to go to the Moon thirty years from now, you would think that the Moon has simply become the Hong Kong of the Solar System.' Space capitalist Jim Benson is another impatient lunar enthusiast: 'I've been waiting for something to happen in space for most of my life ... I'm getting tired of waiting.' Benson runs SpaceDev, one of a number of companies already making millions of dollars from the prospects of lunar exploration. He and his shareholders are speculating on the bright future for projects that exploit the potential of space. He believes that 'space is too expensive to do as a taxpayer-supported programme; we need to find ways to make space pay and then we can do anything that makes a profit, we can go to space and do all those things'.

The obvious things that people could do to make money from the Moon are mining and tourism. And when people start organizing such activities, Benson will be there, to make money: 'When I think of lunar colonies, I think of all the business opportunities ... If SpaceDev can supply the infrastructure, it's going to be a very profitable company.' So confident is Benson of the impending moon rush that he has built new offices in Poway, California, which include a clean room where spacecraft will actually be built and serviced. Benson's company deals with all aspects of the commercialization of space – including proposed mining on asteroids – but the Moon has special significance, as it could become the launch pad for journeys to more far-flung locations. The offices also include a 'mission control' room from which personnel will monitor and control ongoing missions. Benson explains that there will be 'four or five personal computers and monitor screens' inside this unlikely-looking hub of activity. SpaceDev's first project – the Near-Earth Asteroid Prospector that plans to mine the resources of asteroids – was begun in 1997. This is a significant year in the history of space exploration: it was the first year in which global commercial expenditure in space projects outstripped government spending. This situation is likely never to be reversed, and in fact the gap between publicly and privately funded space projects will probably only widen. The Moon is ripe for development and exploitation and, like it or not, the commercialization of space seems to be here to stay.

Don't go there

There are many obstacles and objections to the construction of the first Moon bases and cities. One of the principal objections concerns the economics behind such a programme:

the money spent in setting up a base on the Moon – whether from public or private sources – could be used to feed the millions of starving people here on Earth. If the pattern of Earth-based industry is followed on the Moon, then the economic benefits of lunar exploitation are unlikely to be shared among all people throughout the world. However, many would argue that while industry makes only some people rich, it does bring incalculable benefits to everyone, by developing new technologies or manufacturing goods that then become available to everyone. To keep in check the 'get-rich-quick' mentality that could turn pioneering lunar exploration into a mere money-making exercise, a United Nations treaty was created in 1967, to which several of the industrial nations signed up. The 'Treaty on Principles Governing the Activities of States in the Exploration and Use of Outer Space, Including the Moon and Other Celestial Bodies' included the following requirements on signatory nations:

- 'The exploration and use of outer space, including the Moon and other celestial bodies, shall be carried out for the benefit and in the interests of all countries, irrespective of their degree of economic or scientific development, and shall be the province of all mankind.'
- 'Outer space, including the Moon and other celestial bodies, shall be free for exploration and use by all States without discrimination of any kind, on a basis of equality and in accordance with international law, and there shall be free access to all areas of celestial bodies.'
- 'There shall be freedom of scientific investigation in outer space, including the Moon and other celestial bodies, and States shall facilitate and encourage international co-operation in such investigation.'

So much for the regulation of our futures in space, which may restrain some of the overzealous lunar explorers. There are also innumerable technological obstacles to be overcome in the establishment of lunar bases. The first thing that will be needed is an affordable and reliable method of launching people and supplies from Earth, transporting them across space and landing them safely on the Moon. Once there, lunar pioneers would have to survive on an airless ball of rock: the Moon does not have an atmosphere. Here on Earth, the atmosphere not only provides us with air and water: it blocks some of the sunlight that reaches our planet, reducing potentially carcinogenic ultraviolet radiation. The Moon would afford no such protection from the harmful effects of direct sunlight. Earth's atmosphere also keeps the planet's temperature at just the right level for life to thrive. In fact, the composition of the atmosphere is a direct consequence of the existence of the living things that it supports. Without the atmosphere – in particular, the carbon dioxide and water vapour constantly cycled and regulated by living things – the average temperature at the surface would be about a hundred degrees C cooler. So the Moon is not only airless: it is very cold, too.

The astronauts who have already made it to the Moon could not stay for long. This is mainly due to the fact that they had to take all their food with them – with no atmosphere and no known source of water on the Moon, they could grow nothing. Astronauts have stayed in space for extended periods, in orbit around the Earth, but this was possible because spacecraft would bring fresh supplies of oxygen, water and food. The methods used to maintain the right conditions inside orbiting space capsules can do so for only limited periods of time. Exhaled carbon dioxide is removed from the capsule's air by cylinders filled with a substance called lithium hydroxide, but these cylinders must be changed daily. It would not really be viable to send frequent batches of supplies to a base on the Moon designed to accommodate hundreds or thousands of people.

If the requirements of making it to the Moon and staying alive there are so challenging, how can people be so confident that within their lifetimes there will be permanent bases or even civilizations in this hostile environment?

The first requirement for living on the Moon – a reliable method of transportation – is not yet a reality. Despite the fifty years or so that have passed since humans first propelled rockets into space, the dream of routine, affordable space travel seems no closer. But a new entrepreneurial spirit may be about to change that. As we saw in the previous chapter, the space technology developed in the space race of the 1950s and 1960s was not designed to make space travel routine and affordable. The impressive and exciting achievements of the American and Russian space programmes were driven by political goals as much as – or possibly more than – by scientific ones. But initiatives such as the $10 million X Prize may be the stimulus that private space travel engineers need to design and build the kind of reliable vehicles that will eventually take people and supplies to the Moon.

Assuming that we can make it to the Moon, how will the other requirement be met? How will lunar pioneers sustain themselves if there is no atmosphere? This is perhaps more of a challenge. Even if lunar engineers took billions of tonnes of the atmospheric gases from Earth and released them at the Moon's surface, the gas molecules would soon escape from the weak gravitational pull there, and boil away into space. So if people do ever stay for extended periods on the Moon, they will need to be enclosed in an airtight building, breathing air from an artificial atmosphere. If a leak developed in the building, the atmosphere would escape and the lunar inhabitants would die, so the pressure of the building would have to be monitored constantly.

The good life

Surviving on the Moon is not as simple as just living in an airtight building, however. You would have only the precious resources that you had taken with you. So lunar inhabitants would be unwise to throw their waste away on to the Moon or into space, since otherwise they would soon run out of water and raw materials. Everything would have to be recycled meticulously – residents of the Moon would have to be totally self-sufficient. A lunar base would have to be a self-contained mini-ecosystem, in which all of the resources would have to be carefully monitored and controlled. This requirement, too, may not be as much of a challenge as it seems. What is needed is a large-scale version of a terrarium – an enclosed container used to propagate plants or certain animals in carefully controlled conditions. The terrarium was popular as a form of decoration during the nineteenth century, as well as a being a tool for scientists and plant- and animal-keepers.

In several experiments conducted on Earth with extraterrestrial survival in mind, teams of people have survived in the equivalent of large terrariums for extended periods. The first such sealed environment was 'Biosphere 2' – a steel and glass building covering just over 1.2 hectares of land in the Arizona desert. Four men and four women lived sealed inside Biosphere 2 for exactly two years, from September 1991. Some of the scientists involved in the project had space survival in mind, although Biosphere 2 was set up as an experiment to model the features of Biosphere 1 –

the Earth. The building contains 170,000 cubic metres of air, and has areas dedicated to rainforest, desert and farmland. It also has an ocean, containing 3,800,000 litres of salt water. In similar experiments carried out by NASA – this time with survival in space very much in mind – groups of researchers spent weeks inside a sealed unit nicknamed the 'Chamber', at the Johnson Space Center in Houston, Texas. This steel-walled container is 6 metres in diameter and three storeys high. Scientists on the outside monitored the physiological and psychological condition of the volunteers inside, through video links and medical tests. Future self-contained living spaces like the Chamber could be packed into a spacecraft that was going to the Moon.

The Chamber was developed as part of the Lunar–Mars Life Support Test programme, begun in 1995. It was as long ago as the 1950s when research began into the use of plants for human survival, and NASA took an interest during the 1960s. A tank similar to the Chamber has been in use at NASA's Kennedy Space Center in Florida – to study the production of plants in sealed environments – since 1986. During the first experiment in the Chamber, four volunteers spent four weeks inside. The longest stay inside the Chamber began in September 1997, and lasted for thirteen weeks. The team entered the Chamber with enough air and water to last only one week, and had no supply from outside. So inside, everything was recycled. The water they drank and washed in was obtained by recycling their own urine. The carbon dioxide they exhaled, together with carbon dioxide obtained by incinerating their faeces, was made available to plants that were kept in hydroponic and soil-based tanks. (In hydroponic cultivation, plants grow in soil-less conditions, suspended in nutrient solutions.) As the plants used carbon dioxide, they produced oxygen – which the inmates needed to breathe – and sugars, which the plants needed if they were to grow. Once grown, the plants themselves were a source of food, which could be eaten, and as solid and liquid waste, be recycled once again.

This regenerative approach – using the same raw materials again and again – mimics Earth's own natural recycling system. Some of the plants grown in the Chamber were eaten, but the inmates in this experiment ate mostly tinned and frozen food. In space, it would be uneconomical to take much food with you – an adult human being consumes several hundred kilograms of food each year. So any advanced life-support systems would have to have a facility for growing food. The chief scientist at the Johnson Space Center, Dr Don Henninger, explains that inhabitants of a colony on the Moon could grow plants in the lunar soil – from the regolith. Henninger runs experiments in which he grows plants in 'lunar simulant soil ... a high-titanium basalt'. In this lunar soil, he has grown specially developed dwarf wheat, which grows only about thirty centimetres high. He has also grown potatoes, soya and even some herbs and spices. So you could actually use the soil from the Moon to grow crops. Of course, as Henninger points out, 'Since there is no atmosphere, we'd have to bring it inside some habitation chamber or plant growth area.'

The team inside the Chamber evaluated a diet that was designed for life on the lunar surface, which included fifteen different crops. One of the inmates in the Chamber was Nigel Packham. Though he survived on the diet well enough for the duration of the experiment, he says, 'I'm not sure if taste-wise you'd want to do that for three, four or five years or longer.' As well as having a restricted menu, the chamber provided a very confined living space. It did have a games area and a library, and there were rooms where

Packham and his colleagues could be alone when they wanted to be. But on the whole it was not very comfortable, and there was not much space to move around. Packham says that his experience gave him 'some idea of what a lunar base might be like'. However, the confinement that the inmates felt living in this enclosed container would have been intensified if that container were up on the Moon – where they would not be able to step outside back into normality. The project's psychologist, Al Holland, says, 'Anything you can do to normalize that life is important, which includes the use of plants, not only for physico-chemical regeneration but also for the psychological regeneration that occurs with contact with plants ... If we could have pets on a lunar colony, if they weren't too distracting or too disruptive, that would be even better.'

Don Henninger and his team are planning another experiment, in a much larger chamber, starting in 2005. But the Chamber will have to have other refinements before it is acceptable for an extended stay on the Moon. Although the inmates were not totally self-sufficient, this is well within the capability of the sort of technology being developed at the Johnson Space Center. Eventually, perhaps, a lunar base will look something like Biosphere 2, whose inhabitants really were self-sufficient. The logistics of setting up an installation like Biosphere 2 on the Moon are daunting, to say the least. Not only would the space colonists need to transport thousands of tonnes of raw materials for building their lunar terrarium, they would have to take water and air. The water could be solidified, and the air liquefied to save space, but their lift-off weight would still be the same. To take all we need to set up a Moon base or Moon city would therefore be a mammoth task, and not one that could be undertaken quickly or cheaply.

It would be much easier, cheaper and quicker if, instead of taking all the materials needed to build a base or a city, we could use what is already there on the Moon. Unlike the Earth – which has rich and varied natural resources – the Moon has no oceans and rivers, no wood, no coal or oil. However, it does have a plentiful supply of lunar rocks and dust. We have seen that there are valuable resources on the Moon that can be used to manufacture spacecraft and electronics, so these materials might also be useful for constructing the buildings for a lunar base. Dave Criswell says that you can 'grab a hold of the lunar materials and change them into virtually anything that you are used to producing here on Earth'. His lunar–solar power idea would depend upon this: it would consist of plots with 'solar cells, wiring and microwave generators and reflective antennas; all of that material – or virtually all of that material – is made on the Moon from the lunar dust'.

The Moon consists largely of basalt, which is also common in the Earth's crust, though not deeper down, in its mantle or its core. This fits in with the theory that the Moon was created after a collision of Earth with a huge interplanetary object: only materials from the Earth's crust, not the mantle or core, would have been thrown into space by this collision. Basalt is basically silicon dioxide combined with various metallic elements.

There are two main types of geological region on the Moon: the highlands and the lowlands. The highlands – the lunar mountains and the crater rims – were formed from materials thrown out from the Moon's crust as a result of the impacts of meteorites. And so they are made from the materials of the Moon's crust: basalt high in aluminium. The lowlands – the maria, the 'seas' – are younger areas, formed by the intrusion of molten rock that originated from far beneath the Moon's surface. The molten rock was probably formed by huge meteorite impacts, generally about 3,000 million years ago. Lowland rocks are also largely made of basalt, but with more iron and magnesium and a little less aluminium; a significant amount of titanium is found, too. There is plenty of scope in these two types of geological formation to make strong metal alloys – from the iron, magnesium, aluminium and titanium. To make iron into steel, lunar smelting works would have to be set up. However, there is a snag: steel-making requires carbon and oxygen in addition to iron. There is very little carbon present on the Moon, and huge quantities would have to be taken there. Clearly there is no oxygen gas in the atmosphere – the Moon has no atmosphere. However, oxygen could be liberated simply by roasting the silicate minerals contained in the lunar basalt. This oxygen could provide life support for human inhabitants, too, or oxidizers for rocket fuel. The energy to roast the silicate would come from the reliable supply of solar radiation. The silicate minerals would also be useful for making glass, fibreglass and ceramics. Some of the necessary materials that are not present on the Moon – in particular, carbon – are found on asteroids, and they could be mined from the asteroids and brought to the Moon.

Tireless workers

Naturally, if a base is to be constructed on the Moon, a minutely detailed survey will be needed before any building work can begin. First, though, the construction workers would have to be sure that they were in a location where they could find the right materials. Once all this was done, the lunar minerals would have to be processed, to make the glass and metal alloys from which the buildings could be made. It would be difficult for people to do this, as they would have to live on the Moon during these long early stages. But robots could be sent up to carry out these tasks.

Professor Kumar Ramohalli has clear ideas of the way this would work. He is based at one of the Space Engineering Research Centers set up by NASA at several American universities. At the University of Arizona, Ramohalli has designed several robots for exploring the Moon or Mars, and utilizing the resources there. These include LORPEX (Locally-refuelled Planetary Explorer) and BIROD (the Biomorphic Robot with Distributed Power). LORPEX would manufacture its own fuel from the lunar dust, using solar power; then it would use that fuel in its rocket engines, to make powered 'hops' so that it could explore other regions. BIROD uses some recent developments of robotics, including artificial muscles, which allow it to move around more like a cat than a wheeled robot. Its muscles are powered individually, rather than centrally – just like real muscles, which derive their energy from chemical reactions in the cells of which they are made. It is small enough to fit inside a spacecraft, and could be easily deployed on to the surface of the Moon or Mars.

Robotics is a rapidly advancing field of endeavour. Ramohalli says that 'the Moon will be teeming with intelligent robots during the next thirty years'. Ramohalli – and other researchers – have already made building blocks by baking clay made from lunar-simulant dust. The resulting material, from which the blocks are made, is called 'lunacrete' – the first Moon bases could be fashioned from these lunacrete blocks. Ramohalli has worked out the processes by which a self-contained robot spacecraft could make these bricks. His 'Common Lunar Lander', proposed in 1992, lands softly on the Moon, and its robot arm picks up soil, puts it into a solar oven, and uses the intense heating effect of concentrated sunlight to make bricks and tiles. He believes that a single robot could build twelve huts in six months.

Another robot being developed to play a role in lunar exploration is Nomad, developed jointly by NASA's Intelligent Mechanisms Group and a team from Carnegie Mellon University, Pittsburgh. In a trial conducted in the Atacama Desert, in Chile – over terrain similar to that found on the Moon – this four-wheeled robot travelled 215 kilometres searching out particular types of rock, and operating largely autonomously. The robot had video cameras that gave 360-degree vision, and a sophisticated communications capability, which meant that it could be controlled remotely by scientists more than a thousand kilometres away. It was also intelligent enough to be able to explore and make decisions for itself. Red Whittaker, one of the team who developed Nomad, says that 'Nomad has capabilities to see, to safeguard, to communicate ... it is a continuing persistent evolution of the technologies and the operations and the ideas that are essential to carry us back to the Moon.'

Nomad is equipped to carry out geological experiments, and to work out the nature of and distances to nearby mountains or boulders. For some portions of its trek through the Atacama Desert, the images from the panoramic cameras were displayed on an 11-metre-wide curved screen at the Carnegie Science Center in Pittsburgh. Visitors to the centre could control the vehicle as if they were in it – a technique called telepresence. In the future, telepresence systems may use virtual-reality systems, allowing a user on Earth the chance to 'look around' the Moon and pick things up using a virtual-reality glove. There would be a serious time delay, however, as signals between the Earth and the remote robot would take nearly two seconds each way. The whole robot has a mass of 550 kilograms, which means it could be taken to the Moon relatively easily. One of the clever design features of the Nomad robot is the fact that its wheels retract. The robot is 2.4 metres wide when it is roving – this wide wheelbase gives it stability. So that it can be stowed in the hold of a spacecraft, the four wheels retract under the robot's body, making the robot 60 centimetres narrower. Another member of the Nomad team is Dimi Apostolopoulos, who explains a rather strange attachment to the robot: 'We all have a very intimate relationship with this robot. We are the people who conceived it, designed it, built it – when Nomad is on the Moon, it will be as if one part of ourselves is actually operating up there.'

The idea of robots building lunar bases for human habitation sounds like a scene from a science-fiction film. But if the incentives are there, the technology is sure to follow. Kumar Ramohalli may be right: the Moon really could be teeming with intelligent, industrious robots in the near future. And so, with robots surveying the Moon's surface and then building the initial primitive shelters for humans, the first permanent lunar base could be established within a few years. By living in units like the Chamber, humans could survive on the barren Moon for extended periods. But they would still need to take essentials like water and oxygen – or would they?

Where the Sun don't shine

It has long been assumed that the Moon is totally dry: because it has no atmosphere, and because every part of its surface receives direct sunlight, any ice there would quickly evaporate into space. However, in 1961 three lunar scientists at the California Institute of Technology in Pasadena proposed that water ice might exist at the north and south poles of the Moon. And over the next three decades other scientists developed this idea, suggesting that ice could be deposited in craters by comets that crash on to the Moon. Comets are made of ice and dust – they are sometimes referred to as 'dirty snowballs'. It was noted that there are certain craters at the Moon's poles whose interiors never receive any sunlight. In these shadowed areas, the temperature would never rise above minus 170 degrees C. In 1992 and 1993, in an effort to investigate the possibility of ice on the Moon, radio waves transmitted from Earth bounced off the lunar poles and were detected by Earth-based radio telescopes. The results seemed to suggest that water really might be there.

In 1994 NASA and the US Department of Defense launched a space probe, Clementine, which was supposed to travel – via the Moon – to an asteroid. It never made it to the asteroid because of a fault in its on-board computer software, but it made an astonishing discovery as it flew past the Moon. It shone radio waves on to the Moon's south pole, and the reflected signals were received and analysed on Earth. The results were extremely encouraging – they seemed to indicate the presence of thousands of tonnes of water – until they were shown to be inconclusive. More information was needed to determine whether or not there is water on the Moon. If there is, then lunar survival will be a great deal easier.

In 1998 another tireless worker – this time an orbiting probe called Lunar Prospector – set out to further investigate Clementine's discovery. Lunar Prospector carried a neutron spectrometer, which would measure the energy of neutrons produced at the Moon's surface – these neutrons are expelled when cosmic rays collide with the Moon. The energy of the neutrons depends upon what substances are present on the Moon. So Lunar Prospector could not only search for signs of water, but could also perform a general survey of the distribution of resources on the Moon as a whole. Alan Binder, based at the Lunar Resources Institute in Tucson, Arizona, was the principal investigator on the project. He says, 'We're not just doing science: we're trying to understand the distribution of resources on the Moon which we can then use to build a lunar base, to build a lunar colony.'

Lunar Prospector made a total of 6,800 orbits of the Moon, over a period of eighteen months. And its neutron spectrometer did find fairly conclusive evidence that there were hydrogen atoms in huge quantities in the craters at the Moon's poles. The hydrogen atoms are almost certainly connected to oxygen atoms, in water molecules (H_2O). So Lunar Prospector has confirmed Clementine's results – there really is water on the Moon. The total mass of ice in these craters is not known, but is probably somewhere between ten million and 500 million tonnes. This would be enough to support a colony of about a thousand people for about a hundred years – or much longer with efficient recycling. In a dramatic attempt finally to confirm the existence of water, the Lunar Prospector was deliberately crashed into a crater at the south pole – only when its mission had already overrun and its batteries were running low. The hope was that a plume of debris would be produced by the collision: the 160-kilogram craft was travelling at 6,120 kilometres per hour when it hit. Earth-based telescopes, as well as the Hubble Space Telescope, were watching as the craft crashed into the Moon on 31 July 1999. The telescopes were looking for sunlight reflected off the debris. By splitting that light into a spectrum, astronomers would be able to tell whether water was present.

Despite the earlier positive results, these observations did not find any water in the debris. It is possible that the spacecraft simply missed its target, and hit part of the Moon's surface outside the crater, or that the debris was not carried high enough to be detected – or, of course, that there is no water after all. It may be that the hydrogen detected by Lunar Prospector was an accumulation of hydrogen ions (electron-less hydrogen atoms) from the solar wind, known to be streaming out into the Solar System from the Sun, as described in 'Sun Storm'.

Would-be lunar explorers are very excited about the possible presence of water on the Moon. Not only will water be useful for drinking, washing and agriculture: using solar power, it is possible to separate the hydrogen and oxygen that makes up water, to make fuel for rockets in which to travel back to Earth or farther out into space. Alan Binder believes, 'Everybody likes to imagine themselves being on another planet, exploring the Solar System. This is now at the point where it can happen – it will happen; now that Prospector has found water, we can have a golf course and a swimming pool.'

Many organizations are busy putting forward their ideas of what we can do with the Moon. The first aim is to find out yet more information about it using uncrewed space probes. Several lunar probes have been planned for the near future, including Icebreaker, being developed jointly by a private company, LunaCorp, and the Robotics Institute at Carnegie Mellon University. If their probe is launched, this wheeled robotic vehicle will survey areas around the south pole, looking for water, between June and December 2003. Icebreaker will have a drill and ground-penetrating radar, so that it can look for water in the regolith underneath the surface. The Japanese Institute of Space and Astronautical Science has put together a thirty-year project for a lunar base. The first stage of this project is planned for 2002, when a probe called Lunar-A will be launched to investigate what is underneath the regolith, using seismic surveys. Another proposed Japanese project is LOOM (Lunar Orbital Observatory Mission) a 2,000-tonne space probe that would map out the lunar surface from an altitude of 100 kilometres.

At least two European missions to the Moon are currently being planned. One team, based at the Technical University of Munich, has proposed a low-budget lunar orbiter, called LunarSat, which would fly over the Moon's poles investigating in greater detail the distribution of ice. Other scientific plans that have been put forward include building a huge observatory on the Moon. A telescope that does not look through the atmosphere of Earth – such as the Hubble Space Telescope currently in orbit around Earth – has significant advantages over Earth-based telescopes. Observatories with telescopes based on the Moon rather than in orbit would be easier to update and repair. The European Space Agency (ESA) is among the organizations that plan to set up lunar bases, as part of a programme called the Lunar European Demonstration Approach.

Everyone's going to the Moon

Moon fever is taking hold, with private companies becoming increasingly active in the drive for development. There is LunaCorp, based in Arlington, Virginia; they are running several programmes jointly with the Robotics Institute at the Carnegie Mellon University, including Icebreaker, mentioned above. In another of these joint projects, LunaCorp intends to send two interactive robot rovers to the Moon. People on Earth will be able to sit on 'motion platforms' and experience the same knocks and bumps that the rover is experiencing on the Moon. Some users will be able to control the robot probe, and Internet access will enable anyone with an Internet connection to obtain live pictures from the Moon. The probes will have Internet-ready computer servers built into their wheels, and will communicate via radio like any other probe. This and other aspects of the commercialization of the Moon were explored at the first Commercial Lunar Base Development Symposium, held in July 1999 in League City, Texas. It was organized by the Space Frontier Foundation and funded by the Foundation for the International Non-governmental Development of Space. Things really are happening.

Greg Bennett of Bigelow Aerospace and the Lunar Resources Company gave a talk at the symposium, entitled 'Lunar Cruiseships and Hotels'. The main thrust behind the Lunar Resources Company is the Artemis Project, which aims to establish a permanent, inhabited and self-sufficient base on the Moon within ten years. Once there, its lunar pioneers will explore the Moon to find the best locations to set up mining operations and a permanent lunar colony. The money to do this will come from the initial stages of space tourism: taking people for thrilling rides into space or perhaps around the Moon. In the foreseeable future, say the Lunar Resources Company, the Artemis Project will include travel to and from the Moon in huge spaceliners and a stay at a luxury hotel on the Moon itself. In his work for Bigelow Aerospace, Bennett is involved in visualizing how hotels might work in space. The owner of the company is tycoon Robert Bigelow, who has pledged to support the building of a 100-passenger tourist cruise ship that would orbit the

Moon; he has promised $500 million to the project, before it starts making him money. Bennett can also foresee tours around the original Apollo landing sites, where visitors walk around in transparent plastic-covered walkways.

Patrick Collins, too, feels that the future of the Moon will depend upon space tourism. He believes that the first people to go there are going to be 'adventurers – people for whom it's really their life's ambition; and then it will be the rich who go, once it's really safe; and then if you go there in a hundred years, it will be like Vegas'. Despite the fact that it is in the middle of the desert, Las Vegas is an incredibly rich city, and is growing richer all the time. The money comes from tourism. Collins is certain that the same demand will exist for space-based tourism: 'Frankly, the world is getting smaller and smaller now; there is nowhere you can go that's new; there's nowhere you can go where you are out of reach of a mobile phone. So space is the next frontier for tourism.'

Most people would think that spending hundreds of thousands of dollars to go to the Moon just to avoid your mobile phone is a little excessive, when simply turning it off would be cheaper and easier. However, Collins is probably right about the future of the Moon being dominated by tourism. The Moon draws people towards it for many reasons – space scientists have a million questions; adventure-seekers would find one of the ultimate thrills; artists dream of painting there; and of course people want to mine the Moon's resources. And if people like Patrick Collins and Greg Bennett are right, then tourism is the only thing that can make these things possible. Space tourism, however, will be available only to those people with huge amounts of spare money, who will be very much in the minority. Long-haul air travel suffers from the same inequalities today: the majority of the world's people will never be able to afford to take an intercontinental holiday, and yet some people can travel anywhere in the world and be pampered in luxury hotels. Some people would argue that long-haul air travel would not have been developed if it were not for capitalism in general and tourism in particular. But then, just what value long-haul travel and expensive hotels have to those who will never be able to afford to experience them is another question.

Patrick Collins and his fellow 'space profiteers' are serious. If their visions come true, then perhaps thousands of people will be visiting lunar hotels each year by around 2040. What might a visit to a hotel on the Moon be like?

Even in 2040, your spacecraft has to produce great thrust to accelerate to the speeds necessary to lift you into orbit. So its huge rocket engines start up and lift you into space – as the spacecraft accelerates, your body is pulled up with it. And so you feel yourself being pressed harder against your seat. The downward force is 'g-force': at '3g', you push against the seat three times as hard as your weight would normally do if you were not accelerating upwards. You experience the same thing when a lift begins to ascend inside a building – your legs bend, and you press harder than normal against the floor of the lift. If your spacecraft has windows, you can see the sky darkening outside you as the atmosphere thins. As you move into orbit, the g-force decreases, and eventually you experience less than the 1g that you are used to. In fact, once you are in orbit, your spacecraft and you are both free-falling, in zero-g. You are falling down towards the ground, but you never move any closer to it. This is because you are moving around the Earth in orbit – always at a right angle to your direction of fall – at just the right speed. Once in orbit, you are moving at incredible speeds, but as you gaze out at the Earth, it looks as if you are just drifting slowly. This is simply because you are so far above the Earth's surface, which provides your only visual reference. It does not feel as if you are moving fast either, because at a constant speed your body feels no g-force. In fact, you are now experiencing weightlessness – although you will probably remain strapped into your chair for some time yet.

Your spacecraft acts only as a shuttle to a much larger space cruiser that ferries people between the Earth and the Moon. Your craft docks with the cruiser, and takes this opportunity to refuel. It makes sense to take fuel down to Earth ready for the next flight up into orbit. The fuel taken on board by the shuttle craft has been transported in the cruiser from the Moon, on the return leg of its last trip. It may have been produced by the action of solar power on lunar soil or – if enough ice has been found on the Moon – by splitting water into hydrogen and oxygen. You and your fellow passengers transfer – in zero-g – from the small craft to the large space cruiser, through an extremely well-sealed airlock passage. Your luggage is transferred to the larger craft, too – though the amount you are permitted to carry is very limited. After you board, there is a detailed safety briefing and then you are strapped into your seat again, ready for the push towards the Moon. A carefully calculated rocket thrust sends your space cruiser out of Earth orbit, and into a lunar trajectory.

The space cruiser was built, in orbit, several years previously. In the vacuum of space, it does not need a streamlined shape – as a seagoing cruiser does – if it is to move quickly towards its destination, and so this ship can have any shape. However, most of the modules from which the space cruiser is made are cylinders – because that shape is strong enough to withstand the incredible pressure difference between the inside of the cruiser and the vacuum outside. The cylindrical modules are quite spacious, and there are huge windows through which you can enjoy the view, as Earth grows smaller and you head for the Moon. For the next four days, you enjoy luxury in weightlessness. You can communicate with people on Earth, as the ship is connected to whatever the Internet has become by 2040. You can watch television, play games, or just enjoy the weightless conditions as you travel; there may be a gymnasium, too. Perhaps part of the space cruiser will be set into rotation, so that normal gravity is resumed in this area of the ship. And so after four days' travel, your space cruiser manoeuvres into a lunar orbit and you transfer to another shuttle craft, this time destined for the Moon's surface.

From your vantage point above the lunar surface, you can just make out a source of light near to the Moon's north pole. It is a huge dome made of glass and aluminium, about one kilometre in diameter. Around it, there are fields full of crops inside other huge glass domes. In their experiments in the 1990s, the team at the Kennedy Space Center worked out that about thirty square metres of land would supply each person with food – if genetically modified, high-yield plants are used. This is about half the area of a squash court. Next to the agricultural areas is a collection of scientific laboratories, including several astronomical observatories. About a hundred kilometres from the settlement – beyond the lunar mountains that lie next to the huge dome – are extensive arrays of solar cells and a mining operation. The shuttle craft's rockets fire to slow your approach to the settlement and the craft hovers before landing at the end of a covered tunnel, which extends about six kilometres from the main dome. You transfer from the shuttle craft, through another airlock, into the tunnel. Much of the length of the tunnel is transparent, so you can see the lunar surface, the black starry sky and perhaps the Earth, as you are transported towards the main dome.

Just like a large version of the Biosphere 2 project in the 1990s, the dome is sealed and acts as a self-contained ecosystem. All the resources inside are carefully monitored, although the recycling regime is not as strict as in Biosphere 2 – fresh oxygen can be obtained from the lunar soil, for example, and other essential substances can be mined from asteroids thousands of tonnes at a time and brought to the Moon. Besides, a strict recycling regime may prove uncomfortable for the hotel visitors, who would not want to pay large sums of money to live meagrely during their time on the Moon. The temperature inside the dome is carefully controlled, to avoid those dramatic temperature swings experienced by the exposed surface outside the dome. The glass of the dome is designed to filter out much of the ultraviolet in the sunlight, just as the atmosphere does on Earth, though you can still sunbathe. One of the buildings inside the dome is the hotel; the lighting inside it varies over a twenty-four-hour period, to simulate the cycle of day and night on Earth.

When you arrive in the dome, you are taken to the reception area of the luxury hotel, which is similar to the rich palatial hotels found on Earth. Water falls slowly to the ground in the low gravity of the Moon, so in the hotel foyer there are graceful water features unique to the Moon. There are also trees and shrubs growing all around you – the foyer is grand and welcoming. One of the best things about being on the Moon is the low gravity. For example, people are able to leap as high as their own height on the Moon. There is a range of excursions that you can take to see the sights of the lunar surface – including a tour of the Apollo landing sites, where the lunar pioneers stood some seventy years ago. And of course, there is a gift shop, and you will find a wealth of 'photo opportunities'.

Back to the present

At the moment, this chain of events is a fantasy. However, there is nothing in the story that, in principle, cannot be achieved using technology that exists today or will be developed in the very near future. And there are already serious, costed proposals for a lunar hotel complex in a huge glass dome. In 1998 it was reported that the Hilton Hotels Group had commissioned architect Peter Inston to look into the idea of building a hotel on the Moon. In the same year, Japanese construction company Shimuzu also announced plans to build a lunar hotel. Even if these particular projects are never undertaken, more will follow. Patrick Collins has clear visions about what a lunar hotel might be like: he says that the 'ceilings of the rooms are going to have to be higher, because in the one-sixth gravity on the Moon, when you walk you bounce up and down'. And Greg Bennett supposes that the effects of low gravity could 'change the way you design staircases'. He also believes that in a lunar hotel, you could be 'standing on floors made of the actual rock of the Moon'. That is an exciting prospect – if you can afford it.

The Moon is big business, even though no plans have actually made it off the ground yet. There is already a lunar estate agent, called Lunar Embassy, that sells small plots of land and even city-sized regions of the Moon's surface. Lunar Embassy claims that its operation is legal, and has already formally founded a city called Lunafornia, too. There are plenty of private companies and organizations keen to kick-start lunar development – what about public-sector organizations, or some kind of international regulatory body to govern the future of the Moon? The closest thing is the International Lunar Exploration Working Group, formed in 1995. This is made up of representatives of most of the major space agencies, whose mission is to develop a strategy for lunar exploration, with international co-operation as a priority. Their role may be crucial in helping to define the shape of future developments.

The Moon will always be important scientifically, being our nearest celestial neighbour and consisting of important ancient minerals that may tell us a great deal about the formation of the Solar System. But the Moon will be important to space scientists in another way, too. Because of the low gravity and lack of atmosphere, it is easier to plan, equip and launch spacecraft bound for Mars or Jupiter from the Moon, and it would be much cheaper and easier than launching it from the Earth. So a lunar base may play a central role in our quest to explore the other planets. As we have seen, the interest in the Moon is certainly not restricted to the scientific community. Lunar miners want to dig up huge areas of the Moon, and will need to cover vast areas of the lunar surface with solar cells. And lunar hotels will be set up, if they are guaranteed to make money. Once enough people are working there, bases and hotels may become integral parts of a new civilization.

Alan Binder has a clear vision of the Moon's future: 'Thirty years from now, I can imagine tens of thousands, hundreds of thousands of people living on the Moon ... you could have them stay for ever, and have children, and have their children have children.' For Steve Bennett of the Starchaser Foundation – one of the entrants for the X Prize discussed in the previous chapter – the Moon is 'important for the long-term continued human exploration of the Solar System, because it's filled with all the important things that humans need to live, work and prosper in space'. If a colony grows on the Moon, through lunar exploration and exploitation, then the next logical step would be to set up bases, mines, hotels and colonies on Mars.

The red planet has some features that would make it easier to colonize than the Moon. For example, it has an atmosphere – though not one in which human life could survive. The factors that make Mars uninhabitable at present are its surface temperature and the nature of the atmosphere. These two factors are related: Mars is colder than the Earth not only because it is farther from the Sun than the Earth is (though that is important too). It is also because it has a more tenuous atmosphere than Earth; the atmosphere on our planet prevents energy from leaving the planet as infrared. This is the greenhouse effect, and makes a significant difference to the Earth's surface temperature.

Some long-term plans suggest terraforming the planet: changing the Martian atmosphere to make it like our own. The first step would be to warm the polar ice caps, which are a mixture of water ice and carbon dioxide ice – one method of doing this would be to mount huge mirrors in space that would focus sunlight on to the Martian poles. The water and carbon dioxide would evaporate and form an atmosphere. Through the greenhouse effect, the presence of an atmosphere high in carbon dioxide would warm the planet; then plants could be brought in, changing most of the carbon dioxide into oxygen as they grow, making the atmosphere breathable. This really is possible – in the future, there may be millions of people living on Mars, not confined to sealed domes but living normally in the open air. There is little chance of you setting up home on Mars: this whole process would probably take centuries.

Today, the long-term future of human beings in space is uncertain. Jim Benson of SpaceDev foresees a commercial future in space, filled with many possibilities: 'I think we can accomplish anything that our imaginations allow us – and that the technology is available for – and it's so hard to predict what can be possible; but why not have orbiting businesses and hotels leading to settlements on the Moon, and scientific outposts and even hotels on the Moon and Mars eventually? All of this is possible, and will happen; the question is when?'

With so many problems still to solve on our own planet, the question will also always have to be: 'Why?'

SPACE – THE FINAL JUNKYARD

...searching for solutions...

If you are interested in the future of space exploration, or might want to take a trip into space yourself, there is something you need to know. Space has become a junkyard, full of debris from thousands of space missions. This debris is not merely an unpleasant littering of our space environment: it is a threat to the safety of future space missions. Space junk can damage spacecraft or break through an astronaut's protective spacesuit. It can even be a danger to people back here on Earth.

Floating free

A story from the early years of space exploration will help to illustrate the important features of space debris, some unusual pieces of which were created during the mission. On the morning of 3 June 1965 astronauts Ed White and James McDivitt lay back in their seats in a capsule atop a Titan II rocket. The rocket fired and accelerated upwards. As the rocket climbed, its first stage fell away; the second stage continued into space and pushed the crewed capsule into orbit around the Earth. The average altitude of the orbit was 223 kilometres. In this orbit, capsule and crew were travelling at speeds of more than seven kilometres per second. Soon after the spacecraft achieved orbit, the second stage of the rocket was released. It was also travelling at just over seven kilometres per second and so, after it was released, it continued to orbit with the capsule.

After a few minutes, the crew estimated that the second stage was about a hundred metres in front of their capsule. One of the aims of this mission, called Gemini IV, was to attempt a rendezvous with the second stage of the rocket. This would be important for the impending Apollo missions: astronauts who descended to the Moon's surface in the lunar module would have to ascend again to meet the Moon-orbiting command and service modules. During Gemini IV's attempted rendezvous, McDivitt fired the capsule's rockets, intending to push the craft in the direction of the second stage. Frustratingly, he found that instead of moving closer to his target, he moved above it and behind it. This is because the increase in speed sent the capsule into a higher orbit, which takes longer to complete than a lower one. So firing the rockets to take the capsule towards its goal actually allowed the free-floating second stage to move further ahead relative to the capsule. To achieve rendezvous with an object ahead of it, a spacecraft has to slow down – this drops it to a lower orbit, in which it can catch up whatever it aims to meet. When it has caught up, it moves back into the higher orbit and the rendezvous is complete. All this looks like a graceful, slow-motion manoeuvre. However, it is important to bear in mind that the two orbiting objects are actually travelling at incredible speeds – this is more like a high-speed chase.

The second aim of the Gemini IV mission was just as important. After three revolutions in orbit – each taking about ninety minutes – White was ready for what was probably the most exciting moment of his life: time outside the capsule, attached to it only by a 7.6-metre tether. As well as providing a guarantee that White would not drift away hopelessly into space, the tether included an oxygen hose and other vital links, making it a lifeline in more ways than one. White used a 'personal propulsion unit', or 'zip gun', to move about in space. Pulling the gun's trigger released oxygen under pressure, pushing the astronaut in the opposite direction. The only other way he could have moved away from the spacecraft was by pushing off it – this could have disturbed the orbit.

There is an important point here: when White let go of the capsule to float free, he did not get left behind, even though the capsule was travelling more than seven kilometres per second. The reason for this is that White, too, was travelling at that speed. If you jump from an aeroplane, in the atmosphere, you do get left behind. The difference is due to air resistance. As someone steps out of an aeroplane, they are travelling at the same speed as the aeroplane at first, but moving through the air slows them down, so the aeroplane speeds off without them. The reason the aeroplane carries on moving at the same speed is that its engines provide thrust that pushes it forward despite the air resistance. In orbit several hundred kilometres above the ground, there is virtually no air; there was nothing to slow down White or the spacecraft – and so they orbited together at high speed. In television footage of astronauts on spacewalks it looks as if the spacecraft and the astronauts are drifting slowly, floating almost motionless above Earth. But like the discarded second stage of the Titan rocket, and like White and his capsule, they are actually moving at around seven kilometres per second. This is the main reason why space junk is dangerous – any debris that is left in orbit by a space mission will be travelling at this kind of speed. At seven kilometres per second, a small nut or bolt becomes a lethal projectile that can kill a person or rupture a spacecraft.

During his spacewalk, Ed White was outside for fifteen minutes and forty seconds. While he was above the Pacific Ocean, tethered to the spacecraft, he discarded his outer glove and his helmet visor. These unusual pieces of space junk drifted away from the spacecraft, and probably remained in orbit for several months. Since this first American spacewalk – and the first Russian one three months previously – about a hundred more spacewalks have been taken. During most of them, items have been discarded or have slipped from an astronaut's grasp. Spanners and cameras have been lost, for example. As well as creating their own debris, the crew of the Gemini IV mission may have narrowly missed a deadly collision with another piece of space junk. White and McDivitt reported seeing an unidentified flying object hurtling towards them. Although McDivitt appeared on television many times claiming that the object was a flying saucer, it was probably a piece of debris from another mission.

Although they were travelling at such incredible speeds, White's glove and visor posed no threat to the capsule, as their speed relative to it was very low indeed. In fact, all satellites (apart from those in a polar orbit, described below) orbit in the same direction: from west to east. This is chosen because it is the same direction the Earth itself turns – the Earth's rotation means that a spacecraft is already travelling west–east at 400 metres per second before it even lifts off. Because everything is moving around the Earth in the same direction, head-on collisions between objects in orbit thankfully do not occur. Also, two objects at the same height will be travelling at about the same speed, reducing the risk of collision. Imagine two athletes running in the same lane of a running track; if they are travelling at the same speed, they will not collide, and if there are no other athletes coming in the opposite direction, no one else will collide with them. However, most satellites do not orbit parallel to the equator, so objects in orbit may cross each other's paths, and can then collide. Similarly, if the two athletes cross into each other's lanes, they could collide, but not with any great force.

A similar situation occurs because most orbits are not perfectly circular. Gemini IV's orbit, for example, had an apogee (maximum altitude) of 282 kilometres and a perigee (minimum altitude) of 163 kilometres. Again, there is a chance that objects in different orbits will collide. Most dangerous of all, some satellites orbit over the poles and so cross many other orbits at 90 degrees. Imagine two sprinting athletes in running tracks set at 90 degrees to each other – in this case, the collision could be dangerous. At the speeds that objects orbit in space, it could be catastrophic.

Space jam

While collisions between objects in space are rare, they do happen – and their frequency has increased as more and more objects have been left in orbit. Since the end of the 1950s, nearly 5,000 satellites have been launched into orbit. Only about 350 of them are still working, and many others have returned to Earth. Don Kessler, an expert on orbital debris, says that there are two reasons why debris has accumulated to such a degree: 'One is that people were not aware of what was going on, and second, there's no economic incentive not to leave something in space after you're through with it.' Nick Johnson of NASA's Orbital Debris Research Project estimates that 'the total mass in Earth orbit ... is approximately 5,000 metric tonnes, and is growing at a rate of almost 200 metric tonnes every single year.'

There are more than 8,000 objects larger than ten centimetres across, and countless smaller ones. Among this motley collection are satellites that have either malfunctioned or simply come to the end of their working lives. Then there are rocket bodies – engines and fuel tanks. For objects being placed into a low Earth orbit, usually only one part of the rocket makes it into space – the final stage. However, satellites that are destined to circle the Earth at higher altitudes, such as those that end up in geostationary orbits, arrive in space with two or three. Another source of debris is components of spacecraft such as nuts and bolts and, as mentioned earlier, items dropped by astronauts during spacewalks. The exhaust gases of rockets fired in space can produce huge numbers of tiny particles. Finally, there are fragments of spacecraft produced by break-ups. Many break-ups are the result of explosions, which usually happen because residual rocket fuel mixes with liquid oxygen or because pressure becomes too high inside a satellite's battery.

Collisions involving space debris also lead to break-ups that produce more debris. One of the first satellite break-ups that is attributable to a collision with space junk occurred in 1981, when a Russian navigation satellite called Kosmos 1275 blew up in space. Orbital-debris scientist Darren McKnight explains that 'the interesting thing about Kosmos 1275 is that there was really no energy source on board that should have caused the satellite to break up into 300 pieces of debris ... Mainly, people have come to the conclusion that they think it may have broken up due to a collision with another piece of orbital debris.'

Collisions themselves can sometimes cause explosions in fuel-laden satellites, as well as cause break-up directly. As a result of all these explosions and collisions, there are millions of fragments smaller than tennis balls. Some of the spacecraft break-ups that have created the smaller fragments of space junk were actually planned. From the late 1960s until the early 1980s, the USA and the then Soviet Union conducted tests of 'star wars' weapons that would approach orbiting enemy satellites and fire high-speed pellets at them. The tests created thousands of fragments. Yet another type of debris is coolant leaking from damaged satellites. At present, there are at least thirty Russian satellites leaking liquid sodium and potassium that was used to cool the nuclear reactor on board. At altitudes of between 800 and 100 kilometres, there are clouds of these droplets.

Below an altitude of about 3,200 kilometres, nothing escapes collision with space debris. Evidence of the dangers created by space junk can be seen on the solar panels and other surfaces of satellites recovered from orbit, which show pockmarks of various sizes caused by bombardment by thousands of small particles. By analysing the pockmarks on spacecraft retrieved from various altitudes, researchers have been able to estimate the distribution of space junk in different orbits. The highest concentration of debris is found in low Earth orbits, below about 600 kilometres.

Every piece of space junk is travelling at the phenomenal speed necessary to keep it in orbit. These speeds – several kilometres per second – are called hyper-velocities. When moving at hyper-velocities, an object just one millimetre across can inflict the same damage as a high-speed rifle bullet at close range. A 1-centimetre hyper-velocity object has the same energy of impact as a car travelling at 100 kilometres per hour, or a 1-tonne metal safe dropped from about thirty metres. A 10-centimetre object has about the same energy as the explosion of several sticks of dynamite. Even the droplets of coolant can be dangerous; Don Kessler says that for objects that hit at hyper-velocities, 'whether they're liquid or solid doesn't really make any difference'. Space has become a cluttered and dangerous place to be.

The collision between one of the larger particles and a spacecraft can cause the craft to tumble out of control or stop working altogether. Experiments conducted at the University of Texas show what can happen in one of the worst cases – when a projectile hits a spacecraft at a hyper-velocity. These experiments involve a 'light gas gun', which uses extremely high-pressure hydrogen or helium gas to accelerate small pellets to speeds of several kilometres per second. In one such test, a satellite model made of thick aluminium is placed inside a thick-walled chamber and a plastic pellet is fired at it, at a speed of 6 kilometres per second. Soon after the collision, hyper-velocity scientist Dr Harry Fair enters the chamber to investigate the damage. The model satellite is totally destroyed: the thick aluminium sheets that made the body of the satellite are buckled and are still hot after the collision, and the electronic circuit boards are charred and broken. Fair says that the impact spelled catastrophe for the satellite, and that 'it would be very difficult to conceive of anything surviving after that kind of impact'.

Similar experiments are carried out at NASA's hyper-velocity laboratory at the White Sands Test Facility in New Mexico. The researchers there use light gas guns to fire pellets up to 2.5 centimetres in diameter at 7.5 kilometres per second. The pellets slam into test pieces of various materials, including pieces of multi-layered spacesuit and new types of spacecraft shielding being developed at the laboratory. High-speed X-ray photography highlights just what can happen as a result of hyper-velocity impacts. Frame-by-frame, researchers watch pellets vaporize as they break through a thick aluminium plate, producing a spray of tiny particles. If spacecraft were made of a single layer, then a collision like this would rupture the hull. The particles in the test facility go on to produce pits and cracks in the wall behind the aluminium plate. Multi-layered shielding for spacecraft is being developed at the White Sands facility – each layer absorbs some of the energy of any particle that hits it, and the particles are vaporized before reaching the spacecraft's hull. This research is important for understanding not only impacts involving space debris, but also those involving small meteoroids – natural high-speed pieces of interplanetary rock and dust that can present as much threat as the debris of artificial satellites. When a micrometeoroid vaporizes on impact with a satellite, for example, it produces a plasma – a high-temperature gas of charged particles – that can damage a satellite's electronic circuits.

Whether meteoroid or space debris, there have been several known collisions with spacecraft since the demise of Kosmos 1275 in 1981. In 1996, for example, a French military satellite called Cerise was put out of action when an object the size of a briefcase collided with its stabilizing mast, sending the end of the mast hurtling away. Since the job of the mast was to keep the satellite pointing in the right direction, the satellite began tumbling hopelessly out of control. Operators on the ground have now managed to regain control of the satellite, although its operation has been seriously impaired. Many other satellites have suddenly stopped working for no apparent reason – most of these were probably the result of collisions with space junk. And there have been plenty of known near-misses. In 1999, for example, an American military satellite came within 500 metres of the Russian Mir space station, which had come to the end of its life and been abandoned a few weeks earlier (although it has now been promised a new lease of life by an international investor). During Mir's long and illustrious career in space, the crew had to move into their escape capsule eight times because of potential collisions with orbital debris.

A smash of paint

In one of the early Space Shuttle missions, in 1983, a tiny fleck of paint from another spacecraft hit one of the orbiter's windows, creating a 4-millimetre crater in the toughened glass. This occurred before the space community was totally aware of the dangers of space junk. The Space Shuttle orbiter vehicle is peppered with small collisions during every mission. The windows – costing $40,000 each – must be replaced regularly because of this kind of damage. There is another type of space junk that Space Shuttle crews had to be aware of when they were docking with Mir. Pam McGraw at the Space Shuttle Mission Control explains: 'The Russians tend to throw their garbage out the back door, so when we're approaching Mir, we're flying up through some of that.' In this case, the garbage includes human waste from the toilet facilities of the Russian space station. The waste is contained in capsules, each one an aluminium cylinder about the size of a rucksack. The capsules are deliberately discharged – propelled towards the Earth, so that they burn up re-entering the atmosphere.

When the Shuttle was being designed, in the 1970s, the risk from space debris was not well known. The engineers on the Space Shuttle programme have now installed extra protection against small items of debris. However, larger objects could cause much more serious damage. If you are aboard the orbiter, there is no way you can dodge a chunk of debris using just your senses: neither your senses nor the orbiter will be able to react quickly enough. One of the

Space Shuttle pilots, Scott Horowitz, says, 'You're probably never going to see this particle coming towards you because the closure speed can be in the order of fifty or sixty thousand kilometres per hour – you're just moving so fast.' Ground-based mission control can predict collisions with the larger objects. Horowitz explains: 'We have a whole group of engineers that do nothing but analyse the risk of orbital debris ... and change the alignment of the Shuttle to protect the critical systems from orbital debris hits.' The engineers are able to warn the Shuttle crew thirty-six hours in advance about any chunk of debris that will come within ten kilometres of the orbiter, and adjustments can then be made to the orbit. To reduce the dangers from smaller particles, the orbiter tends to fly with its least vulnerable part – the tail – towards the most likely direction from which space junk will hit.

The odds on a satellite colliding with a piece of space debris cannot easily be calculated – particularly for small debris – but estimates have been made. The probability of collision increases with the size of the 'target' and the length of time it is exposed to the junk-filled environment. Some orbits present a greater risk, too. A person is small compared to a spacecraft, and spends only a few hours at a time in a vulnerable position, so the chances of an astronaut being hit are very small. Various agencies have produced estimates of the probability of collision with space junk. According to the Colorado Center for Astrodynamics Research, the probability that a piece of debris larger than one centimetre in diameter will hit a medium-sized spacecraft during a ten-year period is somewhere between 1 in 100 and 1 in 1,000. So there will be one such hit for every 500 or so satellites. The probability of colliding with smaller pieces of debris is higher, simply because there are more of them. Between 100 and 1,000 pieces of debris less than one millimetre will hit the spacecraft over the same ten years.

NASA has developed a telescope installation specially for tracking objects in orbit. It is based in Cloudcroft, in southern New Mexico, at high altitude and far away from the glare of city lights. The telescope inside the main observatory dome is an example of a relatively new design: a liquid mirror telescope. Nearly all large telescopes, including the one at Cloudcroft, are reflecting telescopes, or reflectors. This type of telescope has a concave primary mirror – shaped like the dish of a satellite aerial. The primary mirror reflects light on to a smaller, secondary mirror, which in turn reflects it on to a light-sensitive electronic component called a CCD (charge coupled device). There is a CCD at the heart of every camcorder and every digital stills camera, which does the same job. When light forms an image on the light-sensitive elements of a CCD, it produces tiny electric charges, which can be amplified to produce a video signal. The signal can be displayed on a television screen or computer monitor, and can be recorded on videotape or as a digital 'movie' in a computer's memory.

Most reflectors have a primary mirror made of silvered polished glass. The primary mirror of the liquid mirror telescope is, as its name suggests, actually liquid: it is made of pure mercury. The mercury is poured into a 3-metre-diameter shallow concave dish, which rotates slowly and continuously – on a cushion of air to avoid shocks and vibrations. This causes the mercury to spread out into a thin, uniform coating on the dish, and it then makes a perfect primary mirror that is more reflective than a glass one. The telescope can point only straight upwards, otherwise the mercury would flow out of the dish. This restriction does not significantly inhibit the telescope's effectiveness, however, as most satellites and their associated debris travel in orbits that are inclined to the equator, and they all pass across the field of view once in a while. During each sighting, an object's speed, direction and distance are logged and analysed. The CCD produces an output that represents the field of view of the telescope.

The CCDs used in telescopes are extremely sensitive, and some can produce up to 200 images every second – eight times faster than in a camcorder. The sensitivity of the CCD in the liquid mirror telescope at Cloudcroft is one of the features that make the telescope useful for tracking tiny pieces of space junk. One of the astronomers who operates the liquid mirror, Mark Mulrooney, says, 'The mirror of the telescope is about 100,000 times larger than your eye, so it can collect about 100,000 times as much light as your eye; there's a CCD camera ... about ten times as sensitive as your eye.' All this means that the liquid mirror telescope can detect objects that would have to shine about a million times brighter for an unaided eye to see them.

The display buzzes with activity, much of it 'noise' created by random signals produced in the CCD. However, the screen shows hundreds of stars, which appear as points of light drifting very slowly across the screen as a result of the Earth's rotation. Most of the stars on the display would be too dim to see with the naked eye, even on the clearest of nights. Every few seconds, a spot of light moves across the star field on the screen. Each of these is an object in orbit. From the speed of the spots of light, Mulrooney and his colleagues can estimate the object's altitude simply by looking at the screen. Looking at a slow-moving blob, Mulrooney comments: 'Ten thousand miles or so, about the size of a desk.' Many of the spots seem to flash on and off as they move across the screen. This indicates that the objects are tumbling. Satellites and pieces of space junk do not give out their own light: sunlight illuminates these objects, and the telescope picks up the reflected light. The flashing signal is due to the fact that the tumbling objects showing up on the display have shiny flat sides and flat solar panels. Mulrooney holds up a 2-centimetre-diameter ball bearing, and says that his telescope can detect such an object at an altitude of 1,600 kilometres.

The liquid mirror telescope was brought into service in October 1996. One of its first jobs was to track the 800 or so large fragments, and some of the countless small fragments, produced by the explosion of a commercial satellite in 1996. The satellite was launched in 1994, and its engine exploded in orbit at an altitude of about 600 kilometres. Information about the orbiting fragments, supplied by the liquid mirror telescope at Cloudcroft, was incorporated into NASA's flight plan for the Space Shuttle mission of February 1997. During that mission, the Shuttle had to orbit to about the same altitude as the cloud of debris from the destroyed satellite, to retrieve the Hubble Space Telescope for repairs. On inspection, the space telescope itself showed considerable damage from orbital debris. At one stage of the mission, the Shuttle orbiter, with the telescope in its bay, had to make a careful manoeuvre to a higher orbit, to avoid a book-sized fragment of the destroyed satellite that could have caused considerable damage.

The liquid mirror telescope is not the only tool that is used to chart and analyse space debris. Several powerful radar systems also monitor the orbiting debris. The radar stations are situated in defence establishments, and their primary function is to keep a watch on military satellites and the threat of incoming ballistic missiles. Increasingly, however, they are being called upon to look out for space debris. The information from radar installations is fed to the US Space Command, which co-habits, in the interior of a mountain in Colorado, with NORAD (North American Aerospace Defense Command). The door to this subterranean site weighs 25 tonnes, and would survive the blast of a hydrogen bomb. Inside the mountain, Space Command collates the information from the radar stations and supplies information to NASA and other space agencies. The radar can detect pieces of space junk even smaller than those observed by the liquid mirror telescope. Both types of installation are necessary, however: some objects reflect radio waves better than light, while others reflect light better than radio waves. Between them, radar and optical observatories can produce an accurate picture of the space debris environment for objects larger than a few centimetres.

In the UK there is another radar installation that tracks military satellites, and which is also increasingly used to track space junk. It is based at RAF Fylingdales on the Yorkshire Moors. The radar here is of a type known as solid-state phased array radar, or SSPAR. It does not have a spinning dish to transmit and receive radio waves, as most radar systems have. It is a huge eight-storey high pyramid with 2,560 small microwave transmitter-receivers arranged in a circle on each of its three faces. The total power output of the radar is equivalent to 25,000 100-watt light bulbs. Microwaves are very short-wavelength radio waves, including one with just the right frequency to agitate water molecules, increasing their temperature – this is why a microwave oven heats only foods that contain water. Any animal that came close to the powerful transmissions of the radar pyramid would be cooked within minutes.

The pyramid shape of the radar gives the team at RAF Fylingdales an all-round view of space, and they can detect orbiting objects that are just a few centimetres across and at an altitude of several thousand kilometres, anywhere above the horizon. Dave Tymon, an expert on orbital tracking who works at RAF Fylingdales, explains the importance of the sort of observations made there: 'We need to be able to protect certain satellites – especially those that are manned – to ensure that no piece of debris hits them; and to do that, we need to know accurately where each piece of debris is.' The ability of the radar at RAF Fylingdales to track and identify objects in orbit is quite amazing. When the French Cerise satellite went out of operation in 1996, its operators contacted RAF Fylingdales to see if the scientists there could work out what had happened. Dave Tymon recalls, 'We ran a simulation of Cerise against all the other pieces of debris and all other satellites over a five-day period.' Of the many thousands of objects in orbit, the team at Fylingdales managed to work out from their simulation that it was a piece of an Ariane rocket launched several years earlier that had hit Cerise.

Objects smaller than one centimetre inflict minor damage that causes degradation of orbiting spacecraft but does not generally threaten entire missions or astronauts' lives. Debris that is larger than a few centimetres can be tracked by the telescopes and radar observatories, and appropriate action can be taken in space. The real danger lies in the intermediate range – objects that are large enough to pose a real threat, but which cannot be detected from the ground. Richard Crowther of the Defence Evaluation Research Agency, in Farnborough, Hampshire, sums up the situation: 'With our radar and telescopes, we can see over 10,000 objects in space; of those, we've seen about twenty break up for unknown reasons, and we believe that they are due to collisions between those objects and objects we can't track.' As more collisions take place, more pieces of debris of intermediate size are created, and space becomes more and more hazardous.

What goes around comes around

You would think that the most obvious first step towards making space safer would be to send fewer objects into space. But there is no sign that the pace of spaceflight development is slowing: there are already more than a thousand satellite launches planned for the next ten years or so. And as hobbyists or teams registered for the X Prize (see 'Day Return to Space') thrust their rockets closer and closer to orbit, there may be even more.

Most of the planned launches are designed to place into orbit a new 'constellation' of satellites that will provide global telecommunications coverage. The first of these networks, Iridium, is already operational (although some of the more than seventy satellites already in orbit have failed and become nothing more than space junk). These satellites function in low Earth orbit, and make it possible for people anywhere on the planet to make telephone calls or connect to the Internet using a low-power transmitter. Until recently, this has been possible only by communicating via satellites in geostationary orbit, more than 35,900 kilometres away, which requires more powerful transmitters. Using satellites in low Earth orbit to communicate will also dramatically reduce the time delay experienced by current satellite communications. The next such network of low Earth orbit satellites is a system called Teledesic – a 300-satellite scheme backed by Microsoft's Bill Gates, which will eventually provide fast, cheap and global access to the Internet.

With more and more objects going into space, the amount of space junk in orbit will increase. However, as we have seen, space junk begets space junk: when one object collides with a spacecraft or discarded rocket fuel tank, it can cause an explosion or fragmentation that produces many more, smaller, pieces. Just as pebbles become sand as the sea crashes them together over thousands of years, some space scientists fear that the space junk currently in orbit will break down into smaller and smaller pieces, gradually producing a shroud around the Earth. This idea is called the Cascade theory, and was proposed by Don Kessler, who says that looking ahead, 'over a period of fifty years, you can have a significant increase in the fragment population'. He can foresee that population eventually forming a ring around the Earth, like those around Saturn. The objects that make up the ring would slowly coalesce, and may even form one huge artificial moon in low Earth orbit. This is the process by which, some astronomers believe, planetary satellites – including our own Moon – formed.

In 1990 space scientists Peter Eichler and Dietrich Rex at the Technical University of Braunschweig conducted a study of space junk. They used powerful computers to work out the likely scenario for the future of orbital debris. They showed that cascade really is possible, or even likely, and that 'critical mass' had already been reached. Since their research was conducted, hundreds of launches have taken place. Other scientists are more optimistic, claiming that the thousand or so planned satellite launches will turn out to be far fewer. They also point out that policies are now in place to reduce the debris-creating capabilities of new spacecraft. Several reports have been produced, with recommendations on how to reduce the impact of launch vehicles and defunct satellites on the space debris population. For example, parts of spacecraft that would normally be discarded – such as lens caps or straps that join the final stage of the rocket to its payload – will be attached to the spacecraft from now on, so that they cannot become space junk. And final-stage rockets themselves, most of which fall back to Earth after a few weeks or months, will be purposely propelled back, as soon as their payloads have been delivered into orbit. The incentive for commercial companies to spend money doing this, say the optimists, will be that in the long run they will save money, as fewer of their satellites will be put out of action by space junk.

But if the likes of Eichler and Rex are right, then the space community must take action to reduce the amount of orbital debris already in space, not simply send up less in the future. Some imaginative solutions have been proposed for doing this, notably by Kumar Ramohalli at the University of Arizona (whose work in robotics featured in 'What Shall we do with the Moon?'). As long ago as 1988, Ramohalli began work on an idea for an orbiting clean-up robot, called ASPOD (Autonomous Space Processor for Orbital Debris). ASPOD is designed to seek out satellites that have gone out of commission, and use a solar-powered laser to slice them into pieces. It would use robotic arms to recover reusable parts such as solar panels, and collect up any other pieces of space junk into a hopper. It would then re-enter the atmosphere, splash down into the sea and be recovered. Another plan that has been proposed to deal with existing space junk is Project Orion, suggested by Jonathan Campbell, a scientist at NASA's Marshall Space Flight Center. Campbell's idea is to use powerful ground-based lasers to knock pieces of debris out of orbit and into re-entry.

However, most space scientists – while they do not take lightly the dangers of space junk – are prepared simply to incorporate better shielding and more space-environment-friendly features into future space designs. The existing problem will go away, they say, as disused or fragmented spacecraft eventually fall back to Earth. The United States Space Surveillance Network has a catalogue of 23,000 space debris objects, of which about two-thirds have fallen back to Earth. According to Richard Crowther, 'Every day we observe at least one object coming back from space to Earth. These range in size from the order of a portable phone to large space stations.' Satellites at high altitudes – above a few thousand kilometres – can continue to orbit for hundreds, thousands or perhaps millions of years without any additional thrust. At these altitudes, there is virtually no atmosphere, so there is nothing to slow down the satellites' orbital speed. The atmosphere is very tenuous at about 200 kilometres, but it does exert a tiny amount of air resistance on objects in low Earth orbit. This slows them down, dragging them gradually lower, into more dense atmosphere, which slows them still further. Eventually, they enter the atmosphere proper and burn up. Space junk that burns up on re-entry appears as streaks of light, like meteor trails.

Meteors – or 'shooting stars' – are those fast-moving streaks of light produced when small pieces of rock from space enter the atmosphere. The pieces of rock (meteoroids) are mostly material left behind by the passage of comets around the Sun. Just as a high-speed collision of space junk with a spacecraft occurs where the orbits of the two objects are inclined with respect to each other, the orbits of comets around the Sun – and therefore of meteoroids – are steeply inclined to Earth's orbit around the Sun. This means that meteoroids approach Earth at great speed – sometimes as much as seventy kilometres per second. As it enters the atmosphere, a meteoroid collides with countless air molecules. This heats its surface – enough to vaporize it. If the meteoroid is small – less than about one millimetre in diameter – this process will continue until there is no more meteoroid left. The destruction of most meteoroids in this way takes only a second or two. The light that produces a meteor trail is actually created by charged particles in the vaporized region around the meteoroid, and not by the meteoroid simply glowing white hot. Larger meteoroids take longer to disintegrate, and leave longer, brighter trails. Meteoroids larger than about the size of a grape can actually make it to the ground without completely disintegrating. The centres of much larger ones can be found as chunks of rock – meteorites – and enormous ones can hit the ground with great force and cause damage. The main differences between space junk and meteoroids entering the atmosphere lie in their composition, shape and speed, but the underlying processes are the same.

Because space junk enters the atmosphere more slowly than meteoroids, its burn-up trails move more slowly, too. Meteors dart very quickly across the sky; a re-entry trail of a hunk of spacecraft looks like a slow-motion version. Size is important: as with meteoroids, small pieces of debris are more likely to disintegrate completely during re-entry than larger pieces. Large items of space junk, such as rocket fuel tanks, do not burn up completely. They can produce trails that last for several minutes. Most satellite bodies are made of aluminium, which is tough and machineable and has a low density, which means that spacecraft made of aluminium will be light for their size. Aluminium is worthy of being used to make space hardware for another reason: it has a relatively low melting and boiling point, so most aluminium objects that return from space burn up completely in the heat of re-entry.

Astronomer Paul Maley spends some of his spare time scanning the skies for space debris and, in particular, re-entry of space debris, using his binoculars. He is most proud of an event he managed to see and record on to videotape in 1984: 'Probably the most unusual and perhaps the most spectacular piece of debris was the re-entry of the Space Shuttle's external fuel tank that I observed from Hawaii ... it began to break up into dozens of pieces that tumbled end-over-end and were flashing and flaring.' The re-entry of large objects like the Shuttle's external tanks creates an exciting fireworks display visible to the naked eye. Many UFO reports have been attributed to the re-entry of space junk into the atmosphere. Some spacecraft are equipped with 'debooster' rockets that can slow them down at the right moment to descend into the ocean, in a controlled re-entry. This is likely to become more commonplace. NASA's 1997 Policy for Reducing Orbital Debris Generation stresses the importance of 'post-mission disposal', for example.

Space agencies can normally predict with some accuracy where objects will land, although this task is more difficult for objects with irregular shapes. The larger or more durable objects that re-enter and make it down to ground level nearly all end up in the sea, since about two-thirds of the Earth's surface is ocean. Occasionally, large pieces of space junk that survive re-entry do hit the ground. One such object was NASA's first space station, Skylab. Launched into orbit in May 1973, Skylab was in use until February 1974 as an orbiting science laboratory, and then was moved into an orbit that would keep it in space until 1983, when the Space Shuttle programme was to begin. However, increased solar activity upset NASA spaceflight engineers' calculations of the air resistance of the upper atmosphere, which began to drag Skylab down noticeably in 1978. More than a year later, in July 1979, the craft was hurtling down to the ground, heading for Australia. The people of Perth, in Western Australia, seemed to be the most at risk, but in the end the surviving fragments of the former space station were strewn across a large uninhabited area.

Another large piece of space debris to make landfall was a fuel tank, with a mass of 260 kilograms, in January 1997. The fuel tank hit Georgetown, Texas, and landed just 40 metres from farmer Steve Gutowski's house. Gutowski recounts the events of that unusual day: 'My wife went out to get the newspaper, she says, "You'd better look out there, boy, there's a dead rhinoceros out there" ... then I looked up to the sky and I said, "There's only one place it could have come down from: heaven."' The fuel tank – from a Delta II rocket – was cylindrical, about three metres long and 1.5 metres high, and had been crushed a little and broken open due to the re-entry and impact. The tank was taken to the Center for Orbital and Re-entry Debris Studies, part of the Aerospace Corporation, in Los Angeles. There it joined a collection of other ex-orbiting objects.

Director of the Aerospace Corporation, Bill Ailor, described the tank: 'You notice this quite prominent hole – we don't know exactly what caused that hole, but it looked like – based on the shape of the metal, the fact that we've seen some high-temperature process going on here – this probably occurred relatively early in the re-entry.' Another of the objects in Bill Ailor's 'space junk museum' was found about thirty kilometres away from Steve Gutowski's farm, by an Airedale terrier called Oliver. What Oliver came across was a beach-ball-sized sphere of titanium. This, too, was part of the Delta II rocket. It once contained pressurized gas that was used to force fuel into the rocket engines. There were three more identical spheres on the spacecraft, and they are probably lying undiscovered in the Texas countryside.

In neighbouring Oklahoma on the same day, Lottie Williams was hit by a small piece of debris, from the same spacecraft, while walking in the park: 'My first instinct was that it was something from a shooting star.' Ms Williams is the only person to date to have been hit by re-entered space junk.

At the Center for Orbital and Re-entry Debris Studies, Ailor and his colleagues analyse samples in a mass spectrometer, which enables them to determine the mixture of chemical elements present. Ailor claims that he can often tell where a satellite was made from the fragments they have recovered, since space agencies in different countries tend to use different alloys. Analyses of re-entry fragments are useful to aerospace engineers, who can check their understanding of the materials they use in the construction of satellites and launch vehicles, and ensure that they do burn up in the atmosphere, or land in a predictable location so that accidents can be avoided.

Break-up of the nuclear family

Re-entry may be useful in reducing the amount of junk in orbit, but it can sometimes create a different problem. More than seventy of the satellites launched into space have carried radioactive material, which can be spread over a large area if the satellites re-enter. In 1970 the first of a series of Russian navigation satellites was launched. These were RORSATs (Radar Oceanic Reconnaissance Satellites), which used radar to track ships at sea. Solar cells would not produce energy fast enough to provide power for the radar, so the satellites carried nuclear generators, each holding about thirty kilograms of uranium-235 fuel. Of the planned thirty-three nuclear-powered RORSATs, thirty-one made it into orbit. All but two of these were moved into 'storage orbits', at around 1,000 kilometres, where they will remain for at least 200 years. Spent nuclear fuel on Earth is normally encased in thick concrete and buried deep underground. Either way, the fuel is highly radioactive, and will not be reasonably safe for thousands of years.

One of the RORSATs that did not make it into the safe storage orbit was Kosmos 954. Launched in September 1977, Kosmos 954 re-entered the Earth's atmosphere – with its nuclear reactor – in January 1978. It broke up and, despite features designed to keep each reactor element in one piece, scattered millions of tiny radioactive pieces over Canada's Northwestern Territory. The scattered material amounted to about one half of the reactor core – the remainder of the core remained intact. Most of the uranium was recovered by a team from the CIA (the US Central Intelligence Agency); some of the pieces recovered from the ground were as small as sugar granules, and the clear-up operation cost $13 million. Four years later, a similar incident with another nuclear-powered Kosmos satellite sent the 30 kilograms of uranium into the Atlantic Ocean. It is the RORSATs still in orbit that are leaking sodium and potassium coolant into space, too.

Another nuclear-powered Russian spacecraft, Mars-96, was at the centre of controversy in November 1996. The craft was destined for Mars but, just one day after launch the final stage failed and it re-entered the atmosphere. According to both the Russian space agency (RKA), and the US Space Command that was tracking the craft, Mars-96 plummeted, its load of plutonium intact, into the sea just off Chile. However, eyewitnesses report seeing a fireball break up into pieces over the Atacama Desert. Patricio Aravena, a Chilean policeman, was one of those who saw the break-up: 'It was ten o'clock at night; I was standing here watering the yard when a bright light appeared in the west. It was very strong ... as it came towards us, it broke into hundreds of bright lights that were still glowing as they vanished over the horizon.' Space Command were not the only ones who were tracking the trajectory of Mars-96. Astronomer Luis Barrera says that he was 'able to plot the trajectory of the rocket carrying the probe; the rocket did explode near to the coast, where some of it landed. But the probe itself carried on to the mountains.' No pieces of debris were found, but the Atacama is a vast, sparsely populated area.

It is not only the Russians who have sent up radioactive materials into space. NASA has used plutonium-238 in electrical generators in about thirty missions. Most of these missions have involved space probes that were to travel to planets farther away from the Sun, where the intensity of sunlight is far less than in the vicinity of Earth. The plutonium is used as the source of power in a 'radioisotope thermoelectric generator', or RTG, which uses a junction of two metals to generate electricity from heat. A small RTG is used in some heart pacemakers, instead of batteries that would have to be replaced. The plutonium in an RTG is in the form of solid pellets of plutonium dioxide. Unlike the Russian nuclear power generator on the RORSAT satellites, an RTG does not use a fission chain reaction to generate heat. Instead, it produces heat by the natural decay of radioactive plutonium. During decay, the nuclei of atoms of plutonium-238 release high-speed electrons – a process called beta radiation. The electrons bump into neighbouring atoms, and this is what heats up the plutonium dioxide. The rate at which the heat is released is just right to power the spacecraft.

There is great concern among many people about the use of plutonium in RTGs. Plutonium-238 does not occur naturally on Earth: it is produced in nuclear reactors. It is poisonous and remains radioactive for many years. In tests, tiny amounts of plutonium inhaled by animals led to cancer. Just a single decaying atom of plutonium inside a living cell can disrupt the DNA inside, and lead to abnormalities in a newborn child. Having said all that, the conker-sized plutonium dioxide pellets in NASA's RTGs are encased in extremely durable iridium and graphite; if they do break, they break into large pieces, not into pieces small enough to be inhaled; and they are extremely heat-resistant and will not vaporize during re-entry or a fire. All these features are designed to reduce the chance of a release of plutonium dioxide. The safety of RTGs was brought up to this standard after an accident during the launch of an American navigation satellite, SNAP-9A, in 1964. The satellite employed a type of RTG called a SNAP (system for nuclear auxiliary power). The satellite blew up over the Indian Ocean and released about one kilogram of plutonium-238 into the atmosphere. The radiation 'footprint' of this accident can still be measured – once the plutonium-238 has released beta radiation, the product of the decay goes on to produce alpha radiation, which is more dangerous.

The largest amount of plutonium ever sent into space is being carried by the Cassini space probe, launched on 15 October 1997 and due to arrive at Saturn in June 2004. It will go into orbit around the planet and, in November 2004, it will drop a small probe on to the surface of Titan, Saturn's largest moon. Saturn orbits the Sun at an average distance of more than 1,420 million kilometres, where the intensity of sunlight is about one-hundredth of that near Earth. So aboard the probe are three RTGs containing a total of 32.4 kilograms of plutonium-238. This mission has been accompanied from the outset by tremendous controversy and has provoked widespread protests. Nonetheless, NASA is confident that its RTGs would survive re-entry and would not scatter plutonium in the atmosphere, and they have another eight missions planned that may use RTGs. Some space probes use radioactive plutonium to keep their instruments warm, rather than to produce electricity. The RHUs (radioisotope heater units) have been used on several previous missions, and NASA plans to use them in space probes that they are considering for missions to Mars and Jupiter and a rendezvous with a comet.

With so much other space junk in orbit, which may cascade into smaller and more numerous fragments, many people think that sending radioactive substances into space is foolhardy. Michio Kaku, a nuclear physicist at the City University of New York, is a prominent voice in the movement against the use of nuclear power in space: 'It's only a matter of time before one of our space probes encounters some kind of space debris in space. I say it's like Russian roulette: sooner or later, you press the trigger often enough, one of our space probes will in fact encounter space debris and spew plutonium on planet Earth.' Whatever the dangers associated with large objects either in orbit or falling back to Earth, space engineers are now engaged in building the largest manufactured object ever to orbit the Earth: the International Space Station.

Action station

By 2004 the ISS (International Space Station) will be up and running. It is one of the most ambitious scientific projects of all time, involving an unprecedented level of international collaboration. The project is being co-ordinated by NASA in the USA, but Russia, Brazil, Canada, Japan and eleven member countries of the European Space Agency are also involved. When fully assembled, the ISS will be 109 metres long and 88 metres wide, and will have a mass of 500 tonnes. It will be four times the size of the current largest orbiting object, the Mir space station. The ISS will consist of modules, connected together in space, and powered by about 4,000 square metres of solar panels that can be turned to face the Sun. It will orbit at an average altitude of 400 kilometres, about the same as the altitude of Mir. Scientists will spend between three and six months there, carrying out a range of experiments in near-zero gravity ('microgravity'). When complete, the ISS will be visible to the naked eye, by virtue of light reflected off its surfaces. In fact, it will be one of the brightest objects in the night sky – brighter than any star and all the planets except Venus.

For such a large object, in low Earth orbit, the chances of being hit by large pieces of orbital debris are significant. In June 1999 a spent Russian rocket came within one kilometre of the two modules that were in place at that time. space-walks around the exterior of the ISS will be routine – more than 160 of them will be conducted during the construction programme alone – and this will increase the risk to the astronauts. In order to be able to calculate the chances of the ISS being hit by debris, NASA put together a test package that members of a Space Shuttle crew attached to Mir in March 1996. The package – called the Mir Environmental Effects Payload – consisted of four briefcase-sized units. One of the units contained an alloy plate on to which small meteoroids and small pieces of space debris impacted. The package was exposed to the space environment until September 1997, when it was retrieved by another Space Shuttle mission. On inspection under a microscope, thirty-eight tiny craters were found, which had been made by hyper-velocity impacts of very small particles. The particles vaporized on impact, and so it is difficult to tell much more than their mass and speed. One of the other units used in the experiment was cushioned, so that it could absorb the impact, and collect the kind of tiny pieces that produced the craters in the first unit. Once this unit was returned to Earth, analysis showed that there was a mixture of micrometeoroids and space junk. The space junk included tiny paint flecks, powdered solid rocket fuel and tiny pieces of aluminium.

The Mir Environmental Effects Packages allowed researchers to work out what kind of dangers the ISS may face, and what kind of shielding it may need. At the Hypervelocity Laboratory in the White Sands Testing Facility in New Mexico, they determine not only what type of shielding to use, but also where it is most needed. Using knowledge of the most likely direction of approach of orbital debris, together with the direction in which the ISS will be facing in orbit, researchers at the laboratory are able to construct computer models of the space station showing the risk of impact. The high-risk, vulnerable surfaces are shown in red, the low-risk, least vulnerable in blue. Justin Kerr, one of the shielding designers for the ISS, says 'If you're a shielding designer for the International Space Station, you want to put your most robust shielding in the front of the spacecraft, in the red areas.'

If the ISS is hit by a large object, it could be a catastrophic collision. The knowledge of the distribution of space junk in orbit will help to give adequate warning, but the ISS is less manoeuvrable than, say, the Space Shuttle. In case of an emergency, there is an escape vehicle for the astronauts – the equivalent of a lifeboat. The experimental version of this vehicle is called the X-38, and has already undergone extensive tests. It is a wingless, lifting body, similar to the X-33 and VentureStar spacecraft described in

'Day Return to Space', but much smaller. It would be deployed quickly and efficiently if an emergency occurs. This vision of escape pods leaving space stations is like something from a science-fiction film. But as long as space remains full of junk, escape may be necessary. If the ISS is hit by a large object, a lot more debris will be created. In the worst case scenario, that might start off the cascade that will fill our skies with tiny pieces of debris. Alternatively, the space station may fall back to Earth. But if it does, will we be able to predict where it will fall?

SUN STORM

...searching for a space weather forecaster...

On a clear night, with the unaided eye, you can see about 6,000 stars. On a clear day, you can see just one – the Sun. Like any of the stars visible at night, the Sun is a huge nuclear furnace radiating enormous amounts of matter and energy into space. The Sun is extremely large – 109 times the diameter of the Earth – and it contains more than 99 per cent of the total mass of the Solar System. And it produces vast amounts of energy, which radiates out in all directions. The Sun is 150 million kilometres away from us – to give you an idea of just how far away that is, consider how long it takes for sunlight to reach us. Light travels 1,000 kilometres through space in just one three-hundredth of a second, and yet sunlight now arriving at the Earth left the Sun a full eight minutes ago.

Despite the immense distance between the Sun and the Earth, life on Earth depends upon the Sun absolutely. Even a small change in the Sun's output of light has dramatic effects on the Earth. But we do not receive only light from the Sun. The 'surface' (outer area of the ball of gases) of the Sun is a seething, turbulent place, and it emits a stream of energetic particles across vast distances. Sometimes, immense blobs of searingly hot gas are ejected, too, during periods of disturbance in the Sun's magnetic field. This is what some people refer to as a sun storm, and the effects of this 'space weather' phenomenon may be felt on Earth. Some of the effects of a sun storm are beautiful: the auroras, such as the northern lights, are much more intense when a sun storm hits. However, there are plenty of undesirable effects too, such as widespread power cuts, satellite malfunctions and hazards to the health of passengers in aeroplanes at high altitude. And the more we have come to rely on technology in our lives, the more of a threat sun storms have become.

Great ball of fire

One of the popular images of the Sun is that it is a huge ball of fire. In one sense, it is. Both fire and the Sun are made of hot gases, which we can see because their high temperature makes them glow. However, that is where the analogy ends. In a fire, the hot gases are released as a result of a chemical reaction. When a candle burns, for example, carbon and hydrogen in the candle wax combine with oxygen from the air to make carbon dioxide and water, and release energy as heat. The products of this reaction – the carbon dioxide and the water – are gases, and are produced at the wick. It is because these gases are hot that they produce light – a process called incandescence. The surface of the Sun is also incandescent, but, in this case, the heat is generated by nuclear reactions, not chemical reactions.

The Sun consists mainly of two gases, hydrogen and helium. About 75 per cent of the Sun, by mass, is hydrogen, and it is hydrogen that takes part in the fusion reactions that power the Sun. When the Sun was born, about 5,000 million years ago, it was a cloud almost exclusively made of hydrogen. Gravity caused this cloud of hydrogen to pull in on itself until the centre of the Sun became dense and cramped, heating the hydrogen gas there. At the high temperatures in the young Sun, the electrons were stripped from their atoms, leaving bare hydrogen nuclei. A gas in which electrons have been separated from their atoms in this way is called a plasma. The contraction of the plasma cloud of hydrogen would have continued had it not been for what happened next.

When the temperature at the centre of the young Sun became high enough, the hydrogen nuclei slammed against each other hard enough to join, or fuse, producing nuclei of helium. This nuclear fusion released huge amounts of heat, and continues to do so. Every second, about 700 million tonnes of hydrogen are converted into about 695 million tonnes of helium. The missing mass is carried away as energy – it was Albert Einstein who first realized the bizarre fact that mass and energy are interchangeable, as summed up by his famous equation $E = mc^2$. The 'E' represents energy, the 'm' represents mass and the 'c' is the speed of light (through empty space). It works out that the loss of 5 million tonnes every second at the Sun's core liberates as much energy as it would take to light 4,000 million million million 100-watt light bulbs. The Sun's energy output at the surface – mainly as visible light and infrared – is equivalent to having 630,000 100-watt light bulbs in each square metre of its surface: no wonder the Sun looks bright. The nuclear fusion releases its energy as gamma rays that heat the Sun's core. The temperature causes the core to expand, fighting against the gravitational collapse. The expansion and contraction are in balance, and the Sun produces heat at a fairly stable rate as it continues to fuse hydrogen. Eventually, the balance will be destroyed – when the Sun begins to run out of its hydrogen fuel. This will be a very long way into the future, and will result in dramatic changes to the Sun, which are discussed more fully in Chapter Five.

Deep in the Sun's core, the temperature is thought to be in the order of 15 million degrees C. At this temperature, the gases in the core radiate huge amounts of energy, largely as gamma rays and X-rays. The intense radiation heats the surrounding layers, which re-radiate the energy to the layers above them. At each successive layer, the temperature is less than the one beneath it – this is because the heat is being transferred outwards in all directions, becoming more spread out in space. The heat transfer by radiation continues outwards. Towards the surface of the Sun, however, the main method of heat transfer is convection. It is convection that is behind the distribution of heat in a pan of water on a stove. The pan is heated at the bottom, and the temperature of the water there increases. This water becomes less dense as it heats up, and it floats to the surface – pushed upwards by colder, more dense water intruding from above. When the water reaches the surface, it cools, and is pushed aside by the water that had replaced it at the bottom of the pan. As it cools still further, it sinks to the bottom again. This churning is called a convection cycle, and it helps to spread heat throughout the water. Due to convection, a pan of water on a stove can be a turbulent place. The same is true of the gas in the outer layers of the Sun. In all, energy generated in the core takes 50 million years to reach the surface by radiation and convection. Even if the nuclear reaction at the core stopped, the Sun would continue shining for many millions of years.

The surface of the Sun is of course vastly more turbulent than a pan of boiling water, as can be seen by the beautiful images produced using large telescopes. These images show how the convection of gases at the surface forms 'cells', each with hot material rising to the surface in the centre and cooler material sinking around it. Together, the cells give the Sun a granulated appearance in these images. The turbulent, granulated appearance of the Sun is due to the twists and turns of its intense magnetic field, which is very different from that of the Earth. Most people are familiar with the shape of the Earth's magnetic field, which is similar to the field of an ordinary magnet. Iron filings sprinkled around a magnet on a tabletop map out the magnet's field. They form lines that loop from one end, or pole, of the magnet to the other. These lines represent the direction of magnetic forces, which align the iron filings between the magnet's north and south poles. The magnet's field is a representation of the direction and strength of these forces. The Earth's magnetic field is very similar to the field around a bar magnet: it, too, loops between the north and south poles. The Sun's magnetic field, on the other hand, is far more complex than the Earth's and is constantly shifting. Sun storms are related to the complex shifting loops of magnetic field lines at and above the Sun's surface.

During a sun storm a huge mass of gas – up to a thousand million tonnes – is ejected from the outermost region of the Sun, the corona. The gases that make up the corona are essentially the same as those making up the rest of the Sun, but for reasons that are not fully understood they are much less dense and far hotter than those at the surface. The temperature at the surface is about 5,500 degrees C, while the temperature in the corona is more than two million degrees C. At this temperature, electrons in all the atoms are stripped from their nuclei – the corona is an extremely hot plasma. Energetic particles from this plasma – mainly high-speed electrons and hydrogen nuclei – are ejected at high enough speeds to escape the Sun's gravity, forming a constant stream from the corona called the solar wind. It has an average speed of about 400 kilometres per second. During a sun storm, unlike the constant stream of the solar wind, plasma is thrown off the corona in individual huge blobs called coronal mass ejections. These blobs are as much as forty times as dense as the solar wind. The frequency of sun storms varies with the activity of the Sun as a whole. During times when the Sun is active, there may be a few ejections every day, while at other times there may be only one every week or so. As they leave the Sun's corona, these ejections expand slightly, and travel out into space.

The coronal mass ejections produced by sun storms travel away from the Sun at extremely high speed. Sometimes they are directed towards the Earth, and take between three and five days to reach us. When a coronal mass ejection meets the Earth, it can produce electrical and magnetic disturbances. A heavy sun storm affects magnetic compasses all over the world, for example. When solar plasma reaches the Earth, it follows the lines of the Earth's magnetic field, because it consists of electrically charged particles. The field lines converge at the Earth's magnetic poles, which lie near to the geographical north and south poles. The electrons and hydrogen nuclei of the plasma travel at incredibly high speeds and, when they enter the atmosphere around the poles, they slam into air molecules, producing an eerie but beautiful glow. This eerie light often takes the form of a waving green-and-red curtain – an aurora. Around the northern hemisphere, the aurora is called the northern lights, or aurora borealis. Around the south pole, it is the southern lights, or the aurora australis. The air molecules become 'excited' – they have more energy – and when they release their excess energy, it is in the form of light.

Dr Nicola Fox, who works at NASA's Goddard Space Flight Center in Maryland, has carried out extensive studies of auroras. She says that the phenomenon is 'completely spellbinding'. There is a faint aurora all year round in the polar regions, because there are always solar particles arriving at the Earth courtesy of the solar wind. But when a sun storm hits, the intensity – and the beauty – increases. Fox explains the different colours present in auroras: 'If you see a lot of green you're seeing a particular excited state of oxygen; if you see a lot of red, you're seeing a different excited state of oxygen.' The green light is more predominant when the particles entering the atmosphere have higher energies than normal – when they have been produced by a sun storm. However, sun storms do more than create beautiful light shows. According to Fox, violent solar events 'can send bubbles of hot plasma streaming from the Sun to the Earth, and here it can cause chaos'.

Solar power cut

At 3.45 am on 13 March 1989, Marie-Claude Bertrand was at home in Quebec: 'All of a sudden, everything just blacked out,' she recalls. She was not the only one whose lights failed that morning, however. Power was lost to about eight million homes, and all Quebec's businesses. Another resident of Quebec, Michael Bailey, looked outside his window when the power cut hit: 'We looked at the city ... we could see different sectors just switching off.'

The power cut was the result of a severe overload on the city's entire electrical supply system. Night workers at the Quebec National Grid Control Center had noticed the surge, and had done their best to cope with it. But the electric current was too powerful, and it knocked out power stations and substations across the grid. It took eight days to get the system back to normal. The surge of current that debilitated Quebec's electricity supply grid was caused by the arrival on Earth of a coronal mass ejection, which had been thrown off the Sun on 6 March. The fast-moving electrically charged particles arriving on Earth induced powerful electric currents in the grid – the energy of the stream of plasma would have been enough to vaporize the Mediterranean Sea. Unfortunately for the residents of Quebec, their city lies on a bed of granite, which is not a good conductor of electricity. If Quebec had been in a different geological location, the ground might have taken the brunt of the electric shock.

During this great geomagnetic storm, other electrical distribution systems were affected across Canada and the USA. A huge transformer was damaged beyond repair in a nuclear power plant in New Jersey, for example. Power cuts were experienced in Sweden, too, as the result of this sun storm, and auroras were observed in regions much farther from the poles than normal – the northern lights were observed as low as New York and London. In assessing the storm, the Oak Ridge National Laboratory estimated that a similar storm in the north-east of the USA could cause damage worth as much as $6,000 million.

The people of Canada were affected by another sun storm five years later, in 1994. This time, two key communications satellites were put out of action. The first satellite went out in the late afternoon. Bruce Burlton is an engineer at Telesat, the company that was operating the satellites. He explains that the satellite 'had spun out of control – it wasn't looking down at the Earth so it couldn't carry communications channels'. Television pictures were lost across much of America and Canada, telephone conversations were cut off, and aeroplanes at several airports were grounded as a result of the loss of the satellite. After five hours of painstaking work, the engineers managed to recover the satellite, stabilize it and return the communications to normal.

However, the disruption was not over: another of Telesat's satellites was put out of action a few hours later. Len Lawson, Telesat's marketing director, had just sat down to relax and watch television when he noticed that the set was not receiving any signals: 'The sports network had gone to noise ... the music station was gone to noise ... the local TV station was gone to noise.' Like the first satellite, this one had also spun out of control – but this time it took five months to recover it. The loss of either satellite would have cost $100 million, which would have meant bankruptcy for the company.

It was a coronal mass ejection from the Sun's corona that had caused the loss of both satellites. The electrons contained within the plasma were absorbed by insulating materials on the satellites, such as thermal blankets that keep the spacecraft warm. When electrons build up and cannot leak away, an electric charge accumulates. This can give rise to an electrical discharge, which can damage a satellite's electronic circuits. Several other satellites have been put out of action in this way. In 1997 AT&T's Telstar 401 satellite was lost during heightened solar activity, cutting transmission on several television channels. And in 1998 more than forty-five million pagers in the USA became inaccessible when the satellite upon which they depended was lost due to a sun storm. All these satellites are in geostationary orbits, at an altitude of about 36,900 kilometres, and are more vulnerable to the effects of sun storms than satellites in low Earth orbit. Those in lower orbits are protected by the Earth's magnetic field, which diverts the plasma away rather like the bow of a boat pushes water aside. As it does so, it produces a teardrop-shaped region of space called the magnetosphere, cushioning us from the blowing solar wind. The magnetosphere is our very own solar windbreak, and it changes shape with the strength of the solar wind.

While sun storms are most likely to knock out the electronic circuits of satellites in geostationary orbits, they can affect satellites in low Earth orbit in a different way. The energy of a cloud of solar plasma released by a sun storm actually causes the Earth's atmosphere to expand slightly when it hits. This increases the drag on satellites in low Earth orbit, which slows them down. When this happens, people on the ground whose responsibility it is to track satellites lose them temporarily. The main organization that tracks satellites in the USA is Space Command at NORAD (North American Aerospace Defense Command) in Colorado. 'When we have a really bad sun storm, we start losing track of the satellites because they're not where they should be,' says Major Gregory Boyette of NORAD. 'During the last serious one I can recall, it took approximately ninety-six hours for us to reacquire our satellites, and figure out exactly where they were, so that we could continue tracking them.' Dr Daniel Baker of the University of Colorado thinks that this could actually be a threat to national security: 'If a country is intent on military mischief and they are sophisticated enough to understand that solar disturbances can have a detrimental effect on our knowledge of what's going on in space, then the worst-case scenario is that they would choose that time to launch some kind of a strike or some kind of military action.'

The expansion of the Earth's atmosphere caused by the arrival of a huge blob of solar plasma can also be helpful. It can cause orbital debris – space junk – to fall to Earth, burning up as it re-enters the atmosphere. During a sun-stormy year in 1991, an estimated 2,000 sizeable items of space junk fell to Earth.

Violent solar activity has also been shown to affect long metal pipelines. Powerful electric currents are induced in these pipelines, which accelerate corrosion. Sun storms can affect the health not only of satellites and pipelines, but of people, too. Astronauts working in space are exposed to the plasma from coronal mass ejections – particularly if they orbit over the poles, where the solar plasma does move close to the Earth because the magnetic field lines converge there. The effects of exposure to high-energy plasma from the Sun are the same as a high dose of radiation: the electrons and hydrogen nuclei can pass into living cells and disrupt the DNA inside. This can lead to cancer. When astronauts are inside a space capsule, they are protected to a certain degree from the flow of plasma. During a spacewalk, however, a spacesuit alone is little protection against extremely fast-moving charged particles. And several Russian astronauts who have conducted spacewalks have since developed cancer and radiation sickness, although it is not certain that the solar wind is to blame. Still, Professor Mike Lockwood at the Rutherford Appleton Laboratory in Didcot, Oxfordshire, says that he 'personally wouldn't be terribly happy about being inside a spacecraft that flew over the auroral regions on a regular basis'. Perhaps we should all think twice about day trips into space and hotels on the Moon after all.

But even on Earth we may not be safe from the Sun's energetic output. Charged particles from the Sun can affect people in high-altitude aircraft, such as Concorde, which cruises at an altitude of 18.5 kilometres. Since it began operating in 1976, Concorde has carried on board a device that detects solar plasma particles. If there is an unusually high flow of the particles, then the pilot drops the aeroplane to a safer altitude. According to NASA's Nicola Fox, 'You usually find the airline companies will switch around the crews on a much more regular basis when we have a violent [solar] event.' The effects of sun storms may even cause health problems at ground level: a doctor in Israel has conducted a study into a possible link between solar activity and heart attacks, the main consequence of which is myocardial infarction. Professor Eliyahu Stroupel, of the Bellinson Medical Centre in Tel Aviv, says that his data showed that 'in years of high solar activity, we have more deaths from myocardial infarction, especially in the older population'. A study in Hungary has even suggested that there may be more car accidents when the Sun is more active.

As increased activity on the Sun can have so many effects here on Earth, it would be advantageous if space scientists could predict when it was about to occur. Professor Mike Lockwood says of the space science community: 'We're very interested in what is going on inside and on the surface of the Sun, because that tells us about the likelihood of events that can wipe out our modern high-tech systems, on which we have become very dependent.' A reliable forecast of the 'space weather' would be very useful to many people beyond the space community, too. The effects of solar activity on communications and electrical power systems on Earth have been known for decades. Since the early 1980s an organization called the Space Environment Center, in Boulder, Colorado, has been issuing warnings to airlines, satellite operators and power companies during times of extreme solar activity. But its predictions of the effects on Earth were often wrong. In fact it was wrong two-thirds of the time (reversing the forecast each time would have increased its accuracy by 33 per cent). During the 1990s there was a great deal of activity aimed at improving the predictions. In 1994, for example, several agencies in the USA got together to consider setting up a co-ordinated effort to improve the forecasting of space weather: the National Space Weather Service. At the same time, there was a growing interest in the Sun in the scientific community. An intercontinental initiative called the International Solar-Terrestrial Program involved proposals for a number of space probes that would increase our understanding of the Sun. This is the latest, very high-tech development in solar investigation – which itself has a long history.

Seeing the light

Today, astronomers have a detailed knowledge of the Sun, although there are still many mysteries still to be solved. Before the birth of modern physics, people had no hope of actually understanding how the Sun provides the heat and light upon which we depend. They could really only notice how it moves across the sky, and wonder what this hot, bright ball might be that rose in the east and set in the west every day without fail. Many ancient cultures held the Sun as the focus of their spiritual celebrations, they built temples or monuments to help them in their sun-worship. Stonehenge seems to be a 5,000-year-old solar observatory, and there are many similar structures throughout the world. The position of the Sun against the stars at the time of a person's birth was thought to be a major influence in that person's life, and many people still believe this to be true. Despite the importance people placed on the Sun, it took a great deal to convince some people that the Earth moves around it, and not vice versa. For example, Galileo was imprisoned by the Catholic church and forced to deny his claim that the Earth moves around the Sun.

The Sun was important to the ancients in many practical ways, as well as spiritual or ideological ones. Its movements across the sky over the period of a year were the basis of the calendar, and it was used to measure the time of day, by observing the shadows it cast on sundials. The Sun was used in navigation, too: by simply measuring the position of the Sun in the sky from a particular location, it is possible to tell that location's latitude. In the sixteenth century, sailors used a cross-staff to make more accurate measurements of the Sun's position. The cross-staff was a stick about a metre long, with a movable cross-piece. A sailor would look along the length of the stick, and point it at the horizon. Then he would move the cross-piece until the top of it was aligned with the Sun (or, at night, the pole star). Many sailors were blinded as a result of their repeated use of the cross-staff.

The first truly scientific investigations into what the Sun might be had to wait until the invention of the telescope. The first astronomical observations using telescopes were made at the beginning of the seventeenth century. It was learned early on that it was not a good idea to look directly at the Sun through a telescope, and soon a safer way to view it was discovered using the magnifying power of a telescope. This involves projecting the visible face (the disc) of the Sun on to a screen behind the telescope's eyepiece. Galileo was among the first to do this. The Sun's disc is still projected in this way today, in large solar observatories as well as by amateurs using small telescopes or even binoculars.

One of the features that is immediately noticeable in a projected image of the Sun is the presence of groups of small dark patches, called sunspots. The early astronomers argued about whether these sunspots are actually on the

Sun's surface – some believed that the Sun, being a celestial body, could not be 'blemished' in any way, and so the dark patches they could see were somehow shadows of something above the Sun's surface. It is now known that sunspots are, in fact, on the Sun's surface. Sunspots were actually noted long before the invention of the telescope, by ancient Greek astronomers – they could see them when they gazed at the Sun through thick clouds or fog. In the first century BC, astronomers in China used the same method to chart the way these dark patches shifted and changed. These observations are far simpler and more consistent when noted through a telescope, and detailed drawings of sunspots were produced by many people from the seventeenth century on. These drawings highlighted how the sunspots moved from day to day, across the face of the Sun. Because sunspots always form in groups, you can follow a particular group as it moves across the Sun's disc. In fact, if groups of sunspots persist long enough, they can be seen disappearing from one side of the disc and reappearing at the other side two weeks or so later. This is because the Sun is rotating, just as the planets are.

Sunspots normally have a lifetime of several days or weeks, and the population of sunspots present varies over time. In fact, the number of sunspots at any time is an important indicator of solar activity. When there are more sunspots, there tend to be more auroras, for example. The number of sunspots, and therefore the Sun's activity, varies over a repeating period of about eleven years. This 'solar cycle' was first discovered in 1843, by the German astronomer Heinrich Schwabe. The solar cycles have been numbered, starting in the 1750s when reliable records of sunspot numbers were first made. The current solar cycle is number 23, and it runs from 1995 to 2006, with a peak in

2000. The last solar cycle was at its maximum at the end of the 1980s, when the violent sun storm hit Quebec's power system. At that time, there was a particularly large group of sunspots – twenty-seven times the average size. There was also an increase in solar flares – another feature of the Sun's turbulent surface, also related to the complex twists and turns of the solar magnetic field.

A solar flare is an eruption of material from the surface of the Sun, which normally takes place along the dividing line between two sunspots. Coronal mass ejections – the really big solar events – are often related to both sunspots and solar flares, but they can happen independently of them, too. The material ejected during one of these violent events comes from the corona, not the surface of the Sun as with solar flares. Sun storms happen all the time – just more frequently and more energetically around the maximum of solar activity. Many people predicted that all the communications satellites in orbit would be knocked out by cataclysmic events at the solar maximum at the beginning of this new millennium – but to date there have been no such disasters.

Not only does the sunspot cycle generally correspond to a variation in the frequency of auroras; it seems to be intimately linked with our weather and climate here on Earth. In the 1890s British astronomer Walter Maunder noticed from historical records of sunspot observations that there were hardly any reported from about 1645 to 1715. This seventy-year period corresponded to a long cold spell, sometimes referred to as the Little Ice Age. During this time, the weather was so cold that rivers froze solid, and in Europe many cities held 'frost fairs' on them. There was another particularly cold period during the first two decades of the nineteenth century; again, this corresponded to a time when solar activity was low. Of course, the climate depends upon so many factors that it is difficult to say for certain that a reduction in the solar activity – indicated by the reduction in the number of sunspots – is necessarily the cause of mini-ice ages; it could be mere coincidence. However, more recently, similar evidence has come to light.

When John Butler joined the Armagh Observatory in Northern Ireland, he was interested in the history of observations made there. The observatory was established in 1790, and detailed meteorological records, as well as records of sunspot observations, have been kept ever since. Deep in the vaults in the passageways underneath the observatory, Butler discovered a cache of these records going back to 1795. Plotting the variations in the sunspot cycle against variations in temperature, he found that the two sets of records matched incredibly well. For Butler, this not only confirmed the findings of Walter Maunder: it has relevance for the debates on climate change today. In the last few solar cycles, activity has been higher than normal – perhaps corresponding to the global warming also attributed to the increase of so-called greenhouse gases in the atmosphere. Butler claims that 'the changes in the Sun are one of the principal causes of the change in temperature of the Earth in recent decades'.

If Butler's assertion that solar activity plays a major role in Earth's weather and climate is true, then scientists will need to find out how this can be so. One mechanism by which it can come to bear on our weather is by affecting the clouds. The idea is that when a sun storm hits the Earth, it can electrically charge the tops of clouds, high in the atmosphere. Clouds are made of water droplets and ice crystals. When a cloud is electrified, electric currents pass through it and this can affect the relative populations of water droplets and ice crystals. Brian Tinsley and his colleagues at the University of Texas are carrying out research into the connection between solar activity and clouds; he explains: 'The electrical currents charge up the droplets at the tops of clouds where they are very cold, and a very small amount of electric charge has the capability to make them freeze.' The more ice crystals there are at the cloud tops, the greater is the likelihood of rain. And with rain, the cloud disappears, allowing the Sun to increase the rate at which it heats the Earth's surface. So the theory goes that sun storms arriving on Earth have the effect of reducing cloud cover, allowing the Earth to warm up.

To test that theory, Tinsley and his team take to the skies whenever auroral activity – and therefore solar activity – is high. They climb to the tops of clouds in an aeroplane that is specially equipped for carrying out high-altitude weather investigation. They sample the cloud directly, collecting water droplets and ice crystals using two metal tubes attached to the aeroplane's fuselage. Detectors in the tubes measure how much electric charge is carried by the cloud particles – to see whether the increased electrical activity leading to the auroras can have effects atmosphere-wide. So far, the results do seem to suggest that there is a greater concentration of ice crystals during periods of intense auroral activity – there may well be a link between sun storms and our weather. Weather is a day-to-day phenomenon, but climate is the average weather conditions over the longer term. Long-term variations in solar activity really could play an important part in climate variations. And solar physicists have confirmed that the actual power output of the Sun varies with the sunspot cycle – so not only do clouds clear, allowing the Sun to warm the Earth more efficiently, but the amount of energy reaching the Earth in the first place is greater, too.

Even today, astronomers are not sure exactly what causes sunspots, or quite how they relate to solar activity. But it is known that they are regions – each normally larger than the Earth – in which the gas is slightly cooler than in the surrounding areas: they are at a temperature of about 4,000 degrees C compared with 5,500 degrees C on the rest of the surface. This explains why they look dark against the rest of the Sun's surface – because they are cooler, they radiate less light. If you looked just at the sunspots, isolated from the rest of the surface, they would be extremely bright. It is also known that sunspots are intimately linked with the activity of the Sun's magnetic field. They form where magnetic field lines are vertical with respect to the Sun's surface – neighbouring sunspots have opposite magnetic polarity, because magnetic field lines are looping out of the surface and back in again. Loops of magnetic field that rise above the surface can form huge tubes of hot plasma that hang thousands of kilometres above the surface. These looped tubes of hot gas are called prominences.

The reason the Sun's magnetic field is so complex compared with that of the Earth has to do with the way the Sun rotates. On the Sun, some parts of the surface take longer to rotate than others. This is not true on Earth: because its surface is rigid, all parts of the world turn around once in twenty-four hours (actually, 23 hours, 56 minutes and 4.1 seconds). However, the situation on the Sun is quite different: regions close to the solar poles take about ten days more than those near the equator to complete each rotation: the period of rotation varies from twenty-five to thirty-five days. The reason for this difference is that, unlike the Earth, the Sun is made of gas. It seems, however, that the inside of the Sun behaves as a rigid ball because of its high density. As it spins, the rigid ball drags the upper, gaseous levels with it. And although the time it takes to rotate once is the same for the whole of this rigid ball, the actual speed at which the surface of the ball moves is quicker at the equator than near the poles. And so it drags the gaseous layer around faster there, too. The same differential rotation is observed on the planets Jupiter, Saturn, Uranus and Neptune, which are also made of gas.

Dr Nicola Fox explains how the differential rotation of the Sun can lead to its complex magnetic field: 'If you imagine twisting up a long piece of string round and round and round, you can see that the configuration is going to become very confused very quickly.' As the magnetic field lines become more twisted, they begin to loop out of the Sun's surface, and – so the theory goes – the result is sunspots, solar flares and a general increase in solar activity. One further oddity of the Sun's magnetism concerns its magnetic poles. While the Sun does generally have a magnetic north pole and a magnetic south pole, because of the complex nature of the magnetic field it sometimes has two north poles or two south poles.

Doing the splits

Astronomers use a wide variety of techniques to study the nature of the Sun's features such as sunspots, solar flares and coronal mass ejections. One of the most important is spectroscopy – the analysis of light by splitting it up into its component colours, to form a spectrum. This technique has its origins in the nineteenth century, although it depends upon a discovery made nearly 150 years previously, by Isaac Newton. In the 1660s Newton was investigating the rainbow-like spectrum produced when white light, such as light from the Sun, passes through a glass prism. Before Newton, scientists had reasoned that the spread of colours in a white light spectrum were somehow introduced by the prism itself. But in a classic investigation – which he referred to as his 'crucial experiment' – Newton used two prisms, recombining the spectrum of the first to produce a spot of white light. During part of this investigation, he blocked off all but the blue light of the spectrum produced by the first prism; in this case, there was neither a spectrum nor a spot of white light produced by the second prism. The colours that recombined to make white light in the first stage of the experiment were the colours present in white light in the first place. Newton realized that white light is actually a mixture of all the colours seen in the white-light spectrum, and that the prism simply split up the light by bending it to different degrees according to its colour. Contrary to popular belief, there are not just seven colours in the spectrum: there is a continuous range of colours, from red to blue.

In the early part of the nineteenth century, German physicist Josef Fraunhofer used prisms to investigate the spectrum of sunlight, and of the light from other stars and from the planet Venus. He noticed that the spectrum of light from these objects was crossed by a number of dark lines, where specific parts of the spectrum were missing. This fact had been discovered before, but Fraunhofer was able to show that the lines had fixed positions in the spectrum, and he catalogued a total of 574 of them. The colour of light depends upon its wavelength – you can fit about 2,500 wavelengths of blue light, but only 1,430 wavelengths of red light, into one millimetre, for example. And so Fraunhofer catalogued not only the 574 dark lines, but also the wavelengths of the missing light.

Fraunhofer and many other nineteenth-century physicists also investigated the spectra of light given out by hot substances, such as metal vapours. Because the light was actually given out by these hot materials, this type of spectrum is called an emission spectrum. These spectra consist of mostly darkness, with just a few bright lines. The bright lines are like a fingerprint of the chemical elements present – you can faithfully identify an element by its emission spectrum. It became clear that the dark lines in the Sun's spectrum corresponded to some of the bright lines observed in emission spectra. Experiments on Earth showed that when you shine white light through a hot gas, you get dark lines, just like those observed in the solar spectrum. The reason for this is that the gas absorbs certain colours – the same colours that it emits. It re-radiates the light that it absorbs, but in all directions, and this is why the spectrum appears darker at those points than across the rest of the spectrum. In fact, a spectrum like that produced from sunlight is called an absorption spectrum. In 1861, armed with this knowledge, German spectroscopists Gustav Kirchoff and Robert Bunsen were able to make the first guess of what elements were present on the surface of the Sun, 150 million kilometres away.

Spectroscopists were able to discover several chemical elements that were previously unknown. The greatest triumph in this endeavour began during the solar eclipse of 1868, when French astronomer Pierre Janssen analysed the emission spectrum of the Sun's corona – which is visible during a solar eclipse because the Moon obscures the Sun's disc. Janssen detected a previously unknown line in the spectrum. Several scientists became involved in the quest to identify the line, and to attribute it to a known element. However, it turned out that the line corresponded to an element that had not yet been discovered. It was named helium, after the Greek word for the Sun, 'helios'. The other main constituent of the Sun, hydrogen, produces several lines across the spectrum. The most important one is red, and is referred to by astronomers as 'hydrogen alpha', or 'Hα'. Between the Sun's surface and the corona is a layer that appears red because there is a great deal of Hα emitted there. This part of the Sun is called the chromosphere (coloured ball). The chromosphere is only about 10,000 kilometres thick, and it is visible, in profile, during a solar eclipse. Solar prominences, thrown up by the Sun's tortured magnetic field, extend from the chromosphere into the corona.

One of the most important developments in solar observation was the spectroheliograph, an instrument that revolutionized the study of the Sun. It was invented in 1889 by American astronomer George Ellory Hale, who was also the first person to discover the magnetic fields associated with sunspots. The spectroheliograph enabled astronomers to produce images of the Sun using particular wavelengths of light. This is the equivalent of putting a filter over a camera to allow only certain colours through. Indeed, astronomers are now able to use a wide range of very specific filters and other instruments that help produce images of objects from the light they emit in a very specific region of the spectrum. Of particular interest were pictures taken in Hα, which enabled astronomers for the first time to see the chromosphere over the disc of the Sun, not only at its edge during a solar eclipse. The chromosphere is alive with interesting features. Across the solar disc prominences appear as long thin trails called filaments, and in parts of the chromosphere above sunspots there are bright patches called plages. The whole chromosphere looks like a translucent ball of worn red velvet.

As well as producing specific, individual wavelengths of light – such as hydrogen's red Hα – hot objects like the

Sun emit light across a continuous range of wavelengths. This is called incandescence (mentioned earlier in connection with a candle flame), the process behind things that glow red hot or white hot. The element of an electric hob heats up to red hot, while the filament of a light bulb is white hot – both are incandescent. The range of wavelengths emitted by an object is characteristic of its temperature. The element of an electric toaster emits only red light and a little orange and yellow light – the low-energy end of the spectrum. An incandescent light bulb emits all the colours of the white light spectrum, including red, orange and yellow light, but also green and blue light. The white light spectrum contains all the colours visible to the human eye, but it is just a tiny part of a far wider spectrum: the electromagnetic spectrum. So light is one of a family of different types of electromagnetic radiation. Beyond the short-wavelength, blue end of the white light spectrum lies ultraviolet, then X-rays and finally gamma rays. Beyond the longer-wavelength, red end are infrared and radio waves. An object that is hotter than the filament of a light bulb emits all the colours of the visible spectrum but also ultraviolet; if it is really hot, it may even emit X-rays or gamma rays. The Sun's surface gives off visible light, but is not hot enough to produce much ultraviolet, and emits virtually no X-rays. The much hotter chromosphere is very strong in ultraviolet, and the corona, at 2 million degrees C, is strong in X-rays.

A wider view

During the twentieth century, astronomers learned how to extend their investigations of the Universe to include these other types of electromagnetic radiation. Radio waves were investigated first: huge parabolic dishes focus the radio waves from distant stars and galaxies on to a receiver. But now there are infrared, ultraviolet, X-ray and even gamma ray observatories, both here on Earth and in space. The main advantage of placing an observatory in space – that observations do not have to be made through the atmosphere – is particularly important for ultraviolet, X-ray and gamma ray wavelengths. Most ultraviolet wavelengths are absorbed strongly by oxygen and nitrogen – which between them make up nearly all of the atmosphere – and much of the rest is absorbed by ozone. Fortunately for animal and plant life, hardly any X-rays and gamma rays make it through the atmosphere. Nearly all radio waves penetrate the atmosphere as if it were not there and, of course, visible wavelengths make it through – otherwise we would not be able to see the Sun, the Moon, the planets and the stars. However, even for wavelengths that make it through the atmosphere relatively unimpeded, it often makes sense to observe from space. For extremely sensitive measurements in the radiowave part of the spectrum, for example, installing a telescope in space means that observations are made out of reach of the 'radio noise' caused by mobile telephones and television transmitters.

The most famous observatory in space is the Hubble Space Telescope, launched in 1990, which has brought a wealth of new astronomical discoveries. It carries instruments for producing images from visible light, infrared and ultraviolet. There is a spectrograph on board, to gather information about the spectrum of distant objects and to take pictures using any particular wavelength of light. Incredible photographs of the stars and planets taken using the spectrograph have brought dramatic new scientific insights, as well as providing the people of Earth with some stunning images of the Universe in which they live.

However, the Hubble Space Telescope does not look at the Sun – its specialities are planets and deep-space objects such as nebulas and galaxies. A number of space-based observatories have been designed to create a better understanding of the Sun, and some of the latest ones are bringing a chance to forecast space weather, perhaps giving us here on Earth more of a chance to prepare ourselves for a sun storm. Beginning in the 1960s, a series of spacecraft called OSOs (Orbiting Solar Observatories) took advantage of the lack of atmosphere in orbit, and gathered information about the Sun's X-ray, ultraviolet and gamma ray emissions. In 1971 the seventh OSO brought the first evidence of coronal mass ejections. X-ray telescopes on board the American space station Skylab were the first to examine the corona in detail in 1973. From the ground, the Sun's surface can be blocked out, by placing a disc of the right size in front of a telescope. But apart from fleeting moments during total solar eclipses, astronomers on the ground had always had to view the corona accompanied by the light of the sky. From Skylab's position in orbit above the atmosphere, it could study the corona against a black sky for the first time. The images produced on Skylab showed that there are huge 'holes' in the corona, which appear as dark patches on X-ray photographs. It has been found that the solar wind from coronal holes is twice as fast as that which comes from the rest of the Sun. The instruments aboard Skylab produced the first clear images of the Sun in the X-ray part of the electromagnetic spectrum – quite a task, since X-rays are absorbed by glass lenses and mirrors. Since 1978, with the launch of the Einstein Observatory, X-ray telescopes have used concentric metal cylinders to focus X-rays on to a detector by glancing reflections, and detailed X-ray images can now be obtained.

Also in 1978, ISEE3 (International Sun–Earth Explorer 3) was launched into space. This particular mission is notable because the spacecraft was flown into a special type of orbit, called a 'halo' orbit. Most space telescopes orbit around the Earth, which means that they regularly pass behind the Earth, losing sight of the Sun. One way to get a constant view of the Sun is to orbit not around the Earth but around the Sun itself, as the planets do. While this would indeed give the spacecraft a constant view of the Sun, it would mean that for months at a time contact with the spacecraft would be lost, as it passed behind the Sun as seen from Earth. This happens with all the planets, too. In a halo orbit, the gravitational forces exerted on a spacecraft by the Sun and by the Earth are equal, and this has the effect of keeping the spacecraft in view of both the Sun and the Earth at all times. This position relative to the Earth is called a Lagrangian point, after French mathematician Joseph-Louis Lagrange, who predicted their existence in 1772.

ISEE3 studied solar flares using an X-ray spectrometer. There were several other missions involving space-based solar observatories during the 1980s, including the SMM (Solar Maximum Mission), launched to coincide with the maximum of solar cycle number 21. The SMM spacecraft carried a range of instruments designed to study the Sun in ultraviolet and visible wavelengths.

Like the 1980s, the 1990s were an exciting time in space-based solar research. In 1990 an ambitious project called Ulysses sent a space probe into an orbit that takes it over the Sun's north and south poles, and extends out as far as Jupiter's orbit, 778 million kilometres from the Sun. Built by the European Space Agency and launched by NASA, the Ulysses spacecraft detects high-speed streams of energetic particles not detectable from Earth. In 1991 it was joined in its quest to discover more about the Sun–Earth system by a Japanese probe, Yohkoh (sunbeam), that carried a range of instruments for analysing the X-ray radiation coming from the Sun. Space-based research into the Sun was hotting up: in November 1994, as part of the International Solar-Terrestrial Program, NASA launched Wind. The instruments aboard Wind act like a space weather vane, and can warn other spacecraft of changes in the solar wind. Another important mission was Cluster, a quartet of satellites designed to study the interaction of the solar wind and coronal mass ejections with the Earth's magnetic field.

Group effort

The idea behind Cluster was to use the data from four satellites together, to create a three-dimensional picture of how sun storms affect the magnetic environment around Earth. The project involved 2,000 scientists from around the world, took ten years of planning and cost £200 million. The satellites were launched aboard an Ariane-5 rocket, from Kourou in French Guiana, South America, in June 1996. Mike Lockwood of the Rutherford Appleton Laboratory says the launch caused great excitement among scientists: 'The whole community, globally, was glued to a screen of some kind to see the launch.' NASA's Nicola Fox was also involved in the Cluster programme: 'As the rocket lifted up, it was just a whole bag of mixed emotions: there was relief that it was finally on the rocket and going up, there was excitement about the type of science it was going to show us.'

The hopes of the scientists who had worked on Cluster were dashed when, just thirty seconds after lift-off, the rocket – and the satellites it was carrying – exploded 3.5 kilometres above the ground, travelling at nearly 860 kilometres per hour. The resulting pieces were scattered over a wide area of the Guianan swamps. Lockwood remembers the effect it had on the people whose work had just been obliterated: 'People were crying around the room, and there was a touch of anger – this should not have happened to the community of scientists who were so looking forward to the data coming in.' To make things worse, the satellites were not insured. The enquiry into the incident found that the problem was with the guidance system, which was using computer software designed for use with an Ariane-4 rocket.

In 1997 there was good news for the scientists involved in Cluster: the satellites are to have a second chance. Using some spare parts from the first mission, the project is going ahead, scheduled for launch in the summer of 2000.

Meanwhile, another spacecraft designed to study the interaction between the Sun and the Earth's magnetic field was launched, without a hitch, in December 1995. Called SOHO (Solar and Heliospheric Observatory), it travels around the Sun in a halo orbit like the ISEE3 spacecraft described above. Aboard SOHO are three sets of instruments, each with a specific task. One set of instruments takes measurements of the composition and speed of the solar wind that blows past it. Analysis of the solar wind, which originates in the corona, should reveal a great deal about how the corona can be so much hotter than the surface beneath it – one of the great mysteries of solar science. The puzzle concerns the fact that the boundary between the surface and the corona seems to disobey a scientific law: if two things are in contact, heat passes from hot to cold, and yet the relatively cool surface is heating the searingly hot corona. This has been pondered over since scientists first discovered the temperature of the corona in the 1940s. There must be some mechanism by which energy is reaching the corona from below the surface, and SOHO may be the one to find out what it is.

Another set of instruments on board SOHO investigates the nature of the corona more directly. One of the instruments is the Extreme Ultraviolet Imaging Telescope, which produces images of the Sun at wavelengths within the ultraviolet range of the spectrum. The telescope acts as a spectroheliograph, because it looks at specific wavelengths produced by the Sun. The wavelengths chosen correspond to particular elements within the corona, and the resulting pictures are awe-inspiring: they highlight the corona, but leave the surface almost totally black, giving these solar portraits a ghostly look. The corona is highlighted in a different way by another type of instrument – a coronagraph, a telescope that has a disc in front of it to block out the Sun's surface. The final set of instruments looks at the Sun's surface itself. They watch the oscillations of the Sun's surface when violent events take place there. It has been known for a long time that the Sun 'rings' like a bell after any kind of disturbance, sending shock waves in all directions across the solar surface and through its centre. By analysing the oscillations, solar physicists should be able to deduce a great deal about the nature of the Sun beneath its surface. This is like seismology, which uses analysis of the oscillations of the Earth caused by earthquakes to investigate the Earth's mantle and core. And so this method of peering beneath the surface of the Sun is called helioseismology.

Another exciting venture using helioseismology to probe the inner workings of the Sun is the GONG project – the Global Oscillation Network Group. This international effort employs six identical helioseismological stations – in California, Hawaii, Australia, India, the Canary Islands and Chile. The fact that they are spread across the globe means that they can provide constant measurements of the Sun – when it is night in one of the centres one or more of the others will then be in daylight. The instruments are designed to detect a particular oscillation of the Sun's surface – one that takes five minutes to complete. The five-minute oscillation promises to reveal new details about the solar interior.

Results obtained by SOHO during 1996 showed dramatic activity on the Sun, producing exciting pictures of coronal mass ejections, for example. Animations made from sets of time-ordered photographs showed the Sun spewing out matter from the chromosphere and the corona. This matter was seen being thrown out at the sides of the solar disc as seen from Earth, so there was no cause for concern, as Mike Lockwood explains: 'Although they're very spectacular to watch, in terms of worrying about the disturbances here on Earth they're not a problem because they're going to miss the Earth.'

SOHO and the other probes were joined in space by another spacecraft in February 1996: Polar. Rather than gazing sunwards as the others do, Polar looks at the Earth's poles, to keep a track of the auroras and disturbances in the Earth's magnetic field.

With all the spacecraft in place, on 7 April 1997, the Sun flared up again. This time, the appearance of the eruption was different. Instead of being directed out in a particular direction, or just at the sides, the Sun blew material off all around, like a halo. This meant that the solar matter was blown off in the direction of Earth, as Nicola Fox explains: 'You could see the circle getting bigger and bigger and this was the sun storm coming towards us.' One of the instruments on board SOHO showed how this violent event made the Sun reverberate, with a powerful shock wave that originated on the side of the Sun facing Earth. This was more evidence that the coronal mass ejection was directed towards Earth. Within three or four days, the matter would reach our planet.

There's a storm brewing

As a result of this observation, the scientific community was on alert, fearing huge magnetic storms on Earth, like those that ravaged Quebec in 1989. The SEC (Space Environment Center) decided that the April 1997 solar event would have no ill effects, and issued no warnings. But the US Air Force was taking no such chances, and it warned military units to expect interruptions in their communications. The reason why no definite forecast could be made is that not all ejections from the Sun have harmful effects on Earth. This is partly because the Earth is such a small target in the vast open space of the Solar System; but even when they do reach Earth, the bubbles of plasma are sometimes rebuffed by the Earth's magnetism, passing harmlessly by. The determining factor is the polarity of the plasma cloud: if it is the same as the direction of the Earth's magnetic field, then the cloud is repelled. If the plasma cloud's magnetic field is polarized in the opposite direction from Earth's magnetic field, the plasma cloud can approach much closer to the Earth, and that is when it can create havoc at ground level.

Despite the SEC's decision not to issue warnings, NASA scientists contacted the media, which broadcast news stories about the sun storm. The telephones of the Space Environment Center began to ring, bringing a stream of questions from individuals and organizations worried about the potential effects. Ernie Hildner of the SEC remembers: 'We had a hospital administrator call up and say, "Of course, we will turn off all our computers, including the medical monitoring computers at the time the storm is supposed to hit" ... Kennedy airport people called up and said, "Should we keep 747s on the ground?"' The telephones were ringing. at NASA, too, as Nicola Fox remembers: 'We had things like, "Will my plants die?" "Should I switch off my computer for the next two days?" "Should I bring my pets inside?"' Everyone made ready, and waited for the storm to disrupt their lives. But this time, the solar plasma swept past the Earth without incident.

In August 1997 NASA launched the Advanced Composition Explorer (ACE). This spacecraft carries a range of instruments, including two magnetometers that determine the polarity of approaching plasma clouds. According to Ernie Hildner, data from ACE would enable the Space Environment Center to make predictions with almost a hundred per cent accuracy one hour in advance. He says that 'customers who are affected by geomagnetic storms will very much appreciate having one hour's notice'. And, since the launch of ACE, the accuracy and sophistication of space weather predictions have increased. In 1999 the Space Environment Center's parent organization, the National Oceanic and Atmospheric Administration, issued a set of 'space weather scales', akin to the Richter scale for earthquakes or the Beaufort scale for wind speed. There are three different scales: one for geomagnetic storms, one for solar radiation storms and one for radio blackouts. Each scale runs from one to five. So, for example, a G5 geomagnetic storm can cause electricity grid systems to collapse and cause problems in tracking satellites and the auroras to be seen as far from the poles as the equator; while a G1 causes mere fluctuations in electrical power supplies and can affect the migration of birds, which depends upon the Earth's magnetic field.

The unprecedented views of the Sun that SOHO, ACE and the other probes supply are beginning to form a clearer picture of how the twists and turns of the Sun's magnetic field can produce the variation in solar output, and how that variation can affect us here on Earth. Having studied the Sun's activity for a number of years, they have provided evidence that the Sun is very active at all times, even during the minimum of the solar cycle. And in 1999 they showed that even during the height of the solar cycle, the Sun can have quiet days. On 12 May, the Wind and ACE spacecraft measured the density of the solar wind as only 2 per cent of its normal value. Without the usual buffeting from the solar wind, the Earth's magnetosphere could expand to five or six times its normal size.

In 1996 scientists based at the Rutherford Appleton Laboratory, who are part of the SOHO programme, discovered a previously unknown feature on the surface of the Sun. During a time when the Sun was particularly 'quiet', they observed flashes of activity that they called 'blinkers', within the granulations created by convection at the Sun's surface. The scientists estimated that about 3,000 blinkers occur at any time across the whole surface of the Sun, each one about the size of Earth and releasing energy equivalent to the explosion of about a hundred million tonnes of TNT. The team believes that blinkers could be the mechanism behind that mystery of solar physics, the heating of the corona. Shortly before the British team discovered blinkers, a team at the Stanford-Lockheed Institute for Space Research in Palo Alto, California, had discovered an effect that they refer to as a 'magnetic carpet'. The scientists describe how relatively small but very numerous loops of magnetic field protrude above the Sun's surface, each bringing as much energy as a large power station would generate in a million years. It is because blinkers and the magnetic carpet occur across the whole solar surface that they are creating excitement among the people studying the Sun. The solar wind spreads out in all directions, throwing particles out at speeds higher than the 500 kilometres per second necessary to escape the Sun's gravity. And to do this, it needs the kind of energy supplied by the extremely high temperatures found in the corona.

With our understanding of the Sun advancing so rapidly, and with ACE and the other solar probes watching the Sun from space, the future of more accurate space weather forecasting looks assured. Perhaps within the next few years, the state of the space environment will be included in our television, radio and newspaper weather bulletins. But according to Ernie Hildner, ACE might be vulnerable to the very thing it is measuring: 'The situation with ACE and sun storms is very like the situation for an anemometer – a wind velocity meter – trying to measure a hurricane: it can be ripped off its moorings, and taken away. Unfortunately, ACE out there helping us measure sun storms about to hit Earth could be damaged by just one of those storms.'

And whether we can accurately predict space weather or not, violent sun storms can disrupt our increasingly technology-dependent way of life. As Professor Mike Lockwood says, 'We could actually find ourselves unable to do things that we've been able to do for most of the past century, because we no longer have the necessary tools to do it; that these things have been removed from us in a stroke.' Many navigation techniques on Earth depend upon a GPS (global positioning system) of satellites. Communications between these satellites and instruments on the ground could be compromised by severe geomagnetic storms. And Dr Daniel

Baker reminds us that sun storms can threaten military communications and satellite tracking: 'There could be a rather catastrophic effect on much of civilization as we know it, if one of these storms occurred at the wrong time in the wrong place.' Perhaps this indicates that we have come to depend on technology in ways that we might regret.

BLACK HOLES

... searching for nature's ultimate abyss ...

Until recently, black holes were hypothetical objects – confined largely to theoretical physics and science-fiction stories. They were a by-product of our increasing understanding of the Universe, a mathematical inevitability, which yet defies our common-sense notions about the nature of space and time. As we shall see in this chapter, a black hole is very strange indeed. But, despite some scepticism, astronomers now in the scientific community, have compelling evidence that these weird objects really do exist in our Universe.

A star is born again

In 1992 a little-known star called V404 Cygni shot to fame in scientific circles when astronomers claimed that it almost certainly was a black hole. Three years earlier a Japanese orbiting observatory, Ginga, had detected a powerful surge of X-rays coming from the star, in the very distant constellation of Cygnus. Along with the X-rays, the star was emitting more visible light than normal, too. It was still far too faint to be detected by the unaided eye, despite growing more than 600 times brighter. Calculations based on the radiation received from the star led to the almost inescapable conclusion that V404 Cygni consisted of a giant star orbiting a black hole. The burst of radiation observed in

1989 – and also, for the first time, in 1938 – was caused by gas heating up as it was pulled off the giant star by the intense gravitational forces of the black hole.

Gravity is the key to understanding black holes. It is the mechanism behind the making of these strange objects, and is also responsible for their behaviour. And it was as an inevitable consequence of the laws of gravity that the concept of black holes originated in the first place. Once you understand that everything in the Universe is affected by gravity and you know how gravity makes things behave, the concept of black holes jumps out at you, begging to be noticed. Even so, the predicted behaviour of these objects is so much at odds with the rest of the physical world that no one could confidently assume that they exist. Since the 1970s more and more evidence has accumulated in favour of the existence of black holes, and astrophysicists now generally accept that they do exist. A number of probable candidates have been discovered, including V404 Cygni.

Before V404 Cygni became a black hole, it was an ordinary massive star – with a mass several times greater than the mass of our Sun. Like the Sun, it formed thousands of millions of years ago from a cloud of interstellar gas and dust – a nebula – brought together by gravity. The force of gravity between any two objects pulls them closer together. When there is a large number of objects – the particles of gas and dust in a nebula, for example – this has the result of bringing everything together towards a common point in space. The Sun began life in this way, as gas and dust from a nebula clumped together in space about 5,000 million years ago. Gravity is tenacious: the force between two objects increases as they move closer together, so, once the gravitational collapse of the solar nebula started, it accelerated, and the clump of matter became more and more dense. This 'protostar' heated up, especially at the centre, where the gases were crushed inexorably. As explained in 'Sun Storm', once the temperature rose sufficiently at the core of this protostar, nuclear fusion was initiated, producing huge amounts of heat. This heat tended to expand the core, and this halted the gravitational collapse. The Sun is in its 'main sequence' – the phase of its life cycle in which the gravitational collapse is offset by the reactions at its core. Eventually, the nuclear reactions will cease, and gravity will take over once again.

As the Sun begins to shrink at the end of its main sequence, its core temperature will increase and, since there is no more hydrogen to fuse into helium, the heat will pass to the Sun's outer layers. Hydrogen in these outer layers will begin to fuse into helium, and this in turn will heat up the centre, where a new set of nuclear fusion reactions may take place. The helium in the core will undergo fusion into heavier elements – carbon first, and then oxygen. At this stage, another force will come into play to fight against the gravitational collapse: electron degeneracy. It was discovered in the 1920s that certain particles, including electrons, resist being squashed into a small space as a result of strange rules that govern their behaviour. The force that results is known as electron degeneracy pressure, and this will prevent any further gravitational collapse, leaving a core in which no more fusion can take place, and that will shrink no more.

And so the Sun will end up as the type of dead star called a white dwarf, composed of carbon and oxygen nuclei, that slowly cools as it sits redundant in space, held up against gravity by electron degeneracy pressure in plasma. Meanwhile, the outer layers, in which there will still be some fusion taking place, will be thrown off into space – first swelling and then ending up as a distended cloud of gas around the white dwarf. In the case of the Sun, the expanding gases will extend beyond the orbit of Venus, and possibly out as far as Earth's orbit. The white dwarf remnant at the centre will have a diameter about equal to the diameter of the Earth and a density about 50,000 times that of gold.

Many stars exist in pairs or groups, held together by their mutual gravity: they are in orbit around one another. If a white dwarf is a member of a pair – or 'binary system' – gravity pulls gas towards it from the outer layers of its partner star. When the gases arrive at the surface of the white dwarf, they may take part in nuclear fusion – the white dwarf temporarily comes alive again – and this results in a sudden brightening of the star as seen from Earth. If the binary system was not visible to the unaided eye from Earth before the eruption, but becomes so afterwards, it will appear as a new star in the sky. For this reason, such an occurrence was given the name 'nova', from the Greek for 'new'. The star called V404 Cygni is a kind of nova, but there is a black hole – and not a white dwarf – in this binary system. Astronomers know that it cannot be a white dwarf, because it is too massive. A star that is much heavier than the Sun cannot become a white dwarf, because due to the greater mass the gravitational collapse can overcome even the electron degeneracy pressure.

When massive stars come to the end of their main sequence, they fuse helium into carbon and oxygen as described above, but fusion does not stop there. The star is hot enough to build up heavier elements, including silicon, sulphur and calcium. This continues until the element iron is produced. When iron fuses into heavier elements, it does not release huge amounts of energy as the previous phases of fusion do. In fact, the iron takes in huge amounts of energy, the star cools and it can no longer be supported against gravity. The outer layers collapse, then are driven out into space accompanied by a huge burst of radiation – which may be visible from Earth as a 'supernova'. Seven such supernovas have been observed in recorded history, the most famous being one that occurred in 1054: it was noted by astronomers in China and Korea and appears on rock paintings in the Americas. This supernova was bright enough to be seen during the day.

Once a star has 'gone supernova', its core may collapse still further, to become a 'neutron star'. Neutrons are found in ordinary matter: together with protons, they make up the atomic nucleus around which electrons orbit. At the centre of a massive collapsing star, however, particles of these three types are crushed together so hard by the gravitational collapse that the positively charged protons and the negatively charged electrons combine, forming more neutrons (which have no electric charge). So, apart from an outer shell of solid iron and an inner core of exotic subatomic particles, the star has become composed almost exclusively of neutrons – hence the name. A neutron star may have a radius of about twenty kilometres and yet have a mass equal to several Suns – its density is a million million times that of gold. A pinhead-sized piece of the material of a neutron star on Earth would weigh as much as a supertanker.

Most neutron stars spin rapidly – a remnant of the rotation of the original star, but at an increased rate due to the fact that the star has shrunk. (This is similar to what happens when spinning ice skaters bring their arms in close to their body to increase their rate of spin.) A neutron star has an intense magnetic field, and this causes protons and electrons around the surface of the star to emit radio waves.

The radio waves form a powerful beam emanating from the neutron star's magnetic poles. As the star spins, these beams sweep across space like a cosmic lighthouse. From Earth, this can be detected as a pulsing source of radio waves – a pulsar.

Most neutron stars collapse no further because of 'neutron degeneracy', which is like the electron degeneracy that halts the collapse of white dwarf stars, only stronger. And yet if the mass of the neutron star is large enough – above about three times the mass of the Sun – even neutron degeneracy will be overcome, and the gravitational collapse will continue unhindered. The result will be an object that keeps on collapsing, with nothing to prevent it crushing itself out of existence. In fact, the object's mass becomes concentrated in ever smaller volumes, and its density increases dramatically. Ultimately, it will shrink to a point of no size – and infinite density. Such a hypothetical point is called a singularity, and its predicted existence is one of the great problems of physics. The gravitational field near a singularity is so strong that even light cannot escape – this is a black hole.

No escape

As explained in 'Day Return to Space', a rocket fired from Earth that attains a speed of 11 kilometres per second has enough energy to escape from the planet's gravitational influences. Any slower, and the rocket will have to go into orbit around Earth or else it will eventually fall back down to the ground. Light is affected by gravity, just like everything else. If you shine a torch straight up into the air on a clear night, the light that makes it through the atmosphere (without hitting air molecules or dust particles) easily escapes the Earth, because it travels at 300,000 kilometres per second – much faster than the escape velocity. If you could shine a torch upwards from the surface of a neutron star, the light would actually be impeded by the intense gravitational field there. It would not be slowed down, however, because the speed of light is the one thing in the Universe that is absolutely constant. This fact was determined from the theory and experiment towards the end of the nineteenth century. So a light beam from a torch travelling towards you or away from you at any speed will still arrive at the speed of light.

Even so, a light beam from a torch at the surface of a neutron star would lose energy as it escapes from its gravitational field. This would manifest itself as a change in the light's wavelength. The wavelength of a light wave would increase as if the light waves were being drawn out. So short-wavelength blue light would be shifted towards the longer-wavelength red end of the spectrum, for example. This effect – called gravitational red shift – has actually been detected in the light coming from extremely dense objects far out in the Universe. What makes this happen is very strange: in a sense, the light really is stretched out, because the gravity near the neutron star actually stretches the space through which the light travels. Put another way, gravity stretches time, so that the time between oscillations of the wave is more than before, again making its wavelength longer. In fact, gravity warps 'space-time' – an inseparable combination of space and time that physicists use to understand the Universe.

The idea of space-time originated with German physicist Hermann Minkowski, in 1908, as an interpretation of Albert Einstein's 1905 Special Theory of Relativity. Before the principles of Einstein's relativity were worked out, physicists imagined that for the sake of equations governing an object's motion, the Universe worked something like a three-dimensional grid. Using Isaac Newton's laws of motion, they could work out how an object would move around this three-dimensional grid, according to how fast and in what direction it starts out and the forces that act upon it. This picture of the Universe assumed that 'now' is a meaningful concept across the entire grid, and that lengths and times are 'absolute' quantities that would be the same for any two people even if they were travelling relative to each other. In this Newtonian Universe, two beacons one kilometre apart and flashing one second apart will be measured as one second and one kilometre apart by anyone who sees them. Common sense tells us that this is what would happen. However, when it was found that the speed of light is constant, relative to any observer, this picture was under threat. In the 1890s Dutch physicist Hendrik Lorentz and Irish physicist George Fitzgerald were separately trying to reconcile the Newtonian Universe with the new discovery about the speed of light. As an attempt to hold on to the three-dimensional picture of things, they proposed that time ran differently for people moving relative to each other, and that lengths were different, too.

This idea was taken on by Einstein in his special relativity theory. According to this, different people travelling at different speeds relative to the two beacons above would see the beacons at different separations – both of time and of space. For one person, the beacons might be in the same place but flash at very different times, while the other sees them flash at the same time but separated by hundreds of kilometres. This is not an illusion – time and space really are different for people moving relative to each other. The separation between the two flashes is constant only when measured in four-dimensional space-time. The concept of space-time is difficult to comprehend, not least because it is impossible to imagine since our senses are restricted to the three-dimensional world that we perceive. However, it is a well-tested idea, and it is a useful way of explaining some of the scientific theory behind black holes, as we shall see.

The concept of space-time was particularly useful when it came to visualizing Einstein's General Theory of Relativity, published in 1916. In this picture, gravity is seen as a warping of space-time rather than a force in the conventional sense. Isolating any one, two or three dimensions makes it possible to picture the effect of gravity. So, for example, imagining three-dimensional space as a flat (two-dimensional) surface, gravity can be thought of as distorting the shape of the surface. This is where the popular image of space-time as a rubber sheet comes from (and we return to this image later in the chapter).

The General Theory of Relativity was a completely different way of interpreting the Universe from Isaac Newton's Universal Theory of Gravitation. Both theories predict the same results for most familiar situations – if Newton's theory was consistently wrong in 'everyday' use, it would not have survived for more than 200 years. Indeed, Newton's theory is still routinely used in many situations, as calculations based on it are simpler and quicker to perform than those of general relativity.

Still, the two theories were very different: Newton's idea was that gravity is a force between any two objects with mass – the force acts across large distances instantaneously. Einstein had to address the fact that his special relativity theory had shown that nothing – not even the transmission of gravitational force between two objects – can travel instantaneously. Everything takes a time to get somewhere else, and the Universe has a speed limit: the speed of light.

All forms of electromagnetic radiation travel at the speed of light – including radio waves, ultraviolet and X-rays. The result of Einstein's incorporating gravity into his relativity theory led to the idea that, rather than being a force, gravity is caused by the distortion of space-time.

To a physicist, the General Theory of Relativity 'feels' right because it is so mathematically elegant. But for such a radical ideological shift to become accepted as a central theory of astrophysics – and of science in general – Einstein's view of things had to prove itself. It has done so amply, by matching prediction to observation in many different situations in which predictions based on Newton's theory failed. One of these situations involves the deflection of light from distant stars as it passes close to the Sun. According to Einstein, light follows the curvature of space-time, and so is deflected as it travels past a massive object like the Sun. This should have the result that the stars from which that light has come appear in a slightly different location in the sky. The effect is very small with an object with the mass of the Sun, but would be much more marked for something like a neutron star. And light can actually be deflected completely around by a black hole, so that it ends up travelling back the other way, or even orbiting the hole a few times before either being sucked in or moving off into space.

The first person to suggest that light might be affected by gravity, in the same way as ordinary matter is, was British clergyman and scientist John Michell in the 1780s. Incidentally, Michell was the first to suggest the existence of objects possessing such a strong gravitational field that not even light could escape. The deflection of light by the Sun's gravity can be observed only by looking at stars very close to the Sun in the sky – those whose light passes very close to the Sun. It is normally impossible to see such stars, because of the daylight and the intense brilliance of the Sun. However, during a total solar eclipse the sky around the Sun is dark, and this makes it possible to see them.

An eclipse in 1919 gave supporters of general relativity a chance to test the new theory. The British physicist Arthur Eddington organized an expedition to Príncipe Island in East Africa, to record accurately the positions of stars close to the Sun in the sky during the eclipse. Comparing the positions of those stars when their light does not pass so close to the Sun – six months earlier, when the same stars were in the night sky – would allow any deflection to be measured. There were two predictions that could have come from Newton's theories. One would predict that light would not be deflected at all as it passes near the Sun, because light has no mass; the other would predict that there would be deflection, but only half as much as that predicted by Einstein's theory. (It is not necessary to go into the details of the mathematics here.)

The 1919 results gave confirmation to Einstein's theory and have been repeated, with more accurate instruments, many times since then. Many other strange predictions of general relativity have been confirmed, too, and the theory that the Universe is made of space-time that is curved around massive objects quickly became accepted. One is the idea that the bending of light around very massive objects, such as huge galaxies or black holes, can become so extreme that it would severely distort the images of more distant objects whose light passes close by them. This effect – an extension of the deviation of light passing close to the Sun – is called 'gravitational lensing'. In the 1970s the first example of a gravitational lens was discovered, and in 1999 ten classic examples of the phenomenon were found by searching through some of the images produced by the Hubble Space Telescope. The existence of black holes is another of the predictions of general relativity, and one of the reasons why people are so keen to find them is to find the ultimate verification of the theory.

When space is pictured as a rubber sheet – using the image mentioned earlier – you can imagine a heavy ball making a circular 'well' around it as it pulls the rubber downwards. Other, smaller, balls placed on the rubber sheet passing near to the heavy ball will move towards the large ball, as if they are being attracted to it. In the same way, according to general relativity, the presence of the Sun in space distorts space-time so that the Earth appears to be attracted towards it. The heavier the ball, the deeper the well and the steeper its sides. A neutron star distorts space so much that only very fast-moving objects can escape from this well. But the gravitational well caused by a neutron star does have a bottom – objects, and light, can emerge from the well if they are travelling fast enough. A black hole, however, forms a bottomless well. You can see why this is troubling to some people: once an object – or even light – travels into a bottomless well, it can never come out again.

Far from a black hole, an object can escape the intense gravitational field. The escape velocity increases as you move closer to a black hole, and the distance from the centre of a black hole at which the escape velocity is exactly equal to the speed of light is called the Schwarzschild radius. It is named after the German astronomer Karl Schwarzschild, who was the first to work out the shape of the space-time curvature around a massive object, using general relativity, in 1916. It was also he who, in the same year, predicted the existence and behaviour of black holes as a consequence of Einstein's theory. From anywhere within the Schwarzschild radius, nothing can ever escape. All the points around the centre of the black hole that are at this distance form the 'event horizon', so-called because no events that happen inside can ever be observed outside.

A black hole defies easy description. Professor Kip Thorne of the California Institute of Technology is one of the pioneers of recent black hole history. He describes it as 'not made of ordinary matter, not made of anti-matter, it's not made of matter at all: it's made of a pure warpage of space and time'. Professor Philip Charles, who was one of those who worked out that V404 Cygni is probably a black hole, says that 'nothing comes out of it, things only go into it; it's that one-way nature that probably upsets people most, causes them to be really disturbed'. And Professor Martin Rees, Astronomer Royal, based at the University of Cambridge, comments on the importance of the singularity: 'In the centre there is a so-called singularity, which is a place where ... everything goes infinite; where, if you do calculations, as it were, the smoke pours out of the computer and something goes wrong. What that means is that deep inside black holes we have a signal that we need some fundamentally new physics to understand what is going on.'

Worrying though the concept of black holes is to some astrophysicists, their existence is now widely accepted. But, what would it be like near to a black hole?

Visiting the death star

Space probes have visited all but one of the planets of our Solar System. One day, a probe will no doubt be sent to the nearest star – a journey that will take tens or perhaps hundreds of years. And in the very distant future, we may visit a black hole – with extreme caution – but for now, this has to be a mission of our imaginations. From our safe vantage point some way from the event horizon, in an extremely rapid orbit around the black hole, we could send a probe down towards – and beyond – the event horizon. As the probe drew nearer to the black hole, it would be violently stretched by the gravitational field. This stretching is a result of what physicists call 'tidal forces'. The closer you are to something, the stronger the gravitational pull. So in ordinary Earth gravity your feet are pulled downwards slightly more strongly than your head, for example, and the difference between the two forces results in a very slight stretching of your body.

The gravitational field around a black hole is vastly more intense than that of the Earth. So the end of the space probe closer to the black hole will be pulled considerably more than the other end. It would certainly not be advisable for a person to venture too close: 'You or I falling into a black hole will get stretched, and our bodies will get mutilated before we even reach the horizon,' explains Kip Thorne. Astrophysicists have coined a memorable term for this effect, as Martin Rees says: 'As you get closer to the centre gravity will get more extreme, you will get squeezed and distorted – spaghettification is the name that some people use for this experience.'

The gravitational field would cause another interesting effect as the probe neared the black hole. The light struggling to escape from the surface of a neutron star is red-shifted because time runs more slowly in an intense gravitational field. And time would slow down for the probe, too – relative to the mother ship. It would appear to move more and more slowly from the mother ship's perspective. At the event horizon, time would be slowed down to a stop: the probe would hang at that location for ever relative to the mother ship. And yet, Kip Thorne tells us, 'If you are inside that spaceship looking back at me, you can wave, you can see me, you may think I can see you, but I can't; you go inside the event horizon you can still see out, you can still see me on the outside, right up to the point when you die.' You would still receive light signals coming in – they have no trouble going in that direction. Nobody knows what happens to objects that fall into a black hole. Once past the event horizon, a probe – if it was still in one piece – would be able to send back no information anyway.

Inside the event horizon, the probe would no doubt be pulled in the direction of the centre – towards the singularity, if such a thing exists. It was Karl Schwarzschild who first predicted the existence of the singularity in 1916. For the next few decades, as physics grew in scope and sophistication, it seemed increasingly likely that black holes did exist and had the kind of properties described above. Until the discovery of the first pulsar in 1967, however, black holes remained a theoretical speculation. As explained earlier, a pulsar is a spinning neutron star – not far removed conceptually from a black hole – and this is why the discovery of pulsars was so important. Many astrophysicists who were unable to accept the strange effects associated with black holes had convinced themselves that a supernova would always throw enough matter into space to prevent ever leaving enough matter to form a black hole or even a neutron star. These things were just too strange to comprehend.

It was British astronomer Jocelyn Bell who was the first to detect signals from a pulsar. She was using a radio telescope to search for other unusual objects called quasars. Studying the hundreds of metres of paper output from the radio telescope, she noticed an unidentified signal: pulses of radio waves with an incredibly regular and very rapid pulsing. After various theories about its origin were dismissed,

Bell wrote the letters 'LGM' next to the trace, to signify the possibility that the signals might be coming from a civilization of 'little green men'. When Bell and her colleagues worked out that this signal could only have been coming from a rapidly rotating neutron star, it was confirmation of the theory of the late stages of the evolution of a star. Bell's discovery of a pulsar confirmed the fact that neutron stars exist, and several more have been discovered; so why not black holes?

It was in the same year as Jocelyn Bell discovered the first signals from a pulsar that the term 'black hole' was coined – by American astrophysicist John Wheeler. But already in 1964 the first serious search for these objects had been undertaken, by Russian theoretical physicist Yakov Zel'dovich. The method used by Zel'dovich has much in common with the approach used by today's black hole hunters. He and his student Oktay Guseinov at the Institute of Applied Mathematics in Moscow scoured the star catalogues looking for binary stars. Their reasoning was straightforward enough: if one of the two stars in a binary system is not radiating much light, and has a mass greater than about three times the mass of the Sun, it is probably a black hole. This conclusion is a result of general relativity and other physical theories applied to super-dense objects: if the invisible partner has a mass less than three times the mass of the Sun, then the object is either a neutron star or a white dwarf. Astronomers had already developed ways to measure the masses of the two partners of a binary system and, from the catalogues, Zel'dovich and Guseinov highlighted five black hole 'candidates'.

In 1966 Zel'dovich, together with another Russian astronomer, Igor Novikov, came up with a giveaway sign that an object might be a black hole or a neutron star. They surmised that a massive, compact object such as a neutron star or black hole would pull huge amounts of gas off a partner star, and that this gas would heat up to enormous temperatures as it fell towards the massive object. At these temperatures, the gas would emit powerful X-rays, which we should be able to detect from Earth. And so Zel'dovich and Novikov suggested looking for binary systems with one partner emitting X-rays and the other emitting visible light. Using X-ray telescopes and optical telescopes together would be a good way of finding black holes or neutron stars in binary systems with ordinary stars. Calculating the masses of the two stars would then confirm whether one of them was a black hole.

In a binary system, stars rotate around a common centre – like two people holding hands, engaged in a whirling dance. If the two people in such a dance weigh about the same, they will both go round at the same speed, about a centre of rotation halfway between them. If one of the people is very heavy, and the other is much lighter, the heavy person will hardly move, while the lighter person will fly around at high speed. In this case, the centre of rotation is much closer to the heavy person. In fact, if you could not see the people, but you could somehow observe their motions, you would be able to estimate their relative masses. In the same way, astronomers can work out the masses of the stars in a binary system by working out how they move around each other.

In 1969 Donald Lynden-Bell, at the University of Cambridge, developed the idea of X-ray emission from black holes or neutron stars suggested by Zel'dovich and Novikov three years earlier. Assuming that gas leaks from the visible star in a binary system, as explained above, Lynden-Bell worked out what might happen as the gas approaches the black hole or neutron star. He found that the gravitational and magnetic fields around a neutron star or black hole would draw the gas into a specific configuration: a rapidly spinning 'accretion disc'. The temperatures near the centre of the disc may be as high as 100 million degrees C – hot enough to produce X-rays, as Zel'dovich and Novikov had proposed. However, in the 1960s, X-ray telescopes were in their infancy. To complicate this further, X-rays do not penetrate the Earth's atmosphere, so X-ray observatories would have to be sent to high altitudes – ideally into space. The first devices to detect extraterrestrial X-ray sources were carried aloft on high-altitude balloons and rockets. They were crude by today's standards, but they did show that X-rays were coming from space, and in 1970 the first true X-ray telescope – called Uhuru – was launched into orbit by NASA.

Swanning around

To illustrate the lengths to which astronomers are prepared to go in search of black holes, consider the story of perhaps the most famous black hole candidate, Cygnus X-1. Uhuru detected a source of X-rays – the first one to be discovered in the constellation of Cygnus (the Swan) – during a systematic survey of the sky. The X-ray source seemed to be very close to a known star, about 6,000 light years distant, with the less than memorable name of HDE226868. Astronomers suspected that there might be a star and a neutron star or black hole in orbit around one another, so they set about determining whether HDE226868 was in orbit around something. They did this by analysing the light coming from it. If the star is orbiting something, and the orbit is side-on or nearly side-on, then it will be moving towards Earth in one part of its orbit and then away from Earth in another part. When the motion is towards the Earth, the light from the star will be blue-shifted, and when the star is moving in the part of its orbit that takes it away from the Earth, its light will be red-shifted. Blue shift is similar to the red shift explained above for light leaving a neutron star, but this time caused by motion, not gravity.

The idea of red shift and blue shift caused by a star's motion is quite simple, and is based on the Doppler effect, which most people are familiar with. The classic example of the Doppler effect is the change in the pitch of a siren as an ambulance speeds past. The siren is more high-pitched than normal when the ambulance is approaching, and is more low-pitched once the ambulance has passed and is receding. The situation with HDE226868 is similar. Like sound, light travels as waves, but in this case the frequency of the waves determines the colour, not the pitch, of the light. In the part of HDE226868's supposed orbit in which the star is moving towards Earth, the light coming from the star will have a higher frequency than normal – it will be shifted towards the blue end of the spectrum. This is the equivalent of a higher pitch in the case of the ambulance. When the star is moving away from Earth, its light is red-shifted.

When astronomers examined the light coming from HDE226868, they found that it did indeed vary from being red-shifted to being blue-shifted – over a regular period of just 5.6 days. The brevity of this orbital period meant that the star must be orbiting something very massive. The astronomers could also discern a slight variation in brightness as the star orbited, too. They attributed this to the fact that the star was elongated, and not spherical. The reasoning behind this is that an elongated star side-on presents a greater surface area and will therefore appear brighter than the same star seen with its bulge facing towards or away from us. This fact led the astronomers to suppose that

HDE226868 was being stretched – by an intense gravitational field. This lent more support to the idea that the object around which the star was moving was a black hole.

Astronomers turned next to the X-ray emissions from Cygnus X-1. They observed how quickly these emissions varied, and found regular variations over a period of a few thousandths of a second. To an experienced astronomer – and for reasons that are not relevant here – this means that the object must be much smaller than any ordinary star. It had to be a very compact object indeed. Putting all the evidence together: there is a massive, compact object that gives out no visible light of its own, but does emit X-rays – presumably from its accretion disc. Cygnus X-1 is indeed a hot candidate for a black hole. Estimates put the mass of the invisible object in Cygnus X-1 at between eleven and twenty-one times the mass of the Sun. If this is true, and if general relativity is correct, it must be a black hole.

The main reason for the large uncertainty in the mass of the heavy object in Cygnus X-1 – between eleven and twenty-one solar masses – is that astronomers cannot tell at what angle the binary system is orbiting. From Earth, even the most powerful telescopes cannot separate the two objects. The light and X-ray emissions appear to originate in the same point in the sky. So all the information about the binary system was inferred from measurements of the variations in the X-ray and light received. The orbit might be nearly side-on; it cannot, however, be totally side-on, because in that case the two objects would eclipse one another. This would result in a different 'light curve', a graph of the light level over a period of days. Astronomers would be able to tell from the light – and the X-ray signals – that the two objects were passing behind each other. The opposite extreme, in which the orbit is inclined at 90 degrees to the Earth, is also ruled out, because in that case there would be no Doppler effect since the star would not be moving towards and away from us. Despite the uncertainty in mass, the fact that the lowest possible mass was well over three times the mass of the Sun meant that Cygnus X-1 was very exciting for astrophysicists.

In 1974 two celebrated black-hole theorists decided to have a bet on whether the source of the X-rays coming from Cygnus X-1 was the accretion disc around a black hole. They were Kip Thorne and Stephen Hawking. Both these scientists believed that black holes do exist. Stephen Hawking, of the University of Cambridge, explains why he went ahead with the bet: 'It was not because I didn't believe in black holes. Instead, it was because I wanted an insurance policy: I had done a lot of work on black holes, and it would all have been wasted if it had turned out that black holes didn't exist – but at least I would have had the consolation of winning the bet.' After enough evidence piled up in favour of Cygnus X-1 being a black hole, Hawking decided to concede the bet. While he was in Los Angeles in 1990, Hawking and his friends broke into Thorne's office, where the hand-written bet was hanging, on the wall, in a frame. Hawking put his thumbprint on the sheet of paper as a signature, and made sure that Thorne received his winnings: 'I had given Kip Thorne a year's subscription to *Penthouse*, much to his wife's disgust.' Hawking's thumbprint has come to symbolize the first real acceptance of the existence of black holes by the astrophysics community.

Black catalogue

Since the discovery of Cygnus X-1, several other very promising black hole candidates have been discovered and analysed, and astronomers have grown increasingly confident that they have definitive evidence of black holes. V404 Cygni is perhaps the most compelling case to date. As explained above, Philip Charles spotted the fact that V404 Cygni is an X-ray source, by looking at the data collected by the Ginga satellite in 1989.

Charles is a black-hole hunter based at the University of Oxford. His quest takes him to powerful telescopes all over the world: the Canary Islands and Hawaii in the northern hemisphere, and South Africa, Chile and Australia in the southern hemisphere. 'Searching for these things is the most wonderful way of going to the frontiers of modern physics ... it's one of the most exciting things a modern scientist can do.' We have seen the lengths to which astronomers must go to analyse their quarry once they have found it – but finding it in the first place is not easy, in a deep sky filled with all manner of other objects. 'When you are looking for a needle in a haystack, you need the needle to shout out and say, "I'm here",' says Charles.

V404 Cygni did shout out, by way of the burst of X-rays detected by Ginga. Since 1989 Charles has been studying this binary system from the ground, at the William Herschel Telescope on La Palma in the Canary Islands. By studying the light and the X-rays from V404 Cygni, Charles and his colleagues concluded that the invisible object in the system must have a mass of at least six times that of the Sun, and probably more like twelve. This is why they are confident that this object must be a black hole. Presently, there are about twenty-five candidates for stars that have become black holes, including Cygnus X-1 and V404 Cygni.

Another member of the black hole hall of fame is GROJ1655-40. This binary system is 10,000 light years away, and appears in part of the sky in the constellation of Scorpio. In 1999 a team led by Rafael Rebelo at the Institute of Astrophysics on the Canary Islands made a remarkable finding related to this black hole candidate. They attempted to determine the gases around the two mutually orbiting objects that make up GROJ1655-40. If there really was a black hole here, it would have been formed from a massive star that would have been through the supernova stage. Would there be any record of a supernova in the gases in that region of space? What they found is remarkable: there were large amounts of the elements oxygen, magnesium, silicon and sulphur – which can only have been produced by a supernova.

However, while the evidence in favour of black holes is becoming ever more conclusive, certain problems still remain. One of them is that the X-ray radiation received from the black hole candidates is lower than that predicted by the theory of the accretion disc from which it emanates. In 1997 Dr Ramesh Narayan of the Harvard-Smithsonian Center for Astrophysics in Massachusetts proposed a theory that could solve the problem: he suggested that the reason that the amount of X-ray radiation is less than expected is that the black hole sucks the gas away inside its event horizon, before it can emit the radiation, so taking its energy with it. This requires an alternative to the generally accepted theory of the formation of the accretion disc. Narayan proposes that rather than forming flat, spinning discs, the gas from the stellar partner of the black hole or the neutron star falls in from all around. This would make it less dense, which would mean that it would retain its energy for longer – until it either hits the surface of the neutron star or moves within the event horizon, never to be seen again. In the case of a neutron star, the gas would still be able to radiate – since there is no event horizon, radiation can still be emitted into space.

To test his idea, Narayan looked at the radiation coming from eleven X-ray binary systems. Six of the systems – including V404 Cygni – were black-hole suspects, while the remaining five were thought to be neutron stars. And what he found is compelling: all six black-hole candidates were lacking in X-rays and the neutron stars were not. Narayan believes that if his theory is correct, then it will provide direct evidence of the existence of the event horizon. It may sound like Narayan has simply made up his theory to fit the facts, but the X-ray spectrum that his theory predicts matches what is observed extremely well.

The central idea

The black holes so far discussed are 'stellar' black holes which, as their name suggests, are created from dead stars. And they are all inside our own Milky Way galaxy. But the matter to make a black hole does not have to have come from a star – anything will do, as long as there is enough of it. For many years, astronomers have wondered whether there might be 'super-massive' black holes lurking at the centres of galaxies. There is enough matter in the centre of a galaxy to form a black hole of immense proportions.

The idea originated as an explanation of the behaviour of so-called 'active galaxies' – a class of deep-space objects, far outside the Milky Way. The first of these active galaxies was discovered, by American astronomer Carl Seyfert, in 1943. Seyfert galaxies, as they are now called, are large spiral galaxies that have surprisingly bright central regions. Discoveries of other types of active galaxies soon followed: radio galaxies in 1946, quasars in 1963 and blazars in 1968. Radio galaxies are huge: they have vast lobes in which electrically charged particles travel at close to the speed of light, emitting radio waves as they do so. The lobes can stretch thousands of light years into space. Quasars (quasi-stellar objects) are incredibly far away – some of them are the most distant objects ever detected. They are also the most powerful objects known, releasing phenomenal amounts of energy. Blazars are similar to quasars: in fact, they may be quasars that point directly towards us.

The first suggestion that super-massive black holes might be the source of power in active galaxies came in 1964, from American astrophysicist Edwin Salpeter at Cornell University and Yakov Zel'dovich in Moscow. Detailed theories to explain the processes behind the production and release of energy from super-massive black holes were developed through the 1970s, and the first strong evidence of their existence came in 1978, as a result of observations made by Wallace Sargent and Peter Young, based at the California Institute of Technology. These two researchers had turned their attention to a galaxy called M87, about 50 million light years away.

Sargent worked out how fast the stars in M87 are revolving around the centre, by studying the red shift and blue shift of starlight coming from around the galaxy's centre; Young estimated the mass of stars in the central part of the galaxy, and therefore the density of matter there. Both of the results suggested that there is as much mass concentrated in the centre of M87 as in about 3,000 million stars with the mass of the Sun – a black hole almost certainly exists there. In 1994, the Hubble Space Telescope was used to look at M87. The images produced were clear enough to show a central disk 500 light years across, made largely of hydrogen gas. Like a CD, there was a 'hole' in the middle of the disc, about a hundred light years across.

The Hubble Space Telescope provided evidence of another super-massive black hole in 1992, when it captured stunning images of a radio galaxy called NGC4261, also a member of the Virgo Cluster. Although NGC4261 is more than 40 million light years away, the Hubble Space Telescope produced a remarkably clear image of the centre of the galaxy, which showed a brown spiral disc inside the galaxy's white fuzzy core. The disc is rotating, and from the speed of the rotation astronomers determined the mass of the small central part of the core. They found that it has a mass more than 1,000 million times that of the Sun – fairly conclusive evidence of a super-massive black hole.

The Hubble Space Telescope has found other examples of what must be super-massive black holes, but perhaps the most compelling evidence came from the Very Long Baseline Array – a collection of radio telescopes across the USA. In 1994, astronomers used the array to examine a galaxy called NGC 4258. They collected radio waves emitted by water molecules near to the centre of the galaxy (as explained in 'The Rubber Universe'), and concluded that the density of the matter at the centre is more than 10,000 times greater than any known star cluster – there is almost certainly a black hole there.

Astrophysicists have long suspected that there may be a super-massive black hole at the centre of our own galaxy. The Milky Way is a spiral about 100,000 light years in diameter, containing about 100,000 million stars. Our Solar System lies about 25,000 light years from the centre of the Milky Way, in one of the spiral arms. If you find the constellation of Sagittarius, you will be looking in the direction of our galactic centre. For decades, it has been known that there is a massive object at the centre of our galaxy. The galactic centre is largely obscured from the view of optical telescopes by interstellar gas and dust, but radio telescopes detected a powerful source of radio waves in the 1950s, named Sagittarius A*. In the past few years, there has been a flurry of activity by astrophysicists trying to find out more about what it might be.

In 1997 astronomers at the European Southern Observatory used the New Technology Telescope to 'weigh' the huge object at the centre of our galaxy. They followed the stars for a year, tracking their motion – which is very slight from our distant viewpoint – and their results suggested that there is an object with a mass about 2.5 million times as much as the Sun, contained in a space less than one-tenth of a light year across at the centre. This is strong evidence indeed to support the idea that there is a supermassive black hole on our galactic doorstep. There is no need to worry about this black hole: however hungry this object might be, we are sufficiently far away, and revolving around the galaxy at high enough speeds, to evade capture.

Perhaps in the future, intrepid astronauts might venture to take a closer look at the object at the centre of our galaxy. Given its extreme size and mass, would this be more dangerous still than a visit to a stellar black hole? Perhaps: near to the centre of a galaxy, stars are ripped apart to feed the huge black hole. Martin Rees takes us on a journey to the centre of the Milky Way: 'If one was to take a journey starting in the outer part of the galaxy and moving in, you'd see very hot glowing gas filling the galaxy, which is being heated by some mysterious powerful central source. You'd find the gas orbiting the disc, swirling at near the speed of light and getting very hot – so hot that it radiates not only light, but very energetic radiation like X-rays.'

Staying at a 'safe' distance and sending a space probe towards the centre, we would watch as the probe descends towards the event horizon, as in our imaginary journey to a stellar black hole. In fact, it turns out that this time the tidal forces are actually less severe: because the black hole is much larger, the curvature of space-time near to the event horizon is much less. The probe would survive passing through the event horizon, and, from its perspective, the light from outside the event horizon would be warped into a smaller and smaller space above the probe. There would be a final crunch as the probe met its fate at the singularity, but no one has any idea what might happen then – and Einstein's general relativity can provide us with no clue.

In addition to stellar and super-massive black holes, astrophysicists have recently discovered what they think are 'middleweight' black holes. In 1999 teams at Carnegie Mellon University and at NASA independently found evidence of this new class of black hole, in the spectrum of the X-rays emitted by the accretion of matter. The objects have masses of between 100 and 10,000 times that of the Sun – an intermediate size between stellar and super-massive. The researchers believe that these newly discovered objects may be the result of a kind of coalescence of many stellar black holes or neutron stars. The black holes they observed were in a 'starburst' galaxy – one in which star formation occurs at a phenomenal rate. Star death also happens quickly in this type of galaxy. The researchers estimate that millions of neutron stars and black holes would have been created in the galaxy they studied during the past 10 million years or so. Perhaps black holes – far from being elusive and rare – are actually extremely common. Each galaxy may have a super-massive one at its centre, many smaller ones forming all the time, and also a collection of intermediate ones.

Going for a spin

There is an important feature of black holes that has not been mentioned so far: the fact that nearly all of them are likely to be rotating. Certainly the super-massive ones at the centres of spiral galaxies will be, since spiral galaxies themselves are rotating. Stars rotate, too, and so when they become stellar black holes they will retain their spin. In fact, in the same way as a dying star gains speed as it shrinks to form a neutron star, a black hole is likely to be spinning much faster than the star from which it formed.

The rotation of a black hole has many interesting possible consequences. One of the most exciting, bizarre and controversial is the possibility of near-instantaneous travel into a rotating black hole and through a 'wormhole' to another point in space-time. The idea goes something like this: the rotation of a black hole may turn the singularity from an infinitesimal point into a ring; the 'well' of space-time inside a black hole is infinitely deep – open-ended – and so could conceivably connect to a similar well elsewhere. Because the singularity is enlarged, you could pass through it without being squashed to nothing. So you could travel through the wormhole and emerge from a 'white hole' at another point in space and time.

However, this theory has now been discounted. Kip Thorne: 'I recall a movie in which the spaceship goes inside a black hole and comes out the other side into another universe. That don't happen ... The fundamental laws of physics say that at the centre of a black hole, there is a singularity – a region in which space and time are infinitely warped, a region where when matter goes into it, matter is destroyed.' And Professor Roger Penrose of the University of Oxford says that 'these things are driven by romantic science fiction ideas ... The scientific evidence is that this is just not going to happen – it's just not possible.'

However, the idea of travel through wormholes may not be dead. Several theoretical physicists, Kip Thorne among them, have developed theories involving wormholes that would have no event horizon around their entrances and no singularities. So, these wormholes would connect objects that are not true black holes, but have similar properties. According to theory, these wormholes would collapse in on themselves unless they were 'held open' by some kind of exotic matter. This matter is as yet undiscovered, but its existence is predicted by various theories of modern physics. Some physicists believe that, in theory at least, such wormholes would allow instantaneous travel across space-time. Russian physicist Igor Novikov believes that travel across huge distances or through time would indeed be possible. Novikov says you would need to create two interconnected objects like black holes, and move them to two different locations: 'One of them can be near our Earth, the other in another galaxy, but this tunnel can be incredibly short – say a few metres ... You can travel from one galaxy to another in, say, a few seconds.'

In 1997 a team at the Massachusetts Institute of Technology and the University of Maryland, led by Dr Shuang Nan Zhang, found a way to measure the spin of a black hole. What they did was to ascertain the distances from the centres of several black holes at which matter could attain a stable orbit. According to general relativity theory, matter that orbits beneath this distance is quickly sucked into the black hole, however fast it is moving. This limiting distance is farther from the centre of a black hole than the event horizon, and varies with the black hole's speed of rotation: the distance from a black hole of its 'last stable orbit' increases with the black hole's rotation.

The result of this is that the accretion disc of a black hole ends farther from the centre of a rotating black hole than from the same black hole if it were stationary. This should have the result that less X-ray radiation is emitted, since material would be able to radiate more energy before being sucked in if the accretion disc is closer to the black hole. Shuang Nan Zhang and his team managed to measure the last stable orbits of several stars, by examining the spectrum of X-rays emitted by the accretion disc. They found that two of them seemed to be rotating much more rapidly than the others. Interestingly, these two fast-spinning stars were also the only two that produced high-speed jets of material. Whatever the mechanism behind the jets, it probably has something to do with the black hole's spin. And so the fact that they occurred only with the two black holes that Shuang Nan Zhang believed were fast-spinning does suggest that his theory might be correct – although more work will need to be done to confirm this.

Theory tells us that there are only three things that we can determine about a black hole: its mass, its rate of spin and its electric charge. The mass can be worked out using the Doppler effect to measure the speed of a black hole's visible partner in a binary system, and Shuang Nan Zhang's method above seems to reveal a black hole's spin. So if astrophysicists can work out a way to measure the electric charge of a black hole, they would be able to know absolutely all of the information there is to know about a particular black hole. This would be satisfying in one way, but utterly frustrating in another. We would still be utterly confounded by the mysteries of what kind of star the black hole originally was, what kind of matter the hole has swallowed and what is going on inside the black hole's event horizon.

A rotating black hole literally drags space-time with it – rather like the way a whirlpool drags water around with it. This creates a region around the 'equator' of the black hole called the ergosphere. Travel within the ergosphere, and you will be dragged around with the black hole. Roger Penrose has devised an interesting thought experiment, in which it is possible to extract energy from a rotating black hole. If a piece of matter breaks up into two within the ergosphere, it is possible for one piece to gain speed by 'borrowing' energy from the other, and to shoot out of the ergosphere. The other piece will be swallowed by the black hole. Penrose himself explains the notion behind the 'Penrose process': 'You could imagine firing a number of particles or bodies into the neighbourhood of the spinning black hole, in the right direction ... one particle carries negative energy into the hole, and that means that the partner has more energy than the one that fell into it. So you can actually extract energy from a black hole using this process.'

Penrose can even see how the idea could be used as a huge power station at some point in the distant future. Penrose's power station is situated at a safe distance from the black hole, outside the ergosphere. The idea is to throw items of rubbish from there into the ergosphere, and as they break up, some are flung out at high speed, taking with them energy from within the ergosphere. The energy of these fast-moving objects could be used to drive some sort of turbine, to generate huge amounts of power. The loss of energy means that the black hole slows down, but so very slightly that you could go on using a black hole as a power station in this way perhaps until the end of time.

Penrose's scheme will never actually be realized: it is just a thought experiment to illustrate one of the strange aspects of black holes. A similar process – in which pairs of objects divide, one going into the black hole and one leaving – is Hawking radiation. This involves a concept of modern physics which, to the uninitiated, is hard to contemplate, but which to a modern physicist is as inevitable as black holes: virtual particles. While black holes are the domain of general relativity, which is used to study very large objects and their effects on space-time, virtual particles spring forth from quantum theory – the branch of modern physics that deals with very small objects such as atoms and subatomic particles.

One of the mainstays of quantum theory is the uncertainty principle – the idea that it is impossible to measure both the position and the speed of a particle with total certainty. One consequence of the uncertainty principle is the fact that there is no such thing as nothing: that even in a total vacuum, pairs of particles fleetingly come in and out of existence. These virtual particles normally annihilate each other within a tiny fraction of a second, and it is therefore not possible to detect them. However, the rate of generation of virtual particle pairs increases as space-time becomes more warped – such as near to a black hole. In 1974 Stephen Hawking showed that at the event horizon of a black hole, pairs of particles would be produced – from nothing – at a phenomenal rate. Some of the pairs would be annihilated, but others would be separated – one particle being sucked into the clutches of the black hole and the other moving off into space.

If virtual particle pairs are separated, each member becomes a real particle. One member of each pair of virtual particles has 'positive energy' and the other has 'negative energy'. The particle with negative energy goes through the event horizon and takes energy from the black hole to 'justify' its existence. The particle with positive energy will shoot off into space, and – according to Hawking's calculations – should be detectable. And so Hawking predicted that black holes should be radiating particles that have been created from nothing, as a result of the uncertainty principle. This whole process may sound unlikely, but has become part of the accepted theory of black holes.

The result of Hawking radiation on a black hole is that the hole gradually loses energy, which radiates away into space. Hawking viewed this as a kind of 'evaporation' of black holes. The rate of black hole evaporation should increase, and the size of a black hole decrease, until the black hole goes out of existence in a flash of energetic gamma rays. The theory is watertight, but no Hawking radiation has yet been detected. This is largely because black holes take in more mass and energy than they radiate, and so are growing, not shrinking. The intensity of Hawking radiation emitted by a large black hole is tiny compared with the radiation from its accretion disc, so the theory is hard to verify in practice.

However, another prediction made by Stephen Hawking concerns tiny black holes left over from the early stages of the Universe, when temperatures and pressures were high enough to cram huge amounts of matter into tiny spaces, smaller than an atom. Given the age of the Universe (see 'The Rubber Universe'), most of these tiny, 'primordial black holes' will each have exploded in a burst of gamma radiation by now. But Hawking has calculated that there may be 300 of them in each cubic light year. When they do explode, it should be possible to detect the gamma radiation produced, and some high-energy physics projects, including the space-based Compton Gamma Ray Observatory, are actually looking for them.

Hawking's theories of radiation and mini-black holes are the result of his attempt to reconcile general relativity and quantum theory – the physics of the very large and the very small. The combination of general relativity and quantum theory is called quantum gravity, and is one of the main preoccupations of modern theoretical physics. Both general relativity and quantum theory are well established, and their results are consistent and have been proved time and time again. For Kip Thorne, a successful theory of quantum gravity would allow physicists for the first time to understand the most elusive and worrying aspect of black holes: 'The singularity is an object at the core of a black hole that is governed not by the ordinary laws of physics, but by the laws of quantum gravity, which we're only just beginning to understand.'

The hunt for more black holes of all sizes continues. The recent discoveries of objects that seem to be black holes have spurred on Philip Charles, one of the most avid black-hole hunters: 'We need to get maybe twenty or thirty so that you can really see the population of black holes we are looking at: do they all have similar masses? Is there a spread of mass? Do they all come from the same sort of star? We can start to work that out once we can get enough of them ... There are things out there that are incredibly compact and they have to be explained; I think that's an exciting prospect.'

THE RUBBER UNIVERSE

...searching for the beginning of time...

The Universe is everything there is: space, time and 100,000 million galaxies. It is larger than we can truly appreciate, and it is growing. The space between galaxies is actually stretching, and the galaxies are flying apart at a phenomenal rate. This suggests that, long ago, the entire Universe was contained in a tiny space that exploded, and has been expanding ever since – this is the theory of the Big Bang. By working out the rate at which the galaxies are moving away from each other, astronomers can make a guess as to when the Big Bang occurred and therefore just how old the Universe is. This sounds straightforward, but is causing problems within the space science community: there is rivalry between astronomers who have calculated different speeds of expansion – but, what is more, some of the most reliable measurements suggest that the Universe is younger than some of the stars in it.

Pulling punches

The person who discovered that the Universe is expanding was American astronomer Edwin Hubble. He studied mathematics and astronomy at the University of Chicago, and also showed great promise as a boxer. But when he graduated in 1910 he put all pursuits aside to concentrate on becoming a lawyer. The fact that he decided to move back into astronomy was to have a huge effect on the world – Hubble changed our view of the Universe for ever. In 1913, at the age of twenty-four, he began studying for a PhD in astronomy, at the University of Chicago. After finishing his studies in 1917 he served in the army during World War I, and then began working at the Mount Wilson Observatory, just outside Pasadena, California. It was there that he made his startling discoveries.

At the time Hubble was completing his PhD, the view of the Universe was already undergoing dramatic re-evaluation. The story begins with another American astronomer, Henrietta Leavitt. In 1907, working at Harvard College Observatory, Leavitt began studying a class of stars called Cepheid variables. These stars vary in brightness over a regular period of days or weeks. Through her painstaking work, Leavitt discovered that the period over which a Cepheid variable star varies depends on its average brightness. Astronomers measure a star's brightness in terms of its magnitude: the brighter a star appears, the lower its magnitude. A star of magnitude 6 is just visible to the unaided eye on a clear night, while the brightest star in the night sky, Sirius, has a magnitude of minus 1.5. During the latter part of the nineteenth century, the measurement of stars' magnitudes had changed from guesswork to a precise science, thanks largely to the introduction of photography.

Astronomers knew that if you could work out a star's actual output of light – its intrinsic luminosity – you could work out how far away it is from Earth. The reason for this is simple: two stars with the same intrinsic luminosity at different distances from Earth would have different magnitudes. In particular, if one star were twice as far away as another, it would appear only one-quarter as bright; three times as far, and it would appear one-ninth as bright, and so on. This is called the inverse square law, since how bright a light source appears depends upon the square of its distance from you. So, move a star four times as far away, and it will appear one-sixteenth as bright.

Alternatively, if two stars were at the same distance but differed in their intrinsic luminosities, they would again have different magnitudes. In this case, a star with twice the luminosity of another would appear twice as bright. However, stars are not all at the same distance, and they do not all give out the same amount of light. What was needed by astronomers trying to work out the distances to the stars was a 'standard candle' – a type of star whose intrinsic luminosity could somehow be ascertained with confidence. Then, by comparing this intrinsic luminosity with the actual magnitude as seen from Earth, a star's distance could be worked out using the inverse square law. Cepheid variables turned out to be just the standard candle that astronomers had been looking for.

Cepheid variables are named after the star 'delta Cephei', which is in the constellation Cepheus. This star is the 'prototype' Cepheid variable – it was the first such star to be studied in detail. Its magnitude varies from 3.6 to 4.3 and back to 3.6 again over a precise period of 5.4 days. Cepheid variables that are brighter than delta Cephei vary in magnitude over a longer period; less bright, and the period is shorter. Leavitt studied around thirty Cepheid variables in the Magellanic Clouds – large fuzzy areas in the sky, visible only from the southern hemisphere. Astronomers realized that all the stars in each of the two Magellanic Clouds were at about the same distance from Earth, but nobody knew how far. Henrietta Leavitt's work showed that there was a definite relationship between the intrinsic luminosities of these stars and the period over which their output varied, but until astronomers could work out how far just one of these stars was, this information was useless.

It was another American astronomer, Harlow Shapley, who worked out the distance to the Magellanic Clouds, and therefore 'calibrated' the Cepheid variables. Shapley was keen to find a way of measuring distances to the stars, particularly because he was troubled by the picture of the Universe as it stood in the early part of the twentieth century. Since the sixteenth century, when Copernicus realized that the Earth orbits the Sun, astronomers had assumed that the Solar System was at the centre of the Universe. Moreover, since the middle of the nineteenth century, astronomers believed that the Universe was about 23,000 light years across and about 6,000 light years thick. This had been worked out by looking at the distribution of stars in the sky; but Shapley was unconvinced. He was particularly troubled by the fact that the misty band of light called the Milky Way was seen only across part of the sky: surely if the Sun and the Earth were at the centre of the Universe, the sky should look the same in all directions.

In addition to the uneven distribution of the Milky Way, there were other problems for the accepted picture of the Universe. One of them involved globular clusters, which were known to be groups of thousands or millions of stars. Like the Milky Way itself, globular clusters were distributed unevenly across the sky, as seen from Earth. By studying Cepheid variables in the globular clusters, Shapley made a startling discovery: most of the globular clusters are tens of thousands of light years away. Together with the fact that they were found in only half of the sky, this challenged the idea that the Universe was only 23,000 light years across with the Sun at its centre. According to Shapley's observations, the Universe was almost ten times larger than people had believed, and the Sun was nowhere near the centre. This was big news, and Shapley became the most famous astronomer of his time.

The other problem for the picture of the Universe in the early part of the twentieth century involved the so-called 'spiral nebulas'. The word 'nebula' comes from the Latin word for 'misty', and was applied to any of many unidentified objects that appear in the night sky, most of them visible only through telescopes. In some, astronomers could make out individual stars, while others appeared as fuzzy, ill-defined patches of light. Some were irregularly shaped, some round, and some were the spirals that confounded the astronomers of the early twentieth century.

To some astronomers, the spiral nebulas seemed to be 'island universes', each a massive star system outside our own. Shapley, however, adhered to the idea that the spiral nebulas are inside our own system of stars, and that therefore the Universe – and not just the galaxy – is a disc about 200,000 light years across. Another American astronomer, Heber Curtis, believed otherwise: he was one of those who believed that the spiral nebulas are outside our own galaxy – although he stuck with the generally accepted idea that the Sun is at the centre of the galaxy. In 1920 Curtis and Shapley staged a well-publicized debate entitled 'The Scale of the Universe', at the National Academy of Sciences in Washington DC. Both men were right on one count and wrong on the other: we now know that the spiral nebulas really are other galaxies far outside our own, and that the Sun is indeed near to the edge of our own galaxy.

New reflections on the Universe

This is the point at which Edwin Hubble's contributions begin to be felt. From 1919 Hubble used a telescope with a

150-centimetre mirror to study the spiral nebulas, in an effort to settle the question of what these objects really are. He thought he could see Cepheid variables inside these nebulas, but could not be sure – he needed to use a more powerful telescope. So, between 1922 and 1924, Hubble used the largest telescope in the world at the time – the 250-centimetre Hooker Telescope, also at Mount Wilson. It had taken nearly six years to grind the concave mirror at the heart of the Hooker Telescope into the right shape. When, in 1930, Albert Einstein and his wife were being shown around the Mount Wilson Observatory, they were informed that the Hooker Telescope was the tool that astronomers were using to work out the nature of the Universe. Einstein's wife Mileva is reported to have remarked that her husband had done that on the back of an envelope.

In 1924 Hubble observed what was without question a Cepheid variable in the largest of the spiral nebulas – the Great Spiral of Andromeda, in the constellation of Andromeda. He charted the variation in the star's magnitude, and worked out that it had a period of one month. From this, he could work out the intrinsic luminosity and therefore the distance of the star. And from this, he could calculate the distance of the Great Spiral of Andromeda. His discovery was startling: he worked out that it is about one million light years away, much farther than Shapley's estimate for the size of the entire Universe. (It is now known that the Andromeda Galaxy, as it is now called, is more than two million light years away.) The conclusion was inescapable: the spiral nebulas were not nebulas at all, but were indeed star systems – galaxies outside our own. In 1926 Hubble devised a system of classification for the galaxies that he observed. He defined three types: spiral, elliptical and irregular.

At about the same time, the American astronomer Vesto Slipher was finding another vital piece of information about the spiral galaxies: the fact that nearly all of them are moving away from us. Slipher worked this out by studying the spectrum of light coming from each of them. The dark lines in the white light spectrum – described in Chapter Four in connection with the spectrum of sunlight – were all there, but were shifted somewhat towards the red end of the spectrum. The red shift of a galaxy is related to the speed at which it is moving away from us, in the same way as the pitch of a moving siren is lowered as an ambulance moves away. This is the Doppler effect, as explained in 'Black Holes'. If a galaxy is moving towards the Earth, its light is blue-shifted, so that the dark lines appear farther towards the blue end of the spectrum.

One of the galaxies that Slipher studied was the Great Spiral of Andromeda. Its spectrum is blue-shifted, and so it is coming towards us. The other galaxies seemed to be receding at phenomenal speeds. Slipher did not know how this fitted into the emerging picture of the Universe – but Hubble did. In 1927 Hubble looked at measurements of distance and red shift for twenty-four of the galaxies studied by Slipher. What he found was remarkable: the farther away a galaxy is, the faster it is receding. In 1929 he made a further discovery: if you divide a galaxy's speed by its distance from Earth, you will always end up with the same number. A galaxy twice as far is moving away twice as fast. The Universe is expanding in every direction.

The number obtained by dividing the speed of a galaxy by its distance from Earth is called Hubble's constant. Hubble's Law – as the relationship between a galaxy's distance and its speed is called – enabled astronomers to measure the distance to faraway galaxies in which individual Cepheid variable stars could not be seen. This was important, since only in those galaxies that lie relatively close can Cepheid variable stars be picked out. So, applying Hubble's law to more distant galaxies, with larger red shifts, astronomers began to realize that the Universe is much, much bigger than they had previously believed.

It would be easy to conclude from Hubble's Law that – since nearly all the galaxies are receding from us – the Earth has a special place in the Universe after all: at the centre. However, this is not the case. To see why, imagine a cake mixture with raisins evenly distributed inside it. In this analogy, the raisins correspond to galaxies, and the cake mixture corresponds to the space between them. Inside an oven, the cake mixture expands, pushing the raisins farther apart; the distance from one raisin to a neighbouring one increases. A raisin twice as far from another will be separated by twice as much of the mixture, and will therefore move twice as far in the same time. Applied to the Universe, the space between the galaxies is expanding. The Earth, the Sun and even the Milky Way are at no place in particular in a vast, expanding Universe.

New world view

Until the time of Copernicus, astronomers knew of five planets, which appear different from the stars because they shift their positions on a nightly basis relative to the star-studded backdrop. In fact the word 'planet' comes from the Greek word for 'wanderer'. When people aimed telescopes at the night sky, from early in the seventeenth century, it was inevitable that their ideas about the Universe would change radically for ever.

The magnification of a telescope is important when studying the Moon, the planets, comets and the nebulas (some of which were later identified as galaxies). But the light-gathering capability of a telescope is just as important, revealing objects much fainter than people can see with the unaided eye. In 1781 the English astronomer William Herschel discovered another planet – Uranus – which is not visible without a telescope. Herschel made many important contributions to astronomy, including being the first to attempt to work out the shape of the Milky Way. As a result of studying the distributions of stars in the night sky, he proposed in 1785 that the Milky Way is shaped like a lens – and he was not far off.

Long before Hubble's revelation that some of the nebulas are galaxies like our own, one of Herschel's contemporaries – German philosopher Immanuel Kant – proposed the same thing in 1755. But neither Herschel, Kant nor any of their contemporaries could unravel the nature of our galaxy or any other until astronomers had at their disposal powerful telescopes, a more sophisticated understanding of light and the technology of photography.

We now know that the Milky Way is a huge disc 100,000 light years across and 2,000 light years thick – not quite as large as Shapley had supposed – and that it contains about 200,000 million stars. Viewed from the side, it would appear as a long white streak, tapering at both ends and with a bulge in the centre. The bulge is about 6,000 light years across at its thickest point. About 200 globular clusters would be visible, dotted above and below the central bulge. Looking down on the disc of the galaxy, the central bulge would appear as a bright oval with long arms reaching out from it and trailing off into space to form a spiral shape.

The whole system is rotating, with the stars closer to the centre rotating faster than those further out – this is the origin of the spiral shape. The Solar System is in one of the

Milky Way's spiral arms – the Orion arm – about 23,000 light years from the centre of the galaxy. Although it is orbiting at a speed of around 240 kilometres per second, it takes about 200 million years to complete each orbit around the galactic centre. The Magellanic Clouds – where the Cepheid variables studied by Henrietta Leavitt are situated – are satellite galaxies, orbiting the Milky Way Galaxy once every 1,500 million years.

The Milky Way is the largest of a group of roughly thirty galaxies called the Local Group. Other members of this exclusive club include the Magellanic Clouds and the Andromeda Galaxy. The Local Group takes up about 125 million cubic light years of space. Travelling from Earth to the Andromeda Galaxy at the speed of, say, a rocket as it leaves Earth's atmosphere – 11 kilometres per second – this would take a staggering 70,000 million years. But the Local Group is just part of a larger cluster of galaxies, called the Local Supercluster. The centre of the Local Supercluster is nearly thirty times as far as the Andromeda Galaxy.

It is largely thanks to the efforts of Shapley, Slipher and Hubble that this picture of the Universe emerged. Cosmology – the study of the nature and the origins of the Universe – owes them a debt of gratitude. But there are many great minds working on the puzzling aspects of cosmology that is the legacy of these pioneers. The very relation that simplified our picture of the Universe – Hubble's Law – brought with it many questions, as yet unanswered. One of them is: 'How did it all begin?'

Hubble's constant – the number you arrive at when you divide a galaxy's speed by its distance from us – is given in units of 'kilometres per second per megaparsec'. A megaparsec is a unit of distance: one parsec is about 3.26 light years, so a megaparsec is about 3.26 million light years.

Hubble arrived at a value of the Hubble constant equivalent to 500 kilometres per second per megaparsec, meaning that a galaxy two megaparsecs (6.5 million light years) away is receding at a speed of 1,000 kilometres per second. Ever since the discovery of Hubble's Law, the numerical value of Hubble's constant has been the subject of debate and controversy. Incidentally, the reason that the Andromeda Galaxy is moving towards the Milky Way rather than away from it is that, being relatively nearby, the mutual gravitational attraction between the two is pulling them together. Only the more distant galaxies are moving away, and for them, Hubble's Law does seem to hold.

In the beginning

Hubble wondered whether the fact that the Universe is expanding means that at some time in the past it was condensed into a much smaller space – the concept we now know as the Big Bang. This idea had already been put forward in 1927, by a Belgian priest and astronomer, Abbé Georges Lemaître. Combining the findings of Shapley, Slipher and Hubble with Einstein's General Theory of Relativity, Lemaître gave the idea of an expanding Universe a firm mathematical footing, showing how it could be explained by the expansion of space, as illustrated by the rising cake analogy above, and not in terms of the galaxies physically flying apart.

Using Hubble's constant – a key to the rate of expansion of the Universe – successive generations of astronomers have worked backwards to estimate how long the Universe has been expanding, and therefore how old it is. Using his own calculation of Hubble's constant during the 1930s, Hubble worked out the age of the Universe to be around 2,000 million years. Geologists had already worked out that the ages of some rocks found on Earth were much older than this, and so Hubble's calculation caused much consternation in the scientific community as a whole.

The Big Bang theory was not the only one that was proposed to explain the creation of the Universe. Its main rival was the Steady State theory. This was put forward by a trio of mathematicians and astrophysicists, Fred Hoyle, Hermann Bondi and Thomas Gold, in the 1940s. It too appealed to general relativity. These three scientists considered it ludicrous that anything could be created by an explosion. However, the only way their theory could account for the fact that the Universe is expanding was to suppose that new matter was being created from nothing all the time. Still, both the Big Bang and the Steady State theories had their supporters, and fierce debates raged.

Also during the 1940s, Russian-born American physicist George Gamow improved the Big Bang theory, proposing that the initial state of the Universe would have been an intensely hot, extremely compact 'fireball' smaller than an atom. Gamow also showed how the Big Bang idea could explain the production of hydrogen and helium in the early Universe. Professor Alan Guth at the Massachusetts Institute of Technology points out that this is one of the strong points of the Big Bang theory: 'The Big Bang theory has to be recognized as being more than just a cartoon image of an explosion from which the Universe came: it really is a detailed mathematical theory. And given this mathematical description of the details, you can actually calculate the rates of different nuclear reactions that would have taken place in the early Universe.'

By the 1950s, the idea that the Universe was once searingly hot and incredibly dense began to be generally accepted by most cosmologists. But the really conclusive evidence in favour of the Big Bang came, by accident, in 1964. American radio astronomers Arno Penzias and Robert Wilson, working at the Bell Telephone Laboratories in New Jersey, were attempting to identify all sources of radio interference, so that communications with satellites could be optimized. When they had eliminated all the known sources of radio waves, they were left with one signal they could not, at first, explain. On further study, and after discussing their findings with Bernard Burke at the Massachusetts Institute of Technology, they realized that they had inadvertently discovered the radiation left over from the Big Bang.

The cosmic background radiation, as Penzias and Wilson's discovery became known, is a microwave signal. Just as the spectrum of light coming from a hot object can reveal that object's temperature – a white-hot object is at a higher temperature than a red-hot one, for example – radiation emitted by any object is characteristic of its temperature. The signal that Penzias and Wilson had detected corresponded perfectly to a temperature of minus 270 degrees C, just under three degrees Celsius above the coldest possible temperature, called absolute zero. Despite its intensely hot beginning, the Universe today is on average a very cold place indeed.

The temperature indicated by the cosmic background radiation fitted almost perfectly with one of the predictions of the Big Bang theory – namely that the intense heat of the early Universe would have produced high-energy radiation. As the Universe expanded, the wavelengths of the radiation would have been drawn out, gradually changing short-wavelength gamma radiation to the long-wavelength microwave radiation that Penzias and Wilson had picked up with their radio antenna. Another way of looking at this change in wavelength is that the Universe has cooled as it has expanded.

And so the Big Bang theory became the only real contender for explaining the creation of the Universe. The fact that the whole Universe began at one point and expanded in all directions means that it has no edge. Furthermore, if the Big Bang really was the moment of creation of space and time, then it would have occurred in every point of the Universe, not in any particular location. However, it is the space between the galaxies that is expanding, not the space within a galaxy or between galaxies that are near each other. Gravity is holding space together in these locations.

Bizarre though the predictions of the Big Bang theory may seem, the theory has been confirmed in many other ways since the 1960s. Perhaps the most convincing confirmation came from a space-based observatory called the COBE (Cosmic Background Explorer). Launched in 1989, COBE mapped the entire sky, making extremely precise measurements of the radiation left over from the Big Bang. Its triumphant year was 1992, when the map was complete. Alan Guth: 'It's in absolutely gorgeous agreement with the predictions of the thermal radiation from the heat of the Big Bang.' Perhaps even more importantly, there were very slight variations in the intensity of the background radiation, exactly as predicted by the Big Bang theory: if the radiation had been perfectly uniform across the entire sky, there would have been no way for the galaxies to have formed in the early Universe; the entire Universe would have been perfectly smooth.

Conflicting measurements

Edwin Hubble's original estimate of the Hubble constant put it at about 500 kilometres per second per megaparsec.

Although cosmologists agree that this is far higher than the actual value, there is still a great deal of disagreement about the actual figure. The values determined between the 1960s and 1990s generally fell into two camps. It was during the 1960s that the rivalry began: one team comprising Allan Sandage at the Carnegie Institution in the USA and Gustav Tammann at the University of Basel, Switzerland, found a relatively low value of 50, while Gérard De Vaucouleurs at the University of Texas and his supporters found a value around 100. Each set of investigators held on avidly to their results, and the debate still rages. The difference between 50 and 100 may not seem too great, but a value as low as 50 suggests that the Universe is perhaps twenty thousand million years old, while a value of 100 suggests nearer eight thousand million years.

Professor John Huchra, at Harvard University, makes it clear that the Hubble constant is very important: 'To a cosmologist, the Hubble constant is perhaps the most important number – in part because it sets the scale of the Universe; it sets the sizes of the things we look at; it sets the luminosities of the galaxies we look at; it sets the age of the Universe. We need to understand that if we really want to talk about the model as a whole and understand where we come from.'

Of the two measurements that must be made for accurate determination of the Hubble constant – the speed at which galaxies are moving away and their distances from Earth – the first is by far the easier. It is calculated using the Doppler effect. The distance of a galaxy is much more difficult to ascertain. The method using Cepheid variables is fairly reliable, but can be used only for nearby galaxies, because individual stars can rarely be observed in very distant galaxies. And, frustratingly, it is the most distant galaxies that are most important in determining the Hubble constant, since those will be the ones unaffected by the gravitational effect of our nearest neighbouring galaxies.

Several different methods for determining the distances to faraway galaxies – and therefore the Hubble constant – have emerged since Hubble first discovered the relationship between galaxies' speeds and their distances. Sandage and Tammann decided to study supernovas – bright outbursts produced during the death throes of aging stars. According to these two cosmologists, all supernovas produce approximately the same amount of light at maximum intensity. This is useful because, as Tammann points out, 'if you can calibrate this luminosity in some galaxy whose distance you know already through Cepheids, you have a standard candle'. Because supernovas are much brighter than Cepheid variable stars, they can be observed in galaxies that are much farther away. It was using this supernova method that Sandage and Tammann calculated their value for the Hubble constant as 50 kilometres per second per megaparsec.

However, there are problems with Sandage and Tammann's method. First, there is disagreement as to whether all supernovas produce the same amount of light. Second, supernovas are rare events. Sandage and Tammann base their measurements on a supernova that occurred in 1937. The intensity of this supernova was measured by astronomers Walter Baade and Fritz Zwicky, by analysing photographic plates that had recorded the outburst. Michael Pierce of Indiana University explains that he and his associate George Jacoby, at the Kitt Peak National Observatory, 'have been analysing the plates using modern techniques, and have found that the supernova was considerably fainter than Baade and Zwicky had claimed ... and this correction alone would revise the Sandage and Tammann value from about fifty to about sixty or sixty-five'.

Another, rather indirect, method of determining the Hubble constant involves measuring the rate of rotation of spiral galaxies. In a rotating galaxy, half the stars are moving towards us and the other half are moving away. Using the Doppler effect, it is relatively simple to work out how fast the galaxy is spinning and, since the rate of spin is related to the mass of a galaxy, the galaxy's mass can be estimated using this method. The mass of a galaxy is related to the number of stars, and the number of stars is related to the total output of light, and therefore the galaxy's intrinsic luminosity. By comparing the luminosity with the galaxy's magnitude, its distance can be worked out. This method tends to find values of the Hubble constant much higher than the value that Sandage and Tammann had worked out.

Michael Pierce, one of the astronomers using the rotation of spiral galaxies to calculate the Hubble constant, says of this technique: 'If one dismisses the idea that somehow the local galaxies are fundamentally different from the more distant ones – and I see no reason to assume such a thing – then I have to accept that the Hubble constant is eighty-five.'

Yet another approach is based on the fact that distant galaxies appear as relatively uniform patches of light, while those that are nearer appear 'grainy' because individual stars can be made out. In the same way, the individual dots that make up a newspaper photograph are visible only when seen close up. So, by studying the variation in light intensity across the surface of a galaxy, it is possible to gain an idea of how far away a galaxy is. This method, developed by John Tonry of the Massachusetts Institute of Technology, seems to be in good agreement with most of the other methods, but cannot be used with the most distant galaxies.

Apart from Sandage and Tammann's value of around fifty kilometres per second per megaparsec, all the other methods so far described have calculated the Hubble constant at higher values. These higher values suggest that the Big Bang happened around twelve thousand million years ago – and the very highest values suggest an even lower age for the Universe, just eight thousand million years. This is a problem for cosmologists, because some stars in the Milky Way Galaxy are known – with some certainty – to be older than that.

Theories of star formation and evolution are well developed, and probably very accurate – as Professor Pierre Demarque of Yale University in Connecticut points out: 'This is very mature physics, compared to what we use in extragalactic physics.' The oldest stars in our galaxy are found in the globular clusters, distributed around the galactic centre. Astronomers are able to work out the ages of globular clusters by examining the proportions of stars within them that have reached the late stages of development. But there are other ways of working out how old stars are.

Professor Michael Rowan-Robinson of Queen Mary and Westfields College, London, says, 'We can determine the ages of these stars not only by studying star clusters, as we do for the very oldest stars, but also by looking at the radioactive elements, similar to how scientists use carbon-14 dating ... Then there's a third method, which is based on white dwarf stars – which is what the Sun will end up as. Because we know how fast these stars cool, we can work out how old these stars are.' Professor Demarque adds another field of study to the list of reasons why astronomers are fairly confident about their understanding of stars: 'We have better ways to probe the interiors of stars; one of these is the new field of helioseismology, which enables us to study the interior of the Sun by studying its oscillation.'

So astronomers are fairly sure that their theories of star evolution are good enough to enable them to be confident that some stars in our galaxy may be as old as 14,000 million years. And yet this is at odds with the higher values for the Hubble constant, which put the age of the Universe as less than the age of the oldest stars. As George Jacoby of the Kitt Peak National Observatory says, 'That is a real conflict. How do we resolve this conflict? Well, I'm not sure how ...'

Help is at hand – from Edwin Hubble's namesake, the Hubble Space Telescope. From the outset, one of its mission objectives was to attempt to settle the debate over the Hubble constant. A team of scientists working on the Key Project on the Extragalactic Distance Scale – in consultation with some of the major players in the debate – set out to determine the Hubble constant to within 10 per cent.

The approach taken by the Key Project team was to check galactic distances found by several different methods against the most reliable method – Cepheid variable stars, for as many galaxies as possible. The superior view of the heavens that Hubble gives astronomers meant that the Cepheid variable technique could be extended to more distant galaxies than was possible before. Cepheid variables in galaxies as far away as sixty-five million light years – in the Virgo and Coma galaxy clusters – were observed with great accuracy, to gauge their luminosities and therefore their distances from Earth.

By the time the Key Project came to an end in May 1999, the Hubble Space Telescope had peered at a total of almost 800 Cepheid variables in eighteen galaxies. The results were encouraging: the team had calculated a value for the Hubble constant of 71 kilometres per second per megaparsec, set pleasingly between the high and low values previously worked out, and leading to an age of the Universe that does not conflict with the ages of the oldest stars.

Far out

All the approaches to calculating the Hubble constant outlined so far – including the research carried out by the Hubble Space Telescope team – are only really useful for nearby galaxies or those at intermediate distances. What is needed for a truly accurate determination of this elusive and all-important number are the distances to the most remote galaxies – thousands of millions of light years away.

Two promising methods are beginning to yield interesting results for galaxies lying at these immense distances across space. The first of these is particularly promising, because it can gauge how far an entire cluster of galaxies is receding. This approach makes use of a phenomenon known as the Sunyaev-Zel'dovich effect. Every cluster of galaxies is shrouded in tenuous but extremely high-temperature plasma (ionized gas). The plasma can affect the cosmic background radiation passing through it. As the radiation meets the gas, it can receive extra energy from the electrons in the plasma, and this affects the spectrum of the radiation. Because the plasma is so tenuous, collisions of radiation and electrons are rare, so the effect is very slight – the radiation is predicted to change by only one part in 100,000.

The Sunyaev-Zel'dovich effect enables astronomers to estimate the size of a galaxy cluster: the greater the size of a cluster, the more chance that the radiation will collide with electrons in the plasma. This done, astronomers can measure how large the galaxy appears in their telescopes, and estimate how far away it is. A project run by NASA's Dr Marshall Joy and Dr John Carlstrom of the University of Chicago has been investigating the effect since 1992, using only microwave signals. More recently, they have been able to compare observations made using ground-based radio telescopes with observations from the orbiting X-ray observatory, Chandra, launched in 1999. Joy and Carlstrom's preliminary values of the Hubble constant are between 40 and 80 kilometres per second per megaparsec but the pair hope to narrow down the range of values.

The fact that light is bent by the gravitational fields around massive objects is the basis for another distant galaxy approach to the Hubble constant problem. As light passes through a large galaxy, it can be bent in the same way as light is bent by a lens. The distance between Earth and a lensing galaxy can be estimated from the effect the galaxy has on the light from even more distant galaxies. In particular, light from the more distant galaxies that takes different paths through the galaxy also takes different amounts of time to reach Earth. Astronomers can work out this difference if something changes in the more distant galaxy – such as a supernova.

Then, the same fluctuation in brightness will be observed at different times and, from this, astronomers can work out the distance of the lensing galaxy. The phenomenon of gravitational lensing is a prediction of general relativity, and is now well tested. But examples of gravitational lenses are few and far between, and using the phenomenon to determine the Hubble constant has yet to prove itself. Preliminary results again cover a range of values, but are pleasingly within the mid-range of values, around sixty-five kilometres per second per megaparsec.

The gravitational lensing and the Sunyaev-Zel'dovich effect seem to point to values for the Hubble constant that fall between the two extremes suggested by the earlier investigators. So did the research carried out by the Hubble Space Telescope's Key Project Team. However, just as everything seemed to be settling down in the debate, another method has found worrying evidence that the larger values of the Hubble constant might just be right after all.

In September 1999 the Hubble Space Telescope was turned to a galaxy named NGC 4258. This spiral galaxy had been used as a benchmark, because it was one of the farthest galaxies whose Cepheid variables could be seen clearly. It was thought that the galaxy was at a distance of about 8.1 megaparsecs (26.4 million light years) from Earth. With the clearest picture yet of the galaxy, Dr Eyal Maoz and his team at NASA's Ames Research Center in California was able to make what he thought was a reliable judgement of the distance of this galaxy, to confirm this figure. However, a new measurement of the distance of NGC 4258 was made using another, more reliable method.

Near to the heart of the galaxy, water molecules are heated by the release of huge amounts of energy by a supermassive black hole and give off microwave radiation at a very precise frequency. The radiation is created by a process much like the way a laser produces light. And just as the word 'laser' is derived from the acronym 'light amplification by stimulated emission of radiation', so the phenomenon observed in NGC 4258 is a 'maser', from the same acronym but with 'light' replaced by 'microwave'. The microwaves are produced in two jets that emerge from the galaxy's core.

By following the way the maser jets move, a collection of radio telescopes on Earth – called the Very Long Baseline

Array – was able to work out how large the galaxy is. Using simple geometry, based on how large the galaxy appears, astronomers could then work out how far away the galaxy is. Their calculations put the galaxy at a distance of about 6.1 megaparsecs (20 million light years). Comparing this highly accurate measurement of distance with Maoz's measurement based on Cepheid variables seems to highlight a flaw in the use of Cepheid variables. If this is correct, then all previous measurements using Cepheids may have been consistently too high by around 12 per cent. And if this is true, then all the previous measurements of the Hubble constant also suffer from that same flaw, since they have all ultimately been calibrated been using Cepheid variables. What this means to cosmologists is that the age of the Universe is about 12 per cent lower than the previous best estimates: perhaps as little as 12,000 million years.

This low value has resurrected the paradox that cosmologists feared when previously confronted with large values of the Hubble constant: stars in our own galaxy are known to be older than the age of the Universe that is suggested by these figures.

Fatal attraction

So at present the world of cosmology seems to be in disarray. Despite some of the most sophisticated tools ever constructed, and a highly developed understanding of the Universe, there is disagreement and paradox. And there is another problem: the effect of gravity. The mass of the Universe has the effect of slowing the rate of expansion, pulling everything together. The more mass there is, the greater this effect. This process is well understood – in terms of general relativity; the trouble is that cosmologists do not know how much mass there is in the Universe. If they did, they would be able to calculate how much the rate of expansion has changed since the Big Bang.

The situation is rather like watching an athlete at only one point during a race and trying to infer from that how long he or she has been running. You might be able to calculate how fast the athlete is running at that point, but he or she will not have been running at that speed for the whole race. It is likely that the speed was higher at the beginning of the race. If you knew at what rate the athlete had been slowing down, then you could produce a reliable estimate of when the race began.

In the same way, even if cosmologists could work out a reliable figure for the expansion of the Universe at the present moment – from an accurate determination of the Hubble constant – they would not necessarily know how old the Universe is. They need to know the extent of the slowing effect of gravity, and they can work that out only if they know how much mass is contained within the Universe. In other words, they can work out only how the expansion has slowed over time – and therefore how old the Universe is – if they can ascertain how much matter there is in the Universe.

According to Einstein's General Theory of Relativity, anything that possesses mass – stars and planets, interstellar gas and dust, and black holes, for example – warps space-time, the fabric of the Universe. This is explained in 'Black Holes'. Gravity between massive objects holds space-time together locally – in galaxy clusters, for example – but the space between clusters of galaxies is expanding steadily in all directions. This expansion is, of course, what the Hubble constant attempts to quantify. But as well as the local-scale distortion of space-time, matter is curving space-time on the scale of the entire Universe. Depending upon how much matter there is – and therefore to what extent the gravitational effects of that matter is pulling space-time back in on itself – the Universe is either 'closed', 'open' or 'flat'.

The behaviours of space-time and the objects within it are based on a branch of mathematics called non-Euclidean geometry. This is a description of space that is different from the descriptions and theorems of Euclid – one of the most influential and prolific mathematicians of ancient Greece. Around 300 BC, Euclid collected a number of mathematical theorems and put them together into a collection of thirteen books called the *Elements*. In an attempt to define fully the nature of shapes and the space they inhabit, Euclid included in the *Elements* a set of ten axioms and postulates, which were supposed to be self-evident truths. Using these as starting points, Euclid went on to set out 465 theorems that helped mathematicians to work out areas and volumes of shapes, for example. The *Elements* was the standard mathematical textbook in many countries of the world for more than 2,000 years.

A classic example of one consequence of Euclid's geometry is the idea that the angles in the corners, or vertices, of any triangle add up to exactly 180 degrees. This is easy to verify for a triangle drawn on a flat surface. But there is a problem if you draw a triangle on a curved surface, such as the surface of the Earth. Imagine a triangle with the North Pole as one vertex and two points on the equator as the others. Since the angle between any line joining the equator to the North Pole is at 90 degrees to the equator, and there are two such lines in this triangle, the angles inside the triangle will add up to more than 180 degrees. The angles inside the triangle that joins Kenya, Papua New Guinea and the North Pole, for example, add up to about

270 degrees. The important feature of this triangle, as opposed to those that Euclid was concerned with, is that it is drawn on to a curved surface rather than a flat one.

Another of Euclid's postulates is the idea that parallel lines will remain at a fixed distance from one another along their entire length. This sounds logical enough, until once again you consider a curved surface. Two people setting out from locations on the equator and travelling due south – in parallel with each other – will converge gradually, and meet at the South Pole. The relevance of this non-Euclidean geometry to the Universe at large is that gravity curves space-time, and therefore causes unfamiliar effects in three or four dimensions, just as these triangles and parallel lines do on a curved two-dimensional surface. As we have seen, for example, the fact that parallel rays of light passing either side of a massive galaxy converge – gravitational lensing – is understood in terms of the curvature of space-time.

The curvature of a surface may be 'open', 'closed' or 'flat'. A triangle drawn on a sphere – a closed surface – has angles that add up to more than 180 degrees. But on an open surface, triangles have angles adding up to less than 180 degrees. An example of such a surface is a saddle. In the same way, the curvature of space-time may make the Universe flat, closed or open. If there is enough matter in the Universe to curve space-time in on itself, then we have a closed universe. If not, then the Universe is either open or 'flat'. One of the main aims of modern cosmology is to discover just how curved the space-time of the Universe is. To do this, they must find out how much mass it contains and how big it is.

An open universe will go on expanding for ever, with the galaxies moving farther apart into eternity. This would be a slow, cold death for the Universe, and would result from there not being enough matter to halt the expansion. A closed universe is perhaps an even more frightening prospect: it will eventually stop expanding, and begin to be pulled in on itself, shrinking back at an ever-increasing rate, eventually reducing in size to nothing. Cosmologists have given this idea a name – the Big Crunch – to relate it to the Big Bang that occurred at the other end of the Universe's lifetime. The other possibility is that space contains just the right amount of matter to stop expanding, but not to contract again. In this case, the Universe would be described as 'flat', with what is called 'critical density'.

A handy analogy may help clarify these complexities: a part-inflated balloon with dots drawn on to its surface. In this analogy, the dots represent the galaxies, while the rubber itself represents the space-time they inhabit. Inflating the balloon makes it equivalent to an expanding Universe: the dots move farther apart in the same way as galaxies are observed to do in the Universe; space-time (the rubber) is stretching, as a result of the Big Bang. In this case, the three dimensions of space are confined to two dimensions in the surface of the rubber. The Universe has no centre, and no edges. Just as a traveller on the 'closed' surface of the Earth can journey around the world and return to where they started – a journey across a closed universe could conceivably bring you back to Earth, after a journey of trillions of years.

An 'open' universe does not contain enough matter to close space-time in this way, and neither does a 'flat' universe. To work out which of these geometrical descriptions applies to our Universe, cosmologists must work out the average density of space. To do this, they must be able to measure the volume of the Universe (or at least large swathes of it) and the total amount of matter contained within it. This sounds like a daunting task – and it is.

Dark secrets

At present, cosmologists have little idea of the total mass present in the Universe. What they do know is that the mass they can observe puts the density of the Universe at less than one-tenth that required to attain critical density (the amount of matter required to halt the expansion of space-time). And yet large-scale observations of the Universe tend to point to a flat universe, which means the density must be close to the critical density. And so there must be much more mass in the Universe than cosmologists can actually observe.

The measurement of the density of the Universe is important not only to predict the future, but to look into the past. The reason for this is that estimates of the age of the Universe (based on the Hubble constant) will depend upon the amount of mass in the Universe. If there is a great deal of mass in the Universe – if the density is high – then the expansion rate will be slowing down quickly. In that case, for the expansion rate still to be as high as it is today, we must be in a relatively young Universe. Conversely, if the density is relatively low, then the Universe must have had a longer time to reach this speed, and we are in a relatively old Universe. In this way, a low-density Universe would help to increase the age of the Universe as calculated using the high values for the Hubble constant found by most experimenters. This would be a relief, as it may just push the age of the Universe beyond the ages of its oldest stars.

It is relatively simple to estimate the overall mass of observable matter in the Universe, by measuring the total output of light from each galaxy, and taking into account the dark 'lanes' in galaxies due to interstellar dust. But cosmologists are sure that there exists a huge amount of matter that cannot be directly observed – what they have called 'dark matter'. Astronomer Heather Morrison at the

Kitt Peak National Observatory in Tucson, Arizona, explains the derivation of the term: 'When we are looking for that matter, we can't account for a significant amount of the mass, so we've had to postulate something we call dark matter to explain it. We call it dark matter because we've never observed any light from it. But it's worse than that – we don't actually know what it is.'

The first clues to the existence of dark matter came in the 1930s, when astronomer Fritz Zwicky studied the motions of clusters of galaxies. Zwicky worked out that, based on the amount of matter that could be seen in these clusters, the galaxies should be flying apart, not attracting one another as they are seen to do. The gravitational influence of some kind of unseen matter is holding these clusters together. Even the mutual gravitational attraction between the Milky Way Galaxy and the Andromeda Galaxy is far greater than can be accounted for by the amount of observable mass. Also, when observing distant stars within our own galaxy, gravitational lensing occurs without anything observable between the Earth and those distant stars.

There are several pet theories to explain the nature of dark matter. Two of them involve tiny particles that would have to exist in great numbers. Both of these types of particle would have to interact very weakly with ordinary matter – otherwise scientists would be able to detect them as they arrive in Earth- or space-based detectors. Two candidates are generally put forward: neutrinos – produced in certain nuclear reactions, and in particularly huge numbers during the Big Bang itself – and WIMPs (weakly interacting massive particles), which were also thought to be produced during the Big Bang. Because of their elusive nature, these particles have yet to be detected in the quantities needed to contribute sufficiently to the dark matter of the Universe.

The gravitational lensing effects of dark matter are blamed upon hypothetical objects called MACHOs (massive compact halo objects). These objects would be large and invisible, and would have high mass – the obvious candidates are black holes. And early in 2000, astronomers based in Australia and in Chile, having used ground-based telescopes as well as the Hubble Space Telescope, announced that they had made the most promising breakthrough in the search for MACHOs so far. The astronomers discovered what they are certain must be two isolated black holes in our own galaxy.

Until now, black-hole candidates had been discovered only by the X-ray emissions from the accretion disc that forms as gas from a binary companion star is pulled off by the black holes' strong gravitational field, as explained in 'Black Holes'. The individual examples were discovered by their gravitational lensing effects on more distant stars. In particular, the Hubble Space Telescope detected that light from distant stars became brighter and dimmer – presumably as it was focused by black holes passing in front of the stars.

The 'gravitational microlensing events' lasted several hundred days each, allowing telescopes based on the ground to study the effects over a long period, while the space telescope moved on to other duties. Detailed calculations highlighted the fact that the unseen objects must have a mass at least six times that of the Sun – too big to be a white dwarf or a neutron star. The existence of these two black hole candidates suggests that isolated black holes may be common in our galaxy – and other galaxies. Larger stars use up their hydrogen fuel more quickly than smaller stars, and it is the larger stars that, according to theory, go on potentially to become black holes. So, given the age of the Milky Way, it seems reasonable to expect that large numbers of massive stars have already burned themselves out and formed black holes. The astronomers who made the discoveries think that isolated black holes could account for a reasonable proportion of the hypothetical MACHOs in the Universe. This may be the first step towards tracking down the 'missing mass' of the Universe.

In one sense, cosmologists are keen to find the perceived missing mass of the Universe, so that the value they calculate for the density of the Universe moves closer to the critical density to which their observations lead them; on the other hand, they would rather have a much lower density, so that the age of the Universe calculated by the best values of the Hubble constant is pushed up somewhat.

The other crucial measurement required for a calculation of the density of the Universe is its size. All we have to go on for that measurement is the observable Universe – bounded by the most distant objects that we can detect. Until recently, the most distant objects observed were quasars – active galaxies whose red shifts are so extreme that they are estimated as being at least 12,000 million light years away. Since light takes time to cross such vast distances, this is not only a very long way away, but also a long way back in time – to the early stages of the Universe. But unfortunately, the accuracy of this estimate of distance is unknown: the measurement is based upon the red shift of these distant galaxies, using the best value for the Hubble constant. This value is in dispute, as we have already seen.

However, in 1999 the Compton Gamma Ray Observatory – an orbiting telescope that can see gamma rays from space – detected events that seem to be occurring at distances that are farther still, and scientists used a different method to calculate their distance. The events are known as gamma ray bursts, and just what causes them is not yet known. However, what is known is that the spread of radiation that we receive in our instruments in orbit around Earth is related to the intrinsic luminosity of the events – the actual amount of energy released. This is similar to the use of Cepheid variables in determining the distances to nearby galaxies.

The ultimate questions to which many people want answers are, 'What is outside the Universe?' and, 'What existed before the Big Bang?' These questions can be posed in another way: 'Where and when does the Universe exist?' Modern cosmology avoids these questions, arguing that time and space were created in the Big Bang – there simply was no time before the beginning of the Universe; and the Big Bang created all space, too, so there is no space outside the Universe. Professor Carlos Frenk of Durham University explains: 'Given that space and time are so closely related to one another, asking the question, "What was there before the Universe?" is exactly the same as asking, "What is there outside the Universe?" – no difference: space was created with the Universe, time was created with the Universe. Time is a concept that is part of the Universe: it did not exist before the Universe came into existence.'

In other words, the Universe is nowhere and 'no-when', since space and time are intrinsic qualities of the Universe itself. Most people are left unsatisfied with such answers, but cosmologists insist that their arguments are watertight, and that we must adjust our common-sense notions of space and time, as relativity theory has shown us.

For now, one of the main questions of cosmology is: 'What is the ultimate fate of our Universe?' We will be able to find an answer only by working out a value for the

Hubble constant as well as determining the density of the Universe as a whole. If there is enough matter in the Universe to halt the expansion and pull space-time in on itself, then it seems that it will all end as a super-dense fireball, just as it began. The Universe could end up as a tiny black hole – a singularity. If there is to be a Big Crunch, it will not happen for many billions of years. By that time, perhaps cosmologists will have agreed on a reliable value for the Hubble constant.

Another cosmological theory, called the Oscillating Universe Theory, says that the Universe could be created again from the singularity that results from the Big Crunch: there could be another Big Bang. So perhaps everything will simply start again.

The Universe is everything there is: space, time and a hundred thousand million galaxies. It is larger than we can truly appreciate, and it is growing ...

AFTERWORD
... looking ahead...

Space science and technology are moving forward ever more rapidly. Increasing understanding of our place in the Universe brings new insights into our origins and our existence, while the ability to leave the confines of planet Earth is opening up new and exciting opportunities for the human race. Less than a hundred years after the first liquid fuel rocket made its faltering leap into the air, we have taken great strides spacewards. People have floated free in orbit and walked on the Moon. There is nothing but time and money stopping us from venturing farther and staying away longer – perhaps even living on the Moon or Mars for extended periods.

Where this will lead in the next twenty, thirty or forty years – and who will go into space in that time – is anybody's guess. In the next few years at least, advances in space exploration are fairly predictable: although the pace of development in spacecraft technology is rapid, it still takes a few years of rigorous testing to move a new spacecraft from the planning stage to production and to its first flight. Building spacecraft is a time-consuming, skilled and risky business and, despite the initiatives of private financiers, there will be no cheap or easy access to space in the next few years. However, the commercialization of space is a growing force, pushing the technology forward and the costs down. Through huge investment and such initiatives as the X Prize, space will no doubt eventually become accessible to many more people over the next few decades. This will satisfy those frustrated with the slow pace of government-led space programmes that have purely scientific goals. The dreamed-of bases or civilizations on the Moon and Mars probably will be realized – perhaps in the lifetime of a reader of this book.

But for the foreseeable future, most of the world's population will be firmly rooted on Earth. At each step along the way to greater accessibility to space, it will be important for us as a species to ask why we want to go there, and who will benefit. Certainly the people who are investing in space futures will benefit – they are shrewd business people who will proceed only if they can be guaranteed a financial reward. Then there are the pioneers, who simply want to push human frontiers ever further forward. It is natural to want to do this, but we must always remain aware of who we are and why we want to go to space.

And it is worth remembering that the potential costs of space travel may be more than just the financial costs of building spacecraft. If space travel becomes routine, what will be the effects on the environment, both on Earth and in space? Will the dreams of the few leave the many behind? And most of all, are we really ready to push out into space when there are still so many problems to solve back here on Earth?

While the future of space travel is fairly predictable, the related sciences of astronomy, astrophysics and cosmology will no doubt continue to be full of surprises. For although theoretical predictions of increasing sophistication are being realized, we uncover more questions and fascinating phenomena that beg for answers. The modern instruments of astronomy can peer deeper into the Universe with increasing clarity, but there will always be more to learn and more to tantalize or to astonish us.

One thing is for certain: in the modern world, space science and space technology are co-dependent – they are locked in a permanent relationship. Space telescopes and space-station laboratories require funding, and they require efficient transportation. The intimate connection between private and public funding of ventures into space – where scientific missions ride on the back of commercial ones, for example – will probably become more common, as the Russian and American governments shy away from the huge investment they once poured into the space race, driven as it was by political goals.

There is one aspect of NASA's space programme that must not be forfeited in the new era of space exploration, and that is education – in the widest sense of the word. The spectacular images of the planets and their moons, of distant stars and galaxies, and of our fiery Sun, appear in books, magazines and newspapers and on television and the Internet. When made freely available to everyone in this way, they surely justify the large sums of money spent on space programmes. As well as being inspirational, space programmes can bring more down-to-earth benefits to everyone through the remarkable advances in science and technology. The costs of mounting space-based scientific operations such as the International Space Station are monumental, of course, and there are many causes that would benefit from such funding. However, there are many other areas of human activity that involve just as much money but that do not bring such benefits – the arms race and the worst excesses of consumer society immediately spring to mind. Another unique feature of modern space exploration is the level of international co-operation that is involved, and here, too, is a reason to continue to venture into space.

There will always be more mysteries to solve in the Universe – the journey towards understanding space will have no end. Along the way, we must not lose our ability to stand and stare at the majesty of it all. Galileo was aware of this when he wrote: 'The Sun, with all the planets revolving around it, and depending upon it, can still ripen a bunch of grapes as though it had nothing else in the Universe to do.'

TIMELINE

The development of space travel

1926 Rocket engineer Robert Goddard launches the first liquid fuel rocket.
1949 V-2-WAC-Corporal rocket becomes the first manufactured object launched into space.
1957 Sputnik I becomes the first manufactured object in orbit.
1958 NASA is formed.
1959 Luna 2 becomes the first probe to hit the Moon.
1961 Astronaut Yuri Gagarin is the first person in orbit.
1965 First spacewalks: Aleksey Leonov (Voshkod 2) and Edward White (Gemini IV).
1968 Apollo 8 is the first crewed spacecraft to orbit the Moon.
1969 Neil Armstrong is the first person to step on to the Moon (Apollo 11).
1973 Skylab is launched.
1976 Two Viking probes land on Mars.
1981 First launch of the Space Shuttle.
1986 Mir space station launched.
1989 Galileo space probe sets off for Jupiter and its moon.
1994 Clementine space probe discovers ice on the Moon.
1996 The X Prize, a competition for private reusable spacecraft, is announced. Also, VentureStar wins X-33 contract.
1997 Cassini probe sets off for Saturn, carrying 32.4 kilograms of plutonium.
1999 First part of the International Space Station in orbit.
2000 First crew board the International Space Station.

The development of astronomy and astrophysics

1543 Nicolaus Copernicus publishes his theory that the Sun is at the centre of the Universe.

1610 Galileo Galilei makes the first scientific observations of the night sky with a telescope.

1667 Isaac Newton publishes his Universal Law of Gravitation.

1838 Friedrich Bessel is the first to measure the distance to a star.

1861 Gustav Kirchoff and Robert Bunsen discover the composition of the Sun using spectroscopy.

1905 Albert Einstein publishes his Special Theory of Relativity.

1916 Albert Einstein publishes his General Theory of Relativity.

1916 Karl Schwarzschild works out black-hole theory using Einstein's general relativity.

1929 Edwin Hubble discovers the law dictating the expansion of the Universe.

1964 Arno Penzias and Robert Wilson discover the cosmic background radiation from the Big Bang.

1967 Jocelyn Bell discovers the first pulsar.

1971 Cygnus X-1 is proposed as a strong black hole candidate.

1974 Stephen Hawking suggests that black holes emit radiation.

1989 COBE (Cosmic Background Explorer) satellite is launched.

1990 Hubble Space Telescope is launched.

1992 V404-Cygni is discovered as most likely black-hole candidate.

1995 SOHO (Solar and Heliospheric Observatory) is launched into orbit around the Sun.

1997 Astronomers at the European Southern Observatory confirm the existence of a super-massive black hole at the centre of our galaxy.

1999 Hubble Space Telescope's Key Project Team concludes, and suggests that the value of Hubble's constant is 73 kilometres per second per megaparsec.

2000 First isolated black holes discovered.

EQUINOX: WARFARE
The Introduction

From virtually the very beginning of his existence, man has employed weapons as a means of attack or defence. During the Stone Age, the humble stone was probably the first object pressed into use as a club and then as a thrown missile, followed by pieces of wood also utilized as clubs and ultimately fashioned into crude spears. The discovery of flint and the exploitation of bones enabled crude axe- and spearheads or knife blades to be fashioned. At the same time, the bow and sling were invented for the launching of shortened spears (which became arrows) and stones. Around 3500 BC saw the development of smelting, a process put to good effect by the Sumerians of Mesopotamia who discovered bronze – the first metal from which effective weapons could be fashioned. In around 1700 BC Egypt was invaded by the Hyksos, who were armed with bronze weapons and bows, and rode into battle in bronze chariots. In addition, they wore armour fashioned from leather with bronze plates attached to protect the vital parts of the body.

The first professional army was formed by Assyria in 1250 BC; it comprised chariot-mounted troops and infantry armed with weapons manufactured from iron. During the war between Persia and Greece of 490 to 479 BC, the disciplined and well-trained Greek hoplite, equipped with helmet and breastplate, a large round shield, short sword and spear, proved more than a match for the Persians. The Greeks developed the use of the phalanx, a tight formation eight men deep, with each man's shield covering the exposed right side of the man to his left. A similar system was subsequently used by the Romans until 390 BC when they introduced the legion, a unit numbering up to 6,000 men. Each legionary was equipped with helmet, armour and shield, and armed with a short stabbing sword and two *pila* (throwing spears). By this time artillery in the form of siege engines had become well developed, for example the ballista, a large mechanical bow firing a long javelin at ranges up to 500 metres, and catapults that hurled heavy rocks. Originally invented by the Sumerians, siege engines were developed further by the Assyrians and ultimately the Romans.

In western Europe, the collapse of the Roman Empire in the fifth century AD was followed by the Dark Ages, which saw a decline in military proficiency. Standing armies comprised poorly trained levies centred round small bodies of professional troops retained by the rulers who reigned following the departure of the Romans and their legions. During this period, the emphasis in weaponry turned to heavy armour, long swords, battleaxes and maces. In about AD 800 heavy cavalry came into vogue and for the following 500 years reigned supreme on the battlefield. The arrival of the longbow, however, spelt the end for heavily armoured knights as was shown at the Battle of Crécy in 1346 when English longbowmen slaughtered over 1,500 French heavy cavalry as well as some 10,000 foot-soldiers and lightly armed levies.

The advent of gunpowder resulted in the development of artillery and, by the middle of the fifteenth century, most armies in Europe were equipped with some form of cannon. Manufactured in iron with short barrels, they were intended for siege warfare and fired solid round balls of stone designed to batter down the walls of castles or defensive emplacements. Handguns had made their first appearance in the middle of the fourteenth century; heavy and cumbersome, they were little more than miniature cannon and possessed little accuracy. The middle of the fifteenth century, however, saw the arrival of the matchlock arquebus, a heavy smoothbore weapon fired from the shoulder with the barrel of the weapon supported by a rest.

The next two centuries saw a number of further developments, especially in the field of hand-held and shoulder-fired weapons, with the arrival of the wheel-lock and flint-lock. Artillery also saw vast improvements and by 1600 had been standardized into four types: culverins, for long range up to 6,460 metres (21,195 feet); cannons, capable of firing a 40-kilogram (90-pound) ball up to some 3,690 metres (12,100 feet); pedreros, which could fire a 27-kilogram (60-pound) projectile up to 1,846 metres (6,055 feet); and mortars, which could send a 90-kilogram (200-pound) shell up to 2,300 metres (7,545 feet) in a high trajectory.

The eighteenth century saw little in the way of radical changes in weapons, with muzzle-loading artillery and small arms continuing in service well into the next century. One of the most significant developments during that period was the introduction into military use of the rifled barrel, which had hitherto been a feature of sporting weapons. At that time the standard military weapon was the smoothbore musket, which was not a particularly accurate weapon at ranges over 50 metres (165 feet). Rifles were, however, capable of producing accurate fire at ranges of between 200 and

300 metres (655 and 985 feet) and first proved their worth during the American War of Independence, when American irregulars used them with great effect against the British. The latter learned the lesson well and in about 1800 introduced the Baker rifle, which was used by the British Army's rifle regiments during the Napoleonic Wars, being replaced by the Brunswick rifle in 1837.

The first half of the nineteenth century saw a number of major advances, including the invention of the first breech-loading weapons. Others included the development of the revolver by American Samuel Colt in the late 1830s, and the first machine-gun, invented by Richard Gatling, which saw service during the American Civil War. The last thirty years of the century saw major advances in the design of automatic weapons, which would have a major impact on military strategy and tactics.

Without doubt, the twentieth century was the most destructive and violent in the history of warfare, featuring two world wars and a large number of regional conflicts. It also saw the development of armaments and weapons systems that would have far-reaching and long-lasting ramifications for almost the entire world, in particular nuclear weaponry which for nearly fifty years posed the threat of mutual destruction by the superpowers.

The First World War was essentially a static conflict fought in the trenches and resulting in huge loss of life on both sides. Artillery was used extensively in support of mass infantry attacks but frequently proved ineffective against well-dug-in troops who, once a barrage had been lifted, opened fire with large numbers of machine-guns. They would cause extensive casualties among attacking troops, whose forays were frequently held up by barbed wire which had also been little affected by shellfire. Among newly developed weapons that made their appearances during the four years of the war were flame throwers, poison gas and tanks. The latter were used first by the British in September 1916. However, at the end of the Battle of the Somme they suffered heavy losses through mechanical breakdown and inexperienced tank crews. The Germans were quick to react to this new threat, developing the first generation of anti-tank weapons.

The First World War was also notable for the introduction of aircraft in a military role. Initially used primarily for reconnaissance and observation, these first unarmed models were equipped with early-generation radios via which they transmitted target co-ordinates in Morse code to the artillery, who thereafter acted on the information. In due course fighters were also developed, seeing extensive service on both sides.

The 1920s and 1930s saw further developments in the field of armour, including the design and development of tanks with rotating turrets containing the main armament. Although Britain, the United States, France, Italy and Russia carried out much experimentation with armoured and mechanized formations, it was the Germans who developed the strategy of *blitzkrieg* and used it with such devastating success during the invasion of Poland in 1939 and that of Belgium, Holland and France in 1940. The ensuing five years of conflict saw considerable research and development of weaponry on both sides, particularly in the areas of armoured fighting vehicles, anti-tank weapons and armour-piercing ammunition.

The Second World War was also notable for the birth of the first strategic weapons in the form of *Vergeltungswaffen* (reprisal weapons), better known as V-weapons, developed by Germany. From June 1944 to March 1945 Britain was subjected to over 8,000 attacks by V1 pilotless flying bombs, the forerunner of today's cruise missile, which were launched from the Pas de Calais on the coast of northern France and from Holland. Popularly known as the 'doodle-bug'. From September 1944 onwards, it was joined by the even more terrifying V2 missile which, although inaccurate, could reach target areas up to about 320 kilometres (200 miles). Travelling faster than the speed of sound, it would arrive completely unheralded until announcing its presence with a massive explosion. The use of V-weapons was indiscriminate. They were aimed at cities and other densely populated areas, and were designed to instil terror into the populations of London and southern England.

Meanwhile, Britain and America were aware of Germany's and Japan's efforts to develop nuclear weapons and thus were determined to pre-empt them by producing their own. The United States, with British assistance, succeeded in winning the race and the dropping of two atomic bombs on Hiroshima and Nagasaki in 1945 hastened the subsequent surrender of Japan.

However, as described in 'A Very British Bomb', after the end of the war Britain discovered that the United States no longer wished to continue the close wartime collaboration and was thus forced to establish its own nuclear weapons programme. Despite major problems and opposition from the United States, it succeeded in doing so and carried out its first test of an atomic bomb in October 1952. Meanwhile, however, treachery on the part of British and American scientists had permitted the Russians to develop their own atomic bomb, which was exploded in August

1949. Development of thermonuclear weapons by the United States, Britain and the Soviet Union swiftly ensued and the years following the end of the Second World War and the beginning of the Cold War saw the start of the nuclear arms race between the superpowers.

Massive resources were devoted by the United States and the Soviet Union to the development of strategic nuclear weapons and the systems to deliver them. Initially, the Soviets relied on long-range heavy bombers for delivery of nuclear weapons, but by the end of the 1950s had also developed intermediate range ballistic missiles (IRBMs). Development of second-generation weapons was by then at an advanced stage and in 1963 intercontinental ballistic missiles (ICBMs), such as the SS-8 Sasin, were deployed. These were followed by third-generation weapons and in the late 1960s by the development of the first Soviet multiple warhead systems known as multiple re-entry vehicles (MRVs). Equipped with an MRV system, the SS-9 Scarp was capable of delivering three warheads, all of which would either attack the same target, to maximize the chance of destroying it, or three individual targets within an area known as the missile's 'footprint'.

In addition to land-based ICBMs, the Soviets also developed submarine-launched ballistic missiles (SLBMs). Development of strategic ballistic missile submarines (SSBNs) followed.

By the end of the 1960s, the United States had also developed a formidable arsenal of strategic nuclear weapons. Like the Soviet Union, during the late 1940s and early 1950s it had developed a long-range bomber, the B-52 Stratofortress, for the aerial delivery of nuclear weapons; but during the 1960s it concentrated on the development of

ICBMs. First among these were the Atlas and the Titan I. Phased out in the mid-1960s, Titan I was followed by the Titan II, a much larger missile with a 9-megaton warhead, which remained the principal US ICBM until the arrival of the Minuteman I. The latter was soon replaced in the late 1960s by Minuteman II, which had a range of 13,000 kilometres (8,000 miles) and a 1.2-megaton warhead equipped with electronic jamming devices and other systems designed to enable it to penetrate Soviet radar-controlled anti-ballistic missile systems. A further enhanced version, Minuteman III, entered service between 1970 and 1975, the first US missile to be equipped with a multiple independently targeted re-entry vehicle (MIRV) system comprising three MIRVs, each equipped with a 170-kiloton warhead.

The United States also developed SLBMs, the first being the Polaris A-1 which became operational in 1960. This was followed two years later by the A2 model and eventually the A3, the latter being the first American SLBM to be equipped with MIRVs. All three models of Polaris were deployed aboard Benjamin Franklin and Lafayette Class SSBNs which at the end of the 1960s were converted to carry a new SLBM: the Poseidon C3.

Both the United States and the Soviet Union also developed tactical nuclear weapons for delivery on to the battlefield by short and intermediate range ballistic missiles fired from mobile launchers. Among those in the American inventory were the MGR-1 Honest John and the MGM-52 Lance battlefield missiles. The latter, which entered service in 1971 to replace the Honest John, could be fitted with a variety of warheads, ranging from one to 100 kilotons. A more advanced weapon was the MGM-31 Pershing 1a IRBM which, with a maximum range of 740 kilometres (460 miles) and a range of warheads of up to 400 kilotons, entered service in 1962. It was replaced in 1984 by the Pershing II which, fitted with a manoeuvring warhead (MaRV), was an exceptionally accurate weapon with a range of 1,800 kilometres (1,120 miles).

The Soviet Union meanwhile produced the FROG-7 unguided battlefield rocket and the SS-1C Scud B SRBM, both entering service in 1965. In 1969 the SS-12 Scaleboard SRBM also became operational. In 1978 the SS-21 Spider entered service as a replacement for the FROG-7. Mounted on an armoured 6 × 6 amphibious transporter/launch vehicle, it had a range of 120 kilometres (75 miles) and carried a 100-kiloton warhead. In 1985, a replacement for the Scud appeared in the form of the SS-23 Scarab.

By the beginning of the 1970s, the Soviet Union had achieved numerical superiority over the United States in terms of operational ICBMs. During the following decade it proceeded to develop further IRBMs and ICBMs with increasing range and enhanced warhead capabilities. In 1975 the first MIRV-equipped ICBMs were deployed, prominent among them being the SS-18 Satan. In addition the SS-20 IRBM was developed and deployed in early 1976. Mounted on a tracked chassis, it could be equipped with a 1.5-megaton warhead and had a range of 2,750 kilometres (1,700 miles). The SS-20 caused considerable concern to the United States and its allies as it had the capability, depending on its deployment location within the Soviet Union, of hitting targets throughout the whole of western Europe and the Middle East.

The Soviets also developed fourth-generation SLBMs which came into service during the 1970s and were operational aboard Delta Class SSBNs. Late 1976 saw the arrival of the SS-N-18 Model 1 which, with a range of 6,500 kilometres (4,000 miles), was the first Soviet SLBM to be equipped with a MIRV warhead system. The most startling development, however, was the Typhoon SSBN which entered operational service in 1983. The largest underwater vessel ever constructed and powered by four pressurized water-cooled nuclear reactors, it was designed to operate within the Arctic Circle from which it would launch its twenty SLBMs against targets throughout the continental United States.

The United States had meanwhile also developed a new SLBM, and in 1979 deployed the Trident I C4 aboard eight Benjamin Franklin and Lafayette SSBNs converted to carry it. In 1989 an enhanced version, the Trident II D5, came into service and was deployed on new Ohio Class SSBNs designed to carry the new missile, which has a range of 7,400 to 11,100 kilometres (4,560 to 6,900 miles) and is equipped with between eight and fourteen MIRVs, each with a 375-kiloton warhead.

By the late 1960s, it had become apparent that neither the United States nor the Soviet Union could hope to win a nuclear conflict as both had the capability to inflict an equal amount of destruction on the other. Thus 1969 had seen the start of the Strategic Arms Limitation Talks (SALT), the aim of which was to halt the strategic nuclear arms race. Further progress on nuclear disarmament began in 1982 with the Strategic Arms Reductions Talks (START), which culminated in 1991 with the START I treaty.

In 1980 the United States and the Soviet Union began negotiations over the elimination of IRBMs, which were defined as weapons having ranges of 1,000 to 5,500 kilometres (620 to 3,400 miles). This resulted in the signing of the Intermediate-Range Nuclear Forces (INF) Treaty of 1987, which also covered SRBMs with ranges of 500 to 1,000 kilometres (300 to 620 miles) and called for the progressive dismantling over a period of three years of a total of 2,619 missiles, two-thirds of which were Soviet.

December 1991 saw the dissolution of the Soviet Union and its replacement by fifteen independent states of which four – Russia, Ukraine, Belarussia and Kazakhstan – still retained strategic and tactical nuclear weapons within their respective territories. In January 1993 the United States and Russia signed an informal agreement, START II, under which ICBMs with MIRV warheads would be destroyed and numbers of strategic warheads limited to between 3,000 and 3,500 by the end of the year 2007. In addition, a limit of 1,700 to 1,750 was placed on warheads deployed on SLBMs. Ratification of the treaty by the US Senate took place in January 1996 but it would not be until April 2000 that the Russian Duma would do likewise.

As recounted in 'Dismantling the Bomb', destruction of nuclear weapons in the United States began in 1991 at a special facility formerly used for the servicing and maintenance of missiles and bombs. The situation was, however, very different in the four former Soviet states, which were virtually bankrupt and unable to fund their respective dismantlement programmes. In early 1991 informal approaches were made to the United States for assistance, and these ultimately led to the establishment of the Co-operative Threat Reduction (CTR) Program, also known as the Nunn–Lugar Program after the two US senators who conceived it. Since then, the United States has committed substantial financial and technical assistance to Ukraine, Belarussia and Kazakhstan in becoming nuclear-free states, and to Russia in helping to overcome the problems of reducing its nuclear weapons arsenal.

Such is the size of Russia's nuclear stockpile, however, and the poor conditions under which much of it is stored, that its security poses a major problem. The United States, well aware of the risk of nuclear material falling into the hands of countries keen to develop their own nuclear weapons, has thus provided funds under the auspices of the CTR Program to improve security measures and finance the construction of special storage facilities. As told in 'Russian Roulette', however, a number of miniaturized tactical nuclear weapons, popularly referred to as 'suitcase bombs' because of the briefcases in which they were installed, were found to be missing from Russian arsenals and their whereabouts have as yet not been confirmed.

At the beginning of the last century, standard infantry weapons in the majority of armies comprised bolt-action rifles and water-cooled machine-guns, while cavalry were mounted on horses and equipped with swords, lances and carbines. Artillery provided support with shrapnel fired by field guns and howitzers at ranges up to 9,150 metres (29,740 feet). Wireless was non-existent and communication was by heliograph, semaphore flags and signal lamp. Aircraft had not been invented and thus air warfare was a dimension yet to be added to the battlefield.

Ninety years later the picture was very different when, during the Gulf War of 1991, Coalition forces in Saudi Arabia engaged those of Iraq a few months after the latter's invasion of Kuwait. As described in 'After Desert Storm', hostilities opened with US Army helicopter gunships destroying two key radars located near the Iraqi/Saudi border, thereby allowing a stream of aircraft to pour through the gap created in Iraqi air defences. Among them were F-117A Stealth fighters, which were invisible to the air defence radars and surface-to-air missile batteries seeking vainly to shoot them down. Meanwhile, warships and nuclear-powered submarines in the Persian Gulf launched salvoes of Tomahawk cruise missiles which subsequently navigated themselves over the 1,000 kilometres (600 miles) of terrain to the Iraqi capital of Baghdad, where they unerringly sought out their respective targets and destroyed them.

During the following ground war, artillery – comprising guns and multi-launch rocket systems, all directed by computerized target data and fire control computers – pounded Iraqi troops at ranges of between 20 and 30 kilometres (12 to 18 miles). When Coalition forces advanced into Kuwait the infantry, armed with self-loading rifles and automatic weapons that could be equipped with night-vision devices enabling them to see and engage the enemy in the dark, were transported into battle in armoured vehicles or by helicopter. They had their own defence against enemy armour in the form of shoulder-fired rockets or medium-range anti-tank guided weapons. In constant radio communication with members of their unit and those of others, they were able to navigate through the most featureless of terrain courtesy of satellite navigation systems. They were supported by main battle tanks featuring computerized fire control systems for their 120-mm guns and equipped with steel-ceramic composite armour providing very high levels of ballistic protection. Meanwhile, helicopter gunships engaged Iraqi armour and infantry with guided missiles, rockets and chain guns, destroying them in large numbers. In the sky above, E-3 Sentry AWACS and E-8 JSTARS aircraft provided information on targets and directed air and ground attacks against them.

Such were the scenes that took place eleven years ago. But what of warfare in the future? Despite the major reduction in the threat of global nuclear conflict between the United States and Russia, the risk of nuclear proliferation has remained, with countries such as Iran, Iraq and North Korea seeking to develop their own nuclear weapons. These and others currently possess short- and medium-range ballistic missiles with ranges up to 3,000 kilometres (1,860 miles) and capable of carrying high explosive or chemical agent warheads.

Furthermore, China possesses approximately twenty CSS-4 ICBMs, all of which are capable of reaching the United States, and is known to be developing two mobile ICBMs. The first of these, designated the DF-31, was tested in August 1999 and intelligence sources judge it will have a range of approximately 8,000 kilometres (5,000 miles). The second missile to be tested between 2000 and 2002 and to possess a longer range. A US National Intelligence Council report, published in September 1999, stated that by 2015 China is anticipated to have tens of ICBMs targeted against the United States.

In order to provide defence against these and similar weapons, a number of counter-measures are currently under development. One such system, mounted in a large aircraft and comprising a high-energy chemical laser capable of emitting massive power, is designed to engage and destroy launched Theatre Ballistic Missiles (TBMs) as they break through the clouds near the edge of the Earth's atmosphere. Further anti-ballistic defence will be provided at even greater altitudes by a US space-based system scheduled to be operational by the year 2010. Comprising six high-energy lasers mounted on satellites, its role will be to engage ICBMs launched anywhere on the Earth's surface.

Meanwhile, it seems that lasers will play an increasing part in conventional warfare, primarily in a defensive role. As described in 'Dawn of the Death Ray', an example of this is a ground-based system, currently nearing completion of development, which is designed to provide defence against short-range battlefield rockets of the Russian 122-mm Katyusha type. Two other systems, fitted to aircraft and armoured vehicles respectively, have been produced to blind the homing heads of infra-red guided missiles launched at them. Further research is currently being conducted into the development of lasers for use as attack weapons for precision strikes against targets with minimum collateral damage to personnel and surrounding areas. It thus appears that they may well become the weapons of the future.

In over 2,000 years of history, man has devoted much of his ingenuity to the development and production of increasingly deadly and destructive weapons, the most dreadful being nuclear weapons capable of devastating entire nations. Despite the partial dismantling of stockpiles of nuclear weapons held by the United States and Russia and the consequent lowering of the risk of global nuclear conflagration, the threat of conflict is still present in several parts of the globe. It seems that man is unlikely ever to learn from his mistakes and all too easily forgets the lessons of the past. Perhaps it would be as well to remember the words written in 1965 by an eminent military historian, the late Brigadier Peter Young, which are still very relevant thirty-five years later: 'We live in a technological age. But it is an age in which people whose interests are opposed still strive to solve their problems by force, though their methods may be those of the economist, the politician and the diplomat. Perhaps two world wars have bred a brand of statesman capable of keeping the lid on Hell. It does not seem likely, for we are not especially skilful in selecting our masters. Let us therefore remember the words of Santayana: "He who forgets his history is condemned to relive it."'

A VERY BRITISH BOMB

On 6 August 1945 a United States Army Air Force B-29 bomber dropped an atomic bomb over the Japanese city of Hiroshima; three days later, on 9 August, another was dropped over Nagasaki. Such was the massive loss of life and devastation that five days later, on 14 August, Japan surrendered unconditionally.

The United States had won the race to develop an atomic weapon and in so doing had brought the Second World War, which had lasted almost six years, to a conclusion. The Americans could not have done so, however, without the powerful impetus of Britain and vital input from British scientists, who had been the first to demonstrate that an atomic bomb was possible. A small group of the latter later played crucial roles in the highly secret Manhattan Project, the codename of the American atomic weapon development programme.

Britain had been involved in nuclear research since before 1940, and had been very much in the lead during that period. It was a British scientist, Professor James Chadwick, who had discovered the neutron in 1932 (and was awarded a Nobel prize for the discovery in 1935). In late 1938, in Berlin, Otto Hahn and Fritz Strassmann discovered the phenomenon of nuclear fission in uraniam bombarded by slow neutrons. The full import of it was not understood, however, until shortly afterwards when Lise Meitner and her nephew, Otto Frisch, both refugees from Nazi persecution, interpreted this phenomenon as nuclear fission (a term borrowed from biology by Frisch). The practical possibility of a fast chain reaction resulting in an atomic bomb was recognized in Britain in 1940 by Frisch and Professor Rudolf Peierls, the latter also a refugee from Nazi Germany. In February that year, they produced a short but convincing memorandum stating that if uranium-235, which comprises 0.7 per cent of natural uranium, could be separated from the rest of the material, it would be possible to produce a bomb using only a small amount. Furthermore, although separation would be difficult, uranium-235 had the potential to make a very powerful weapon. Up to that time, it had been calculated that the amount needed to bring about an explosive reaction was impracticably large to produce, and would mean a bomb too large to be carried by any aircraft.

Peierls and Frisch foresaw the long-term hazards of radiation and fall-out, and concluded their three-page memorandum by stating that they thought it likely that the Allies would never wish to develop such a bomb. It was intrinsically a weapon of mass destruction that would inevitably cause much civilian death and suffering. The sole reason for their bringing this to the attention of the government was their fear that Germany might already be working on the development of atomic weapons, and their conviction that the only effective deterrent would be for the Allies to possess their own. The Germans were indeed carrying out nuclear research for military purposes and during 1940 the British received warnings from France and other sources to that effect.

The memorandum had the desired effect. It soon reached the highest levels of government where it resulted in the establishment of the high-powered committee, codenamed MAUD, which in July 1941 produced two detailed reports. The first examined the principle of an atomic weapon, method of fusing, probable effect, the preparation of material and the resources, including time and cost, that would be required to produce a uranium bomb. It stated that 25 pounds (over 11 kilograms) of uranium-235 would be needed for an explosive yield of 1.8 kilotons, equivalent to 1,800 tons of TNT. The second report examined how uranium could be used to fuel a reactor that could be used not only as a source of power but also to produce a new fissile material, 'Element 94', later named plutonium. Plutonium-239 would be an even better explosive than uranium-235.

Copies of both reports were despatched to the United States where the scientific establishment, which at that time was concentrating solely on research into nuclear fission for peaceful purposes, was slow on the uptake. It should be remembered that at this point the Americans were not participants in the war against Germany, and thus there was no incentive for research for military purposes. Meanwhile, unknown to all those involved, copies also found their way to Moscow courtesy of a Soviet agent inside the British government. Although the culprit was never officially identified, the finger of suspicion would later be pointed at John Cairncross, a young civil servant who at the time of the MAUD report was working in the private office of Lord Hankey, the cabinet minister responsible for scientific matters.

On the strength of these reports, the British government decided to proceed with research and development of an atomic bomb. Indeed, the decision was taken at the highest level, by the Prime Minister, Winston Churchill, himself, and the level of secrecy was such that not even the War Cabinet was consulted about it. Responsibility for the project was given to the Department of Scientific and Industrial Research, which in the autumn of 1941 formed the Directorate of Tube Alloys. This was responsible for overseeing research carried out by a number of establishments including the Cavendish and Clarendon laboratories at Cambridge and Oxford respectively, as well as laboratories at the universities of Bristol, Birmingham and Liverpool.

By the latter part of 1941, the British were ahead of the United States in uranium research and thus had much to offer a combined Anglo–American project. But when approached by the Americans to pool resources, they responded coolly, citing fears over security and making it apparent that they wanted to restrict collaboration to exchanges of information. The situation changed dramatically, however, on 7 December 1941 when Japanese aircraft attacked ships of the US Navy's Pacific Fleet at its base at Pearl Harbor, Hawaii, sinking four battleships, damaging a number of other vessels and destroying 180 aircraft. US casualties totalled over 3,400, of whom more than 2,300 were killed. On the following day the United States declared war on Japan and, within days, on Germany and Italy.

Now that they were combatants, the Americans lost no time in beginning work on their own atomic weapon programme. Swiftly allocating considerable resources to it, they pushed ahead independently and such was the speed of their progress that by June 1942 they had surged ahead of the British. By this time the latter had changed their minds and were seeking close collaboration on the lines previously proposed by the Americans, who now, however, no longer regarded British input as essential to their programme. When a team of scientists, led by the head of the Tube Alloys directorate, Wallace Akers, visited the United States early in 1942, they found that the Americans were already overtaking them in development of processes for producing fissile material. Furthermore, they discovered that uranium-235 was no longer the material most likely to be used in nuclear weapons, plutonium by then being the Americans' first choice.

The increasing American lead inevitably resulted in a corresponding reduction in British bargaining power with regard to Anglo–American collaboration, and by the late summer of 1942 it became apparent that the likelihood of integrating British and American research efforts was fading fast. Moreover, the free exchange of information, maintained since July 1941, appeared to be coming to an end. Among those heading the Manhattan Project, including its chief General Leslie Groves, the attitude was that the British were asking for access to technology to which they had contributed little or nothing. Furthermore, doubts were being expressed among such quarters as to whether Britain should be allowed to possess nuclear weapons after the war. The situation was not helped by the fact that in September 1941 Britain had, with the knowledge of the Americans, signed an agreement with Russia for the exchange of scientific information, albeit restricted to a list of subjects approved by the United States. When President Franklin D. Roosevelt, who had previously been unaware of the agreement, was apprised of it, he endorsed proposals for restrictions being placed on exchange of information on atomic energy between the United States and Britain.

In mid-January 1943, while attending a conference with Roosevelt in Casablanca, Winston Churchill raised the subject of Anglo–American collaboration with the President's personal aide, Harry Hopkins, who assured him that the situation would be rectified by Roosevelt on his return to the United States. Nothing happened, however, and repeated communications from Churchill to Hopkins on the subject were either fobbed off or ignored.

The period of January to April 1943 saw a virtual cessation of communication between the British and American scientific establishments as they waited for the two leaders to resolve the matter. In Britain morale was very low as it became apparent that the Americans were no longer interested in collaboration. Despite the formidable cost of some 69.5 million pounds and the resources required, including 20,000 men and 500,000 tons of steel, serious consideration began to be given to the idea of Britain going it alone in building diffusion and heavy-water plants and a reactor, with the ultimate aim of producing a British bomb.

In May 1943, however, the Americans gained the monopoly of the supply of uranium and heavy water from Canada. Discovered in the early 1930s, heavy water (D_2O), also known as deuterium oxide, is used as a moderator to slow down fast-moving neutrons in order to increase their chances of hitting a nucleus and thus maintain the process of a controlled chain reaction. Later that month, Roosevelt and Churchill met again in Washington for the Trident Conference to determine future operations against Germany and Japan following victory in North Africa. The subject of Tube Alloys collaboration was raised again by Churchill and the President agreed to a resumption of exchange of information, while also concurring that the development of nuclear weapons should be a joint project. Once again, however, nothing happened.

In July the situation improved with a visit to London by Dr Vannevar Bush, chairman of the National Defense Research Committee, and the US Secretary of State for War, Henry L. Stimson. During a meeting with Stimson, Churchill succeeded in ironing out difficulties and misunderstandings relating to post-war concerns about commercial exploitation of nuclear research. It was made clear to Stimson that the British government had never attached any importance to industrial exploitation. At the same time, the Americans revealed that serious offence had been caused by the British bypassing those heading the Manhattan Project and appealing direct to President Roosevelt. The upshot of the meetings during this visit was that the obstacles to the resumption of Anglo–American collaboration were removed and a draft agreement produced. It later transpired that while Bush and Stimson were in London, Roosevelt had decided, on the advice of Harry Hopkins, to honour the undertakings he had made previously to Churchill. The President had cabled Bush in London, instructing him to reinstate full exchange of information with the British.

In August 1943 Churchill and Roosevelt met in Quebec for a conference codenamed Quadrant, at which future plans for the invasion of France and operations in the Pacific and south-east Asia were laid. On 19 August the Quebec Agreement, partly based on the draft agreement produced by the British in July, was signed by the two leaders, laying the foundations for Anglo–American collaboration in atomic weapon development for the rest of the war and ostensibly thereafter.

Four months later, in mid-December 1943, the first of a contingent of some forty British scientists left England for the United States. They were led by Professor James Chadwick, who subsequently established a close relationship with General Leslie Groves, the chief of the Manhattan Project, despite the American's reputation for being an anglophobe. Among the contingent's other members were Professor Rudolf Peierls, who later became head of the Manhattan Project's theoretical physics division; a physicist, Dr Klaus Fuchs who, like Peierls, with whom he had worked at the University of Birmingham, had fled to Britain from Nazi Germany; and William Penney, a young professor of mathematical physics. Twenty of the contingent subsequently went to Los Alamos, the American nuclear weapons laboratory located 96 kilometres (60 miles) north of Albuquerque in the mountains of New Mexico. The remainder were dispersed among other parts of the project elsewhere – with the exception of Hanford in Washington State, the site of the production reactor, from which they were excluded. While the British group was small, it was of extremely high quality and proceeded to make a disproportionate contribution to the Manhattan Project.

On 19 September 1944 Winston Churchill and President Roosevelt signed the Hyde Park Agreement, so-called because it was the name of the President's home in New York State. In this, the two countries agreed that knowledge of nuclear weapons research should be restricted to Britain and the United States, and that once a weapon had been developed it might after 'mature consideration' be used against the Japanese who should be warned that such bombing would be repeated until they surrendered. The deal also stated: 'Full collaboration between the United States and the

British government in developing tube alloys for military and commercial purposes should continue after the defeat of Japan unless and until terminated by joint agreement.'

In July 1945, however, a general election took place in Britain and the Labour government of Clement Attlee was elected to office. Attlee, despite having been Deputy Prime Minister during the war, now became fully aware for the first time of Britain's involvement in nuclear weapon research. In October, at a meeting at ShellMex House in London, the decision was taken by a small group of senior ministers within the new government to establish the Atomic Energy Research Establishment (AERE), which would have a wide-ranging brief into all aspects of nuclear research.

After Japan's surrender, and the end of the war, Britain naturally expected to continue in its nuclear partnership with the United States, under the terms of the Quebec Agreement and the Hyde Park Agreement, and at the beginning of 1947 an approach was made to the Americans for technical assistance in constructing a nuclear reactor. The British, however, received a rude shock when it was made clear to them that there was to be no further collaboration.

In May 1946 the United States had passed the Atomic Energy Act, commonly known as the McMahon Act (as it had been sponsored by Senator Brian McMahon), which established the US Atomic Energy Commission and imposed a mantle of secrecy over all aspects of nuclear energy, backed up by draconian laws banning the divulging of information of any sort relating to the subject. Furthermore, it declared that there would be no sharing of information about nuclear technology with any other nation. Part of the blame for the inclusion of Britain within the terms of the McMahon Act can be laid at the door of the tight secrecy surrounding the Quebec Agreement of August 1943 and the Hyde Park Agreement of September 1944, of which Congress had no knowledge. Senator McMahon was later reported as stating that if he had known about the British contribution to the Manhattan Project, the act would have been drafted differently and Britain excluded from its provisions.

In addition to the McMahon Act, further problems arose. Even after that Act, the Combined Policy Committee and Combined Development Trust – two bodies set up under the Quebec Agreement to procure uranium ores and allocate them – continued to function for some time. However, there were serious tensions and the British, who played an important part in worldwide surveys and procurement, had a hard struggle to obtain the annual allocations they needed. They now requested their share to be shipped to them but the Americans resisted this, concerned that there was insufficient ore for their own production. Furthermore, the Americans regarded nuclear technology as being of a cataclysmic or 'end of the world' nature, sole possession of which would render the United States invulnerable to its enemies. At this juncture, some members of the US Senate discovered the existence of the Quebec Agreement. They perceived it as a violation of national sovereignty, which had long been regarded as sacrosanct – particularly among the more conservative elements in US politics – and were vociferous in their opposition to any continuation of nuclear collaboration with Britain.

Despite his dismay at the withdrawal of American collaboration, Prime Minister Clement Attlee was of the firm opinion that without nuclear weapons Britain would be left isolated and powerless in the face of the growing threat of the Soviet Union in Europe, and thus would have to develop her own independently of the United States. He strongly believed that international control of nuclear power should be exercised through the United Nations (UN) but the UN Atomic Energy Commission, which he had been largely instrumental in establishing, failed to reach an agreement over a plan for international control. In the light of this and the McMahon Act, Attlee felt that Britain must acquire the capability to produce nuclear weapons.

The entire programme was contained in the Ministry of Supply, which provided a large and powerful infrastructure together with much experience, gained during the war, in managing research and production establishments and running large scientific projects. Within the ministry, a new Department of Atomic Energy (DATEN) was formed, taking over all the responsibilities of the Directorate of Tube Alloys.

To establish the programme, however, Attlee had to engage the services of four particular men. The first of these was Lord Portal of Hungerford who, as Air Marshal Sir Charles Portal, had been Chief of the Air Staff during the war. A trusted and proven Whitehall operator, he was a practical man as well as being politically minded. He was considered ideal for the task of resolving the problems of obtaining the resources required by the planned nuclear weapon development programme.

In January 1946 Professor John Cockcroft was appointed to head the newly created AERE, which was subsequently located at Harwell, on the border of Berkshire and Oxfordshire; work on the site began in April. Cockcroft was a Nobel Prize winner who had been at the Cavendish Laboratory before the war and had, with Ernest Walton, been the first to split an atom artificially. During the war he had worked in Canada at Chalk River, an Anglo–Canadian–French research establishment that worked on the design and construction of nuclear reactors. After taking up his appointment, Cockcroft proceeded to recruit some of those with whom he had worked at Chalk River.

The British bomb would be based on plutonium and therein lay a problem. During the war, the British scientists working on the Manhattan Project had never been granted admission to the plutonium production plants at Hanford, in Washington State, and were given no information concerning the construction of reactors or plutonium production processes. Nor were they permitted access to the uranium separation plant at Oak Ridge in Tennessee. Unlike uranium, plutonium does not exist in nature – it has to be manufactured, atom by atom, by placing uranium in a nuclear reactor. As the Americans had discovered, the magnitude of the entire Manhattan Project was almost unimaginable. In 1945 it matched the US automobile industry in terms of industrial investment, scale of operations and numbers of people employed. Its cost was estimated at 2 billion dollars, which was equivalent to the cost of sending the first man to the moon.

During the war, Britain had almost exhausted its supplies of ammunition: a crisis that could have had fatal consequences. The situation had been saved by the inspired leadership of one man, an engineer named Christopher Hinton. In January 1946 the government appointed him to take on yet another gargantuan task, that of producing plutonium for the British bomb.

One of the problems facing Hinton was the shortage of qualified engineers. At that time, Britain was facing the massive task of post-war reconstruction and engineers in particular were needed to build up the country's industrial base again, repair the damage caused by six years of war and revive export trade. Furthermore, the number of qualifying engineers had been reduced to almost nil during the war, and so it was extremely difficult to recruit suitably qualified personnel. It was principally for this reason that, during the month following his appointment, Hinton chose to establish his Atomic Energy Production Division in the north of England, at Risley in Lancashire, rather than in the south where he considered he would never find any engineers at all. On starting to look for likely candidates, however, he encountered another problem: as the project was being conducted under the auspices of the Ministry of Supply, all recruitment had to be carried out through the Scientific Civil Service (SCS), whose pay levels were not competitive with those being offered by industry. Furthermore, those scientists and engineers seriously interested in working for the SCS and attracted by nuclear research all tended to want to work at Harwell, which offered more attractive career prospects.

Hinton thus turned to some of the engineers who had worked under him during the war; once again secrecy was paramount and those approached could be told nothing at first. Before long, fourteen individuals gathered at Risley to receive an initial briefing from Dennis Ginns, an engineer who had worked at Chalk River in Canada and who gave his new colleagues a basic introduction to the principles of nuclear energy. As his lecture progressed, those listening to him realized that they were being asked to use chain reactions to release the huge amounts of energy stored in the nucleus according to Einstein's famous equation $E = mc^2$. They also concluded that such a system could provide a possible source of energy for the future; in addition, it was also a possible means of producing the most devastating type of bomb that the world had ever known.

Having undertaken to deliver sufficient plutonium for the first bomb by 1952, Hinton knew that there was no time to be lost and set his team to work, subjecting them to a punishing schedule while he looked for an isolated site for his plutonium factory. Eventually he selected Sellafield in Cumbria on the coast of north-west England. The site was subsequently christened Windscale and, when further land was taken over for additional reactors, the new area was named Calder Hall. Ultimately, the entire area was renamed Sellafield.

Hinton initially wanted to build pressurized graphite-moderated gas-cooled reactors but on advice from Harwell, which advised that design and construction would take too long, decided instead on an air-cooled design in which air passed over the reactor, cooling the uranium fuel rods before being discharged via very tall chimneys. His decision to opt for an air-cooled rather than water-cooled system was based on two facts. Firstly, there was the problem of finding adequate supplies of very pure cooling water. Secondly, there was the risk inherent in using water as a coolant, especially in graphite-moderated reactors. If a reactor overheated and the water turned to steam, the steam would not remove heat so efficiently from the reactor core which would then overheat even more. A further problem, in a 'loss of coolant accident', was that some or all of the moderating effect of the water, slowing down the nuclear reactions, would be lost and a runaway reaction could occur with the risk of a hydrogen explosion (such as that which was to occur at Chernobyl in April 1986). It was for that reason that the Americans had sited their reactors at

Hanford in remote areas. In Hinton's eyes, therefore, graphite water-cooled reactors were unsuitable in a small, highly populated island such as Britain.

This decision brought Hinton into direct conflict with his superiors in the Ministry of Supply, who were convinced that as the Americans were using graphite water-cooled reactors, then Britain should follow the same proven route in order to be able to produce plutonium as soon as possible. Angered at what they saw as intransigence on his part, they were at one point prepared to discipline him but he stood his ground. Adhering to his choice of air-cooled reactors, he opted for a design with tall chimneys 125 metres (410 feet) high through which air would be blown to cool the reactor.

Hinton was advised by scientists at Harwell and by meteorologists that this system would be quite safe because if any air was contaminated while passing through the reactor, by the time it emerged from the chimneys at 410 feet it would be too dispersed to contaminate the atmosphere. Subsequently, however – and by which time building at Windscale was well under way – Harwell had second thoughts, principally as a result of a visit paid by its director, Professor John Cockcroft, to the United States. There, he heard rumours that problems had been experienced with particulate emissions from the chimney of a reactor contaminating the ground in the area. (It later transpired that Cockcroft was mistaken, as the emissions were from a chemical plant at Oak Ridge in Tennessee.) Once made aware of this possible threat, Hinton decided that the chimneys would have to be fitted with filters. However, the chimneys themselves were already half-built and so, with no time available for any delay in the already very tight schedule, the hastily designed huge filter galleries had to be constructed 122 metres (400 feet) up in the air, very near the top of the chimneys. The difficulties of regular maintenance, as well as construction, ultimately proved formidable.

Meanwhile, a further problem had arisen with the high-purity graphite blocks from which the reactors themselves were being constructed. During the machining, Harwell carried out research on the material and discovered that when graphite is heated and irradiated with neutrons, it expands unidirectionally. This would result in distortion of the blocks and narrowing of the channels, preventing insertion of the fuel elements and limiting the life of the reactor to only two years or so. The blocks thus had to be redesigned and in the event worked satisfactorily.

Another problem facing Hinton was that there was no time to assemble a pilot chemical plant; a full-scale one had to be built straightaway. Furthermore, once the manufacturing progress began and the radioactive uranium was beginning to go through the process, it would be impossible for anyone ever to go inside the building to make alterations. The building itself would be vast: 60 metres (200 feet) high and surmounted by a 60-metre (200-foot) chimney.

In March 1946 work began on building a factory at Springfields, near Preston in Lancashire, where uranium ore would be crushed and purified to a very high degree, eliminating any impurities that would otherwise poison the reactor, and turned into uranium metal which was cast into bars. These would be sent either to Windscale in the form of fuel elements for irradiation in the reactors there, or to another facility at Capenhurst, in Cheshire, where the uranium would be fed through a gaseous diffusion plant that gradually separated out the uranium-235 isotope comprising 0.7 per cent of the total in natural uranium, a complex chemical process.

Meanwhile, responsibility for development of the bomb itself had been given to the fourth member of that quartet of remarkable scientists: William Penney. As mentioned earlier, he had been a member of the British contingent working at Los Alamos. One of Britain's team of leading nuclear scientists, he had gained three doctorates by the age of twenty-five. After studying physics at the Imperial College of Science and Technology at the University of London, and at Cambridge University, he had taught at Imperial College from 1936. During the war, he had carried out research for the Ministry of Home Security on the effects of bombing and for the Admiralty on the 'Mulberry' artificial harbours used in the D-Day landings in Normandy in June 1944. On being sent, in December 1943, to the United States, he had been appointed a principal scientific officer at Los Alamos, and was the sole member of the British contingent to witness the bombing of Nagasaki, viewing it from an accompanying aircraft in the company of an RAF officer, Group Captain Leonard Cheshire VC. A few days later, he was sent to visit the city itself to assess the effects of the bomb damage and measure the effects of its blast. He was thus among the first to see the appalling effects of an atomic explosion.

Such was Penney's standing with the Americans that they wanted him to continue working with them on their post-war nuclear development programme. Before conducting the first of a series of test explosions in July 1946 at Bikini Atoll in the Pacific, they asked the British government if they could borrow him because of his expertise in the effects of explosions. Accompanied by a small team of other

British scientists, he played a vital role in these tests, precisely gauging blast and impact effects at different locations.

Penney had intended to return to academic life after the war and had been offered a professorship at Oxford at about the same time as he was approached to develop a British atomic bomb. According to accounts later given by some of his former colleagues, he was concerned about the moral implications of the bomb, although he was convinced of its value as a deterrent. He came under intense pressure from those at senior levels of government who saw him as essential for the success of the project and, after lengthy consideration, gave in and was appointed Chief Superintendent of Armaments Research.

Interestingly, Penney was the only one of the forty British scientists who had worked on the wartime Manhattan Project to be employed on the British bomb itself, although a number took up places at Harwell and elsewhere. There were various reasons for this: some had decided to stay in the United States, while others decided to work at Harwell, which was part of the programme, or return to academic posts held before the war.

In January 1946 Penney took up his post as head of the large armaments research establishment at Fort Halstead, on the Downs south of London. In May 1947 he received his directive concerning the development of an atomic bomb and set about recruiting a team of scientists. During the summer of 1947 he assembled some thirty or so suitably qualified individuals at the Woolwich Arsenal, which was part of his fiefdom. Addressing them in the library, the curtains drawn for security and an armed guard outside the door, by all accounts he opened with the dramatic statement: 'Gentlemen, the Prime Minister has asked me to make an atomic bomb and I want you to help me.' The information that Penney proceeded to divulge was a complete revelation to those gathered together that day. Although it was two years since the end of the war, only a certain amount of information had been released about the bombs dropped on Hiroshima and Nagasaki. This had maintained that both were relatively crude devices based on what was called the 'gun method': this in essence comprised a gun-type barrel assembly containing two sub-critical pieces of uranium-235. One was fired into the other, the latter being precision-machined to accept it. At the moment of impact, the entire mass of uranium became supercritical and initiated a chain reaction.

What had not been revealed was that the bomb dropped on Nagasaki had been of a different type, an implosion-initiated device in which a spherical core of plutonium was squeezed by high explosive into one supercritical mass, initiating the chain reaction. The squeezing effect had to be instantaneous and uniform over the whole surface of the sphere and this was achieved, as Penney was able to tell those listening to him that day, by a system developed at Los Alamos during the war: the use of multi-point detonation and explosive lenses. Simultaneous firing of a number of detonators produced inward pressure, squeezing the sphere uniformly. Meanwhile, similar in principle to a spectacle lens that corrects the passage of light so that it strikes the eye in the correct spot, each explosive lens corrected the inward travelling pressure wave until it formed the correct shape that produced the perfect squeeze.

Among those approached by Penney was John Challens, an electronics specialist who in 1936 had joined the research department at the Master General of the

Ordnance's department at Woolwich; he had spent the war working on guns and rockets, much of it at a rocket range at Abercore in western Wales. Challens accepted and was subsequently responsible for the task of developing the electronic firing system for the new bomb.

Another recruit, who worked under Challens, was Eddie Howse who during the war had worked on radar at Malvern and Rugby. In 1948 he was posted to Fort Halstead, where his first task was to design a pulse transformer without knowing the reason for doing so. He laughed when remembering those early days: 'The secrecy never worried me because I went straight from college to working on radar and took the Official Secrets Act at that stage and after that, everything I did was secret. When I first met my wife-to-be, the only question I asked her was whether her father was a member of the Communist Party. She said no and I said, "Well, that's all right, I can marry you then."'

Other recruits to Penney's team included mathematical physicist John Corner; Roy Pilgrim, an expert in blast measurement who had worked with Penney at the American Bikini Atoll test; metallurgist Graham Hopkin; electronics and instrumentation experts L. C. Tyte, Charles Adams and Ieuan Maddock; explosives experts Ernie Mott and Bill Moyce; health and safety specialists David Barnes and Geoffrey Dale; chemist Dai Lewis; as well as radiochemist Frank Morgan. Recruiting for the team was difficult and took a great deal of time. As John Challens recalled: 'I spent a lot of time on interviewing boards. The real problem was you couldn't tell the candidate what you wanted him for. And a lot of people weren't prepared to take you on trust.'

All candidates underwent a lengthy series of interviews during which they were told nothing about the work involved or the appointments for which they were being considered. Eventually, however, a team was assembled and started work. This was subsequently augmented by a team of Royal Air Force (RAF) officers headed by Wing Commander John Rowlands. Its role was to ensure that the bomb was designed in such a way that the RAF could store, service, transport and, if necessary, use it. Furthermore, the team would ultimately have to advise the Air Ministry on the construction of buildings for secure storage and servicing of such weapons, as well as specialist equipment and training of RAF personnel in handling of them. All science graduates, members of the team were attached to each of the departments established by Penney.

Secrecy was paramount and Penney and his scientists worked in a virtual state of quarantine, as Clement Attlee had made it clear that he did not want anyone knowing that Britain was developing its own atomic weapon. This placed severe limitations on the scientists when, for example, they were talking to colleagues at Woolwich Arsenal who were not involved in projects connected with the nuclear bomb programme. At the same time, efforts to obtain resources were hampered by the fact that key people could not be approached as they were not permitted to know the reason for such requirements. This inevitably prevented the programme from enjoying the priority it had been accorded. An example of this was that Lord Portal was issued by Attlee with a series of letters of authority which would have greatly facilitated the obtaining of resources, but could not use them as he was unable to divulge the existence of the programme.

Short of personnel, resources and technical information, the entire programme was racing against the clock to keep to the deadline set by the government. In order to try and alleviate the problems under which it was labouring, Clement Attlee decided to approach the Americans in a final attempt to enlist their help.

Unlike the United States, which had greatly profited from the Second World War, Britain was virtually bankrupt. Despite their so-called 'special relationship', the British had been forced to buy or lease every ship, aircraft, tank and weapon obtained from the United States, which itself did not enter the war until 1941. By this time Britain had been standing virtually alone in opposing the might of the Axis powers. The country was on its knees after six years of suffering and privations. Attlee later summed up the reasons for Britain's parlous state and how to rectify it: 'We alone, of all the nations, went through two great wars. We stood alone. For that fight we put everything in that we had. We sacrificed all our wealth overseas. We converted all our industries and that is why we are left in this position today. The only way out is by greater output of all the things that we need. And that means harder work.'

At the end of the war, in order to finance its desperately needed programme of reconstruction, Britain had negotiated a 7.5-billion-dollar loan from the United States. By 1947, however, the remaining amount available was 500 million dollars and so the country was close to bankruptcy. At that time the US Secretary of State, George Marshall, had proposed the Marshall Plan, which would fund the rebuilding of Europe and Britain. As mentioned earlier the Americans were, however, determined to exercise a monopoly on nuclear technology in general. Now they decided to force Britain to surrender its share of the uranium ore and forget about the Quebec and Hyde Park Agreements.

Refusal to do so would result in withholding of further financial aid. This demand was made in negotiations in December 1947, during which it was also made clear to the British that there would be no further sharing of technological information (although this was in any case forbidden by the McMahon Act) unless they complied.

Horrified and angered not only by the United States' refusal to continue sharing nuclear technology but also by the outrageous use of such blackmail, the British had no choice but to agree to the Americans' demands. These negotiations did, however, lead to a *modus vivendi* a month later which continued the arrangements for joint procurement of uranium supplies and for co-operation in nine areas of nuclear research. Despite this, however, Clement Attlee was even more determined that Britain must have its own atomic bomb.

One of Attlee's other major concerns was that the press would discover the existence of the programme and publicize it. Such was the degree of security that the government was unable to issue a D-Notice, the mechanism by which the voluntary co-operation of the press is obtained over matters of national security. In May 1948 it was decided instead to issue a guarded statement about Britain producing an atomic bomb, in return for which it would expect the press to comply with a D-Notice that would block any further discussion of the subject. In the House of Commons, in reply to a question concerning development weapons put forward by a primed backbench member of parliament, a minister issued a statement declaring that research and development was of a high priority and that development of all types of weapons, including atomic ones, was taking place. At the precise moment that the statement was being made, editors throughout Fleet Street were being issued with copies of a D-Notice stating that they could not engage in any investigation, reporting or discussion of any sort relating to the development of British atomic weapons.

As explained earlier, in an atomic bomb there are two ways of achieving an atomic explosion: the gun and implosion methods. In a Hiroshima-type bomb, two sub-critical pieces of uranium-235 in a gun-like assembly were shot together so that they suddenly formed a supercritical mass and exploded. This would not work in a bomb using plutonium-239, which is more reactive and liable to predetonate, resulting in an ineffective partial explosion. Thus the implosion method was used for the Nagasaki-type plutonium bomb; a sphere of plutonium was enclosed in a much larger sphere made up of carefully shaped charges of conventional high explosive which, when detonated, compressed the plutonium-239 so that it became supercritical and an atomic explosion occurred. The implosion method was more efficient but extremely difficult to design.

In the United States during the war, one of the British scientists who helped to solve many of the problems in achieving success with implosion techniques was Dr Klaus Fuchs. As mentioned earlier, he was a German émigré, who had studied physics and mathematics at the universities of Leipzig and Kiel. Forced to flee when the Nazis came to power in 1933, he made his way to Britain where he studied at the University of Edinburgh from which he received his doctorate. In 1940, however, along with thousands of others classed as 'aliens', he was interned and subsequently transported to Canada. By the end of the year, however, he had been released and returned to Britain. In the spring of

1941 he was invited by Professor Rudolf Peierls to join him at Birmingham University. Naturalized British in 1942, Fuchs was employed in the Tube Alloys programme until late 1943, when he was recruited for the Manhattan Project and sent to the United States where he contributed a great deal to the theory and design of the plutonium bomb. On returning to Britain in the summer of 1946, he became head of the theoretical physics department at Harwell.

At this time, it was suspected that the Soviet Union was in the process of developing an atomic bomb and Britain's Secret Intelligence Service (SIS) was devoting considerable resources to gleaning sufficient intelligence to be able to estimate a production timescale. Despite the somewhat shabby treatment accorded to Britain by the United States over nuclear matters, the two countries were still co-operating over matters of intelligence. In 1948, the SIS supplied the Americans with reports based on deciphered Soviet intercepts relating to atomic weapon experiments. The United States meanwhile mounted an operation, codenamed Project Snifden, using specially equipped US Air Force WB-29 aircraft to conduct airborne monitoring around the periphery of Russian airspace in the Soviet Far East while ostensibly carrying out meteorological survey flights.

On 29 August 1949 the first Soviet atom bomb test took place and it was not long before a WB-29, on a flight in the area of Kamchatka, detected a cloud of radioactivity. A few days later, an RAF Halifax aircraft collected samples of radioactive fall-out over the Atlantic.

The evidence was irrefutable but many in America, including some scientists and General Leslie Groves, generally held the Soviet Union in contempt and were certain that the Soviets would not be able to develop a bomb for many years. The only exception was one non-nuclear scientist, a Nobel laureate named Irving Langmuir, who had a high regard for Soviet scientists and had predicted that the Soviets would be in possession of a bomb within five years of the end of the war. President Harry S. Truman, who was not a technically educated man, refused to believe that the Russians could have tested a bomb and persisted in holding discussions with his staff during which he put forward the theory that the explosion could have been caused by a reactor blowing up because of faulty design.

Even when presented with definite intelligence confirming that it was a bomb and that it had been of a similar design to the implosion type (nicknamed 'Fat Man') dropped on Nagasaki, he insisted that the Atomic Energy Commission officials presenting the information to him sign the document, confirming that they believed the information to be true. On 23 September 1949, Truman announced to the world that the Soviet Union had exploded an atomic bomb.

In view of previous predictions by both the SIS and the newly formed Central Intelligence Agency (CIA) that the Soviets would not possess an atomic weapon before 1953, the revelation of the test was not only a shock to both organizations but also a possible indication of treachery on the part of persons unknown. In the United States the Federal Bureau of Investigation (FBI) swiftly mounted a major operation to determine whether Russia's unexpected triumph was in any way connected to Alan Nunn May, a British scientist who had worked on the Tube Alloys programme in Britain before being transferred to Chalk River in Canada.

In September 1945 a defector from the Soviet embassy in Ottawa, a cipher clerk named Igor Gouzenko, had revealed that Nunn May had passed information to the Soviet Union while working at Chalk River. Nunn May had been recruited by Colonel Nikolai Zabotin, an officer in the Soviet military intelligence service, the Main Intelligence Directorate of the General Staff, better known by its acronym GRU. Zabotin, who operated under the diplomatic cover of military attaché, had contacted him not long after his arrival at Chalk River and thereafter contact had been maintained through another GRU officer, Pavel Angelov. During the following two and a half years, Nunn May paid a number of visits to other establishments involved in atomic research, including the Argonne Laboratory in Chicago. Indeed, security officials there became suspicious of the frequency of his visits and eventually refused his further requests for access.

By the beginning of 1946 Nunn May had returned to Britain, where he took up a post at London University. By then he was attempting to sever contact with the Soviets but in March he was arrested and confessed everything. On 1 May 1946 he was sentenced to ten years' imprisonment. It later transpired that he had been an avowed Communist while at Cambridge University but, inexplicably, had not been vetted before being recruited for the Tube Alloys project.

While the FBI sought to confirm whether there had been any traitors among those who had worked on the Manhattan Project, American cryptographers of the US Armed Forces Security Agency (AFSA), later the National Security Agency (NSA), were decrypting wartime Soviet cable transmissions between the United States and Moscow. This was part of a major codebreaking operation called 'Venona' which began in 1946 and by mid-1947 revealed evidence of a massive Soviet espionage operation in the

United States during the war. Shortly after President Truman's announcement, the AFSA decrypted two intercepts: one of these was a summary of a Manhattan Project document while the other identified the author and source as the British scientist Dr Klaus Fuchs.

Investigation of Fuchs's background by the Security Service, popularly known as MI5, revealed that he was a member of a family well known in Germany as active Communists. Having joined the German Communist Party in 1930, he had fled Nazi Germany for Britain three years later on the party's orders, his left-wing sympathies being recorded on his arrival and passed to the Security Service. Further evidence of his membership of the Communist Party came a year later when the Security Service received a letter from the Chief Constable of the police in Bristol; he had been informed by the German consul in the city that Fuchs was living with a family of suspected Communists and that he was a Soviet agent. Bearing in mind the biased source of this report, this information had been ignored.

A second report on Fuchs had been received from a Security Service informer within the German refugee community who had revealed Fuchs's Communist connections and his activities in Germany before 1933. The Security Service, heavily involved in investigating subversive organizations at the time, had not checked further on Fuchs, but the officer in charge of its Communist Section had written a report stating that he was more likely to betray secrets to the Russians than to the Germans and had advised that he should not be allowed any more access to secrets than was strictly necessary.

Fuchs was thus accorded a low security rating but unfortunately the Security Service was unaware of the exact nature of his work at the time and did not communicate its recommendations to his employers. Moreover, for unexplained reasons, it did not do so when he was transferred to the United States in 1943 to work on the Manhattan Project; when completing an American security questionnaire about him, it merely stated that he was 'politically inactive and unobjectionable'.

Fuchs returned to Britain and his work at Harwell. Before this, the Security Service had carried out a five-month-long investigation into him but found nothing. Had it checked with the Canadian security authorities, it would have discovered that one of his associates while interned in Canada was a well-known German Communist named Hans Kahle, who was a suspected Russian agent and who in 1945 was spotted in East Germany working for the Soviets.

Following the decryption of the Venona intercept, Fuchs's telephone was tapped and his mail intercepted. He was questioned but revealed nothing until 24 January 1950, when one of his interrogators suggested that he might be permitted to remain at Harwell if he admitted to any wrongdoing. This ploy worked and he confessed, revealing that he had started supplying information to the Soviets from late 1941, while working on the Tube Alloys project under Rudolf Peierls at Birmingham University. He had contacted the leader of the underground German Communist Party in Britain, Jürgen Kuczynski, for assistance in passing on information to the Soviets, and Kuczynski had put him in contact with Simon Kremer, a GRU officer based at the Soviet residency in London. Fuchs was subsequently handed on to another GRU officer named Ursula Beurton, the sister of Jürgen Kuczynski. Operating under the codename of 'Sonya', and living in the Oxfordshire town of Chipping

Norton, her cover was that of a Jewish refugee from Germany named Mrs Brewer. Beurton was to remain undetected in Britain until 1947, when she was questioned by the Security Service after being named by a Soviet intelligence officer, Alexander Foote, who had defected to Britain earlier that year. She succeeded, however, in convincing her interrogator, William 'Jim' Skardon, of her innocence and thus was not placed under surveillance. Two days later she disappeared and surfaced some years later in East Germany, having been made an honorary colonel in the Red Army in recognition of her achievements on behalf of the GRU.

After Fuchs's arrival in the United States in late 1943, control of him as an agent had been transferred from the GRU to the Soviet intelligence service, the KGB. His controller was Anatoly Yakovlev, a KGB officer acting under the diplomatic cover of a Soviet vice-consul in New York, with whom he communicated via a Swiss-born courier named Harry Gold whom he first met on 5 February 1944. Based in New York until the latter part of 1944, Fuchs passed a considerable amount of information to the Soviets before being posted to Los Alamos in New Mexico. It was not until February 1945, when he had returned to New York on holiday, that he had been able to re-establish contact with Gold.

Following his return to Britain in 1946 and taking up his new post at Harwell, Fuchs had continued to work for the Soviets, passing information to his KGB controller in London. It appears, however, that by the time of the period leading up to his arrest his enthusiasm for such work had waned, possibly due to the fact that he had by then developed an increasing respect and affection for his adopted country.

Immediately after confessing, Fuchs was arrested and on 2 February 1950 charged with espionage. His trial took place on 10 February, lasting exactly one and a half hours; he was convicted and sentenced on the same day to fourteen years' imprisonment. At the time, the political ramifications of his treachery were far-reaching and caused huge embarrassment to Britain. It appeared that a major lapse in British security had not only compromised the Americans but also enabled the Soviet Union to develop its own atomic bomb and thus achieve apparent parity within a very short time.

In fact, Fuchs was not the only Soviet agent working inside the Anglo–American nuclear weapon programmes. At least three other members of the Tube Alloys project in Britain, one identified only by the letter 'K' and the other two by their Soviet codenames of 'Moor' and 'Kelly', had reportedly passed information to the Soviets. Meanwhile, as already mentioned, the Soviets also succeeded in penetrating the Anglo–Canadian nuclear research project at Chalk River in Canada through their recruitment of Alan Nunn May and others. In the United States an American scientist working on the Manhattan Project, identified only by the codename 'Mar', was recruited by the Soviets in April 1943 and by December of that year had passed on details of the construction of reactors, cooling systems, production of plutonium and protection against radiation. Others were also recruited: one was a scientist in the radiation laboratory at Berkeley, in California, while another worked in the Manhattan Project's metallurgical laboratory at Chicago University. A third was a construction engineer who passed on information about plant and equipment used in the project.

In addition to Fuchs, the Soviets had two other agents within Los Alamos, both recruited in November 1944. The first was a machinist, an Army sergeant named David

Greenglass who was recruited via his wife, the sister of a woman named Ethel Rosenberg. The latter, along with her husband Julius, an engineer in the US Army's Signal Corps, headed a ring of Soviet spies in the United States before both were arrested in May 1950, together with Greenglass and another conspirator named Morton Sobell, shortly after Harry Gold had been apprehended. Brought to trial in March 1951, both were subsequently convicted of espionage and sentenced to death. They were executed at Sing Sing Prison on 19 June 1953. Greenglass, who appeared as the chief prosecution witness, was sentenced to fifteen years' imprisonment while Harry Gold and Morton Sobell were each sentenced to thirty years'.

The second Soviet agent in Los Alamos was a brilliant nineteen-year-old physicist named Theodore 'Ted' Hall who, it is believed, was the first to reveal the secrets of the implosion method to the Soviets. Both he and Fuchs are thought to have independently provided Moscow with the plans of the American bomb and to have given the date of the first test at Alamogordo which was scheduled for 10 July 1945 (although in the event it was delayed for six days by bad weather). Hall subsequently supplied the Soviets with the results of the test and such was the mass of detail in the information supplied by him, Fuchs and the other Soviet spies within the Manhattan Project that the first atomic weapon exploded by the Soviet Union was an exact replica of the 'Fat Man' bomb produced at Los Alamos.

Meanwhile, Christopher Hinton and his team at Windscale were experiencing further problems with the reactors, caused by faulty information supplied by Harwell on graphite, neutron flux and filtration. The core of each reactor comprised a precision-machined block of extremely pure graphite pierced with holes in which rods of uranium were placed. When enough rods were placed close together, a spontaneous nuclear combustion began. The fission chain reaction would in time consume some of the uranium-235 atoms contained in the fuel and convert some of the uranium-238 atoms to plutonium. The graphite block, while providing the structure containing the uranium rods, also acted as a moderator, slowing down the neutrons and thus increasing the probability of their hitting an atomic nucleus. However, successful operation all depended on a sufficient quantity of neutrons being released in the fission process.

Tom Tuohy was manager in charge of the reactors, or 'piles' as they were called, and it was his task to carry out all the necessary measurements leading to criticality, as he later recalled: 'When uranium was loaded and a measurement was necessary, it didn't matter what time of the day or night – I was in there with my second-in-command, taking the necessary measurements to go up to criticality. If the reactor was going to produce enough plutonium for a bomb, the chain reaction would have to start when a very specific amount of uranium fuel had been loaded.' The amount that had been forecast was approximately 42 tons; in the event, it proved to be 102 tons.

Until then, Penney had assumed that Windscale would not only produce the plutonium but also carry out the difficult process of forming it into the sphere which would form the core of the bomb. With the plant experiencing such problems, however, he realized that he would have to set up another facility to manufacture the core. This was built at the new weapons site at Aldermaston, in Berkshire, to which High Explosive Research (the early code name for the atomic weapon development programme) had begun to move from Fort Halstead in April 1950. Penney and his team were well aware of the hazards of plutonium and developed safe methods for handling it. Reputedly the most dangerous of metals, it is silvery grey in colour and oxidizes quickly on exposure to air, producing a very fine-particled oxide. If this is inhaled and lodges in the lungs, is ingested and lodges in other parts of the body or gets into the bloodstream through an open wound, it can cause serious problems leading to cancers.

A specially designed building housed the area in which the material would be handled, with special ventilation systems and corridors designed to ensure that no leakage could occur from it. Personnel working in it wore pressurized suits with integral breathing systems; initially manufactured from heavy black neoprene rubber, these were known as 'frog suits' as they resembled the diving suits worn by frogmen and indeed were developed from such.

By now it was 1951, and the general election of that year saw the Conservative Party returned to power with Winston Churchill once again taking up office as Prime Minister. Hitherto he had been unaware of the existence of the post-war nuclear weapon development programme which had so far cost 100 million pounds; such was his enthusiasm when he learned of it that he endorsed October 1952 as the date for the testing of Britain's first atomic bomb. During the following months, diplomatic efforts were put in hand to pave the way for the tests to take place in the area of the Montebello Islands off the north-west coast of Australia.

By the early summer of 1951, construction of the two reactors at Windscale had been completed and the loading of the fuel elements took place. Just as the reactors were about to be started up, however, devastating news arrived from Harwell. The theoreticians had suddenly realized their original calculations were incorrect and that the piles would in practice operate much less efficiently than planned. The reactors would simply not make plutonium quickly enough to meet the deadlines. A major redesign would involve years of delay and was out of the question. A radical solution was needed.

The Harwell calculations had missed one crucial factor: many of the neutrons that should have been colliding with uranium nuclei and converting them to plutonium would instead simply be absorbed by the aluminium casings surrounding the uranium fuel rods. The casings each had fourteen cooling fins. The engineers calculated that if these fins could be trimmed very slightly, the reduction in aluminium would bring the reactor back to up to full power. But the reactor contained 36,000 uranium fuel assemblies, a total of 504,000 fins. Undaunted by the magnitude of the task, the engineers built special rigs and worked round the clock to clip 1.6 millimetres (a sixteenth of an inch) off every single fin. At the same time, small graphite 'boats' were produced at Windscale's workshops to narrow the fuel channels in the reactor and increase the reactivity in the pile. Within three weeks, the entire process had been completed and the fuel reloaded into the reactors. The measure worked: the first reactor went critical in October 1951 and the second in June the following year.

The process of producing plutonium starts when sufficient uranium is inserted into a reactor to start a fission reaction. A chain reaction begins when a uranium atom splits and ejects a number of neutrons which then either split other atoms or are absorbed by them. The process is regulated by the use of control rods, made of neutron-absorbing material, which can be motored into the pile core gradually or slowly withdrawn. By controlling the process the number of fissions can be maintained at a steady level. The fuel elements have to be left in the reactor for a sufficient length of time to produce the desired plutonium-239 but not for too long or the result will be plutonium-240 which is not an efficient explosive.

With the problem of production solved, Christopher Hinton and his team started to manufacture plutonium. Although there were still fifteen months to the test date, much remained to be done. Firstly, the reactor had to be run long enough for plutonium to be produced in the fuel elements, which then had to be extracted from the reactor, dissolved and put through the separation plant to separate out the small amounts of plutonium from the large quantities of leftover uranium and other by-products. Thereafter, the plutonium would have to be purified, turned into metal and machined into hemispheres to form the bomb's core.

By August 1952, despite the continual setbacks that had plagued it, Windscale had produced sufficient plutonium to meet the deadline set for the test, which was codenamed 'Hurricane'. Slugs of plutonium, each in its own sealed cylindrical container, were packed into a drum which was sealed and loaded aboard an Army truck on which it was despatched to Aldermaston. On its arrival, the drum was checked for gamma radiation before being unloaded and wheeled on a trolley into Building A1, where it was taken into one of the newly constructed laboratories.

Building A1 was designed for the manufacture of plutonium cores and consisted of three elements. The operating corridor comprised a glove box, a stainless-steel structure equipped with a 12-millimetre-thick (1/2-inch) Perspex window, allowing engineers and other operatives to see what they were doing, and sealed apertures equipped with arm-length rubber latex gloves to enable them to carry out their work from the outside while fully protected. Fully airtight, it was filled with inert argon gas to prevent oxidation of the plutonium. Behind it was a large stainless-steel room, measuring some 6 metres (20 feet) wide by some 18.5 metres (60 feet) long, from which the rear of the glove box could be accessed by the engineers and from where modifications to equipment could be carried out or new equipment installed. This room was subject to contamination and thus only personnel wearing 'frog suits' could enter it. Above these two areas was a floor containing the banks of high-efficiency filters. As the filters would need changing periodically, this floor was also considered to be a contaminated area where 'frog suits' needed to be worn. In addition to these three areas, there were adjacent laboratories providing support for the main laboratory. Nearby was the complex's other building which housed administrative offices and other laboratory facilities.

Up until now, the scientists at Aldermaston had only been working on experiments with small pieces of plutonium. With the arrival of the material from Windscale, they were able to start producing the hemispheres that would comprise the core of the bomb. The pieces of plutonium were placed inside the glove box and weighed before being passed to the next stage where they were cut up and placed in a crucible for melting prior to casting. At this point a blue halo-like flame developed over one of the quantities of molten plutonium, and it was feared that a critical reaction was developing which would have resulted in death or injury to those working in the immediate vicinity. Fortunately, however, the flame (which was later thought to have perhaps been caused by some impurity in the argon gas in the glove box) died away shortly afterwards and the process continued with the plutonium being poured into casts. Once they had cooled, the plutonium hemispheres were broken out of their moulds and on the following day placed in dies and pressed into shape, a process that removed any fissures and cavities on the surface of the metal. Any excess material was then trimmed off with a small lathe before each hemisphere was clad in gold foil, which was then cold-welded. All these processes took place at different stations within the glove box.

A problem occurred, however, when two frog-suited engineers attempted to bring out the completed hemispheres from the contaminated stainless-steel room. Entrance and exit was via a venturi (a short tube inserted into a wider pipeline), which was used to maintain airflows, and then through two heavy steel doors that sealed off the chamber. On this occasion, however, both doors jammed and the two men, together with the two hemispheres in containers, were trapped inside. One door was eventually opened but the other refused to move. Eventually, some two hours later, the water seal pit at the bottom of the door was drained, allowing a gap of about 45 centimetres (18 inches). Through this the plutonium hemispheres could be passed out in their containers, and the two engineers could escape. Afterwards, an escape hatch was installed to cater for any similar mishap in the future.

By now time was running short and the decision was taken to despatch two plutonium cores, comprising four hemispheres, by air to the test site in the Montebello Islands, off the coast of north-west Australia, a journey of some 16,000 kilometres (10,000 miles). The task of escorting them was given to Wing Commander John Rowlands, the head of the RAF team. Packed in four sealed steel containers, one hemisphere in each, the plutonium cores and a neutron initiator were driven in two unmarked furniture vans from Aldermaston to the RAF base at Lyneham in Wiltshire, accompanied by Rowlands, another RAF officer and a scientist, Bill Moyce, travelling in a staff car. On arrival, the entire consignment was loaded aboard a Hastings transport aircraft and, together with the three men, set off on its journey. A number of contingencies had been taken into consideration, including the possibility of the aircraft crashing; in that event Rowlands, his accompanying officer, Moyce, and the aircrew would have been required to bale out with the four containers, which had been designed to float. Meanwhile, to distract any unwelcome publicity, a decoy aircraft carrying William Penney and a set of dummy containers had taken off at the same time as Rowlands's aircraft and flew to Australia.

As it happened, the journey proved uneventful as the aircraft flew via Cyprus, Sharjah in the Persian Gulf and Ceylon (now Sri Lanka) to Singapore. There the consignment, the two officers and Moyce were transferred to a Sunderland flying boat and flown to the Montebello Islands. On 18 September the aircraft landed on the Montebello lagoon, taxiing to a position alongside HMS *Plym*, the River Class frigate which contained the bomb and inside which it would be exploded.

Three months beforehand, while Aldermaston had been producing the bomb's core, a Royal Navy task force had assembled at the naval base at Chatham, in Kent. This comprised an aircraft carrier, HMS *Campania*, which would act as flagship; two landing ships, HMS *Zeebrugge* and HMS *Narvik*, carrying a detachment of Royal Engineers and their heavy plant and equipment; a third landing ship, HMS *Tracker*, which would be the health control ship; and the River Class frigate HMS *Plym*, complete with a specially designed and extensively equipped weapon room in which the unarmed bomb would be transported and ultimately exploded. All the vessels were specially converted and fitted out in complete secrecy at shipyards at Birkenhead and Chatham and on the Clyde.

The flotilla would have to provide complete support for all personnel involved in the tests. They would be based for three months on the islands, which were uninhabited and had no resources whatsoever, not even fresh water. In addition to food and accommodation, transport would have to be provided to ferry scientists to different locations on the islands, where a vast range of equipment would be sited to record the effects of the explosion.

Meanwhile, much preparatory work had been taking place on the islands with generous help from Australia. The Australian Army provided support with civil engineering and construction tasks, assisted by Royal Engineers who sailed from Britain in February 1952. From October 1951 onwards, the Royal Australian Navy (RAN) and the Australian Weather Bureau had begun meteorological observations. The RAN also surveyed and charted the islands and their surrounding waters, and sited buoys and moorings. In addition it would assist in ferrying passengers and supplies, provide a weather vessel and patrol the area before and after the test. Moreover, three Australian scientists would take part in the test, one of them being Professor E. W. Titterton, Professor of Physics at the Australian National University, who would participate as an expert in telemetry. Ultimately, it was agreed that Professor L. H. Martin, the Defence Scientific Adviser, and W. A. S. Butement, Chief Scientist at the Australian Department of Supply, would join him as senior observers.

During the weeks prior to the flotilla's departure, all the equipment required for the test, including the bomb itself minus its plutonium core and neutron initiator, was loaded aboard the vessels along with a large contingent of scientists, physicists, engineers, mathematicians, doctors, chemists and botanists, each of whom had a positive role to play in the test. In June 1952, led by HMS *Campania*, the task force set sail for the Montebello Islands.

The voyage took eight weeks, during which much work was carried out checking, testing, repacking and stowing equipment ready for use after arrival. Conditions on board the ships were overcrowded and uncomfortable and some of the scientists suffered badly from seasickness. On 8 August HMS *Campania* and the rest of the flotilla arrived and anchored in an area east of the Montebello lagoon; thereafter preparations for the test proceeded apace. On the same day, the Australian government declared the Montebello Islands and a surrounding area of 64 kilometres (40 miles) radius a prohibited area for safety and security reasons.

It was predicted that the proposed underwater detonation, if it achieved an explosive power similar to that of the Nagasaki bomb, would result in a fireball rising high into the atmosphere and taking with it a vast column of water and sand, weighing 10,000 to 100,000 tons, as well as fission products, bomb constituents and many tons of steel from the ship, some of which would be vaporized. Due to the huge weight of water drawn up into the column, fall-out would be produced quickly and within a few thousand yards from the site of the explosion. Most of the radioactivity would be dispersed in the sea around the islands, and would be of high levels only in the area of the explosion and for a short duration afterwards.

Meteorological conditions, particularly wind direction and strength, would be crucial to the safety aspects of the test, particularly with regard to the Australian mainland which lay 75 kilometres (47 miles) to the south-east of the islands. Wind conditions on the west coast of Australia had been studied for two years, and it had been discovered that of ten days in October when other conditions, such as tides, were suitable, there would probably be no more than three when the wind would be blowing away from the mainland. Meteorological forecasts would be provided by the RAN weather vessel and by Royal Australian Air Force (RAAF) bases at Pearce Onslow and Port Hedland.

The safety criteria for protecting the mainland, the flotilla and the site of the main island base specified acceptable wind directions and speeds at surface level and altitudes of 1,525 metres (5,000 feet) and 9,145 metres (30,000 feet). They also dictated that no air up to 1,525 metres (5,000 feet) must reach the mainland within a period of ten hours after the explosion, this allowing for heavy particles of fall-out borne by winds of low strengths to fall into the sea. Following the test, airborne monitoring of atmospheric radioactivity would be carried out at high and low altitudes by RAAF aircraft fitted with detection equipment.

All was in place for initial test rehearsals on 12 and 13 September, a full rehearsal taking place six days later on 18 September, the day of the arrival of Wing Commander John Rowlands's party and the two plutonium cores. On 22 September, William Penney also arrived and final preparations for the test intensified.

Penney and his scientists had designed the bomb so that the initiator and core could be kept separate from it until the last possible moment. It had been decided to position the bomb on a ship, below the waterline, as part of an additional test to determine the effect of a nuclear bomb, planted by a hostile force, exploding aboard a vessel in a major port such as Liverpool or the London. On 2 October, the day before the test, the ship's complement were disembarked, less a small skeleton crew comprising an officer and a small group of sailors who would take the target vessel to the test location. Also remaining on board were Wing Commander John Rowlands and a team of scientists, led by John Challens, who were responsible for carrying out the final procedures for arming the bomb. They were accompanied by Eddie Howse, whose overall responsibility within the project had been the design of the firing circuits but who on this occasion was responsible for liaising between the firing control team on one of the Montebello Islands and those aboard the warship.

By the day before the test, HMS *Plym* was stationed at her predetermined location. It was Wing Commander Rowlands's responsibility to assemble the plutonium core, which he did by using a portable vacuum glove box to extract the two hemispheres and the marble-sized polonium-beryllium neutron initiator, known as an 'urchin', from their respective containers. He had previously conducted a number of criticality experiments at Aldermaston, using a special machine which, surrounded by monitoring equipment, had brought two hemispheres together approximately 2.5 micrometres (a thousandth of an inch) at a time until they were touching. In the event of criticality beginning, a release mechanism would have dropped one hemisphere immediately away from the other.

Having inserted the initiator into a recess in one of the hemispheres, Rowlands carefully assembled the entire plutonium sphere and screwed it into the 'gauntlet', a device for holding the sphere so that it was located precisely in the centre of the bomb. The final component was a high-explosive cartridge which was screwed on to the gauntlet before the entire assembly was slowly lowered, using a small hoist, into the bomb and secured in position. By this time it was just after midnight in the early hours of 3 October; firing of the bomb was only a few hours away.

Wing Commander John Rowlands then disembarked and was transferred to HMS *Campania* as the team of scientists under John Challens continued with arming the bomb. This was not a particularly easy task as the HMS *Plym*'s hold was very limited in size, while the bomb measured 1.5 metres (5 feet) in diameter. They first had to insert all the thirty-two detonator units, each the size of a small coconut and containing two 'exploding wire' detonators. The weapon's initiation system, comprising two firing circuits, had a large degree of redundancy built into it so that if there was a detonator failure on one circuit, the detonator on the other circuit would take its place. Once the thirty-two detonator units were all in place, a total of sixty-four firing cables had to be connected up. Once that lengthy and tedious process was complete, all but John Challens, Eddie

Howse and three members of the ship's crew disembarked from the vessel.

The final task was to connect the two firing circuits to the bomb's power units, which were in turn connected to banks of batteries. As Challens connected the power units to the batteries, Howse was in communication with the control centre, which was monitoring power voltage readings transmitted from the ship to the control centre by telemetry. Acting as a link, he advised Challens on each reading until all were correct. Once all connections had been made, three switches were pressed and Challen inserted a master plug which completed all the circuits in the bomb ready for firing. At approximately 4 a.m., Challens, Howse and the three remaining crew then disembarked from HMS *Plym* in pitch darkness to a launch that took them to one of the islands 10 kilometres (6 miles) away, where an underground control room and bunkers had been constructed by British Army sappers.

Challens carried with him a second master safety plug which, on his arrival ashore, he delivered to the control room. This was inserted into the firing system and all was ready for the test to take place. Meanwhile, all personnel changed into special clothing and took their places in the control room and bunkers. At 9.15 a.m. local time, the firing switch was pressed and three seconds later a massive column of water rose swiftly and silently into the sky, the top of it forming into a huge cloud. Almost half a minute passed before the noise of the explosion, which had the force of 20 kilotons, reached the ears of those who emerged from the bunkers, watching the cloud as a tremor shook the ground around them. Shortly afterwards, it was announced that the test had been a total success.

Despite all the difficulties, Britain had succeeded in developing and exploding its own atomic bomb. Three weeks later, however, the United States exploded the first thermonuclear weapon, a hydrogen bomb device that was a thousand times more powerful than the British weapon. The Americans' decision to proceed with its development without delay stemmed from their belief that Klaus Fuchs had also provided the Soviets with information concerning thermonuclear technology. By 1957 Britain, fully committed to being a nuclear power, had also tested its own hydrogen bomb. This was undoubtedly the principal factor that led to the United States deciding that the wartime 'special relationship' was worth reviving after all.

DISMANTLING THE BOMB

For almost fifty years the world lived in the shadow of the threat of nuclear conflict. With the end of the Cold War, it appeared that the threat of mutual destruction might have been avoided. In the wake of arms reduction, however, a major problem arose: how to dispose of many thousands of nuclear weapons.

The nuclear arms race had its early beginnings during the Second World War when, as described in 'A Very British Bomb', first the United States and then the Soviet Union began their respective atomic weapons development programmes. On 2 August 1939 Albert Einstein, who six years earlier had fled Nazi Germany to continue his research, wrote to President Franklin D. Roosevelt, informing him of a major development in his research which appeared to confirm the theoretical predictions of his famous Relativity Theory. Among its many radical ideas, it predicted that matter could be converted into energy and thus made into a bomb. Einstein's formula stated that: '$E = mc^2$ in which energy is put equal to mass multiplied by the square of the velocity of light showed that very small amounts of mass may be converted into a very large amount of energy.' Following the invasion of Poland during the following month, Einstein and another physicist, Leo Szilard, warned the US government of the peril that would threaten the world if Nazi Germany were allowed to win the race to produce an atomic bomb.

In 1941 an American committee reviewing atomic research carried out to date reported that it would take approximately four years to produce a nuclear explosive and that it would be eighteen months before a chain reaction in natural uranium was achieved. Furthermore, the committee estimated that it would take at least three to five years to produce sufficient weapons-grade uranium.

Five months after the United States' entry into the Second World War in December 1941, the decision was taken to proceed as quickly as possible with the Manhattan Project: the production of a nuclear weapon. In August 1942, the US Army was tasked with overall responsibility for the Manhattan Project and, in October, physicist Robert Oppenheimer was appointed as director of Project Y, the team of scientists who would design the actual bomb itself. During the 1920s, Oppenheimer had carried out research at the Cavendish Laboratory at Cambridge University, which was renowned for its pioneering work on atomic structure.

In 1943 the Manhattan Project's weapons division was established at Los Alamos, the site selected in November of the previous year by Robert Oppenheimer who had spent part of his childhood there. In the laboratory, Oppenheimer and his team carried out a programme of experiments with uranium and another radioactive material, plutonium.

Uranium is an abundant material which forms approximately two parts per million of the Earth's crust. Discovered in 1789 by Martin Heinrich Klaproth, who named it after the planet Uranus, it is the heaviest-known natural metal with a density of 19.04 gm/ml (that of lead is

11.34 gm/ml). It is a hard, dense element, silvery-white in colour, malleable and capable of being polished to a high degree although it tarnishes quickly on exposure to air.

Uranium ore is mined and then leached, subsequently being recovered by solvent extraction and roasting. This produces a crude concentrated material called 'yellow cake', which is dissolved in hot nitric acid, subsequently undergoing purification and calcination to form uranium trioxide. A further process of hydrogen reduction produces uranium dioxide, which can then be reduced further with calcium to produce metallic uranium in a powdered form. In order to produce metallic uranium, uranium dioxide is hydrofluorinated into UF_4 which is then mixed with magnesium metal filings and compressed before being baked at 900°C. A reaction between the UF_4 and the magnesium takes place, after which the molten uranium metal is poured into moulds.

Several types of uranium can undergo fission but uranium-235 does so more readily and emits a greater number of neutrons per fission than other such isotopes. A small amount of uranium in any assembly, known as a sub-critical mass, cannot undergo a chain reaction as neutrons released by fission are liable to leave without striking another nucleus and causing it to fission. If a further quantity is introduced, it increases the chance of released neutrons causing another fission as the latter are forced to traverse more uranium nuclei. There is a thus greater likelihood that a neutron will hit another nucleus and split it.

As we have seen in 'A Very British Bomb', plutonium is a synthetic element produced by irradiating non-fissile uranium-238 with neutrons. Fissionable, weapons-grade plutonium is plutonium-239, which has a critical mass only one-third of that of uranium-235. The element was first produced in mid-1941 at the University of California at Berkeley. It was already known, however, that the isotype plutonium-239 would be highly fissile.

As explained in 'A Very British Bomb', there are two methods of achieving an atomic explosion: the gun and implosion methods. By April 1945, it was recognized that sufficient uranium-235 would not be forthcoming for a test of a gun-assembly bomb until the beginning of August. However, there would be enough plutonium-239 for testing of an implosion-assembly device in early July and another in August.

The first plutonium test bomb, weighing 2 tons, was assembled at a solitary ranch house in New Mexico and on 14 July 1945 was hoisted to the top of a 30-metre (100-foot) high US Forest Service watchtower at a location known as the Trinity Site, 192 kilometres (120 miles) south of Albuquerque. In the early hours of 16 July a final warning flare was fired and at 5.29 a.m. the last firing circuit was connected and the bomb detonated. The energy yield of the explosion was equivalent to 21,000 tons of TNT. Such was the devastating sight and effect that Robert Oppenheimer was moved to quote the sacred Hindu text, the Bhagavad-Gita: 'I am become death, the shatterer of worlds.'

Three weeks later, at 8.15 a.m. on Monday 6 August, a single US Air Force bomber, a B-29 called the *Enola Gay*, flew over the Japanese city of Hiroshima and dropped a bomb, nicknamed 'Little Boy', of the untested gun-assembly uranium type. It exploded at an altitude of 580 metres (1,900 feet) over the city with the force of some 15 kilotons (15,000 tons of TNT), instantly causing appalling destruction to the surrounding area of 10 square kilometres (4 square miles).

About 66,000 people died instantly in the blast and a further 69,000 were injured. By the end of that year, the total loss of life in Hiroshima had risen to 140,000.

Three days later another B-29 bomber, *Bock's Car*, flew over the city of Kokura but bad visibility prevented the crew establishing the aim point. Flying on to Nagasaki, the aircraft dropped a plutonium bomb, nicknamed 'Fat Man', which exploded at 11.20 a.m. local time at a height of 500 metres (1,650 feet) with a force subsequently estimated to be equivalent to 21,000 tons of TNT. About 39,000 people were killed and 25,000 injured by the explosion, while almost half the city was destroyed. On 14 August, Japan surrendered.

Despite the end of the Second World War, the development of nuclear weapons continued with US scientists improving their efficiency and deadliness. The Americans decided, however, that they would keep their technology to themselves and, as recounted in the previous chapter, passed an Act of Congress outlawing any further disclosure of nuclear technology, even to the British who had been closely involved especially in the development of 'Fat Man'.

As already mentioned, the principal elements involved in the manufacture of nuclear weapons are plutonium-239 and uranium-235. The latter is, however, very costly and difficult to produce and for that reason plutonium-239 soon became the preferred element in the manufacture of nuclear weapons.

The heart of a nuclear bomb is known as a 'pit' (the American word for the stone in a peach or cherry) and comprises a sphere of fissile material (plutonium-239 or uranium-235) encased in a shell manufactured in a non-nuclear material such as stainless steel. During the functioning of a boosted nuclear weapon, a mixture of tritium and deuterium gas is injected into the sphere, which is then very tightly compressed by means of a high explosive, resulting in critical mass, and then energy being released by fission reactions.

Within a micro-second of detonation, the chain reaction causes the pressure inside the pit to reach millions of pounds per square centimetre and the temperature tens of millions of degrees. As the atoms split, energy is released in the form of X-rays that leave the bomb at the speed of light. The flash sets buildings, trees and people on fire before the explosion has even been heard. Air surrounding the bomb absorbs the X-rays and heats up, becoming visible. The huge increase in pressure creates a shockwave, which punches outwards from the centre of the explosion. Close behind are winds of over 480 kilometres per hour (300 miles per hour) which shatter and destroy everything within thousands of metres of 'ground zero'. A giant fireball of hot air begins to rise, sucking up thousands of tons of rock and earth while forming itself into the shape of a mushroom. The cloud will climb to over 15,000 metres (50,000 feet), a swirling mass of hot and highly dangerous radioactive particles – the fall-out. This will drift with the wind over thousands of kilometres before decaying and becoming extremely diffused.

Despite the United States' efforts at keeping the technology secret it was not long before it fell into the hands of the Soviet Union, whose physicists had been actively engaged in nuclear and atomic research during the 1930s. In February 1939 the Soviets discovered that the United States and Britain were researching the possibilities of nuclear fission, but any further Soviet research was terminated abruptly by the German invasion of Russia in June 1941. During early 1942 the absence of any articles on nuclear fusion or fission being published in western scientific journals, which had hitherto published them on a regular basis, heightened Soviet suspicion that the Allies were conducting secret nuclear research for military purposes. The physicist Georgy N. Flerov wrote to Joseph Stalin, urging that the Soviet Union should begin building an atomic bomb without delay.

During the following year, on the orders of Stalin, the Soviet nuclear physicist Igor Kurchatov, who had previously been the director of the nuclear physics laboratory at the Physico-Technical Institute in Leningrad, began a research project. By the end of 1944, some hundred scientists were working under Kurchatov. On 17 July 1945 Stalin attended the Potsdam Conference where he met other Allied leaders, including President Truman who commented to the Russian leader that the United States had just developed a 'new weapon of unusual destructive force'. On his return, Stalin ordered that work on the Soviet bomb project be stepped up. On 7 August, the day after the bombing of Hiroshima, he placed Lavrenty Beria, head of the NKVD, the Soviet secret police, in overall charge of it.

As described in 'A Very British Bomb', the Soviets had penetrated the wartime American nuclear weapons programme. On 29 August 1949, they carried out their first nuclear test of a bomb at Semipalatinsk, 2,880 kilometres (1,800 miles) south-east of Moscow. As already recounted, Britain followed suit on 3 October 1952. The global nuclear arms race had begun.

It was not long, however, before the United States and the Soviet Union achieved the next quantum leap in the development of nuclear weaponry: the thermonuclear or hydrogen bomb.

The principal difference between atomic and hydrogen bombs is essentially that of fission and fusion. In an atomic bomb, energy is released when heavy atomic nuclei (uranium-235 or plutonium-239) fissions. A hydrogen bomb utilizes the energy released when atomic nuclei of tritium and deuterium (isotopes of the light element hydrogen) fuse into heavier ones. A hydrogen bomb consists of two components: a primary (fission) and a secondary (fusion). The explosive process begins with the detonation of a fission bomb (the primary component); this creates the intense pressures and high temperatures (measuring several millions of degrees) that are necessary to set off the process of fusion in the secondary component. This entire process takes place within a fraction of a second.

A hydrogen or thermonuclear explosion is vastly more powerful than that generated by an atomic bomb. The blast effect is in the form of a shockwave that spreads outwards at supersonic speed, causing total destruction within a radius of several miles. The flash is of such an intense whiteness it can blind observers at a distance of several kilometres, while at the same time igniting fires. Fall-out contaminates the atmosphere, water and soil for many years. The explosive yield of thermonuclear weapons is also many times greater. While that of atomic bombs is calculated in kilotons, one kiloton being the equivalent of a thousand tons of TNT, the yield of thermonuclear devices is measured in megatons (equating to one million tons of TNT).

The first US thermonuclear bomb was tested on 1 November 1952 at Enewetak Atoll with the Soviet Union and Britain following in August 1953 and May 1957 respectively. China, which tested its first atomic bomb in 1964, carried out a thermonuclear test in 1967 while France, which had exploded its first atomic bomb in the Sahara in 1960, did likewise in 1968.

Such were the subsequent developments in thermonuclear weapons that within a decade the Soviet Union had exploded the largest bomb ever detonated on the face of this planet. It possessed 5,000 times the power of the bomb dropped on Hiroshima and its explosive yield was estimated to be the equivalent of 60 million tons of TNT. The power of a 60-megaton device is such that if dropped on a city anywhere in the world, it would destroy the entire city, most of the suburbs and start fires as far as 50 kilometres (30 miles) from the point of explosion, its fall-out meanwhile killing millions of people downwind.

The following years saw the deadly logic of MAD – Mutual Assured Destruction – with the superpowers realizing that none of them would start a war which none of them would win. Ironically, this proved to be one of the cornerstones of global peace as it ultimately led to negotiations over nuclear disarmament between the United States and the Soviet Union. The first tentative halt to the nuclear arms race came in 1963 with the Partial Test Ban Treaty. This was followed by the Strategic Arms Limitations Talks (SALT) which were initiated by President Lyndon B. Johnson in 1967 and ultimately resulted in the signing of the SALT I and SALT II agreements in 1972 and 1979 respectively.

The most important agreements within SALT I were the Treaty on Anti-Ballistic Missile (ABM) Systems and the Interim Agreement and Protocol on Limitation of Strategic Offensive Weapons. The ABM Systems treaty limited both the United States and the Soviet Union to one ABM launch site, and the number of interceptor missiles to 100, while the Interim Agreement and Protocol froze the numbers of intercontinental ballistic missiles (ICBMs) and submarine-launched ballistic missiles (SLBMs) to then current levels for a period of five years during which further negotiations for SALT II would be conducted. Both agreements were signed in Moscow on 26 May 1972 by Presidents Richard Nixon and Leonid Brezhnev. The ABM treaty was subsequently ratified by the US Senate on 3 August 1972.

Negotiations for SALT II lasted seven years, agreement being hindered by problems over asymmetry between US and Soviet strategic missile arsenals: the Soviet Union possessed missiles with large warheads while those of the United States were smaller but more accurate. Other problems relating to definitions, new developments and methods of verification also had to be resolved before final agreement could be reached. It was eventually signed on 18 June 1979 by President Jimmy Carter and Brezhnev, and submitted for ratification by the US Senate shortly afterwards, SALT II placed limits on a number of weapons systems: ICBMs and SLBMs which could be equipped with multiple independently targeted re-entry vehicles (MIRVs); long-range bombers capable of launching nuclear missiles; and strategic launchers. Each country was permitted a total of 2,400 systems.

During the following year, however, tensions between the United States and the Soviet Union, following the latter's invasion of Afghanistan in 1979, resulted in President Carter withdrawing the treaty from the Senate. Thereafter both sides observed the conditions of SALT II on a voluntary basis and in 1982 began further negotiations under the new heading of Strategic Arms Reduction Talks (START).

The purpose of START was to reduce the American and Soviet nuclear arsenals. Between 1983 and 1985, there was considerable interruption. In 1985, however, negotiations were resumed and in July 1991 an agreement was reached under which the Soviet Union would reduce its total number of warheads from 11,000 to 8,000 with the United States reducing its stocks from 12,000 to 10,000. Reductions were also made in both sides' numbers of ICBMs, SLBMs, long-range bombers and mobile launchers. In May 1992, following the collapse of the Soviet Union, the United States conducted negotiations separately with the former Soviet states of Russia, Belarus, Ukraine and Kazakhstan. This led to an additional agreement under which all sides would adhere to the terms of the 1991 START treaty and the latter three states would either destroy all their nuclear weapons or deliver them to Russia.

In 1991 the United States and Russia began dismantling their nuclear arsenals. Between them, they agreed to destroy over 50,000 nuclear weapons. The destruction process started off with much publicity, the world's media watching as missiles and bombs were fed into massive hydraulic presses and crushers or packed with high explosives and blown up. In reality, however, what the television cameras were recording was merely the destruction of empty casings; there was no mention of what was happening to the plutonium 'pits'. The truth is that these were posing a major unforeseen problem with regard to disposal.

When plutonium was first produced in the United States for use in nuclear warheads, little or no thought was given to the ultimate problem of its disposal when such weapons were dismantled at the end of their operational lives. In the words of Victor Rezendes, Director of Energy and Science at the General Accounting Office, 'From day one when we first produced plutonium in this country, we never had an option for its disposal. The notion always was that we were at war, the production of nuclear warheads was the key and the most paramount thing for this country to achieve. The disposal option was always considered something that would be done down the road.'

While US scientists were rather belatedly turning their attention to this problem, heavily guarded convoys of vehicles carrying nuclear weapons were already heading for central Texas from missile bases throughout the United States. Their destination was, and continues to be, a massive nuclear storage depot at Pantex, outside Amarillo. For over forty years, Pantex had been the final assembly centre for the majority of the United States' nuclear weapons. Now its products, unused and no longer required, were returning to be dismantled.

Operated by the US Department of Energy, it is America's only nuclear weapons assembly and disassembly facility and as such is probably the most secure and heavily guarded installation in the world. Located on the high plains of the Texas Panhandle, 27 kilometres (17 miles) north-east of Amarillo, it occupies a 16,000-acre site just north of Highway 60 in Carson County. Originally constructed in 1942 by the US Army Ordnance Corps for the loading of artillery shells and bombs, it was converted in 1950 for nuclear weapons assembly, testing, quality assurance and repair. Ultimately, its operations were extended to include disassembly, retirement and disposal of nuclear weapons, fabrication of chemical high explosives and high-explosive development work in support of research establishments. Currently, Pantex's task is to dismantle between 1,500 and 2,000 nuclear weapons per year.

Those entering or leaving the complex are searched thoroughly by heavily armed guards. Closed-circuit television cameras maintain constant surveillance around its perimeter and throughout the different sectors comprising the entire installation. Ground radar and seismographic detectors pick up anything moving around the perimeter, even Texas jackrabbits attempting to burrow under the razor-wire fences. Throughout the sensitive areas of the installation, tall poles are sited to prevent unauthorized helicopters from landing while armoured vehicles, equipped with machine-guns, patrol the nuclear storage areas.

On arrival at Pantex, nuclear warheads are taken to deep within the installation where they are unloaded and moved to a transit area. There they are subjected to a detailed check: scanning for any damage, cracking or corrosion and ensuring that their safety mechanisms have not been tripped. If no problems are found, the weapons are moved to massive blast-protected bunkers, nicknamed 'Gravel Gerties', for dismantling.

All weapons are first stripped down to their constituent parts, which total some 6,000 in all. The highly complex arming mechanisms are then removed; in the case of bombs this is followed by the parachute assembly which is designed to allow a delivering aircraft the chance to fly well clear as the bomb descends to the predesignated height at which it will explode. Finally the 'physics package', containing the plutonium warhead, and its high-explosive casing, is swiftly removed from the remainder of the weapon and transferred immediately to high-security areas within the installation.

More sophisticated weapons such as Tomahawk cruise missile nuclear warheads are subjected to further tests. Inside their steel casings, the warheads are surrounded by highly radioactive tritium gas which is used as the booster necessary to produce the fusion explosion. Any leakage of the gas would be extremely dangerous; to test it for such, each warhead is placed in a powerful vacuum chamber. As the air in this is pumped out to within a few millionths of an atmosphere, the pressure differential will force the gas to leak from even the most microscopic cracks and be detected.

Gold and silver connectors are extracted from warhead assemblies for recycling, while all plastic and metal components are removed and crushed before being placed in permanent storage as low-level radioactive waste. The plutonium pit in each weapon is, however, left intact. Indeed it cannot be crushed, burned or destroyed in any way as no technology exists for its disposal. There is no alternative but for it to be placed in highly secure storage.

Each 'pit' is placed inside a steel barrel fitted with a cellulose fibre liner and sealed. Barrels are then stored above ground in concrete bunkers originally constructed to store conventional munitions during the Second World War. They are stacked in racks by a sophisticated hydraulic loader specially designed to protect the operator, each barrel then being bar-coded. Thereafter, laser readers travel up and down along the rows of stacks, checking that the correct numbers of barrels are still in place and that there are no excessively high radiation readings which would indicate damage to any of them. Forty-ton concrete blocks are positioned in front of the bunkers' doors to secure them. The pits are currently destined to remain sealed inside their barrels stored in the bunkers for the foreseeable future. It was expected that by the year 2000, 20,000 pits would be in storage – each one more powerful than the bombs dropped on Japan in 1945.

Pantex has, however, experienced problems over transportation of weapons to the installation and suffered an escape of tritium gas from a bomb. In addition, a nuclear weapon was accidentally dropped while being dismantled. The result of all these problems was a scaling-back of the rate at which warheads were being dismantled and consequently during one year only 63 per cent of the target figure for that period was achieved.

Contamination of the seventeen US Department of Energy installations that have been involved in the dismantling of nuclear warheads also poses a major problem. Initially gauged at tens of billions of dollars, the estimated cost for decontamination was subsequently put at 300 billion dollars – a figure now thought to be probably too low. Indeed, it is almost impossible to arrive at a firm figure as the technologies necessary for decontamination do not exist. Some of the installations will thus remain contaminated for centuries.

As described earlier, plutonium was first produced in 1941 as a by-product of uranium. As with all radioactive materials, it has a 'half-life' which is defined as the period of time required for half a gram (a fiftieth of an ounce) of radioactive material to decay completely to the point of being totally harmless. All the plutonium on this planet is approximately fifty years old but its half-life is over 24,000 years. Total decay will take over 250,000 years.

The principal problem with plutonium is the emission of an alpha particle which is a helium nucleus. Very heavy and slow moving, it does not easily penetrate other materials: a layer of dead skin cells will stop an alpha particle. While plutonium outside the human body is thus not a major concern, it becomes one if ingested or inhaled, because the alpha particle causes ionization of cells in the body which can mutate into cancerous forms and other materials. Tests have shown that only eighty-millionths of a gram of plutonium is required to cause cancer.

There are an estimated 200 to 250 tons of plutonium either in nuclear warheads or removed from them. In addition, there are approximately 1,000 tons of highly enriched uranium associated with nuclear weapons worldwide. As large numbers of missiles and bombs are dismantled, the problem facing governments and scientists is the long-term disposal of the highly enriched uranium and plutonium extracted from them.

There are four optional means for disposal of such materials. First, recycling for use in nuclear fuel – a method advocated by many people. Second, vitrification and burial in specially constructed storage areas. Third, transmutation into isotopes which cannot be used in the manufacture of weapons. Last, send the material into space and dump it in the sun.

While the West regards plutonium extracted from weapons as waste and concerns itself with developing disposal technology, Russia takes a different viewpoint, regarding it as a fuel for use in the generation of nuclear power. Minute quantities of weapons-grade plutonium are mixed with low-grade uranium to produce a fuel called mixed uranium–plutonium dioxide (MOX) which is used in pellet form in fast-neutron breeder reactors. This is, however, a difficult and costly procedure and would require massive investment on the part of the Russians to enable them to convert their reactors to use MOX. Furthermore, it would not dispose of the plutonium completely as the spent fuel would have to be reprocessed when extracted from the reactors – the remaining plutonium would have to be removed, mixed with uranium and then returned to the reactor as MOX once again. Thus the MOX elements would have to be recycled a large number of times in order to dispose of the plutonium, making this impractical as a single-cycle method of disposal.

If the sole criterion is the prevention of plutonium being used in the manufacture of weapons, the most favoured method of disposal is vitrification, which is already used in the disposal of high-level waste from nuclear reactors. The plutonium is contaminated with radioactive poisons to prevent any attempt to recover it in the future and then fused with silica to produce molten borosilicate glass which is sealed into steel canisters. Each canister is rigorously checked for leakage by remote-controlled probes before being moved to a permanent storage area. However, this method has only been used for the disposal of reactor waste and so far no one has been able to vitrify weapons-grade plutonium successfully. When and if they succeed in doing so the material will have to be interred for 250,000 years in a highly secure site. Furthermore, those responsible for doing so will have to be certain that the material encasing the poisoned plutonium does not deteriorate and eventually fall away, leaving an accumulating deposit of pure plutonium.

The option of transmutation has also been given serious consideration. In the mid-1990s, research began on the use of linear accelerators to destroy plutonium. The process would involve taking hydrogen atoms (comprising protons with electrons spinning round them) and stripping off the electrons. The protons are fired into a linear accelerator and their positive charge pushes them forward on the front of an electric field, in a similar fashion to a surfer riding a wave. By the time they reach a point three-quarters of the way down the accelerator's tube, the protons have accelerated to almost the speed of light. At the far end of the tube they smash into a block of lead, producing a shower of neutrons that are absorbed into the plutonium which breaks up into short-lived, more stable elements. The results of initial experiments looked promising, but at the time even the most optimistic forecasters estimated that it would be a minimum of fifteen years before transmutation of plutonium by linear acceleration would be a viable proposition.

Meanwhile, research was also conducted into the final option of launching consignments of plutonium into space and dumping them in the sun. Consideration had to be given to the possibility of launch failures and the risk of a massive catastrophe involving a cargo of several tons of plutonium. Research was carried out into the development of crash-proof containers capable of withstanding impacts at the highest velocities possible according to the laws of physics. Like transmutation, however, it is a long way from being a viable process.

In the meantime, plutonium pits continue to be stored in barrels in the absence of any international policy for their disposal. Under the terms of the SALT II agreement of June 1979, limiting the numbers of intercontinental ballistic missiles, submarine-launched ballistic missiles, heavy bombers and strategic launchers, the total number of warheads in the United States and Russia was restricted by the year 2003 to no more than 3,500 each, of which 1,750 could be deployed on submarines. Although SALT II was never ratified, due to renewed tensions between the two countries following the Soviet Union's invasion of Afghanistan, its arms limitations were observed voluntarily by both countries. During the 1980s the nuclear arsenals of the United States and the Soviet Union exceeded the SALT II limits by a factor of ten, thus posing a massive problem of disposal requiring colossal expenditure. In 1994 it was estimated that the cost of dismantling one nuclear warhead was between 30,000 and 100,000 dollars, with tens of thousands waiting to be dismantled.

While the United States could afford to devote considerable resources to dismantling its nuclear weapons, this was not the case with Russia and the former Eastern Bloc countries which were suffering, and continue to do so, from unstable economies and a lack of hard currency. Evidence of the lack of funding for the Russian nuclear industry could be seen in the virtually unguarded and heavily contaminated radioactive waste sites outside Murmansk, the nuclear reactors badly in need of repairs and modernization and a nuclear ballistic missile submarine lying on the bed of the Atlantic Ocean with its warheads still aboard.

In 1991, as the Soviet Union began to disintegrate, senior figures in the Russian political and military establishments became concerned over the threat to the security of the vast quantities of nuclear weapons located throughout Russia and what were fast becoming former Soviet states. They turned to the United States and in particular to two members of the US Senate, Senators Sam Nunn and Richard Lugar, requesting their assistance in safeguarding Russia's nuclear arsenal.

While Nunn and Lugar were receptive to the Russians' approach, recognizing the dangers posed by the risk of nuclear weapons falling into the wrong hands, the same could not be said for the rest of the US political establishment and the country as a whole. The United States was understandably glad to see the end of the forty-five-year-long Cold War which had cost it dear, and was reluctant to commit itself to large-scale assistance to its old foe. Indeed, the House of Representatives had already thrown out a proposal for financial assistance to the tune of one billion dollars for the former Soviet Union. Furthermore, 1991 was a presidential election year and the presidential candidates, George Bush and Bill Clinton, were concentrating their attentions on vote-winning domestic rather than foreign issues.

Despite the odds stacked against them, however, Nunn and Lugar took up the challenge and succeeded in winning other members of the Senate to their cause, developing a plan for the collaborative dismantling of the former Soviet nuclear arsenal. Remarkably, they succeeded and in November 1991 the Co-operative Threat Reduction (CTR) Program, also known as the Nunn–Lugar Program, was passed by the Senate, subsequently being approved by the House of Representatives and signed off by President George Bush.

The CTR Program had five objectives. First, it was to assist Russia in carrying out reduction of its strategic arms to levels in accordance with the terms of the START treaty; this was to be carried out with the establishment of weapon dismantling and destruction projects. Second, it was to assist in reduction and prevent proliferation of nuclear weapons and fissile material by enhancing the safety, security, centralization and control of nuclear weapons throughout the former Soviet Union; the dismantling of weapons would result in large quantities of fissile material which would thereafter have to be securely stored and controlled. Third, assistance was to be given to Kazakhstan and Ukraine in the elimination of weapons and delivery systems infrastructures to be limited under START; this would entail establishment of facilities for dismantling missiles, silos, heavy bombers and their weapons.

The Program's fourth objective was to assist the former states of the Soviet Union to eliminate and prevent proliferation of biological and chemical weapons by helping them to achieve compliance in accordance with the 1993 Chemical Weapons Convention; this would involve the elimination of a biological weapon production facility in Kazakhstan and provision of assistance to Russia in the destruction of its chemical weapon stocks as well as the elimination of its production facilities. The CTR Program's final objective was to promote demilitarization and military reform throughout the new states of the former Soviet Union, assisting the armed forces of each state in their transition to western lines, as well as to reduce proliferation threats by providing assistance for improvement of border security and controls designed to prevent illicit movement of material and technology related to nuclear, biological and chemical weapons.

Initial progress was slow, primarily because of Russian reluctance or refusal to allow access to nuclear facilities and because it took time before sufficient funds became available; but by the following year the Program was gaining momentum. A number of major engineering projects were established in Kazakhstan, Belarus, the Ukraine and Russia with American companies carried out the task of dismantling weapons and delivery systems. Among the latter were well-known names such as Hewlett Packard, Raytheon, Lockheed-Martin, Westinghouse, Bechtel, Allied Signal, AT&T and Caterpillar.

In June 1996 the Ukraine was declared totally free of nuclear weapons. A special facility had been built for the dismantling of missiles, with warheads being removed and missiles extracted from silos after which they were broken up. The silos themselves were then blown up and the entire area ploughed and turned over to agriculture. A total of 111 SS-19 Stiletto lightweight ICBMs were dismantled, 144 SS-19 silos were destroyed and 1,900 nuclear warheads were returned to Russia. Fifty-five SS-24 medium ICBMs and fifty-one silos were also destroyed, their 460 warheads having been removed beforehand and deactivated. The CTR Program also provided for the building of 261 houses at the Pervomaysk ICBM base and 605 apartments at the Khmelnitsky as part of the demobilization of six regiments of the Strategic Rocket Forces of the former Soviet Union.

In addition, a number of Tu-95 Bear and Tu-160 Blackjack long-range bombers were also dismantled, the latter being a variable-geometry aircraft of similar appearance to the US Air Force's B-1 bomber but larger. With a range of 14,000 kilometres (8,700 miles), the Blackjack could carry up to twelve Kh-55MS (NATO designation AS-15 Kent) cruise missiles, each of which carried a 200-kiloton nuclear warhead and had a range of 3,000 kilometres (1,865 miles).

In November, Kazakhstan also became completely rid of nuclear weapons by the CTR Program: 104 SS-18 Satan heavy ICBMs and 1,400 nuclear warheads had been returned to Russia. Produced in six variants (Models 1 to 6), each with a differing warhead configuration, the SS-18 was the largest missile in the Soviet armoury and indeed in the world, having a maximum range of 16,000 kilometres (nearly 10,000 miles). The highly accurate Model 4 was capable of carrying up to ten multiple independently targeted re-entry vehicle (MIRV) 500-kiloton warheads. With a further 204 being stationed in Russia, it was estimated that the Model 4 force alone had the capability of destroying 65 to 80 per cent of the United States' ICBM silos, using two warheads against each.

In addition, 147 (mostly SS-18) silo launchers, missile control centres and test silos at Zhanghiz-Tobe, Derzhavinsk, Semipalatinsk and Leninsk were dismantled and destroyed. Meanwhile, thirteen vertical test boreholes at the Soviet nuclear weapons test facilities at Balapan were sealed; in the Degelen Mountains, a massive complex of 181 test tunnels and shafts was also closed and sealed.

Besides large numbers of missiles, Kazakhstan also held a huge amount of weapons-grade uranium produced for use as fuel for a new type of submarine developed by the former Red Navy. The uranium, in the form of fuel pellets, had been produced at the Ulba Metallurgical Plant at Ust-Kamenogorsk. The development programme had proved a failure and had been abandoned, but by that time a vast amount of uranium fuel had been produced. No longer required, it was consigned to the vaults of the plant and forgotten. In 1992, following the creation of the Republic of Kazakhstan, the existence of the fuel came to light and the Kazakh government turned to the Americans. An initial investigation by the US team revealed a quantity of just over a thousand containers of uranium-235 enriched to 90 per cent – sufficient to manufacture fifty nuclear weapons. Aware that countries such as Iraq and Iran would be keen to acquire the fuel for use in their own nuclear weapons development programmes, the Americans decided to move quickly.

On 8 October 1994 a highly sensitive operation, codenamed Project Sapphire, was launched. Three giant C-5 Galaxy transports of the US Air Force flew a team of thirty-one specialists and 130 tons of equipment, including a mobile nuclear laboratory, to Ust-Kamenogorsk. Here, 600 kilograms (1,320 pounds) of uranium-235 were stored and there was concern that it might be stolen. During the following six weeks, the team packed and sealed the uranium in special stainless-steel containers. In addition, it packed some 30 kilograms (66 pounds) of uranium oxide powder, a quantity of spent fuel rods and other material. By the time the task was complete, the entire consignment comprised fifty steel drums ready for shipment by air to the United States. On 20 November the team, accompanied by the uranium, flew back to the United States where the entire consignment was transported to the US Department of Energy's Y-12 plant at Oakridge in Tennessee.

The United States took similar measures three years later when the former Soviet state of Moldavia announced that, because of its parlous financial situation, it was putting up for sale twenty-one of its air force's MiG-29 Fulcrums, of which fourteen were Fulcrum-C variants capable of carrying nuclear weapons. This caused considerable anxiety in Washington, where there were fears that nations such as Iran might attempt to acquire the aircraft and thus a nuclear delivery capability. Furthermore, the US Air Force was keen to lay its hands on the Fulcrum-C in order to discover more about it and, in particular, its helmet-mounted display system and AA-11 Archer air-to-air missiles. In June 1997, a deal was struck between Moldavia and the United States, the latter purchasing 500 AA-11 missiles and a large quantity of spares in addition to the twenty-one aircraft. In October, a fleet of US Air Force C-17 Globemaster transports flew the Fulcrums, missiles and spares from Moldavia to Wright Patterson Air Force Base in Ohio.

In Belarus, meanwhile, a total of eighty-one SS-25 Sickle lightweight mobile ICBMs, together with all the infrastructure for their support and maintenance, were removed to Russia and the country was declared free of nuclear weapons in November 1996. However, any further work under the CTR Program was terminated as a result of gross violations of human rights by the tyrannical regime of Belarus's Communist dictator, President Alexander Lukashenko, who had come to power in 1994.

It was in Russia, however, that the CTR Program carried out the majority of its work and is continuing to do so. A high priority was the collection from throughout the former Soviet Union of some 30,000 tactical nuclear weapons. Another was the destruction of SS-18 Satan ICBMs, these posing the foremost threat to the United States. Under the terms of the START II Treaty, missiles equipped with MIRV warheads were banned. Russia was required to reduce its arsenal of SS-18s by a total of 254 by the beginning of 2003 and the CTR Program became heavily involved in assisting in carrying out that requirement: destroying rocket motors, providing facilities for the disposal of rocket fuel, dismantling warheads and storing fissile material. During the period up to December 1999, 116 SS-18s, 119 SS-11 Sego, ten SS-17 Spanker, thirteen SS-19 and thirty submarine-launched ballistic missiles (SLBMs) were destroyed, along with fifty missile silos and forty-two heavy bombers. In addition, 100,000 tons of missile liquid propellant were also eliminated.

Dismantling of warheads obviously raised the question of secure storage of fissile material. Existing Russian facilities, inadequately guarded and possessing little in the way of security systems, were totally unsatisfactory. Consequently, in October 1992, the United States signed an agreement with Russia for the design and construction of the Fissile Material Storage Facility, half of the cost of the facility being borne by the United States. Designed by the US Corps of Engineers, it was originally intended to be located at Tomsk but was eventually sited at Mayak in Siberia, just over 1,280 kilometres (800 miles) east of Moscow. This was also the site of the designated RT-1 nuclear reprocessing plant. Construction began in the autumn of 1994 and the first stage of the facility is due for completion in 2002; it will accommodate plutonium removed from over 6,000 weapons.

In addition, the United States supplied over 26,000 fissile material containers for use at Mayak; 117 railcar conversion kits to enhance rail transport of nuclear weapons; 150 containers and 4,000 Kevlar ballistic blankets for security and ballistic protection of nuclear weapons during transit; emergency support equipment (including five mobile response complexes) in case of accidents involving nuclear weapons; and computer systems and training to enhance Russian capability for accounting and tracking of nuclear warheads.

Checks to ensure that CTR Program assistance is being used properly are carried out through a system of audits and examinations (A&E). During October 1999, a total of seventy-six A&Es were carried out in Russia, Ukraine, Kazakhstan and Belarus, evaluating assistance which had been provided at a cost of 599 million dollars out of the total of 2.1 billion dollars allocated to the CTR Program. Detailed reports on the results of these audits are submitted to the US government and Congress.

A major area of concern was the large number of nuclear submarines and other vessels lying inactive since the demise of the Soviet Union. In October 1996 a report in the British *Daily Telegraph* described fifty-two decommissioned nuclear submarines of Russia's Northern Fleet rotting at anchor off the Kola Peninsula, among them seventeen Victor and November Class vessels laid up at the remote naval base of Gremikha in the eastern part of the peninsula, and others at the Sevmorput naval yards at Murmansk and Polyarny. In March 2000 a report by the Bellona Foundation, the Norwegian organization established in 1986 to monitor nuclear safety issues related to the former Soviet Union military complex, was published in its *Nuclear Chronicle*. The report quoted the Russian Ministry of Nuclear Energy (Minatom) as stating that thirty first- and second-generation Northern Fleet vessels, all of them still containing spent fuel, were in danger of sinking and that leakage from the primary cooling circuits in their reactors had been detected. Having been moored for up to fifteen years, their hulls were badly corroded and were no longer watertight. Similar problems were also reported to exist at several naval bases and yards belonging to the Pacific Fleet.

While such vessels may no longer constitute a military threat, they pose a major one to the environment. As stated by the Bellona Foundation in a report in August 1998, the submarine branch of the Soviet/Russian Navy has had a chequered history with regard to nuclear safety. Between 1961 and 1998, there were several accidents on submarines in which a total of over 500 people died, the most serious being caused by fires or severe damage to nuclear reactors through cores overheating, with release of radioactivity occurring in some cases. In three instances, the accidents were so severe that the vessels sank; all of them involved submarines of the Northern Fleet. One of these took place in

October 1986 in the Atlantic Ocean, north of Bermuda, when the Yankee Class submarine K-219 suffered an explosion in one of her missile tubes, possibly following a collision with an American submarine. The vessel was forced to surface after a fire broke out in the missile compartment and water starting leaking into the vessel. Eventually, the crew had to abandon ship and she subsequently sank, her two reactors and sixteen missiles ultimately posing a hazard to the environment.

The most publicized accident involving a Soviet submarine took place in the Barents Sea 480 kilometres (300 miles) off the coast of Norway, west of Bear Island. On 7 April 1989 the Red Navy Mike Class submarine K-278 *Komsomolets* suffered a fire that broke out in her stern and spread to other compartments. The 6,400-ton vessel, the sole forerunner of a new class, possessed a titanium hull designed to enable her to operate at depths of 1,000 metres (3,280 feet),and was equipped with an advanced type of reactor that propelled her at speeds faster than those of any other submarines in existence. The *Komsomolets* surfaced but the internal pressure caused by the fire and high-pressurized oxygen resulted in her hull rupturing and she sank to the seabed 1,685 metres (5,528 feet) below, with the loss of forty-two lives, including that of her commander.

Initially, it was considered that the wreck posed only a small threat to the marine environment surrounding it but during the following years a report stated that there was a risk of leakage by 1995 from the reactor and two R-84 nuclear torpedoes, containing a total of 4 kilograms (9 pounds) of plutonium, with which the vessel was armed in addition to her conventional weapons. According to a subsequent Norwegian report, the motor of one of the R-84s had been damaged in the accident and leaked fuel which, on making contact with salt water, had produced ammonia. There was a danger of the gas reacting with other fuel aboard the sunken vessel and causing an explosion. In the summer of 1994, an expedition was carried out to survey the wreck and seal off the areas of high risk. Signs of plutonium leakage were detected and it was discovered that one of the R-84 warheads had broken open and released 22 grams (about 1 ounce) of plutonium. The expedition succeeded, however, in plugging some of the holes in the submarine's hull.

Although the subsequent levels of radioactivity emanating from the *Komsomolets* were low, Norwegian scientists were concerned by plankton and other organisms consumed by fish becoming contaminated and entering the food chain. Fishing had to be cut back in the area due to minor contamination levels and the perceived threat of further, more extensive contamination. This in turn had a major effect on Norway's fishing industry; its revenues from sales of fish to Europe and Russia were threatened.

Three options were considered for resolving the problem of the *Komsomolets*. The first, that of raising the entire vessel, was soon dismissed as being too hazardous as well as costing in the region of an estimated 1 billion dollars. The second option was to raise the bow section containing the two R-84 torpedoes; this was also rejected as being too dangerous since it was considered that leakage could have rendered the weapons unstable. The third and most feasible option was to hermetically seal the entire submarine in a gel, manufactured from the shells of crustaceans and containing 2 per cent chitosan, which would, it was envisaged, bind radionuclides more efficiently than concrete. This would, however, be only an interim measure as eventually the warheads would have to be removed due to the plutonium-239 in them having a half-life of 24,000 years. The sealing of the submarine was scheduled for the summer of 1995 and, according to the Bellona Foundation, took place at some point thereafter.

A total of some 250 nuclear-powered submarines were built by the Soviet Union; by the time of writing, in mid-2000, 183 of them, from both the Northern and Pacific Fleets, have been laid up for one or more of three reasons: in accordance with START treaty requirements, the vessel's age or lack of facilities for proper repair and maintenance. Of those, 120 still contain spent fuel in their reactors. The Russian Navy does not possess the proper infrastructure nor facilities onshore for the removal of spent fuel from its submarines. Instead, it uses a fleet of twelve tankers and a further twelve ships and barges to transport and store spent fuel and liquid radioactive waste. Most of the tankers are, however, over twenty-five years old and the majority of the other support vessels are no longer in a functional state and were themselves due for decommissioning by the end of the 1990s.

Their desperate state was illustrated during 1999 when two incidents took place involving vessels of the Pacific Fleet. In August, there was a leak of liquid radioactive waste aboard a support tanker based at the Zvezda shipyard in the city of Bol'shoy Kamen; in December, one of the Pacific Fleet's four ageing storage barges caught fire while loaded with 100 submarine spent fuel assemblies. In order to improve the situation in the Northern Fleet, eight nuclear-powered icebreakers operated by a civilian organization, the Murmansk Shipping Company, from its Atomflot base at Murmansk, were pressed into service as support vessels.

Following the collapse of the Soviet Union, the increased decommissioning of submarines inevitably exacerbated the problems of storage and ultimate disposal of radioactive waste and spent nuclear fuel. Lacking the necessary infrastructure and facilities, Russian efforts at storage were makeshift, spent fuel being stored in the reactors of decommissioned vessels and ashore in obsolete storage facilities and transport containers. At the end of February 1999 the Bellona Foundation reported in *The Nuclear Chronicle* that the most dangerous site was at Andreeva Bay in the Kola Peninsula, 64 kilometres (40 miles) from the border with Norway, where 21,000 spent fuel assemblies were stored in containers outdoors.

The report also mentioned that similar situations existed at the Gremikha naval base and at the Pacific Fleet base at Shkotovo. According to a report by Igor Kudrik in *The Nuclear Chronicle*, by the year 2000 the Russian Navy would be storing almost 105,000 spent fuel assemblies from its Northern and Pacific Fleets, these representing between 228 and 238 submarine loads as each vessel was equipped with two reactors, each containing 220 to 230 fuel assemblies. In addition, the Murmansk Shipping Company is currently storing 3,500 spent fuel assemblies, removed from its eight icebreakers, while the Northern Fleet is doing likewise with seven submarine reactor cores. None of these can be reprocessed in the normal way by the RT-1 plant at Mayak: the fuel assemblies are of a zirconium-coated type and the cores are from Alpha Class submarines which were equipped with liquid metal cooled reactors. A further 3,130 zirconium-clad spent fuel assemblies are also stored in the support vessel *Lotta* moored at Atomflot icebreaker base at Murmansk.

A further problem exists at Murmansk where a nuclear cargo vessel, the *Lepse*, is also moored at the Atomflot base. During the period 1962 to 1981, the vessel was used as a support ship for the Murmansk Shipping Company's ice-breakers. Since 1981, however, it has been used as a floating store for 624 spent fuel assemblies, some of which are damaged with the fuel jammed inside, salvaged after an accident aboard the icebreaker *Lenin*, the remainder being from Northern Fleet submarines. Some of the assemblies are zirconium-clad and cannot be reprocessed at Mayak's RT-1 plant.

In 1992 the CTR Program initiated a further programme for dismantling nuclear ballistic missile submarines as required by the terms of the START I treaty. Four locations were initially considered for the work to be carried out: Nerpa in the Kola Peninsula, Zvezdochka in Archangel County (south-east of the peninsula), Zvezda in the far east of Russia and the nuclear reprocessing plant at Mayak. In 1996, after a further review of the project, the decision was taken to construct special dismantling facilities at Nerpa and Zvezdochka.

During 1997, however, it became apparent that the Russian government was unable to pay the shipyards to carry out the dismantling work. The CTR Program therefore contracted directly with the shipyards and in 1999 signed contracts for dismantling seven vessels: one Yankee and six Delta class strategic ballistic missile submarines (SSBNs). Of those, one was already dismantled, three had spent fuel removed and the remaining three were defuelled during 1999 in accordance with the contracts.

At the time of writing, the immediate objective of CTR's submarine dismantling programme was the destruction of a total of thirty-one SSBNs: one Yankee, twenty-six Deltas and five Typhoons. The destruction of the latter will remove a major threat to the United States. Powered by four pressurized water-cooled nuclear reactors, the Red Navy's six Typhoons are the largest underwater vessels ever constructed with each carrying twenty SS-N-20 Sturgeon submarine-launched ballistic missiles (SLBMs), each equipped with ten MIRV 100-kiloton warheads. The SS-N-20 has a range of 8,300 kilometres (5,160 miles) which would have allowed a Typhoon to fire its missiles from within the Arctic Circle and hit targets anywhere within continental United States.

The submarine-dismantling aspect of the CTR Program was initially concerned only with the destruction of SSBNs, with priority being given to relatively new vessels. During 1997, however, the Russians raised the subject of general purpose nuclear-powered submarines. In Russian naval vernacular, the term 'general purpose' would normally cover submarines armed with torpedoes, depth charges and anti-ship cruise missiles. In the context of nonproliferation and disarmament, however, the Russians applied it to all laid-up or decommissioned submarines not covered by the CTR Program.

By 2000, some 110 such vessels will have been laid up, comprising submarines of the Alpha, Charlie, Echo, Hotel, November and Victor classes. In February 1999, an official of the US Department of Energy stated that the United States would be launching a programme for the dismantling of general-purpose submarines once cost and project assessments had been made. In a speech given in December 1999 at a conference at the Centre for Nonproliferation Studies at the Monterey Institute of International Studies, Senator Richard

Lugar emphasized the importance of the destruction of the cruise-missile-equipped vessels once the dismantling of ballistic missile submarines was complete. If funding is granted by Congress, this additional part of the submarine-dismantling program may begin during the period 2002 to 2003.

Measures have also been initiated for dealing with the problem of spent fuel. The Arctic Military Environmental Co-operation (AMEC) organization, established in 1996 by the United States, Norway and Russia with the role of focusing on environmental hazards posed by military activities in the Arctic, designed and built a 40-ton steel–concrete cask, with an operating life of fifty years, for local storage and transportation of spent fuel assemblies. In 1998, US Vice Secretary of State Strobe Talbot announced a project to design and construct fifty giant 80-ton casks, but such is their size and weight that there have been misgivings over the manageability of such containers.

According to a report by the Bellona Foundation, however, around 430 of the AMEC 40-ton flasks would be required to defuel a total of some 150 laid-up submarines, ninety of them currently moored at different locations in the Kola Peninsula. The use of casks, however, is only a partial solution to the storage problem as sites for them have not yet been prepared. A number of locations at Andreeva Bay, Gremikha, the Nerpa shipyard and the Kamtchka Peninsula in the far east of Russia are under consideration but no firm decision has been made to date. Furthermore, problems have arisen in the form of a dispute with the Russian State Nuclear Regulator (GAN) which, according to the Bellona Foundation, has refused to license the casks, claiming that production models would not be manufactured to the same standards as the prototype which passed all the GAN tests. These criteria included drop-tests and exposure to fire and water, designed to ensure that the spent fuel in six containers inside the cask remained intact.

It appears that part of the problem lies in the long-running dispute between the Defence Ministry and GAN: the latter for many years has been attempting to gain control of naval nuclear sites such as the nuclear waste facilities in the Kola Peninsula. At the same time Minatom has entered the fray by trying to strip GAN of any influence on matters concerning nuclear safety at Minatom and Defence Ministry sites. A total of 100 casks were due to be manufactured at a plant at Izhora, near St Petersburg, at a unit cost of 150,000 dollars, with the CTR Program bearing the cost of the first twelve casks. Should GAN be proved correct in its allegations over production standards, the CTR Program would be forced to cancel its order, although Minatom has already stated that in such circumstances it would pay for the casks. Nevertheless, the dispute has highlighted the safety issues surrounding spent fuel, and the problems facing western organizations and governments that wish to have the question of liability settled in order that they are not made responsible for any damage caused by a nuclear accident involving any equipment manufactured or financed by them.

September 1998 saw the completion of a liquid waste processing facility at the Atomflot base at Murmansk. The project, established in 1994 by the United States, Russia and Norway, upgraded existing facilities, increasing capacity from 1,200 cubic metres (1,570 cubic yards) of liquid radioactive waste per annum to 5,000 cubic metres (6,540 cubic yards). The facility was designed to handle not only the waste generated by the Murmansk Shipping Company's eight icebreakers but also all that produced by the Northern Fleet's warships.

Other international projects to solve the problems caused by dismantling submarines have included the funding by Japan of a liquid radioactive waste processing facility for the Pacific Fleet in the Russian far east. Designed by the American company Babcock & Wilcox, it was constructed at the Amur shipyard and subsequently tested at Vostok before entering service at the Zvezda shipyard in late 1998. Like the AMEC casks, however, the project has encountered problems, with the Russians now maintaining that it would be too expensive to operate and that it would be unable to cope with the quantities of waste produced.

Under the auspices of the Commonwealth of Independent States Co-operation Programme (CISCO), the European Union has also participated in a number of projects dealing with the potential environmental hazards, caused by pollution from nuclear spent fuel and radioactive waste. Apart from several regional studies, it has been involved in two specific projects. The first of these was a feasibility study into the removal and subsequent management of the 624 spent fuel assemblies currently stored in highly unsatisfactory conditions aboard the support vessel *Lepse*. The other project was a feasibility study into the options for storing large amounts of spent fuel at the RT-1 processing plant at Mayak. Unfortunately, however, the Russians dismissed the findings, which showed that the cost of a new dry storage facility would be less than that of upgrading the existing wet storage on which construction had ceased several years previously.

Norway has been involved in the majority of projects and has played a leading role in putting forward a number of initiatives. These comprise emptying and closing the spent fuel storage facility at Andreeva Bay; carrying out a study for the management of the fuel and the construction and operation of a facility for temporary storage; establishment of a temporary spent-fuel storage facility at Mayak; construction and operation of a specialist vessel for transportation of spent fuel; construction and operation of four special railcars for transportation of spent fuel; the upgrading and operation of a temporary spent fuel storage facility at the Zvezdochka shipyard in Sverodvinsk; delivery of a mobile liquid radioactive waste treatment facility for use at Murmansk; resolving the problems caused by the support vessel *Lepse*; and upgrading the radioactive liquid waste treatment facility at the Atomflot base at Murmansk.

International efforts aimed at resolving the major potential environmental hazards have, however, been hindered by a number of obstacles within Russia itself. Contributing nations naturally required certain safeguards, similar to those that protect the CTR Program, before they were prepared to release funds for projects. Among these were ratification by the Russians of taxation and customs exemptions, as well as nuclear liability agreements. In addition, they needed a plan from the Russians as to how they intended to deal with the whole question of management, storage and disposal of radioactive material. Finally, there was the problem of obtaining access for international experts to sites at which projects would take place. Solutions to all these problems were slow in appearing although there has reportedly been progress in the last year or so.

The situation has not been improved by a recent chilling in relations between the United States and Russia, caused by a growing disparity in power as Russia became increasingly reliant financially on the West. The situation was exacerbated by other factors, including Russia's financial collapse in 1998, Bosnia, the crisis in Kosovo, the enlargement of the North Atlantic Treaty Organisation (NATO), corruption scandals in Russia itself and the war in Chechnya, where the West became increasingly concerned at the scorched-earth tactics adopted by Russian forces.

While the CTR Program has succeeded in reducing the threat of nuclear conflict through the destruction of strategic weapons and delivery systems in accordance with the terms of the START treaty, the threat of proliferation still remains. As described in the next chapter, countries such as Iran and Iraq are currently seeking to develop their own military nuclear capability by attempting to obtain materials and technology from within the former Soviet Union.

RUSSIAN ROULETTE

The end of the Cold War saw a rapid deterioration in the former Soviet Union's arsenal, which rapidly fell into disrepair. Ironically, this resulted in a greater risk than ever of missiles being launched, not as an act of war but either by accident or without authority.

Just after dawn on 25 January 1995, the world came closer to a nuclear holocaust than at any time since the Cuban missile crisis of 1962. A Russian radar early-warning station in the Kola Peninsula detected a swiftly moving object, measured at over 1,500 kilometres per hour (932 miles per hour), approaching north Russian airspace from the direction of the Barents Sea, an area always treated very seriously by the Russians as US Navy strategic ballistic missile submarines (SSBNs) were known to operate there. During the following five minutes there was mounting alarm as the operators at the radar station flashed details of the fast-approaching threat up the chain of command to a senior officer on duty, via a special terminal system called 'Krokus', stating that they suspected it to be a missile from an American submarine. They were under considerable pressure: seven years earlier a young German, Matthias Rust, flying a light aircraft, had penetrated Soviet air defences and landed in Moscow's Red Square. There had been severe repercussions and so on this occasion the radar operators, despite fearing a reprimand for sounding a false alarm, were concerned to avoid suffering the same fate as their predecessors.

Doubtless thinking along the same lines, the senior officer, a general, decided that it would be prudent to pass on the alert to higher levels rather than subsequently be held responsible for initiating a catastrophic holocaust. He transmitted it via 'Kavkaz', a complex system comprising cables, radio links, satellites and relays that represented the heart of the Russian command and control system. Kavkaz communicated the alert to three portable command and communications systems called 'Chegets'; these are contained in three briefcases that at all times accompany the Russian President, the Defence Minister and the Chief of the General Staff. Similar in principle to the nuclear 'football' that permanently accompanies the President of the United States, the Chegets allow instant communication between the three leaders and the General Staff's national command centre. President Boris Yeltsin, Defence Minister Pavel Grachev and the Chief of the General Staff, General Mikhail Kolesnikov, were soon in communication.

Meanwhile, the radar operators reported that the object was separating and immediately it was assumed to be a Trident II D-5 SLBM deploying its MIRV warheads. Certainly, it appeared to be displaying the characteristics of the much-feared American missile, which can carry up to eight W-76 or W-88 warheads with explosive yields of 100 kilotons or 475 kilotons respectively.

Orders were flashed to the command posts of Russia's Strategic Rocket Forces, ordering them to go to a state of increased readiness to launch; at the same time, similar orders were also transmitted to the commanders of the Russian Navy's ballistic missile submarine fleet. At that time, despite the end of the Cold War and the deterioration in her armed forces, Russia still maintained forty-five strategic ballistic missile submarines. During the next four minutes, commanders waited for the next order in the sequence which would lead to their missiles being launched but, eight minutes after the alarm was first raised, the objects faded from Russian radar screens as they fell into the sea; a major crisis had been averted.

The incident had been caused by the launching of a Norwegian rocket as part of a scientific research programme into the Northern Lights. As the elements of the rocket's first stage had fallen away, they had resembled Trident SLBM MIRV warheads heading south into Russian territory. The Russians had, however, been warned of the launching beforehand. On 21 December 1994, the Norwegian Foreign Ministry had sent a letter to neighbouring countries and the Russian Foreign Ministry, advising that between 15 January and 15 February Norway would be launching a Black Brant XII, a four-stage research rocket as part of a collaborative project with the US National Aeronautical and Space Administration (NASA), the latter having supplied the engines.

The issuing of such a warning letter was a familiar procedure for the Norwegians, who had launched 607 rockets since 1962. On this occasion, however, the letter had become lost in the slow-moving bureaucratic system of the Russian Foreign Ministry. An official had taken down the details but for some unexplained reason the information had not been passed on to the Defence Ministry and ultimately to the forces manning Russia's early-warning system. This oversight, however, was just one factor that had caused the crisis. There was a much larger, more fundamental reason.

During the forty-five years of the Cold War, the Soviet Union expended vast resources in a massive defence programme under which it amassed a huge arsenal of nuclear weapons totalling in 1991 an estimated 11,159 warheads. As part of a massive military infrastructure which covered the Eastern Bloc, the Soviets established a two-tier early-warning system known as 'Kazbek' and comprising seven ground-based Dnestr and Dnepr 'Hen House' radars which provided total coverage of the airspace around Soviet territory.

In 1982, nine military satellites were launched to detect launches of US land-based ICBMs: these were positioned in high-elliptical orbit, looking down at an angle and searching for infra-red emissions that would signal the heat of missile exhausts. The satellites were, however, inclined to stray from orbit and had to be replaced frequently. The system was designed with a certain amount of built-in redundancy, seven satellites being sufficient to cover the entire United States. Information on any launch would have been transmitted down to a network of radar vessels, constantly criss-crossing the oceans and seas on the Earth's surface, which would in turn have relayed it to the headquarters of the General Staff in Moscow.

During the late 1980s a second system, comprising four satellites in geostationary orbit, was also deployed. Its exact role has never been ascertained but it is thought to have been tasked with covering gaps in the coverage of the nine high-elliptical orbit satellites. The geostationary satellites' ability to cover the oceans patrolled by American and British SSBNs has never been confirmed but Paul Podvig, of the Center for Arms Control, Energy and Environmental Studies in Washington, believes that at least one of them, Cosmos-2224, is able to do so. But despite the addition of the geostationary satellites gaps still remained. During every twenty-four hours the high-elliptical orbit system was blind for two periods of six hours and one hour respectively. Even with the assistance of one of the geostationary satellites, there was still a period of three hours each day when there was no Soviet satellite surveillance at all.

This highly sophisticated early-warning system, however, fell into disrepair during the 1990s with the radars, which had been built in the 1960s and 1970s, reaching the end of their operational lives by the late 1990s. They were due to be replaced by 'Daryl' and 'Volga' type radars with further early-warning stations being constructed to ensure 360° coverage of Soviet airspace. Two-thirds of the construction of the new stations had been finished by the early 1990s but the subsequent collapse of the Soviet Union not only prevented the work being finished but also left some of the radars outside Russia in newly independent states which no longer wished to accommodate them: at Mykolayiv in the Ukraine; at Mingacevir in Azerbaijan and at Balqash in Kazakhstan. One of the key locations was in Latvia from where it covered the areas over the North Atlantic and North Sea where American and British SSBNs were believed to be on patrol. In May 1995, however, following their independence from the Russian Federation, the Latvians destroyed a newly constructed radar station, thus opening a gap in radar coverage to the west only partially covered by satellites, a number of which had ceased to function; by 1999 only three high-elliptical orbit and two geostationary satellites were still operational. Another radar station was built in Belarus to compensate for the one lost in Latvia, being completed in late 1998 and early 1999.

Partially blind, the Russians were thus forced to assemble a relatively primitive backup system which provided far less accurate data. In addition to some of the satellites no longer being operational, most of the radar ships were decommissioned. Such were the gaps in coverage that in 1999 American researchers were able to map 'corridors' from the Pacific and the Far East along which missiles could be launched into the heart of Russia.

The fact that Russia's nuclear deterrent force now depends on an unreliable early-warning system has very serious ramifications given that Russia still maintains its Cold War era policy of 'launch-on-warning': launching its missiles when those of the enemy are still airborne. This policy was originally designed to prevent the former Soviet Union's nuclear weapons being destroyed by a pre-emptive strike. Russia currently relies heavily on launch-on-warning because of the high level of vulnerability of its forces. Due to lack of resources to keep their strategic nuclear forces at a constant state of high readiness, the Russians can no longer disperse their mobile and submarine-launched nuclear weapons into forests and oceans. Currently, only one or two of the Russian Navy's SSBNs are deployed at sea at any given time and perhaps only one regiment of nine mobile missiles is dispersed from its base. Consequently, the bulk of the Russian nuclear arsenal is based in its silos, parked in its peacetime bases or moored in naval docks where it is very vulnerable.

While it may have been valid during the Cold War, during which the Kazbek system was functioning efficiently, the launch-on-warning policy is now highly dangerous and the consequences could be nothing short of catastrophic, leading to the destruction of major urban centres in other nuclear powers such as the United States, Britain, France and China. One typical Russian missile, such as the SS-18 Satan ICBM or SS-N-20 SLBM, reportedly would cause at least ten times the devastation of the bomb dropped on Hiroshima in 1945.

Such is the risk of a missile launch resulting from a wrongly identified threat that the West has been forced to take the problem very seriously, despite most Russian military analysts claiming that the chances of such an occurrence are very slim. This is based on the premise that Russia's 3,000 remaining nuclear missiles and launch systems are equipped with safety systems to prevent a mistaken launch. According to these analysts, no Strategic Rocket Forces missile could be launched without the receipt of special codes transmitted by the National Command Authority.

Authority to launch is transmitted via the Cheget briefcase systems accompanying the Russian President, the Defence Minister and the Chief of the General Staff who are the three individuals responsible for making the decision to launch Russian missiles and who as the official holders carry the key to each case. Each Cheget case will respond only to the handprint of its respective official holder who must enter his own password before the system can be activated. Once access is granted, contact is established with the General Staff command centre. Using the integral telephone, each of the three official holders can communicate with military advisers. Any degree of launch can be ordered, ranging from a single missile to an all-out nuclear strike. Once it has been agreed between the President, the Defence Minister and the Chief of the General Staff, the authorization code is transmitted by the national command centre. The order to launch cannot be transmitted from one of the cases alone; it must be confirmed by the other two.

The Cheget system is designed to provide a guarantee against an accidental launch or one based on mistaken data, while at the same time reducing delays in launch authorization to a minimum. It was introduced in 1983 following the deployment of American SSBNs equipped with the Trident I C-4 SLBM in Norwegian waters in 1979. With a Trident taking only ten minutes to reach Moscow, as opposed to the thirty-minute flight time of an ICBM launched in the United States, Soviet leaders needed to be able to respond to a threat more swiftly than had previously been the case.

The requirement for such a rapid-response communication system was highlighted in 1983 when, at just past midnight on 26 September, alert alarms sounded in a secret bunker at Serpukhov-15, the strategic nuclear weapons command bunker from which the Soviet Union monitored its high-elliptical orbit satellites watching US land-based ICBMs. In command of the bunker that night was Lieutenant Colonel Stanislav Petrov. As he sat in the commander's chair, one of the satellites reported that a nuclear missile attack was under way; the computer analysed the information and indicated that five Minuteman ICBMs had been launched from their silos in the United States. Petrov reported the alert to his superiors at the early-warning system's headquarters; they in turn passed it to the General Staff's national command centre which was responsible for contacting the Soviet leader, Yuri Andropov, and with whom the latter would consult concerning the launch of a retaliatory attack.

Fortunately, Lieutenant Colonel Petrov chose to disbelieve the information, deciding that it was a false alarm and advising his superiors as such. His decision was based on the fact that there was no indication of any threat from the Kazbek early-warning system's radar stations, which were controlled from a different command centre; furthermore, having previously been told that a nuclear attack would be massive, he guessed that any American attack would have involved a far larger number of missiles. His decision resulted in a nuclear holocaust being avoided but his subsequent treatment at the hands of his superiors did not reflect any recognition of that. Following the incident he was investigated and questioned at length, his interrogators unsuccessfully attempting to make him a scapegoat. Although he was not punished for his actions that night, neither was he rewarded; he eventually retired to life as a pensioner living outside Moscow.

The false alarm was subsequently traced to one of the satellites which had picked up the sun's reflection off some clouds and had interpreted it as a missile launch. The incident was particularly dangerous as it took place during a period of tension between the Soviet Union and the United States; only a few weeks earlier, Soviet aircraft had shot down the Korean Air Lines Flight 007. It was later described by a former CIA analyst, Peter Pry, as 'probably the single most dangerous incident in the 1980s'.

In the eyes of the Russians, American and British SSBNs still pose a threat and thus they have retained the Kazbek system. In the event of an attack a ten-minute countdown would begin, during which Russian political and military leaders would have to determine how to respond. At every stage, stringent safeguards are supposed to be in place to prevent a mistake or a bad decision made under pressure.

At nine minutes to impact, operators in the radar early-warning centre try to confirm that blips on their screens are missiles. At eight minutes, contact is established with senior officers at the national command centre who activate the three Chegets, alerting the three official holders. A minute later, the three leaders are in contact with the radar early-warning centre, confirming that an attack is in progress, and discuss what action should be taken. At six minutes, the special communication circuit is switched on, connecting the headquarters of Strategic Rocket Force formations with regiments equipped with silo-based missiles as well as others with mobile missile launchers and missile-carrying trains. At the same time, commanders of Russia's Northern and Pacific Fleets are also alerted and order the commanders of their SSBNs to go to a state of immediate readiness.

By five minutes, the three Russian leaders will have had to come to a decision and transmitted their orders to the national command centre, which in turn transmits the unblocking codes to missile and submarine commanders. At three minutes to impact, missiles are prepared for launch. Under the safety procedures, officers in the field or at sea must confirm that the orders received by them are genuine. At two minutes, missile commanders use keys to activate their missile systems and enter the unblocking codes. At one minute to impact, they await the final orders. If it does not arrive, the missiles will not be launched. If it does, firing buttons are pressed.

The safety procedures carried out at every stage are designed to prevent an unauthorized or mistaken launch but there is now evidence that the chain of command can be circumvented. According to one report, Strategic Rocket Force commanders have stated that it is now possible for a junior officer to launch a missile without authorization from the national command centre.

As recounted in the previous chapter, since the collapse of the Soviet Union and the breaking away of some independent states, all nuclear weapons previously based in the Ukraine, Kazakhstan, Belarus and other states were withdrawn to Russia where they were stored at locations all over the country. Some have since been dismantled and destroyed under the auspices of the Co-operative Threat Reduction Program but, according to official figures released on 30 July 1999, Russia still possesses a large number of strategic nuclear weapons comprising just over 7,000 warheads carried by ICBMs of the Strategic Rocket Forces, SLBMs of the Russian Navy and bombers of the Russian Air Force. In addition, it possesses between 6,000 and 10,000 tactical nuclear weapons.

During the Cold War, the Strategic Rocket Forces were considered the premier service in the Soviet Union. Originally formed in 1946 as a special purpose brigade of the Supreme High Command Reserve, in 1960 they were designated the 24th Guards Division of the Strategic Rocket Forces (Raketnyye Voyska Strategicheskogo Naznacheniya – RVSN). On 7 May 1960 the RVSN was elevated to the status of an armed service in their own right, on a par with the other four Soviet armed services: the Army, Navy, Air Force and Air Defence Forces. All strategic nuclear missiles with a range of over 1,000 kilometres (621 miles) were thereafter assigned to the RVSN while tactical nuclear weapons remained the responsibility of the Army, Navy and Air Force. By 1962, the year of the Cuban missile crisis, the

RVSN comprised 110,000 men and, as the Soviet Union continued its massive programme of nuclear weapon development in its efforts to seek parity with the United States, thereafter expanded further. By the early 1970s, the Soviet Union had achieved superiority in numbers of nuclear weapons, by which time the RVSN numbered some 350,000 men.

Control of all RVSN formations was, and continues to be, exercised directly by the Supreme Commander-in-Chief via the national command centre of the General Staff and the RVSN's own main headquarters, using a multi-level network of command posts on twenty-four-hour monitoring watch. Twelve thousand personnel are responsible for maintaining weapons and communications systems in a fully operational state and, in the event of an alert, reacting instantly by bringing missiles to an immediate state of readiness for launching.

The aftermath of the collapse of the Soviet Union, however, has seen large-scale deterioration in the majority of the Russian armed forces. While the RVSN is still supposed to be first and foremost among the Russian armed forces, it has nevertheless suffered. Previously, officers and soldiers were well paid and such was the standing of the RVSN, based on the importance of their role, that their commander-in-chief took precedence over his counterparts in the other four armed services. The situation is now very different. Officers often go without pay for many months and when it is received, it is pitifully low. Like the rest of the Russian armed forces, budgets within the RVSN are tight and bills frequently go unpaid; it is common for the heating in missile bases to be switched off, requiring personnel to work in very low temperatures. At some bases, the underground command posts are crumbling and suffer from flooding. Many soldiers have been forced to moonlight as taxi drivers and, in extreme circumstances, to beg. In some instances, officers in the RVSN have vented their anger by going on strike. In the late 1990s, the ever-increasing degree of hardship resulted in suicide rates increasing throughout the Russian Army to unprecedented heights.

According to Dr James Thompson, a psychologist at University College London who has spent several years investigating the effects of low morale in the Russian armed forces, 'You have to work in the hope that everyone else is going to do their job properly and that you'll be paid at the end and there's someone evaluating your progress, to praise you when you do well. When that breaks down, people tend to go into their own little groups, with their own agendas. They will see what they can get away with, in testing the system. You will believe that no one is going to inspect you, if you fool around. Nuclear weapons and people have always been a bad mix, but when those people haven't been paid, are badly managed and are within a system which has totally lost morale, then there is a danger of unauthorized actions and indeed, those could culminate in an unauthorized launch.'

Further problems have been caused by RVSN units being understrength. With a current total strength of some 100,000, of which 50,000 are conscripts, the RVSN is at approximately 85 per cent of its establishment. This has resulted in officers of all ranks having to carry out spells on alert duty more frequently, on average a total of 130 twenty-four-hour periods per year. Of RVSN officers, 99 per cent have a degree in engineering and over 25 per cent of non-commissioned personnel are volunteers; the remainder are conscripts carrying out two years' service, more than half of whom have no secondary education.

One of those who has dared to voice his concern over the deteriorating situation is Colonel Robert Bykov, a former officer in the RVSN who has become increasingly anxious about the lack of safeguards for launching missiles of which he has extensive knowledge. During the 1960s he was involved in developing and improving them, carrying out tests in remote areas of Siberia. Currently semi-retired, he now investigates problems within the RVSN and his findings have made him very unpopular, so much so that he fears for his life and with good reason – in 1994 an attempt was made on it. After receiving a tip-off that some secret documents had been left in a briefcase, he arranged to meet two other officers and a close friend, a journalist named Dima Kholodov, at the location where the briefcase had been left for them. Kholodov arrived before the others and opened the briefcase. Unfortunately, it contained a bomb which exploded, killing the journalist; the principal suspects were believed to be RVSN officers who resented Bykov's team investigating their affairs. Since then Bykov, who has continued his investigations, has lived behind steel doors at his apartment and has tried to stay out of trouble.

Such was his concern over the current state of the RVSN, however, that Bykov continued to speak out. In 1997, he wrote an article published in a Russian newspaper, *Komsomolskaya Pravda*, in which he disclosed that Russia's nuclear forces were in danger of falling apart, stating that equipment and systems frequently ceased to operate properly and that the central command systems could no longer be relied upon, periodically going into 'loss of regime mode' in which it would refuse to transmit commands. Even more worryingly, he also revealed that there had been reports of individual missile silo systems switching automatically to combat mode, although he maintained that the main command system would prevent an accidental launch. Following the publication of his article, Bykov was investigated by the Federal Security Service.

Bykov's article had repeated similar claims made shortly beforehand by Defence Minister General Igor Rodionov. In January 1997, increasingly concerned at the continuing disintegration of Russia's strategic nuclear forces, he attempted to contact President Boris Yeltsin, but he was ill and unable to see Radionov. He resorted to writing a letter, voicing his anxiety and informing the President that command and control systems, including the early-warning system and the underground command posts, were disintegrating. He advised Yeltsin that the reliability of the control systems could no longer be guaranteed and that the situation might soon be reached whereby Russian nuclear weapons could no longer be controlled. When Rodionov was finally granted an interview with Yeltsin, he was rebuked for writing the letter.

Almost immediately afterwards, Rodionov was dismissed and replaced by General Igor Sergeyev, commander of the RVSN, who subsequently acknowledged that the command and control systems were ageing and that the RVSN was suffering major problems, including shortages of trained manpower and lack of housing for up to 17,000 of its officers.

During the heyday of the Soviet Union, the RVSN always enjoyed top priority in terms of funding and personnel. With the arrival of Mikhail Gorbachev, however, funds for maintenance and upgrading were slashed drastically and inevitably equipment and systems suffered as a result. Ageing computers developed faults and began to malfunction. The command, control and communications network was beyond its intended lifespan and was in desperate need of overhauling and replacement. There were reports of an increasing frequency of false signals transmitted for no apparent reason, playing havoc with the computer systems.

As mentioned earlier, there were also incidents of highly automated systems switching to combat mode of their own volition within a matter of seconds, behaving as though they had received launch orders and beginning to execute programmes leading to the firing of weapons. Fortunately, no missile has so far accidentally entered the final stage of combat mode; the absence of the unblocking codes transmitted from Moscow system was designed to prevent this happening. Those codes are required to permit activation of switches which complete launch circuits.

Colonel Robert Bykov maintains, however, that the system is far from foolproof and that bored operators have interfered with the safeguards. Having studied the launch systems, they have become totally familiar with them to the extent that they discovered aspects of which even the designers were unaware. By rewiring the circuits, they were able to circumvent the key switches and complete the launch circuits. Consequently, there have reportedly been instances where some officers could feasibly have carried out unauthorized launches.

According to Bruce Blair, a former officer in the American nuclear forces and now an academic at the Brookings Institution in Washington specializing in research on Russian nuclear weapons, this is entirely possible. 'There are ways to circumvent safeguards; the question is: how much time and impunity are offered to an aberrant unit to do these sorts of things? I'm reminded of the situation in Ukraine which essentially could have taken control over strategic missiles based on its territory and, by the Russian General Staff's own estimation, bypassed the existing locking devices, the safeguards on those missiles, within a matter of days to weeks. So clearly, all of these safeguards only work for a period of time.'

In 1996 the Russian Defence Ministry formed a department to investigate any instances of troops discovered to be tampering with missile launch circuit systems. The ministry's principal concern was that if a missile was launched, there was no way it could be destroyed or rerouted in mid-flight. In order to reduce the risk of an unauthorized launch, the ministry imposed an additional safeguard whereby at least two, in some instances three, men were required to launch a missile. Two separate keys had to be turned before the firing button was pressed. According to Robert Bykov, however, this measure proved ineffective in practice. If one man was asleep or left his position, the remaining man could easily operate the system and carry out an unauthorized launch.

In 1994 Russia, the United States and Britain took the major step of detargeting their ICBMs and SLBMs. Missile computers were given a new programme, the Zero Program. If any missile was launched accidentally or without authorization, it would land in the middle of the sea. At least, that was the theory. According to Bruce Blair, the detargeting agreement was entirely cosmetic and symbolic, and had absolutely no effect on the combat readiness of Russian and US forces. Nor did it reduce the risk of unauthorized or accidental use of nuclear weapons.

The truth is that original programmes, containing Cold War targeting co-ordinates, have remained in Russian missile system computers. In the event of a member of the RVSN deciding to launch a missile without authorization, the system can be switched from the Zero Program to the Cold War targets within seconds. Some Russian military analysts fear it may even switch of its own accord. According to Alexander Pikayev, a member of the defence committee of the Duma, Russia's lower house of parliament, irrespective of whether or not a launch is authorized, if a missile is launched the computers would switch from the Zero Program to the Cold War target co-ordinates and thus the weapon would fly against its targets anyway.

In 1997, however, it became apparent that an even more serious threat exists in the form of a device which is potentially far more dangerous than any strategic or tactical nuclear weapon. During the Cold War, both the Soviet Union and the United States developed Special Atomic Demolition Munitions (SADM) for use by special forces. In essence these were miniature thermonuclear bombs, each comprising a plutonium core surrounded by high explosive which, when exploded, as described in 'Dismantling the Bomb', produces an extreme and uniform increase in pressure leading to a nuclear explosion.

In the case of the American device, the W-54 Special Atomic Demolition Munition, it was developed between 1960 and 1963 and some 300 were deployed from 1964 until 1988. Based on the W-54 warhead, it had a variable explosive yield of 0.01 to one kiloton. This was the same warhead as used in the Davy Crockett tactical nuclear weapon system which used a 120-mm or 155-mm recoilless rifle to launch a nuclear projectile designed for use against Soviet troop formations. Weighing just over 23 kilogrammes (51 pounds), the W-54 was the smallest and lightest implosion-type fission bomb ever deployed by the United States. Designed for sabotage attacks against enemy command centres and strategically vital points on lines of communications, such as bridges, tunnels, and dams, the W-54 SADM weighed less than 74 kilograms (163 pounds) and was carried in a special case containing the cylindrical warhead, a code-decoder and firing unit. A larger unit, the Medium Atomic Demolition Munition (MADM), was also developed; weighing less than 182 kilograms (400 pounds), it was in service from 1965 to 1986.

The W-54 SADM's Soviet counterpart reportedly had an explosive yield of one kiloton, sufficient to devastate part of a city and kill up to 50,000 people. If such a device was detonated in the centre of London, for example, it would destroy all buildings within a 500-metre (1,640-foot) radius, killing up to 20,000 people within that area. Up to a radius of one kilometre, there would be more destruction and approximately 50 per cent casualties. Within hours, prevailing winds would carry nuclear fall-out as far as the M25 motorway, requiring London to be evacuated. Unlike the W-54 SADM, the Soviet model was built into a briefcase and, activated by a key, incorporated a timer that could be set for periods from under an hour to several days.

In May 1997 General Alexander Lebed, Russia's former Security Council Secretary, paid a visit to Washington. During a closed meeting with Congressman Curt Weldon, Chairman of the House of Representatives' National Security Committee, he disclosed that eighty-four of 132

SADM-type weapons could not be accounted for by the Russian Defence Ministry. In September, during an interview broadcast on the American television programme *60 Minutes*, and on another with the Russian news agency Interfax, he stated that special atomic demolition munitions had been produced for use by Spetsnaz troops of the Main Intelligence Directorate (GRU) of the General Staff and that he had first been made aware of their existence during his period of tenure as Secretary of Russia's Security Council. Lebed concluded by stating that such weapons, possessing no safety systems of the type incorporated into other tactical nuclear weapons, were ideal for use by terrorists.

Lebed's statement about the existence of Russian SADMs was immediately dismissed by Moscow, which had always denied the existence of such devices, stating that all Russian nuclear weapons were strictly accounted for and stored under the tightest possible security. US State Department officials meanwhile adopted a similarly sceptical attitude, stating that there was no evidence to support Lebed's claims.

Days after Lebed's revelations Vladimir Denisov, who had served as his deputy on the Security Council, suggested during an interview with Interfax on 13 September that all tactical nuclear weapons such as SADMs had been withdrawn to central storage facilities while adding, 'It was impossible to say the same about former Soviet military units which remained on the territory of the other states in the CIS.' He also stated that, 'There was no certainty that no low-yield nuclear ammunition remained on the territory of Ukraine, Georgia or Baltic States or that such weapons had not appeared in Chechnya.' This appeared to support a remark made by Lebed in one of his interviews to the effect that the majority of the GRU's Spetsnaz units were deployed along the former Soviet Union's borders and thus some SADMs may have remained in former Soviet states after the USSR's collapse.

On 22 September, Lebed received support from Professor Alexei Yablokov, a former adviser to President Boris Yeltsin on environmental affairs, who published a letter in the newspaper *Novaya Gazeta* confirming the existence of the SADMs, which by this time were being popularly referred to as 'suitcase bombs', and stating that he had met the scientists who had designed them.

Two days later, however, the newspaper *Pravda* published an article in which a spokesman for Minatom dismissed both Lebed's and Yablokov's claims, repeating earlier government statements that all Russian nuclear weapons were stored under strict control. On 25 September Lieutenant General Igor Valynkin, chief of the Defence Ministry's Twelfth Main Directorate (which, better known as the Twelfth GUMO, is the organization responsible for the storage and security of all tactical and strategic nuclear weapons), also publicly refuted Lebed's and Yablokov's claims. In an interview with journalists, he claimed that all Russian tactical nuclear weapons had been withdrawn to special central storage sites controlled by the Twelfth GUMO to ensure that they did not fall into the hands of terrorists. Valynkin went on to admit that it was technically possible to produce a miniature warhead but denied that the Soviet Union or Russia had ever produced such a device.

Further denials of the existence of such weapons followed from other Russian government departments, including the External Intelligence Service (Sluzhba

Vneshney Razvedki – SVR) which succeeded the KGB's First Chief Directorate, all of whom denied any knowledge of such devices. The head of the National Centre for the Reduction of Nuclear Danger, Lieutenant General Vyacheslav Romanov, also added his voice in denying the existence of such weapons, as did General Lebed's successor as Security Council Secretary, Ivan Rybkin, who stated that no documents relating to SADMs had been found in the council's records.

All such denials were, however, largely undermined on 27 September by a programme broadcast by the ORT television network which recounted how small nuclear devices, with yields ranging from 0.01 to 0.35 kilotons, had been manufactured for geological and oil exploration purposes, and had been used in Kazakhstan during the mid-1970s. In addition, the programme revealed that small nuclear weapons had also been developed for military use but had been returned to Russia in the early 1990s.

On 6 October, General Lebed repeated his claims at an international conference in Berlin, stating that he believed that SADMs had been manufactured in the Soviet Union and that he had been unable to investigate the locations of those missing before being dismissed. On that same day, however, President Yeltsin signed a number of amendments to the Russian Federation Law on State Secrets, classifying all information concerning military nuclear facilities. On 31 October, Professor Alexei Yablokov publicly threatened to release details of the nuclear 'suitcases' unless the President replied to a letter that he had sent him four days earlier.

At the beginning of November, the Kremlin implicitly admitted that SADMs did exist and it was subsequently revealed that in 1996 General Lebed, appointed Security

Council Secretary by President Boris Yeltsin after withdrawing his challenge to the latter for the presidency, had indeed been tasked with mounting an investigation into the number of such weapons manufactured and their whereabouts. A special commission had been formed on 23 July 1996 to check whether such weapons were in the Russian Army's nuclear arsenal, interview any specialist troops trained in their use, and investigate whether similar weapons could be manufactured illegally. Headed by Vladimir Denisov, it included representatives of Minatom and the Russian security services, and by September had reached the conclusion that no SADMs were stored in Russian Army arsenals. Lebed's report, submitted to President Yeltsin, had confirmed the existence of such weapons and stated that only forty-eight of them were accounted for. He had not, however, had the opportunity to pursue his investigations further. On 18 October, both he and Denisov had been ousted from the Security Council and no efforts had been made to investigate further the matter of the missing SADMs.

On 6 November, following the Kremlin's admission, Professor Alexei Yablokov was summoned by the Russian Defence Council and co-opted to draft a presidential decree co-ordinating efforts to locate all 'compact nuclear weapons', bring them under secure control and arrange for their subsequent destruction as soon as possible. Not only was this a further tacit admission that the weapons existed; it also confirmed that they were not under secure controls and thus could be a major security risk.

Meanwhile, following his dismissal by President Yeltsin as Secretary of the Security Council, General Lebed had left Moscow for Krasnoyarsk, a huge province in Siberia, where he was subsequently elected governor. In April, when interviewed for this programme, he was asked to confirm that eighty-four SADMs were missing. Refusing to talk while being filmed, he subsequently revealed off-camera that his public statement about the missing SADMs had led to his being investigated by Russia's State Prosecutor for alleged disclosure of state secrets; if he was found guilty, it would be the end of his political career and thus any further comments made by him would only reinforce the case against him. When Congressman Weldon was interviewed in May 1998, he was quite clear about what Lebed had said to him in 1997: 'He said his job was to account for 132, as the Chief Adviser for Defence to Boris Yeltsin. He said he could only find forty-eight. We were startled. We said, "General, what do you mean you can only find forty-eight?" He said, "That's all we could locate, we don't know what the status of the other devices were, we just could not locate them."'

In September 1998, further confirmation of the existence of the Russian SADMs came when a former colonel in the GRU, the highest-ranking officer ever to defect from the organization, testified in closed hearings before the US Congress. During his testimony, he reportedly stated that he personally had identified and reconnoitred locations for the emplacement of SADMs to be used in the event of war. While he had no knowledge of any of the weapons having been smuggled into the United States, he conceded that it was possible given that a number of them had disappeared from Russia's inventory of tactical nuclear weapons. The colonel reportedly revealed that the weapons could be armed only by trained specialists, the process taking some thirty minutes, and that they would self-destruct if opened improperly. He also disclosed that the SADMs would have been smuggled into the United States in a similar fashion to drugs, by light aircraft or boat, or landed on the coast by submarine. Members of GRU Spetsnaz units would have retrieved them and subsequently concealed them close to their intended targets, one such location being the Shenandoah valley in northern Virginia, only a short distance from Washington. The colonel reportedly also disclosed that during the 1962 missile crisis, SADMs had been stockpiled in Cuba without the knowledge of the Cubans.

During 1999 there was considerable further speculation over the allegedly missing SADMs but no definitive information on their whereabouts. In his interview in May 1998, Congressman Curt Weldon also recounted details of a meeting in December 1997 with Defence Minister General Igor Sergeyev: 'I went to Russia on my thirteenth trip out of fourteen or fifteen that I've taken, last December, and I requested, besides my other meetings, a meeting with the Defence Minister ... And I said to General Sergeyev, after a wide range of topics that we discussed in a session that lasted well over an hour, I asked him specifically: "One, did you build small atomic demolition munitions, as we suspect you did? Two, do you know where they are? And three, have you destroyed them all?" And to me he said, "Yes, we did build them, we are in the process of destroying them, and by the year 2000 we will have destroyed all of our small atomic demolition devices, the so-called nuclear suitcases."'

As Congressman Weldon went on to point out, while he believed General Sergeyev to be sincere, there was no proof that the Russians had in fact located all the SADMs that had been built: 'The key question is not just to make sure that they're destroyed but were there others? Are there others that have not been accounted for? Perhaps not just to the Ministry of Defence, perhaps built by the Ministry of

Atomic Energy, a very powerful entity in Russia. Or built by, or for, the KGB, a completely separate entity, and responsible for all of the intelligence work of Russia. We just don't know the answer.'

The 'suitcase bombs' affair inevitably focused western attention on the issue of nuclear security in Russia. As mentioned earlier, within the Defence Ministry responsibility for security of nuclear weapons lies with the Twelfth Main Directorate, the Twelfth GUMO. Shortly after the end of the Second World War, the Soviet Council of Ministers formed the First Main Directorate with the role of co-ordination of all work on atomic projects. Two years later, a specialist department was created by the Defence Ministry to study US nuclear weapons and their employment. Following the testing of the first Soviet nuclear bomb in 1949, the First Main Directorate and the Defence Ministry department were amalgamated to form a directorate charged with the centralized testing, stockpiling and operation of nuclear weapons.

This new organization was the forerunner of the Twelfth GUMO, which currently maintains large depots in which are kept all tactical and strategic nuclear weapons retrieved from former states of the Soviet Union or withdrawn from service with the Russian armed forces. These are highly secure sites guarded by special Twelfth GUMO units. In addition, the directorate also possesses specially trained troops responsible for transporting nuclear weapons which currently are perceived as being particularly vulnerable to attack from criminal groups or terrorist organizations while in transit.

Although security of nuclear weapons is considered relatively effective, that situation could change. The Twelfth GUMO has also been affected by Russia's severe economic problems which, as mentioned earlier, are severely affecting its armed forces. An indication of this occurred on 5 September 1998, a month after Russia's economic collapse, when five members of a Twelfth GUMO detachment guarding nuclear sites at the port of Novaya Zemlya, Russia's principal nuclear testing facility, killed a senior NCO and, having taken another NCO hostage, attempted to hijack an aircraft. After seizing further hostages, they demanded to be flown to Dagestan in the Caucasus but were subsequently overpowered and disarmed by special forces troops of the Interior Ministry. A month later the head of the Twelfth GUMO, while seeking to allay fears over a breakdown of security at its storage depots, admitted that his troops were not being given any priority over pay and that some were receiving food for the winter instead of their salaries.

Following the incident at Novaya Zemlya, there were two further occurrences of breakdowns of discipline involving military personnel at nuclear facilities: on 11 September a young member of the crew aboard an Akula Class nuclear attack submarine went berserk with a hammer and a submachine gun, killing seven people before locking himself in the vessel's weapons compartment where nuclear, as well as conventional, torpedoes were stored. He was reportedly killed when special forces troops stormed the submarine, although a spokesman for the Northern Fleet later claimed that he had committed suicide. On 20 September, a senior NCO of a detachment of Interior Ministry troops guarding the RT-1 nuclear reprocessing plant at Mayak turned his AK-47 assault rifle on other members of his unit, killing two and wounding another before escaping with his weapon.

The lack of security at civilian facilities, where fissile material is stored, has also given cause for concern. In the words of a CIA report published in September 1998, 'Russian nuclear weapons-usable fissile materials, plutonium and highly enriched uranium, are more vulnerable to theft than nuclear weapons.' According to a more recent report, 'The Next Wave – Urgently Needed New Steps to Control Warheads and Fissile Material' by Matthew Bunn (the Assistant Director of the Science, Technology and Public Policy Program in the Belfer Center for Science and International Affairs at Harvard University's John F. Kennedy School of Government), and published in April 2000 by the Carnegie Endowment for International Peace, the former states of the Soviet Union are estimated to be in possession of approximately 1,350 tons of weapons-usable nuclear material, comprising highly enriched uranium and plutonium, of which 700 tons is contained in nuclear weapons and 650 tons in various other forms. While all the nuclear weapons and 99 per cent of the weapons-usable material are in Russia, quantities significantly large to pose threats are currently stored in civilian nuclear facilities in the Ukraine, Kazakhstan, Belarus, Latvia and Uzbekistan where serious security problems exist. As quoted in Bunn's report, in March 1996, the Director of Central Intelligence, John Deutch, in his testimony before the Permanent Subcommittee on Investigations of the US Senate's Committee on Governmental Affairs, stated that weapons-usable nuclear materials were more accessible than at any previous time due mainly to the collapse of the Soviet Union and the increasingly deteriorating economic problems throughout the region.

During the 1990s there were a number of reported incidents involving the theft of nuclear material from within Russia. During 1992, 1.5 kilograms (over 3 pounds) of weapons-grade highly enriched uranium was stolen from the Luch nuclear research facility at Podolsk, near Moscow, and in July 1993 1.8 kilograms (4 pounds) of 36 per cent enriched uranium were stolen by two sailors from the Andreeva Guna base in the Kola Peninsula, not far from the border with Norway. Subsequently caught and arrested, they stated that they had been ordered by officers to steal the uranium; those investigating the case suspected the possible involvement of an organized crime syndicate.

One highly publicized case involved the theft in 1993 of 4.5 kilograms (10 pounds) of uranium-235 from the Sevmorput naval shipyard, one of the Russian Navy's principal nuclear fuel storage facilities for its Northern Fleet. On 27 November a naval officer, Captain Alexei Tikhomirov, climbed through one of the many holes in the wooden fence surrounding Fuel Storage Area 3-30 and broke the padlock securing the rear door of the building containing nuclear submarine fuel assemblies. The door was connected to an alarm but it had been corroded by the salty sea air and was useless. The three guards supposedly patrolling the area were middle-aged women armed with pistols: not only were they too nervous to use their weapons but were scared of carrying out their rounds at night for fear of being raped. Once inside, Tikhomirov succeeded in breaking off the enriched uranium from three fuel assemblies with the aid of a hacksaw. Minutes later, he left the area in a vehicle driven by an accomplice and headed for the house of his father, a retired captain in the Red Navy and former commander of an SSBN. There he concealed the uranium in the garage.

The theft was discovered on the following morning but only because Tikhomirov had left the padlock lying on the ground in front of the door, to be found by one of the guards doing her rounds. During the following months Tikhomirov's accomplice, a former captain named Oleg Baranov, attempted to find a buyer for the uranium for which he and Tikhomirov were asking 50,000 dollars; among those he reportedly approached was an organized crime group. In June 1994 Tikhomirov's younger brother, Dmitry, a lieutenant on board a nuclear refuelling maintenance vessel who was also involved in the plot as he had briefed his brother on where to find the fuel assemblies, took it upon himself to try and sell the uranium. He asked the help of a fellow officer, who reported the conversation to a senior officer. Shortly afterwards both Captain Alexei Tikhomirov and his brother, along with Oleg Baranov, were arrested and the uranium recovered.

All three men had been driven to such a desperate measure by poverty-induced despair. Alexei Tikhomirov's salary as a captain was equivalent to £140 per month – if he was fortunate enough to receive it. Subsequently tried and found guilty, he was sentenced to between three and five years in prison while Oleg Baranov received three years. Dmitry Tikhomirov received clemency as he had not actually taken part in the theft but was discharged from the Navy.

Following this, up until early 1996 there were five attempted thefts of nuclear submarine highly enriched uranium fuel from Northern Fleet storage depots in the area of Murmansk and Archangel. A Northern Fleet military prosecutor was reported as stating that an organized crime group operating in Murmansk and St Petersburg had made approaches to naval officers, offering between 400,000 and 1 million dollars per kilogram (just over two pounds) of highly enriched uranium. In January 1996, 7 kilograms

($15^1/_2$ pounds) of highly enriched uranium were reportedly stolen from the Pacific Fleet base at Sovietskaya Gavan. Subsequently 2.5 kilograms ($5^1/_2$ pounds) of the material appeared 5,000 miles (8,000 kilometres) away in the hands of a metals trading company in the Baltic city of Kaliningrad.

In November 1993, as recounted by Andrew and Leslie Cockburn in their book *One Point Safe – The Terrifying Threat of Russia's Unwanted Nuclear Arsenal*, an even more serious theft took place at Zlatoust 36, a city in the Urals and one of the centres of Russia's military nuclear complex. Located there was the weapons assembly facility producing MIRV warheads for the ICBMs and SLBMs of the Strategic Rocket Forces and the Navy respectively. Security at the facility, which normally only permitted access to the storage bunkers under strict supervision, had grown lax and on this occasion two men working in the facility succeeded in purloining two warheads and smuggling them out of the plant in the back of a truck. Driving into the city, they headed for a garage where they concealed their highly dangerous booty. It was not until three days later that the warheads were discovered to be missing and the alarm raised. A check of the personnel who had access to the warheads soon pointed the finger of suspicion at the two culprits and shortly afterwards they were arrested and the warheads retrieved. Both were subsequently tried and imprisoned, the entire affair being covered up. The motive for the theft was never revealed although there was speculation that the men might have stolen the warheads for a potential buyer.

In May of the following year, two people were arrested in St Petersburg in possession of 2 kilograms (about $4^1/_2$ pounds) of uranium-235 enriched to 98 per cent; while investigations produced no positive evidence, it was suspected that the material had originated from a nuclear plant east of the Urals. In the same month, 10 May, 5.6 grams (less than a quarter of an ounce) of 99 per cent pure plutonium-239 were found in the garage of a retired mechanic, Adolf Jaekle, in the German village Tengen. The discovery was made by chance during a police investigation into Jaekle, who was suspected of involvement in counterfeiting. When questioned, he maintained that he had obtained the plutonium from a Swiss businessman in Basle, known to have commercial connections in Moscow. The businessman denied any involvement in the affair.

It later transpired that the German intelligence service, the Bundesnachtrichtendienst (BND), suspected that the material had been smuggled out of Russia by representatives of a Bulgarian company called Kintex, which was one of four organizations licensed by the Bulgarian government to trade in arms. Kintex was known to have links with Iraq and its dictator Saddam Hussein. During their search of Adolf Jaekle's house, the Federal Criminal Bureau (BKA) reportedly found documentary evidence of contact between him and Iraqi agents. The BND apparently also suspected that Iraqi agents had arranged payment for shipment of the plutonium to Switzerland, believing that funds for such had been deposited in a bank in Zurich. The affair took an even more serious turn when Jaekle claimed that a further 150 grams (6 ounces) of plutonium had possibly already been smuggled out of Russia and cached in Switzerland, information which was passed swiftly to the Swiss authorities. On 8 August six men, a German and five others from the Czech Republic and Slovakia, were also arrested in connection with this affair. Adolf Jaekle was tried in November the following year and sentenced to five and a half years' imprisonment for smuggling weapons-grade plutonium into Germany.

In August 1994, another well-publicized incident took place when German police and officers of Bavaria's State Criminal Bureau (LKA) met Lufthansa Flight 3369 as it landed at Munich airport and arrested a Colombian, Justiniano Torres Benitez, and two other men. At the same time, they seized a case containing a canister holding over 560 grams ($1^1/_2$ pounds) of mixed oxides of plutonium and uranium, of which 408 grams (just over a pound) comprised plutonium dioxide. According to 'Plutonium, Politics and Panic' published by Mark Hibbs in *The Bulletin of the Atomic Scientists*, the major part of the rest of the material comprised uranium dioxide dominated by depleted uranium. Approximately 87 per cent of the plutonium was fissile 239, the remainder was plutonium-240.

The first the Russians knew of the affair was when it appeared in the international press and on television; instead of notifying them, the Germans had leaked the story to the press, declaring that the material was Russian. The story was much hyped by the international media which proceeded to indulge in an orgy of scaremongering, claiming that further amounts of plutonium and highly enriched uranium were being smuggled into Germany where they were being bought by representatives of terrorist groups and countries such as Iraq, Iran and North Korea.

In the event, the entire affair was subsequently revealed to be a 'sting' operation mounted by the BND and the LKA whose agents had approached Torres Benitez in Spain with a request to sell them 4.5 kilograms (10 pounds) of plutonium for 250 million dollars. The latter had agreed to supply one kilogram for 100 million dollars and had subsequently provided a 0.5 gram sample. Thereafter he had travelled to Moscow, returning on 10 August with his two companions and the canister of fissile material. Although the German authorities claimed that the operation had been a success, others disagreed, pointing out that the operation had resulted only in the arrest of three smugglers and that its primary purpose, identification of the source of the plutonium, had not been achieved.

In December 1994, acting on a tip-off from Interpol, police in Prague seized two canisters, containing 2.7 kilograms (6 pounds) of 88.8 per cent enriched uranium-235, from a car parked in a street in the Czech capital. Analysis found the material to comprise almost 88 per cent uranium-235 and 11 per cent uranium-238 with traces of uranium-234 and 236, minor isotopes not found in fresh fuel, suggesting that the material had been irradiated in a reactor and then reprocessed in a plutonium separation plant.

By late January 1995 three men, a Ukrainian, a Belarussian and a Czech, had been arrested in connection with this affair. It transpired that the uranium had been stolen and smuggled out of Russia in to Prague where it had been awaiting a buyer when it was seized. The Czech authorities were unable to determine its source despite attempts made to identify its 'fingerprint', the unique signature left on fissile material by the process of enriching or burning. No two nuclear processing facilities are identical and thus leave their individual signatures on every batch of uranium, in theory enabling its source to be identified. As the material was in the form of uranium dioxide, some experts in the West suspected that it had been stolen from a plant producing naval reactor fuel. A large number of Soviet submarines were powered by uranium dioxide fuel which allowed the high burn-up required for lengthy operations without the requirement for refuelling. While the majority of Soviet submarine fuel was enriched to 30 per cent and 60 per cent, being produced at Vladivostok, Serodvinsk and Severomorsk, a certain amount was enriched to over 80 per cent.

In March 1997, a group of criminals were arrested in the town of Rubtsovsk, south of Novosibirsk in western Siberia near the border with Kazakhstan, in possession of a quantity of uranium-235. More members of the group were apprehended at Berdsk, another town in the same area where they were planning to hand over the uranium to prospective buyers who were in fact undercover agents of the Novosibirsk County police. Yet again, during investigations into the theft, efforts to trace the source of the uranium proved unsuccessful. Later in the year, a Russian team visiting the abandoned Sukhimi research centre in the Abkhazia region of Georgia discovered that 2 kilograms (4.4 pounds) of 90 per cent highly enriched uranium had disappeared from the facility which had been abandoned in 1993 during the Abkhaz civil war. At the time of writing, they are still missing. In September, Russian police seized 3.8 kilograms (8.4 pounds) of stolen uranium-238 and 2 kilograms (4.4 pounds) of red mercury oxide concealed in a house in Ivanov, a town in the north Caucasus. At the same time, several members of a gang were arrested; they had been attempting to sell the material to prospective buyers in Moscow, the Baltic States and elsewhere. It had been stolen from the nuclear research centre at Sarov, formerly known as Arzamas-16, from where a container of fissile material had previously disappeared in 1994.

In the latter part of 1998, a group of workers at one of the Minatom facilities in the Chelyabinsk region, which included the nuclear weapons design laboratory at Snezhinsk, the processing plant at Ozersk and the nuclear weapons assembly and dismantling facility at Trekhgorny, attempted to steal 18.5 kilograms (40 pounds) of radioactive material later described by Russian officials as being suitable for use in the assembly of a nuclear weapon.

In November of that year, an employee at a leading Russian nuclear weapons design institute was arrested for spying for Iraq. It was precisely to prevent such occurrences that the CTR Program has also channelled resources into defence conversion programmes which includes employment by the International Scientific and Technical Corporation, an organization formed under CTR auspices, of some 24,000 scientists previously involved in the Soviet defence establishment. The purpose of this was to prevent scientists being tempted by offers from regimes such as those of Iran, Iraq and North Korea who were keen to acquire their expertise for their own nuclear weapon development programmes.

During the last decade, a considerable amount of evidence came to light that certain countries were seeking to develop a military nuclear capability, notably Iraq, Iran, Libya and North Korea. During the Gulf War in 1991, Iraq succeeded in concealing much of its programme of weapons of mass destruction from Coalition bombers and cruise missiles: the nuclear complexes at Tarmiya and Ash Sharqat were damaged but not destroyed while the most important facility, at Al Atheer, remained intact. Information from an Iraqi defector, a leading nuclear physicist, and satellite photographic surveillance subsequently revealed the presence of calutrons (electro-magnetic systems for enriching uranium) which were buried by the Iraqis pending visits by UN inspection teams and retrieved thereafter. These were later destroyed but by 1995 it was evident that the Iraqis had changed tack and were seeking weapons-grade uranium and plutonium, thus dispensing with the need for enrichment facilities.

Iraq used its covert procurement networks to source material and equipment, despatching agents throughout the former Soviet Union. In 1995 it was successful in obtaining 120 missile guidance gyroscope systems removed from Russian Navy SLBMs. Three years later, the CIA warned that Iraq was seeking to purchase either fissile material or complete nuclear weapons.

Iran has also been active in searching for nuclear weapons and technology in the former Soviet states. In 1992 Iranian agents were spotted visiting nuclear facilities, in particular in Kazakhstan where they appeared at Ust-Kamenogorsk, Semipalatinsk and Aktau with a list of items they required for the Iranian nuclear weapons programme. Kazakhstan was by then, however, unwilling to upset the United States and the West with whom it was fast establishing close links and thus turned down their approaches. Nothing daunted, Iran turned to the Russians, dangling as bait a multi-million-dollar contract to complete the construction of a nuclear power station, containing two 1,000-megawatt light-water reactors, at Bushehr situated 750 kilometres (466 miles) south of Teheran on the coast of the Persian Gulf.

During 1994 Viktor Mikhailov, the Minister of Nuclear Energy and head of Minatom, carried out a number of visits to Iran. On 8 January 1995 he signed a contract for the Bushehr project with Iran's Atomic Energy Organization, which was headed by Reza Amrollahi, the Iranian Vice-President who was also responsible for his country's nuclear weapon development programme. According to Andrew and Leslie Cockburn in their book *One Point Safe*, the contract was for the construction of 'Block No. 1' at the nuclear power facility at Bushehr. Two months later, however, a copy of a protocol relating to the contract was passed to a member of the US Embassy in Moscow, revealing that among the facilities to be supplied by the Russians were a gas centrifuge and a desalination plant. This caused alarm in Washington, as a gas centrifuge is a system used for enrichment of uranium while a nuclear-powered desalination plant of the type supplied by the Russians produced, as a by-product, plutonium of a very high level of purity.

Approaches to the Russians, and Mikhailov in particular, to try and persuade them not to proceed with the deal proved fruitless. When the news of the sale of the two reactors was made public, however, there was a storm of protest led by Professor Alexei Yablokov who, despite Mikhailov's statements to the contrary, knew that plutonium could be extracted from spent fuel from such plants. Yablokov investigated the details of the sale and discovered the protocol, which resulted in his publishing an article in *Izvestia*, giving full details. This gave the Americans the opportunity to publicize the issue which had hitherto been kept tightly under wraps. In May President Bill Clinton, equipped with a comprehensive report on the entire Iranian military nuclear development programme, attended a summit in Moscow during which he discussed the matter with President Boris Yeltsin. The latter refused to halt the sale of the two reactors but agreed to block that of the gas centrifuge which, he agreed, could be used in the production of nuclear weapons.

In August 1995, however, following the visit of an Iranian delegation led by Reza Amrollahi, the Russians signed a second contract with Iran for the supply of two 400-megawatt reactors for the Neka nuclear research complex in northern Iran which is believed by western experts to be part of the Iranian military nuclear research programme. In September, it was reported that China was constructing a calutron at the nuclear research establishment at Karaj, some 160 kilometres (100 miles) north-west of Teheran. The same report disclosed that large numbers of Russian scientists were working at Iran's eleven nuclear research establishments, of which only five were open to international scrutiny.

While the idea of Iraq and Iran possessing a military nuclear capability is one nightmare scenario for the West, that of terrorists laying their hands on them is another. There have been numerous reports that Al Qaida, the Islamic fundamentalist organization headed by Osama bin Laden, has long been seeking to acquire nuclear weapons through its contacts in the Ukraine and other former states of the Soviet Union. Indeed, as mentioned in Matthew Bunn's report 'The Next Wave', the United States has issued a federal indictment of bin Laden and his organization, charging that he and others have on various occasions attempted to acquire components for nuclear weapons.

Another group known to have sought nuclear material and technology is the Japanese cult Aum Shinrikyo, which was responsible in March 1995 for the sarin nerve agent attack on the Tokyo subway. The group had recruited large numbers of members in Russia, among them scientists from Moscow State University and members of staff of the Kurchatov Research Institute in Moscow which possessed relatively large quantities of weapons-usable highly enriched uranium. It is known that a senior member of the cult, Kiyohide Hayakawa, carried out many visits to Russia to procure weapons and other equipment.

Such was the growing concern over the poor physical security and careless accounting at nuclear sites in Russia that the United States instituted a series of programmes to improve the security of, and accounting procedures for, nuclear weapons and fissile material in the former Soviet Union. These included a material protection, control and accounting (MPC&A) Program, funded by the US Department of Energy and designed to improve security and accounting of all nuclear materials at sites throughout former Soviet states; the construction at Mayak of a special secure storage facility, funded by the Department of Defense, for fissile material extracted from dismantled weapons and spent fuel assemblies; provision of equipment for upgrading of security of nuclear weapon storage facilities and transport; and programmes for the removal of weapons-usable nuclear material from sites considered to be vulnerable. In addition, the Department of Defense and other US agencies such as the Department of Energy, Federal Bureau of Investigation and the US Customs Service are assisting former Soviet states in improving the capabilities of their own law enforcement agencies to combat the smuggling of nuclear materials.

Meanwhile, the thefts of fissile and radioactive material have continued in various parts of the former Soviet Union. In September 1999, six people were arrested in Vladivostok while trying to sell 6 kilograms (13 pounds) of uranium-238 to undercover police officers posing as prospective buyers; the material was subsequently reported as having been stolen from a shipyard in the far east where

Russian Navy nuclear submarines were serviced. During the following month, two people were arrested by police in western Ukraine after being caught smuggling 24 kilograms (53 pounds) of uranium from Krasnoyarsk County in Russia. During February 2000 three men were arrested in Alma Ata, the capital of Kazakhstan, after being found in possession of 530 grams of uranium (nearly 1½ pounds); in the city of Ussuriysk two military officers were apprehended while attempting to sell 20 grams (about an ounce) of radioactive strontium. The end of February saw police arrest a group of five men in Donetsk, eastern Ukraine, who were trying to sell twenty-eight containers of radioactive strontium-90. At the beginning of April, customs officials in Uzbekistan stopped a truck near the border with Kazakhstan and seized ten steel containers of unidentified radioactive material; the vehicle and driver were Iranian but the latter maintained that his destination was not Iran, but Pakistan.

Despite the efforts of the United States and other countries to reduce the dangers from Russia's nuclear stockpile, its size and infrastructure are such that there is still a very long way to go before its hazards are removed completely. Meanwhile, deteriorating safeguards still pose the threat of an unauthorized missile launch, while poor standards of security and lax accounting in many civilian nuclear establishments are such that nuclear material is still vulnerable to theft. While such conditions exist, the threat of a nuclear holocaust is still present.

AFTER DESERT STORM

High technology and new weapon systems played principal roles during the Gulf War, enabling Coalition forces to overcome those of the Iraqi dictator Saddam Hussein within forty-four days.

At 4.45 p.m. local time on 16 January 1991, at a US Airforce (USAF) base at Khamis Mushayt in Saudi Arabia, pilots and ground crew of the 37th Tactical Fighter Wing were preparing their forty-two F-117A Stealth aircraft for a raid in the early hours of the following day on the heavily defended capital, Baghdad. Possessing no guns or radar, the F-117A relied solely on technology to protect it during its mission. Coated in radar-absorbent material, the aircraft's skin would soak up much of any energy aimed at it, its unconventional shape deflecting any further beams away from the interrogating transmitter. Indeed, the radar signature of the F-117A is the same as a small bird or large insect.

Its existence officially denied until six years after it entered service in 1982, the F-117A is a single-seat high-subsonic aircraft powered by two General Electric F404 engines located above the fuselage within deep recesses to minimize their infra-red signature. Costing 45 million dollars, it is equipped with sophisticated navigation and attack systems integrated into a state-of-the-art digital avionics suite designed to increase its mission effectiveness and reduce pilot workload, with detailed planning for missions into high-risk target areas being carried out by an automated mission planning system.

Its primary role being low-level precision strike, the F-117A is armed with air-to-ground missiles and laser-guided bombs carried in two internal weapon bays. Weapons carried by F-117As during the Gulf War included the GBU-27 improved 2,000-pound bomb and the Mk.82 500-pound bomb, both fitted with the Paveway III laser guidance system. The aircraft is equipped with a forward-looking infra-red (FLIR) system producing high-definition images on a monitor in the instrument panel, enabling the pilot to view and identify a target. He then switches to a second downward-looking infra-red system, boresighted with a laser target designator, locks the laser on to a target and at an appropriate point releases his laser-guided bombs which, as will be explained later in this chapter, home in on the infra-red energy reflected from the target.

At 2.35 a.m. local time on 17 January, 966 kilometres (600 miles) away in the Persian Gulf, another high-technology weapon was about to be launched against Saddam Hussein, this time an unmanned RGM-109 Tomahawk cruise missile. Ten minutes later on board the cruiser USS *Bunker Hill* and other warships, night turned to day with flame erupting from the vessels' decks as Tomahawks burst out of their launchers, hurtling up into the air before deploying their small, stubby wings and heading away into the night towards Baghdad. Meanwhile, UGM-109 Tomahawks burst forth from the sea, having been launched from submarines below the surface. Just over fifty minutes later, the night was filled with explosions and balls of flame as the missiles impacted on their targets, the sky illuminated by glowing balls of tracer and the muzzle flashes of anti-aircraft weapons as Iraqi gunners fired wildly into the air, desperately attempting to bring down the invisible enemy. Such was the intensity of the anti-aircraft fire that eyewitnesses later described it as an 'awesome fireworks display'.

To the F-117A pilots flying through the storm of anti-aircraft fire on that first night of the war, it seemed inevitable that they would be hit – but in the event all survived unscathed, having successfully completed strikes against command posts and communications centres as well as radar and surface-to-air missile sites. Air Vice Marshal Bill Wratten of the Royal Air Force, deputy commander of British forces during the Gulf War, later commented, 'I think everybody was well pleased with the success of the Stealth fighter, the F-117. It did all that was expected and hoped of it. It went in, as far as we know, unseen and came out unseen. The very early waves were targeted against the Iraqi communications systems for obvious reasons. Now the fact that he [Saddam Hussein] stopped broadcasting at the precise time on target of the first wave was reasonable indication of that first wave being successful.'

As Colonel John Warden of the USAF's War Fighting Unit later pointed out, 'Within the first ten minutes of the war, there had been attacks on the telephone system, the electrical system, on all of the sector operations centres, the national air defence operations centres, the associated intercept operations centres, and on significant command posts at strategic and operational levels. This happened across the entire breadth of the country and it happened, for practical purposes, simultaneously. Once we had made the first attack, he [Saddam Hussein] was doomed because we had imposed strategic paralysis from which he did not recover.'

Colonel Warden's observations on that first night's attacks on Baghdad were all too accurate. By the end of that first night of air strikes, the Iraqis realized they were deaf, dumb and blind. Stealth technology had not only received its baptism of fire but had also dramatically proved its worth in more than one way. In addition to knocking out the entire Iraqi command and communications system in Baghdad, the Stealth aircraft had also proved that they were capable of achieving the same results as large numbers of conventional aircraft requiring considerable support.

As Lieutenant Colonel David Deptula, a USAF war planner, explained: 'What Stealth allows you to do is to obviate the need for large force packaging that you need with conventional aircraft assets. You don't need all the suppression of enemy air defences aircraft, what we call SEAD, to go along with Stealth to protect them. That allows you to free up a whole series of assets to strike a wide array of targets simultaneously. An example of this took place early on in the war when we had a conventional attack package consisting of four Saudi Tornadoes, and four A-6s. Since it was relatively early on in the conflict, these were escorted by four F-4Gs designed to suppress enemy air defences, five EA-6Bs, electronic combat jamming aircraft, and twenty-one F/A-18 Hornets carrying high-speed anti-radiation missiles to take out surface-to-air missile sites. A total of thirty-eight aircraft to get four bomb droppers across one target. At the same time, we had twenty-one F-117s attacking thirty-eight targets. That is the value of Stealth – it allows you to accomplish much more than if you only had conventional aircraft.'

Dawn on 17 January revealed to the Iraqis the appalling scale of destruction and havoc wreaked by the F-117As and cruise missiles during that first night. Thereafter the barrage continued remorselessly. In the Persian Gulf, more of the 1.3 million-dollar missiles lanced upwards from launchers on the battleships USS *Missouri* and *Wisconsin* and headed for Baghdad. Guided by its terrain-following radar, each missile followed its individually pre-programmed route as it flew towards its own target.

The Tomahawk is a 6-metre (20-foot) long missile powered in the initial stage of its flight by a rocket that launches it from its tube on board a ship or submarine. Thereafter, a small turbofan cuts in and the missile proceeds to fly at a speed of approximately 885 kilometres per hour (550 miles per hour) on a course towards its target, initially steered by an inertial guidance system that uses gyroscopes and sensors to govern changes of course and speed. Once it reaches a coastline, another more precise system, called TERCOM, takes over and navigates by scanning the surrounding landscape at pre-set intervals, measuring altitude readings and comparing them to pre-stored data in the computer's memory. Flying at altitudes between 30 and 90 metres (100 to 300 feet), the missile follows the contours of the terrain. As it nears its target, another guidance system takes control. Called the Digital Scene Matching Area Correlator (DSMAC), it digitally photographs the target area and compares the image to one already stored in its memory, resulting in final adjustments being made to the Tomahawk's course as it begins its approach to the target.

Such were the heavy defences in and around Baghdad that only cruise missiles were employed in attacks on the city during daylight. Aircraft were, however, used during the day on targets elsewhere, principally against Iraqi positions in Kuwait and near the Saudi border. As dusk fell on 17 January and the aircraft returned from their first day of operations, preparations were being made for a second night of attacks. Among the aircraft employed on nocturnal raids was the venerable F-111, an aircraft that had been in service with the USAF for twenty-five years. Although elderly, it carried an array of sophisticated bombs, precision-guided munitions popularly known as 'smart bombs'. Costing 80,000 dollars apiece, these are conventional bombs fitted with a laser-seeking sensor, a simple guidance mechanism and steering fins on the nose. They are, however, over four times more accurate than their conventional counterparts.

A 'dumb' bomb, as conventional weapons are now called, will fall along a parabolic trajectory governed by a combination of forces: gravity, wind resistance and the forward velocity of the launching aircraft. However, dumb bombs are also affected by other factors, principally winds, air currents, aerodynamics and structural asymmetrics, which cause them to deviate from their intended path. Such factors would result in half of a stick of four dumb bombs dropped from an altitude of about 3,000 metres (10,000 feet) missing the aiming point on the ground by up to one-eighth of a mile, an area of lethality known as the circular error probability (CEP). During the Second World War, bombs dropped from high altitudes on Germany in daylight raids by USAF B-17 Flying Fortresses landed as much as a mile from their targets, resulting in a CEP of 1,015 metres (3,300 feet). This meant that 9,000 bombs would be needed to hit a small target area measuring 18.5 metres (60 feet) by nearly 31 metres (100 feet). Almost 1,000 B-17s would have been required to drop such a massive amount of ordnance, a number beyond the capabilities of any contemporary air force in terms of both aircraft and aircrew.

It was during the Vietnam War that tests were conducted with laser-guided bombs. Since then, advances in technology have resulted in munitions of increasing precision and such is the accuracy of current weapon systems that a bomb can be guided precisely to the required point of impact. Such a high level of accuracy, producing devastating results with the minimum of collateral damage, is achieved by the attacking aircraft illuminating a target with the invisible beam of a laser emitting coded pulses of infra-red light. In effect, the target acts as a laser lighthouse, reflecting the infra-red energy back into the sky in the form of a cone. When first developed, the system was classified as top secret and codenamed Pave Tac.

A typical attack carried out by two F-111s would see the Pave Tac aircraft arriving at the target area with the weapons officer's radar and the Pave Tac infra-red detector set observing it. The laser provides information about the target, providing information on range to it, which allows the pilot to position his aircraft to provide the best opportunity for the sensor on the other aircraft to 'see' the target. The weapons officer switches to a video view of the target as seen by the Pave Tac system and maintains the laser's lock on the target. Meanwhile, the bomb from the second aircraft has been released and is flying down into the centre of the cone of infra-red light, at the point of which is the target. The system is so precise that it allows destruction of the target with minimal collateral damage and loss of life within the immediate surrounding area.

On the night of 17 January, the second wave of air strikes on Iraq took place but in the early hours of the following day Iraq struck back, launching eight ballistic missiles against Israel.

Until then, the Coalition forces in Saudi Arabia had not taken seriously the threat of Iraq using its missiles which were based on a modified version of the Scud-B short range ballistic missile (SRBM). Developed in the 1960s by the Russians, the Scud had the capability to deliver conventional high-explosive, chemical or nuclear warheads. Not a particularly accurate weapon, it was intended for use against large targets such as major storage dumps, marshalling areas and airfields. The initial version, designated by NATO as the SS-1B Scud-A, was mounted on a tracked chassis, being raised into the vertical position from a small platform. A larger version, the SS-1C Scud-B, was subsequently produced. With a maximum range of 280 kilometres (174 miles), it was carried in a protective casing on a large MAZ-543 eight-wheeled transporter that also accommodated all the necessary support elements for the missile.

Iraq's involvement with ballistic missiles had begun during the previous decade when, while at war with Iran, it had invested heavily in missile research and development, using technology, equipment and components acquired from the West. The main missile development centre was Saad 16, located outside Mosul in northern Iraq, which was constructed in strict secrecy by a consortium of European countries, notably Austria and West Germany. A web of Swiss, West German and Austrian companies provided high technology equipment and systems covertly sourced from Britain, Europe and the United States.

In 1984 Iraq joined forces with Egypt and Argentina to develop a missile. Designated Condor-2, it was a two-stage solid-fuel-powered missile with a payload capacity of 500 kilograms (1,100 pounds) and a maximum range of 1,000 kilometres (620 miles). Meanwhile, Iraq purchased a number of Scud-Bs from the Soviet Union and proceeded to modify them by reducing the 1,000-kilogram (2,200-pound) warhead to 500 kilograms (1,100 pounds) and enlarging the size of the fuel tanks in order to increase the maximum range to 600 kilometres (373 miles). The modified missile, redesignated 'Al-Hussein', was subsequently used with devastating effect against Iranian cities in the final stages of the Iran–Iraq War of 1980–8.

During 1987 Iraq additionally acquired over 120 Badr-2000 missiles from Egypt, also using them to blitz Iran. These were modified Scud-Bs too, having a reduced payload of 275 kilograms (605 pounds) and a range of 600 kilometres (373 miles). Meanwhile, Egypt shipped three prototype Condor-2 missiles to Iraq for trial use against Iran and in the following year supplied sixteen more. It was estimated that Iraq used approximately 360 missiles during the war with Iran, launching some 200 over a period of forty days between February and April 1988.

In December 1989 the Iraqis tested an intermediate range ballistic missile (IRBM), claiming it had a range of 2,000 kilometres (1,243 miles). Designated Al Aabed, it comprised a three-stage type powered by five Scud boosters and carried a 750-kilogram (1,650-pound) warhead. In November 1990 they claimed to have tested another, designated Al Hijara, also with a range of 2,000 kilometres. The appearance of these two missiles caused concern as it meant that the Iraqis possessed the capability to reach Israel, Saudi Arabia and the Gulf States.

The eight Scuds that landed in the area of Tel Aviv on 18 January fortunately caused only limited damage, although sixty-eight people were injured. However, the inaccuracy of the missile was irrelevant in such circumstances; with a target as large as a city the size of Tel Aviv, a Scud would cause damage wherever it fell. The lack of physical damage also mattered little as the political impact of the attack was considerable and sent ripples reverberating around the world. Indeed, this had been the precise objective behind the use of such an imprecise weapon. While on this occasion the missiles had been armed with conventional high explosive, it was all too obvious that in further attacks against Israel or Coalition forces the Iraqis might use chemical warheads. It was well known that they had used chemical weapons during the Iran–Iraq War; on 16 March 1988 they had done so against the Kurdish town of Halabja, massacring its occupants.

It was clear that the Iraqi missiles, whether or not they were carrying chemical warheads, had to be stopped and the main hope of doing so lay in the US Army's MIM-104 Patriot Tactical Air Defence Missile System, units of which were rushed by air to the Gulf to join the Coalition forces, some being deployed to Israel.

Developed as an anti-aircraft weapon during the late 1970s, the Patriot was modified in the mid-1980s to provide defence against ballistic missiles also. The 5.38-metre (17.5-foot) long missile is a single-stage solid-fuel rocket-motor powered type giving a speed of Mach 3. Weighing 1,000 kilograms (2,200 pounds) and carrying a 100-kilogram (221-pound) high-explosive fragmentation warhead, it has an effective range of between 65 and 68 kilometres (40 to 42 miles) and an altitude ceiling of 24,000 metres (78,740 feet).

Each Patriot unit is equipped with eight launchers, which also serve as containers for transportation and storage, each launcher having four missiles. The system is mobile, each four-missile launcher is mounted on a wheeled trailer. The system's fire control unit comprises the MPQ-53 phased array radar which performs IFF (Identification Friend or Foe) interrogation, target acquisition, tracking and missile guidance; the MSQ-104 engagement control station which accommodates the system's fire and operational status control computer; and the MJQ-24 power plant.

Once a target has been detected, the Patriot is launched and is initially steered to the target by the missile's guidance system, which turns it towards the target as it flies into the radar's beam. Thereafter the system's computer directs the missile to the target until it becomes semi-active as its own radar receiver steers it to the point of interception.

During the following weeks of the Gulf War, much publicity was given to the Patriot which was perceived, along with the F-117A Stealth fighter, cruise missiles and laser-guided bombs, as yet another example of the West's technical supremacy. The incoming Al-Hussein missiles were mostly detected at a range of some 112 kilometres (70 miles) and were engaged at 16 and 32 kilometres (10 and 20 miles). Television audiences were able to watch and marvel as Patriots soared upwards into the night skies over Israel and Saudi Arabia, intercepting the Iraqi missiles with thunderous explosions and impressive balls of flame. In fact, those attempting to shoot down the Iraqi missiles faced an extremely difficult problem.

At the point of launching, an Al-Hussein missile weighed just over 5 tons, most of which comprised liquid propellant. At the top of its trajectory it would reach an altitude of over 150 kilometres (93 miles) before heading down towards its target; at this point it would be detected by a Patriot battery's radar whose high-speed computers would immediately calculate its trajectory velocity and potential impact point. With the missile within range at under 68.5 kilometres (42.5 miles), a Patriot would be launched and in theory would intercept the Al-Hussein and destroy it. However, in this instance the theory was wrong.

The original Scud-B was a large slow missile, providing an easy target. The Al-Hussein, however, not only had increased range but also greater speed, travelling at approximately 8,000 kilometres (5,000 miles) per hour, and thus intercepting it was far more difficult. A further problem was that if the missiles re-entered the Earth's atmosphere at an angle, which they appear to have done on a number of occasions, unequal aerodynamic forces caused their bodies to break up. The large pieces of debris falling to earth created solid tracks on the Patriot radars, which assumed they were missiles; on the first night that they were used in Tel Aviv, twenty-eight Patriots, costing 17 million dollars, were launched at the debris of five Al-Hussein missiles.

Once it was realized that the missiles were breaking up, the Patriot radar operators quickly learned to distinguish between debris and the warheads and to engage the latter. As was later pointed out, however, the missile debris falling earthwards at a speed of some 800 kilometres per hour (500 miles per hour) could still kill or certainly cause considerable damage.

Saddam Hussein's objective in launching his missiles against Israel was, however, political rather than military. He was attempting to draw Israel into the war, putting pressure on the Arab elements within the Coalition. His hope was that they would withdraw from it rather than fight on the same side as Israel against an Arab nation. Patriot was deployed to Israel to prevent just such a situation arising and thus, from a political point of view, proved a success.

While the Iraqis' fixed-missile launch sites were attacked and destroyed by air strikes during an early stage of the war, the mobile launchers continued to cause a problem. British and US special forces were given the task of hunting them down but they were difficult to locate, the Iraqis having dispersed them widely, concealing them from airborne and satellite surveillance under bridges and highway flyovers. Emerging to launch their missiles, the transporter/launchers would return to hiding where they would rendezvous with support vehicles and reload. A subsequent study by the Chief of Staff of the USAF, General Merrill McPeak, estimated that between 18 January and 26 February 1991 the Iraqis launched forty missiles against Israel and forty-six against Saudi Arabia.

Location and destruction of the mobile launch vehicles became a top priority for the Coalition in its efforts to prevent Saddam Hussein achieving his aim of drawing Israel into the war. The answer to the problem of locating them was found in the airborne element of JSTARS, the acronym for the US Army/USAF Joint Surveillance Target Attack Radar System, which at that time was still under development.

Designed to carry out airborne surveillance from a stand-off distance of 250 kilometres (155 miles), JSTARS is capable of detecting, locating and tracking fixed and mobile targets on the ground over an area of more than 20,000 square kilometres (7,720 square miles) in all weather conditions, providing commanders with near real-time data. Installed in modified Boeing 707-300Bs designated E-8C STARS, the system's main element is the

Northrop/Grumman AN/APY-3 high performance, multi-mode airborne radar that detects, locates and classifies ground targets. The system is interface with the US Army's Ground Station Module (GSM) via a special data link, the Surveillance Control Data Link (SCDL), to transmit and receive target data. With a ceiling of 12,500 metres (41,000 feet), the E-8C aircraft's operating altitude is 10,700 metres (35,000 feet). Its maximum range is 4,480 kilometres (2,800 miles) and it has an operational endurance of eleven hours without refuelling. The number of crew varies between twenty-one and thirty-four.

On board the E-8C aircraft, radar operators are positioned at consoles that incorporate displays on which are electronically generated maps of sectors of the areas under surveillance, showing roads, rivers, forests and other topographical features. Overlaid on to these are displays of yellow dots, each of which represents a moving object. These are assembled over a period of some twenty to thirty minutes to provide a pattern from which the operator can try to assess what is happening in his sector of responsibility. If he detects organized movement or something that is indicative of, for example, a column of moving vehicles, a synthetic aperture radar picture will be taken of that sector to identify the items detected. Such methods proved highly successful in that they resulted in the detection and location of Iraqi assembly areas and, through the use of imagery analysers, the types of vehicles in them.

Initially JSTARS was used to concentrate on Iraqi armoured formations but once the Iraqis commenced their ballistic missile attacks, it was thereafter used to assist in locating the mobile launchers. On several occasions, the system located launch sites and was able to direct aircraft to them, resulting in several cases in interdiction before launches occurred.

JSTARS' downlink capabilities proved especially valuable, enabling surveillance imagery seen in the aircraft to be transmitted simultaneously to ground units. Early in the war, in January 1991, elements of three Iraqi divisions crossed the border in three columns from Kuwait into Saudi Arabia. The easternmost column, comprising over forty tanks and a number of armoured personnel carriers of the 5th Mechanized Division, advanced on the coastal resort and oil town of Khafji 12 kilometres (8 miles) south of the border. It was deserted at the time, its 15,000 inhabitants having been having been evacuated as the town was within range of Iraqi artillery located in the south of Kuwait.

At 8 p.m. on the night of 29 January, the Iraqis encountered Saudi and Qatari troops together with a screening force of US marines of the 1st Marine Division. By 9.30 p.m. air support had arrived, including an E-8C JSTARS which during the following four days of fighting proved invaluable in defeating the Iraqi attack, detecting and tracking the movements of the Iraqi forces and passing information to commanders on the ground while also directing 100 sorties by F-15 aircraft, and forty by F-16s, against ground targets. At one point, it observed a column of enemy armour heading from the border town of Wafra towards Khafji, and directed further air attacks against it; the Iraqi armour never reached the town. During the latter part of the battle, B-52 bombers were diverted for strikes against JSTARS-designated targets.

While JSTARS provided such invaluable support to ground forces, the E-3 Sentry AWACS (Airborne Warning And Control System) did likewise for Coalition aircraft, providing all-weather surveillance, command, control and communications support.

The E-3 Sentry is a modified Boeing 707-320, easily recognized by its rotating 9-metre (30-foot) diameter radar dome mounted above the fuselage. With a ceiling of over 8,840 metres (29,000 feet) and endurance of over eight hours before refuelling, it is equipped with a radar subsystem capable of surveillance over land and water from the Earth's surface right up to the stratosphere. It will detect low-flying aircraft at ranges of over 320 kilometres (200 miles) and medium and high altitude targets at greater distances. Working in combination with an Identification Friend or Foe (IFF) subsystem, the radar can identify and track friendly or hostile aircraft by eliminating ground clutter. Other subsystems comprise communications, navigation and computers for data processing. Radar operators, stationed at consoles which show data in graphic and tabular format, carry out functions including surveillance, identification, weapons control, battle management and communications.

The radar and computer subsystems are capable of acquiring and displaying broad and detailed battlefield data as events occur, including position, tracking and status information relating to aircraft and vessels, both friendly and hostile, which can be transmitted to major command and control centres on land or at sea. When acting in support of air-to-ground operations, the E-3 Sentry AWACS can supply information required for reconnaissance, interdiction and close air support for ground forces, while also supplying commanders of air operations with information essential for control of the air battle. Furthermore, as part of an air defence system, it can detect, identify and track hostile aircraft, directing fighters to intercept them.

During the Gulf War, E-3 Sentry AWACS aircraft were among the first to deploy to Saudi Arabia, thereafter establishing twenty-four-hour radar surveillance on Iraq. Flying over 400 missions in over 5,000 hours on-station, they provided surveillance and control for more than 120,000 sorties flown by Coalition aircraft while also providing up-to-the-minute information on enemy deployments for senior commanders.

By the third week of Operation Desert Storm, the Coalition had air superiority and thus virtual freedom of the skies. This resulted in the decision to intensify bombing raids over Iraq and in so doing to call on another piece of technology which, like JSTARS, was still under development: the Thermal Imaging Airborne Laser Designator (TIALD). Fitted to the Tornado GR1 aircraft of the Royal Air Force, TIALD represented the very latest in 'smart' electro-optical precision guidance systems.

Contained in a pod fastened to the belly of an aircraft's fuselage, TIALD comprises an infra-red thermal imager (TI), a television (TV) sensor, a laser rangefinder/designator, stabilized optics and an automatic video tracking system. A steerable sensor head permits the common sightline of the TI, TV sensor and laser rangefinder/designator to be directed on to targets by either the operator or the aircraft's avionics systems. The TI has the capability to penetrate smoke and haze, the infra-red telescope providing a wide field-of-view, which is used initially for target acquisition, and a narrow one for recognition of the target and aiming of the laser beam. A further 2× and 4× magnification can be achieved through use of an electronic zoom. The TV sensor is equipped with a narrow field-of-view lens and is used to provide images when conditions are such that the contrast in the TI image is adversely affected or when countermeasures are being employed.

The role of TIALD is to steer air-dropped laser-guided munitions directly to their targets. Prior to a mission, target co-ordinates are entered into the aircraft's main computer and en route to the target the operator will boresight the radar to that position. At a distance of some 32 kilometres (20 miles) from the target, using the system's slave mode, he will direct TIALD to observe in the same direction as the radar; and at a range of 24 kilometres (15 miles) will select the narrow field-of-view to magnify the target. Having recognized it, he will use a manual control device to place an aiming graticule precisely over the target, thereafter engaging the video tracking system which will automatically maintain the beam from the laser designator on the target. In normal circumstances the beam will stay switched on for thirty seconds until bomb impact, remaining on the first target for only a few seconds before being switched to a second target to direct bombs which have been released and are already dropping towards it.

On 19 December 1990, only a month before the beginning of hostilities in the Gulf, the British Ministry of Defence decided to accelerate production of TIALD. The programme was given top priority and less than two months later the system entered service with the RAF wing based at Tabuk, in Saudi Arabia. In early February 1991 five modified Tornado GR1s, two TIALD pods and ten crews were deployed to guide in the bombs of fourteen other Tornado GR1s. Three days after they arrived in the Gulf, the first TIALD-led sorties were carried out against fourteen airfields and thereafter continued throughout the rest of the conflict. Targets included five strategically important bridges and fourteen airfields throughout Iraq, with individual targets comprising hardened aircraft shelters, runways, fuel storage areas, ammunition and bomb dumps and aircrew facilities. A total of 229 direct hits by TIALD-directed laser-guided bombs was recorded.

The true value of laser-guided munitions in the Gulf War is illustrated by the fact that of the 88,500 tons of bombs dropped by Coalition air forces during the conflict, only 8 per cent (7,080 tons) were precision-guided. That percentage, however, destroyed over 50 per cent of targets.

By the middle of February the Iraqi air force had been systematically destroyed, or had fled for sanctuary to Iran, and Coalition commanders were thus able to turn their attention to the impending ground war. Bombardment of Iraqi formations now became the responsibility of long-range artillery and in particular another newly developed system, the Multi-Rocket Launch Rocket System (MLRS).

The MLRS is a rocket artillery system mounted on a high mobility stretched M2 Bradley chassis and crewed by three men: commander, driver and gunner. The M270 launcher, containing twelve rockets loaded in two six-round pods, is an automated self-loading unit equipped with a self-aiming system and a fire control computer. Rockets can either be fired individually or in ripples of two to twelve in less than sixty seconds, the fire control computer ensuring that accuracy is maintained by re-aiming the launcher between rounds.

MLRS fires a number of different projectiles. The standard MLRS surface-to-surface rocket contains 644 M77 anti-personnel bomblet munitions, which are dispensed in mid-air and are armed in mid-air as they fall to earth, a drag-ribbon system ensuring that they land correctly. One MLRS salvo from a launcher can dispense up to 8,000 bomblets in less than one minute at ranges of up to 32 kilometres (20 miles). Other MLRS munitions include the Extended Range Rocket which, with a range of up to 45 kilometres (28 miles), carries 518 improved bomblets; the AT2 Rocket, which dispenses twenty-eight anti-tank mines; the Reduced Range Practice Round which has a range of between 8 and 15 kilometres (5 and 9 miles); and the Army Tactical Missile System (ATACMS), which saw service during the Gulf War. The ATACMS carries 950 cricket-ball-sized M74 munitions to ranges of over 165 kilometres (102 miles). There are three further variants of the ATACMS, designated Block I, II and IIA, which are equipped with thirteen BAT sub-munitions, each comprising an unpowered glider equipped with acoustic and infra-red sensors for target detection and terminal guidance respectively. The maximum range of each variant is 140 kilometres (87 miles). A new guided munition, the GMLRS, which will have a range of 70 kilometres (43 miles) and which will be equipped with a global positioning system (GPS) and an inertial guidance system, is due to enter production in 2004.

The fire-control computer permits fire missions to be carried out manually or automatically. Co-ordinates and other instructions are received from a battery command post, target data being transmitted direct into the computer which proceeds to present the crew with a series of prompts to arms and fire a pre-selected number of rounds. Multiple fire mission sequences can be programmed and stored in the computer. Rocket reload pods are carried in a tracked support vehicle, the launcher being equipped with a boom and cable hook assembly for reloading that can be controlled by a single soldier using a portable boom control device.

For the Iraqi troops, being at the receiving end of MLRS barrages was terrifying. They called the rockets 'black rain', a term referring to the high-explosive bomblets which burst over them, showering them with white-hot metal fragments. Almost 10,000 rockets were fired by a total of 189 MLRSs. At the same time, the Iraqis were also subjected to barrages by conventional artillery that engaged targets at ranges of over 24 kilometres (15 miles). By day and night, Coalition guns pounded Iraqi positions, 'softening them up' in preparation for the advance into Kuwait.

On the night of 24 February, the Coalition ground forces began their advance. Once again, technology played its part by enabling them to do so under cover of darkness. Armoured vehicles were equipped with third-generation image intensifiers and infra-red thermal imagers incorporated into their periscopes and weapon sight systems. Commanders and drivers of vehicles wore night-vision goggles that enabled them to observe the desert around them, while infantrymen were equipped with image intensifiers fitted to their personal and crew-served weapons, allowing them to be used at night.

Image intensification is a process of enhancement of light by a factor of up to 40,000, enabling items to be observed under conditions of darkness which otherwise would be invisible to the naked eye. Having no signature, image intensifiers are passive devices used for surveillance and target acquisition, being produced in a variety of configurations which include: night-vision goggles (either helmet-mounted or strapped to the user's head with a harness), used by aircrew, drivers or personnel equipped with weapons fitted with infra-red aiming projectors; individual weapon sights, mounted on rifles, machine-guns and other weapons operated by a single man; crew-served weapon sights, mounted on anti-tank guns and similar types of weapons operated by more than one man; night observation devices, large devices employed for long-range surveillance, often with photographic or video systems attached.

In the case of hand-held or weapon-mounted devices, first-generation image intensifiers were large and heavy, while their effectiveness was considerably reduced in urban environments where street or vehicle lights and other forms of white-light illumination caused the automatic gain controls to close down. The development of second-, third- and ultimately fourth-generation devices saw a marked improvement in performance with considerable reductions in size and weight as well as less susceptibility to white light.

Thermal-imaging devices operate by detecting changes in temperature and produce graphic images sufficiently detailed that operators are able to identify objects. Thermal imagery can penetrate fire, smoke, rain and foliage, thus making it ideal for surveillance and target acquisition purposes. Like image intensifiers, thermal imagers are passive devices but are vulnerable to electronic countermeasures and can be deceived by infra-red decoys.

Leading the British 1st Armoured Division were the 4th and 7th Armoured Brigades, whose armoured regiments led the way in their Challenger 1 main battle tanks (MBT), which ultimately were credited with destroying up to 300 Iraqi tanks without suffering a single casualty from enemy action. Entering service with the British Army in 1982, the Challenger 1 was based on the Shir 2, a tank originally developed for supply to Iran during the 1970s but which was never delivered due to the Iranian revolution in 1979. Powered by a Rolls Royce Condor CV-12 1,200 brake horsepower diesel engine, with an estimated maximum range of 500 kilometres (310 miles) and giving a maximum road speed of 56 kilometres (35 miles) per hour, the Challenger was equipped with hydropneumatic suspension providing improved stability across country.

With a crew of four, its turret and hull were manufactured in Chobham ceramic and steel composite armour which provides a high level of protection against the majority of battlefield weapons, including anti-tank guided weapons equipped with high-explosive anti-tank (HEAT) warheads. Its fully stabilized main armament was the Royal Ordnance L11A5 120-mm rifled gun, firing armour-piercing fin-stabilized discarding sabot tracer (APFSDS-T) and high-explosive squash head (HESH) ammunition. Secondary armament comprised two 7.62-mm machine guns: one mounted coaxially with the main armament and the other on the commander's cupola on the turret. The Challenger was equipped with a computerized fire control system, which incorporated a laser rangefinder, and featured a thermal observation and gunnery system known by its acronym of TOGS.

Following the Gulf War, Challenger 1 was replaced in 1992 by Challenger 2, which entered service in July 1994. Although possessing the same hull and running gear as Challenger 1, it is equipped with a new type of turret featuring new thermal imaging and fire control systems and designed so that it can be further enhanced with battlefield information control and navigational systems. The commander's position features a gyrostabilized fully panoramic sight, with a laser rangefinder and thermal imager, and a day/night periscope. The gunner has a similarly equipped gyrostabilized primary sight, a coaxially mounted auxiliary sight and a day/night periscope. Both hull and turret are clad in Chobham Series 2 composite armour and use Stealth technology to reduce radar signature. The main armament features the Royal Ordnance LS30 120-mm gun developed by the Royal Armaments Research and Development Establishment (RARDE) at Fort Halstead; featuring a chromed rifled bore, it is capable of firing projectiles at far higher velocities than the L11A5 fitted to the Challenger 1 and producing greater penetration of armour. The co-axially mounted machine-gun is a McDonnell Douglas 7.62-mm electro-mechanically driven weapon with adjustable rate of fire.

Following the Challengers were the Warrior infantry combat vehicles of the armoured infantry battalions. Manufactured by GKN Sankey and designed to fight its way on to an objective or provide fire support for dismounted infantry, the Warrior entered service in 1988 and eight battalions were subsequently equipped with it. With a hull constructed of all-welded aluminium armour, the vehicle is powered by a 550 brake horsepower Perkins Condor CV-8 diesel engine with an estimated maximum range of 660 kilometres (410 miles) and giving a maximum road speed of 75 kilometres (46 miles) per hour. The steel turret is power-operated and is equipped with a 30-mm Rarden cannon, a co-axially mounted Hughes EX-34 7.62-mm calibre chain gun and two quadruple-barrelled smoke grenade dischargers. Both the commander's and gunner's cupolas are equipped with Pilkington day/night sight systems and periscopes for all-round vision when closed-down. Operated by a crew of three comprising commander, gunner and driver, the Warrior accommodates a section of eight fully equipped infantrymen who deploy from the troop compartment via a power-operated door at the rear. The vehicle is not equipped for the section to fire their weapons from it but the roof of the troop compartment is equipped with periscopes and double hatches. In addition, the vehicle features a nuclear biological and chemical (NBC) warfare protection system.

Variants of Warrior include: a command post vehicle; an artillery forward observation vehicle equipped with additional radio communications and surveillance systems; an anti-tank vehicle equipped with the Milan medium range anti-tank guided missile; and a recovery vehicle.

Leading the US Army and US Marine Corps armoured formations was the M1A1 Abrams, an improved version of the M1 MBT. Designed during the 1970s and first entering service in 1980, this highly sophisticated tank also features composite armour similar to the Chobham type used on the British Challenger. In 1988, work began on the development of depleted-uranium armour for use on the M1A1. Although such armour would add a further 2 tons to the 63-ton weight of the M1A1, having a density of two and a half times that of steel, it would provide a considerably enhanced level of protection.

Operated by a commander, gunner, loader and driver, the M1A1 is powered by a 1,500 brake horsepower Avro-Lycoming AGT-1500C gas turbine engine with an estimated range of 465 kilometres (289 miles) and a maximum road speed of 68 kilometres (42 miles) per hour. When an NBC overpressure protection system is fitted, maximum range is reduced considerably to 204 kilometres (127 miles). The tank's extensive equipment specification includes a deep-water fording kit, a position location reporting system and a digital electronic fuel control unit.

The main armament was initially a 105-mm gun but this was replaced in 1986 with the M256 120-mm smooth-bore weapon developed by Rheinmetall of West Germany and capable of firing M8300 High Explosive Anti-Tank Multi-Purpose Tracer (HEAT-MP-T) and APFSDS-T, the latter including a depleted uranium penetrator providing high penetration of armour. Secondary armament comprises a co-axially mounted M240 7.62-mm machine-gun, a similar weapon for use by the loader and an M2 12.7-mm heavy machine-gun equipped with a 3× sight and located on a powered mount on the commander's cupola in the turret.

The M1A1's gunner's position is equipped with a sight system incorporating a thermal imager which displays an image of the target together with a measurement of its range, accurate to within 10 metres (33 feet), which is measured by an integral laser rangefinder. The range data is automatically transferred to a digital fire control computer which automatically produces a solution that is also based on other data including wind velocity (measured by a sensor mounted on the turret), the bend of the gun (calculated by the main armament muzzle reference system), lead angle and the cant of the vehicle (provided by a pendulum static cant sensor also located on the turret). The only data entered manually is the type of round to be fired, the temperature and barometric pressure.

The commander's position is fitted with six periscopes, giving a 360° arc of observation and a stabilized thermal viewer which provides him with a day/night surveillance capability, automatic cueing of the gunner's sight and a second fire control system which enables him personally to fire the main armament. The driver's position is equipped with three periscopes, and an image intensifying periscope for use at night or in poor visibility, giving him a 120° arc of observation when driving the vehicle with his hatch closed down.

Ammunition is stored in armoured containers separated by sliding doors from the crew compartment, which is similarly separated from the vehicle's tanks by armoured bulkheads. Protection against flash or fire within the vehicle is provided by a halon gas fire suppression system designed to activate automatically within two milliseconds. In addition, the M1A1 is also equipped with an NBC warfare protection system comprising overpressure clean-air conditioning, chemical agent detection and radioactive monitoring systems. Each member of the four-man crew is also equipped with his own NBC suit and respirator.

A total of 1,178 M1A1s, transported by sea from Europe and fitted with NBC protection systems, took part in operations during the Gulf War, together with 594 M1A1 Heavy Armour (HA) tanks fitted with depleted uranium armour. All performed very satisfactorily with a high degree of reliability, few cases of mechanical failure being reported subsequently. They were opposed by a large number of Iraqi tanks, the majority of which had been supplied by the former Soviet Union. Of these, 500 were T-72 MBTs which, in spite of some advanced features, proved no match for the M1A1 with its capability of producing accurate fire while traversing rough terrain at speed. M1A1s spearheaded US assaults on Iraqi lines of defence, engaging enemy tanks at ranges of up to 4,000 metres (13,123 feet) primarily with APFSDS ammunition. The Iraqis were more often than not dug in to reduce their signature and this reduction in mobility resulted in almost 50 per cent of their number being knocked out by aircraft before Coalition forces had even crossed the border into

Kuwait. During the war, only eighteen M1A1s were withdrawn from service due to battle damage and nine suffered damage from mines; all were repairable.

During the same year as the Gulf War, the first M1A2 was delivered to the US Army. An initial fifteen M1A2s were built, followed by sixty-two. Thereafter the Pentagon placed orders for the upgrading of approximately 1,000 M1s to M1A2 configuration. Meanwhile, Saudi Arabia and Kuwait ordered 315 and 218 M1A2s respectively.

Although armed with the same 120-mm smoothbore gun as the M1A1, the M1A2 is equipped with enhanced navigation, surveillance and fire control systems. It is envisaged that ultimately all M1A2s will probably be fitted with depleted uranium armour. In addition, a further upgrading, called the System Enhancement Programme (SEP), has been introduced to enhance the M1A2's capabilities even further. This essentially comprises an upgrade to the tank's computer system, providing an operating system designed to allow further future enhancement, improved processors, increased memory capacity, high resolution colour flat-panel displays and an increased degree of user-friendliness via a soldier–machine interface. A major enhancement is the integration of a second-generation forward-looking infra-red (FLIR) sight which will replace the existing thermal imager system and the commander's thermal viewer. A fully integrated variable power (3× to 6× wide field and 13×, 25× or 50× narrow field) target acquisition/sighting system, it is designed to enable both the gunner and commander to acquire and engage targets more swiftly and with greater accuracy.

Other SEP enhancements include an under-armour auxiliary power unit capable of producing all the electrical and hydraulic power required during surveillance operations, when the tank's main engine is switched off, and of charging the vehicle's batteries. In addition, a thermal management system has also been included; this ensures that the temperature within the crew compartment is maintained under 95° and that of electronic units under 125°, ensuring maximum operability in extreme conditions.

Like its British counterpart, the Warrior, the M2 Bradley Infantry Fighting Vehicle is designed to provide mobile protection for armoured infantry and fire support for dismounted operations. It features a hull manufactured from welded aluminium reinforced in certain areas with spaced laminate armour. Operated by a crew of three – commander, gunner and driver – it carries a six-man infantry section in its troop compartment in the rear from which weapons can be used via firing ports in the hull. Designed to keep up with the M1A1/M1A2 Abrams MBT, the Bradley is powered by a Cummins VTA-903T 600 brake horsepower turbo-diesel engine with a maximum range of 483 kilometres (300 miles) and giving a maximum speed of 66 kilometres (41 miles) per hour. The vehicle is amphibious, being equipped with an inflatable pontoon fitting to the front and sides of the vehicle with propulsion, at a speed of just over 6 kilometres (4 miles) per hour being provided by the tracks.

The vehicle's main armament is the McDonnell Douglas M242 Bushmaster 25-mm chain gun which is capable of firing either armour-piercing (AP) or high-explosive (HE) ammunition, selection of either being available via operation of a switch. Capable of defeating the majority of armoured vehicles, including some MBTs, the Bushmaster has a maximum range of 2,000 metres

(6,560 feet), depending on the type of ammunition in use, and can be fired either semi-automatically or automatically. The standard rate of fire is 200 rounds per minute. Mounted coaxially is an M240 7.62-mm machine-gun.

For dealing with heavy armour, the M2A1 version of the Bradley is equipped with two TOW anti-tank guided missiles carried in a housing on the left-hand side of the turret. Reaching a speed of almost Mach 1 during flight, the highly accurate TOW is equipped with a large shaped-charge high-explosive warhead reportedly capable of destroying any known armoured vehicle. Launching of the missile, however, has to be carried out while the vehicle is stationary, and the reloading of the launcher housing is carried out by the infantry section in the rear of the vehicle via a special hatch.

A variant of the M2A1 version is the M3 Bradley Cavalry Fighting Vehicle which differs externally from the M2 in only one way – the absence of firing ports in the hull. With the familiar crew of three (commander, gunner and driver), the M3 carries two scouts, additional radio communications equipment and either Tube-launched, Optically tracked, Wire-guided (TOW) or Dragon missiles.

During the Gulf War, a total of 2,200 Bradleys were deployed in Saudi Arabia and Kuwait. Acting in close concert with US Army tank units, they followed up close behind and engaged Iraqi infantry and armour. According to one report, more enemy armoured vehicles were destroyed by Bradleys than by M1A1 MBTs. Tragically, however, some fell victim to friendly fire by US aircraft which attacked and destroyed a number of vehicles.

Following the Gulf War, upgraded versions of both vehicles were developed and produced. The M2A2 featured additional appliqué, steel armour to enhance ballistic protection, with provisions being made for the use of explosive reactive armour tiles designed to defeat shaped charge warheads. Subsequent developments resulted in the A3 upgrade which included a vehicle control and operation system designed to control and automate some crew functions and to enable transmission, reception, storage and display of digital data; an improved target acquisition and engagement system and an independent thermal viewer for the commander, both second-generation FLIR; and a GPS satellite navigation system.

Although the opening phase of the Gulf War was conducted by the Coalition air forces, the very first shots were fired by a unit of the US Army. In the early hours of 17 January 1991 two USAF MF-53J Pave Low III special operations helicopters, newly equipped with satellite navigation systems, guided eight AH-64A Apache attack helicopters of the 101st Airborne Division (Air Assault) towards the Saudi border and two key Iraqi air defence radar sites. Shortly afterwards, the Apaches attacked both installations with Hellfire missiles, destroying them both and punching a large hole 32 kilometres (20 miles) wide in the Iraqi air defence system, enabling Coalition aircraft to carry out their bombing raids deep inside Iraq.

A total of 274 Apaches were deployed during the Gulf War. Before the start of ground operations, they attacked enemy armoured units while also carrying out reconnaissance and attack missions deep into Iraqi-held territory. Once Coalition forces advanced into Kuwait, Apaches provided close air support against Iraqi armour, ultimately being credited with destroying over 500 tanks and hundreds of armoured personnel carriers and soft-skinned vehicles.

Introduced into US Army service in 1984, the Apache is a multi-role combat helicopter designed to operate by day and night, and in adverse weather conditions. Powered by two General Electric T700-701C turboshafts giving a cruise airspeed of 233 kilometres (145 miles) per hour, it has a combat radius of 261 kilometres (162 miles) and a flight endurance of an hour and 48 minutes which can be extended by the use of an external 1,044-litre (230-gallon) fuel tank. It is flown by a team of two positioned in tandem, the pilot's position being located behind that of the co-pilot/gunner.

With an all-metal semi-monocoque fuselage, the aircraft is designed with maximum survivability and crashworthiness in mind. The crew compartment and other vital areas are protected with armour manufactured from boron carbide bonded with Kevlar which provides protection against weapons up to 12.7-mm calibre. Blast shields, which provide ballistic protection against weapons up to 23-mm calibre, act as bulkheads between the pilot and co-pilot's positions, thus ensuring that at least one will survive a strike by a single round.

Nicknamed the 'Flying Tank', the Apache is equipped with a formidable array of weapons controlled by the Target Acquisition Designation Sight (TADS) system which allows the co-pilot/gunner to acquire and engage targets by day and night and in adverse weather. Mounted in a turret in the nose of the aircraft, the TADS system comprises a day TV system fitted with a zoom lens, a forward looking infrared (FLIR) unit, a direct view optic (DVO) sight unit providing wide (18°) and narrow (3.5°) fields of view, a laser spot tracker and a laser rangefinder/designator (LRF/D). Mounted in its own turret above the nose is the Pilot Night

Vision Sensor (PNVS) which enables the pilot to fly the aircraft in all conditions.

Both TADS and PNVS are interfaced with the Integrated Helmet And Display Sighting System (IHADSS) which projects images on to 50-millimetre (2-inch) square monocular displays mounted on the crew's helmets. In the case of the pilot, key flight data, such as airspeed, radar altitude and heading, is superimposed on to data from the PNVS FLIR. To acquire and designate a target, the co-pilot/gunner merely has to observe a target, place the graticule in his helmet-mounted display on it, select a weapon and fire. The system is designed so that the pilot and co-pilot/gunner can interchange images, both being able to fire weapons and, in the case of the pilot being wounded or killed, the co-pilot/gunner being able to fly the aircraft.

First among the Apache's weapons is the McDonnell Douglas M230 30-mm chain gun, located under the aircraft's nose, which can be used against troops, soft-skinned vehicles and area targets. With a rate of fire of 625 rounds per minute and a range of 6,005 metres (19,700 feet), it fires both M789 high-explosive dual-purpose (HEDP) and M799 high-explosive incendiary (HEI) ammunition. The same types of targets can also be engaged with the Apache's Hydra-70 70-mm unguided rockets, seventy-six of which are carried in four M261 launchers mounted on four pylons located in pairs on either side of the fuselage. With a maximum range of 5.5 kilometres (nearly 3½ miles), the Hydra-70 can be fitted with different types of warhead ranging from the M151 High Explosive, for use against personnel, to the M261 High Explosive Sub-Munition, containing nine M73 sub-munitions designed for attacking light armoured or soft-skinned vehicles.

The Apache's foremost anti-armour weapon is, however, the laser-guided AGM-114 Hellfire missile, equipped with a hollow charge warhead, of which the AGM-114C, AGM-114F and AGM-114K versions are currently in service. The AGM-114C features an improved semi-active laser seeker while the AGM-114F is designed for use against reactive armour, having dual warheads. The AGM-114K Hellfire II missile also features an advanced dual warhead system, electro-optical countermeasures hardening, reprogrammability for adaptation to different threats and mission requirements, a semi-active laser seeker, an improved target reacquisition system to recover laser lock-on if lost, and a programmable autopilot for trajectory adjustment.

Once fired, Hellfire missiles can be controlled either from the launching aircraft or 'handed off' either to another aircraft or to forward observers on the ground equipped with a laser target designator which illuminates the target and guides the missile to it. The latter method has the advantage of allowing the aircraft either to remain concealed by terrain or depart the area immediately after launching a missile.

The current most advanced version of the Apache is the AH-64D Longbow Apache, a remanufactured and upgraded aircraft which features a number of enhancements which include: updated T700-701C engines, a fully integrated cockpit, digital communications, integrated GPS/inertial navigation unit, enhanced doppler velocity rate sensor, a fire control radar target acquisition system and the fire-and-forget AGM-114L Longbow version of the Hellfire missile.

The AN/APG-78 fire control radar (FCR) is a multimode millimetre wave sensor with its antenna and transmitter mounted on the aircraft's main rotor head. It is capable of detecting, classifying and prioritizing targets on the ground and in the air. In addition, the FCR will detect obstacles and provides assistance to the air crew when flying in adverse weather conditions. The FCR can be operated by either the co-pilot/gunner or the pilot, the latter for example using it to search for airborne targets while the other uses TADS to engage others on the ground. The system can also be used in conjunction with TADS, whose electro-optics can be used to identify targets detected by the radar. The FCR also has a threat detection capability through its Radio Frequency Interferometer (RFI) which detects and pinpoints with a high degree of accuracy radars in the search and acquisition mode, in particular when they have locked on to the Apache. The RFI contains a extensive 'library' of pre-programmed threat signatures which enables it to recognize the type of radar and weapon system threatening the aircraft.

The AGM-114L Longbow Hellfire RF missile initially accepts targeting data from the FCR but will thereafter respond to guidance from the co-pilot/gunner's TADS or from another aircraft. It is capable of operating in two modes: lock-on before launch, which is used to acquire static or mobile targets at short range prior to launching; or lock-on after launch, in which the missile acquires and locks on to stationary targets at extended ranges shortly after leaving the aircraft.

The Longbow Apache also has the capability of carrying air-to-air missiles for engagement of air targets. At the time of writing, trials are currently being conducted by the US Army with two missiles: the Raytheon Air-to-Air (ATAS) Stinger and the Shorts Starstreak. The ATAS Stinger is an infra-red homing fire-and-forget missile based on the well-proven man-portable, shoulder-fired surface-to-air missile (SAM) designed to engage aircraft at low altitudes. It is equipped with a super-cooled, two-colour passive infra-red seeker which is extremely sensitive to both infra-red and ultra-violet wavelengths, giving it the ability to avoid electrical and optical countermeasures. Also possessing an IFF system, the ATAS Stinger has enhanced range and manoeuvrability and an all-aspect engagement capability.

The Shorts Starstreak undergoing trials is a helicopter-launched version of the man-portable shoulder-fired close-range SAM designed to provide defence against helicopters and ground attack aircraft. The missile comprises a two-stage solid-propellant rocket motor, a separation system and a forward unit containing three high-density darts. Each of the latter is equipped with a thermal battery, guidance and control system, steering fins and a high-density penetrating warhead. On firing, the missile is boosted clear of the launcher by the rocket motor's initial stage which separates as the second stage ignites, the missile accelerating away and reaching a speed of over Mach 4. Canted nozzles cause the missile to roll, resulting in the deployment of stabilizing fins that provide aerodynamic stability in flight. As the second stage burns out, the three darts are automatically detached and separate, their individual warheads being armed in the process, maintaining a high degree of kinetic energy as they home in on the target. They are guided independently by a laser beam transmitted by the auto-tracker in the Apache's TADS, which scans the target in the form of a grid, the darts homing in on the geometric centre. A delay fuse is initiated as each dart impacts, allowing it to penetrate the target before exploding.

During the Gulf War, much of the target data for Coalition air, ground and naval forces was supplied by remotely piloted vehicles (RPVs), alternatively known as unmanned aerial vehicles (UAVs). These carried out a total of 522 sorties, lasting 1,641 hours, before and during Operation Desert Storm itself.

The principal RPV used by US forces in the Gulf War was the Pioneer. Initially developed in Israel and manufactured in the United States, it weighs 210.45 kilograms (463 pounds) and is 4.3 metres (14 feet) in length. Launched by pneumatic catapult and powered by a 26-horsepower two-stroke twin-cylinder rear-mounted engine running on 100-octane aviation fuel, it has a range of approximately 160 kilometres (100 miles) and a flight duration of five hours. Equipped with auto-pilot, navigation and radio-link data communications systems that enable it to operate in either programmed or manual control modes, it carries a number of high quality video sensors for conducting reconnaissance and surveillance tasks in adverse environments and under battlefield conditions. These comprise gyro-stabilized high resolution television and FLIR systems for operations during the day or at night, or in conditions of reduced visibility.

The Pioneer is controlled by a ground control station (GCS), either ship-borne or land-based, which directs it throughout the duration of a mission. Incorporating advanced electronics for mission planning and execution, the GCS comprises three bays manned by two operators: the Pilot Bay contains all the controls and displays necessary for control of the RPV; the Observer Bay provides control facilities for all the on-board sensors; and the Tracking Bay displays the RPV's location based on data provided by a tracking communication unit (TCU). Located separately from the GCS, the TCU contains a communications system providing a highly sophisticated jam-resistant data link, with a range of 169 kilometres (100 miles), for both control of the RPV and for real-time downlink of video and telemetry transmissions from the on-board sensors. In a land-based system, the TCU can be remotely located up to 1,000 metres (3,280 feet) from the GCS, being connected with the latter via a fibre-optic link.

During the preflight, launch and recovery stages of a mission, a portable control station can be used by the pilot to control the RPV, enabling the GCS to concentrate on other tasks. In order to enable commanders in the field to receive immediate results of an aerial reconnaissance of areas of operations, remote receiving stations can be located at their headquarters, providing them with the most up-to-date information.

The Pioneer saw shipborne service during the Gulf War with the US Navy, and on land with US Army and US Marine Corps units. RPVs were launched from the battleships USS *Missouri* and USS *Wisconsin*, proving highly successful in the roles of target selection, gunfire support, battle damage assessment and maritime interception operations. In the case of the latter, two Iraqi high-speed craft were intercepted and destroyed as a result of information provided by a Pioneer. In the surveillance role, Pioneers pinpointed Iraqi coastal surface-to-surface missile batteries and air defence batteries, as well as identifying over 300 vessels and detecting movements by major Iraqi armoured formations. They flew a total of sixty-four sorties lasting 213 hours, providing support for naval gunfire in a total of eighty-three fire missions.

On shore, a US Army UAV platoon was assigned to VII Corps and its Pioneers proved invaluable in flying a total of forty-six sorties over 155 flying hours during which they sent back video coverage of a large number of targets including enemy armour, artillery, convoys, fortifications and command posts. Such were the numbers of targets identified that there were simply too many to be engaged rapidly enough.

Three US Marine Corps UAV companies were deployed during the Gulf War and their Pioneers were used in the reconnaissance, surveillance and target acquisition (RSTA) role. Once targets such as troops, armour, artillery, surface-to-air missile (SAM) sites, bunkers and supply depots had been located and identified, the information was used for directing gunfire or close air support on to them. Such was the quality of the material transmitted by the Pioneers that on one occasion a SAM site was not attacked as the video clearly showed that it was a dummy. During the entire war, US Marine Pioneers flew 323 missions during a total of 980 flying hours.

Such was the success of the Pioneer during the Gulf War that it was deployed aboard US Navy warships on operations during the mid-1990s in Haiti, Somalia and Bosnia. In the case of Bosnia, a Pioneer was launched on 13 October 1995 from the US Navy amphibious assault ship USS *Shreveport* to conduct an examination of the damage caused by Serb shelling of Mostar where a United Nations safe-haven was located. Thereafter, it carried out reconnaissance missions in which it gathered intelligence on the state of roads, bridges, tunnels and villages which had been reported as badly damaged by shelling. The Pioneer was able to show that much damage had been repaired and that roads and bridges were once again operational. A similar mission was carried out over the port of Dubrovnik and towns further inland on the following day.

The featureless desert terrain over which the Coalition forces advanced and fought during Operation Desert Storm was such that maps were of little use in navigating over the sandy wastes. The difficulties facing commanders of vehicles and units are well illustrated by one British artillery officer when describing his own experience of the high speed advance into Kuwait: 'You've got to imagine from the point of view of my job: you're in a vehicle bumping up and down, you've got the flat featureless terrain, you've got your headsets on with your infantry commander talking in one ear and your artillery commander speaking in the other – it is very hard to cope with all that and read a map as well.'

The key to troops navigating accurately around the desert lay in the fact that the heavens are criss-crossed by a network of satellites forming the Global Positioning System (GPS). Each satellite broadcasts a continuous signal, giving the time and its precise position, which is picked up by receivers installed in ships, vehicles and aircraft, or hand-held by personnel, which use it to calculate their respective positions on or above the Earth's surface.

Operation Desert Storm was the first large-scale operational military use of GPS and it proved invaluable. Not only were commanders of armoured formations able to navigate accurately during the day and at night while moving at speed, and to report their locations when required to do so, but logistic support units, bringing up fuel, ammunition and rations, were also able to rendezvous with the desert with frontline units. As Lieutenant Colonel Peter Williams of the Royal Artillery later commented, 'Certainly,

I ignored my maps, threw them away and relied totally on my satellite navigation device. It was such a brilliant system which enabled us not only to determine where we were at any given time but also to key in where we wanted to go. If we wanted to put in any waypoints, it would actually navigate for us – it would say left a bit, right a bit, the whole way there. The only trouble was, very occasionally, for about half an hour in the morning and evening the satellites would obviously be out of orientation and you would get a dreaded display saying "GPS bad" – gloom all round because it meant that it wasn't going to tell you where you were and wasn't going to help you at all. So, the show tended to grind to a halt during those periods.'

There is currently more than one GPS system in operation. The one developed by the US Department of Defense, during the eighteen years prior to the Gulf War, is called Navstar and comprises two segments: Space and Control. The Space Segment nominally comprises a constellation of twenty-four satellites that orbit the earth over a period of twelve hours, repeating the same track and configuration over any point approximately every twenty-four hours. There are six orbital planes, with four satellites in each, equally spaced 60° apart and inclined at approximately 55° to the equator. Between five and eight satellites are visible from any location on the Earth's surface at any one time.

The Control Segment, called the Operational Control System, comprises a master control station (MCS) located at Schriever Air Force Base near Colorado Springs, where it is operated by the USAF's 2nd Space Operations Squadron, and a network of five passive monitoring stations. The first of these is also located near Colorado Springs, the other four being on the islands of Hawaii, Ascension Island in the South Atlantic, Diego Garcia in the Indian Ocean and Kwajalein in the Pacific. These track the satellites twenty-four hours a day, passing real-time data to the MCS which analyses it to determine whether there have been any changes or malfunctions. Any navigational or other instructions from the MCS are transmitted once or twice a day by radio to the satellites via three ground-based uplink antenna systems located on Ascension Island, Diego Garcia and Kwajalein.

The Navstar system provides two services. The Precise Positioning Service (PPS) is available only to US and Allied military forces, and authorized civilian agencies. These organizations are equipped with the necessary cryptographic systems and keys necessary to decode the signals which provide a degree of accuracy of 22 metres (72 feet) horizontally and just under 28 metres (92 feet) vertically. Other civilian users have access to the Standard Positioning Service (SPS), which is available worldwide at no cost to the user and without restrictions. Intentionally degraded by the US Department of Defense, SPS provides a degree of accuracy of only 100 metres (328 feet) horizontally and 156 (510 feet) metres vertically.

The Gulf War was the first major conflict that saw extensive use of high technology. On the Coalition side, computers were used at all levels, ranging from hand-held devices computing mortar fire data to the highly sophisticated systems in Washington and London being fed with up-to-date intelligence information to produce 'game plan' scenarios and proposed solutions. In the final analysis, however, it was always the senior commanders who had to take decisions.

As General Sir Peter de la Billière, commander of British forces during the Gulf War, later stated: 'The computer feed-outs that we then had were obviously information that we looked at very carefully and considered. But let's be quite clear, computers don't run wars – human beings run wars and you must be very careful to make sure that all the information you get from computers is no more than just another piece of information. At the end of the day, you the commander have got to take the decisions and make judgements based on a wide range of information, including that provided by computers.'

DAWN OF THE DEATH RAY

Despite the end of the Cold War and the current programmes of partial nuclear disarmament being conducted by the United States and Russia, there is still a potential threat from ballistic missiles in certain regions. Although reduced in size, Russia's intercontinental ballistic missile (ICBM) arsenal still poses a major threat to the West, and it is estimated that by 2015 it will still possess as many missiles as its economy will permit, albeit well within the limitations laid down by the START treaties. Furthermore, China will by that time probably possess small but significant quantities of ICBMs capable of reaching the West.

In September 1999 the US National Intelligence Council's Office for Strategic and Nuclear Programmes produced a report outlining emerging ballistic missile threats in other regions. Foremost among those nations believed to be capable of producing ICBMs within the next fifteen years is North Korea, whose Taepo Dong 2 space launch vehicle reportedly could be converted to an ICBM capable of carrying a warhead payload of several hundred kilograms with sufficient range to reach the United States. Similarly, the report states that Iran, with Russian assistance and technology, could test a North Korean type ICBM, capable of delivering a payload of several hundred kilograms to some parts of the United States, by the end of this decade. Likewise, Iraq is reported as possibly being able to test a similar weapon by 2010 although this would depend on the degree of foreign assistance available.

Meanwhile, in southern Asia and elsewhere in the Middle East there has been a proliferation of short-range (SRBM) and medium-range (MRBM) ballistic missiles with maximum ranges of under 1,000 kilometres (621 miles) and 1,000 to 3,000 kilometres (621 to 1,864 miles) respectively. Pakistan possesses M-11 SRBMs and Ghauri MRBMs, both supplied by North Korea, while India has developed its Prithvi SRBM and is currently testing the Agni MRBM. In May 2000, it was reported that Syria had purchased 300 Scud-D SRBMs and twenty-six mobile launchers from North Korea. With a range of 708 kilometres (440 miles), this weapon would enable the Syrians to hit all areas of Israel from deep within their own territory.

Such weapons could inflict considerable damage on Western forces or interests, as was illustrated in 1991 during the Gulf War when Iraq launched its Scud missiles against Coalition forces and Israel. According to the National Intelligence Council report, missiles with conventional high explosive have been used in a number of regional conflicts during the last twenty years.

It is for these reasons that the United States in particular is developing new defensive weapon systems, based on powerful high-energy lasers, to counter the ballistic missile threat not only within the continental United States but in areas where US and Western interests or forces come under threat.

History has it that the first recorded use of light on the battlefield occurred over 2,000 years ago during the Peloponnesian Wars in ancient Greece when troops on one side used the concave faces of their highly polished bronze shields to direct the rays of the sun into the eyes of the enemy. The story goes that such was the painfully dazzling effect of the concentrated mass of light that victory was assured at a stroke.

At its focal pinpoint, laser light is capable of an intensity greater than that of the sun and can unleash terrifying force. One example is a laser developed in the United States, the most powerful of its type in the world, which produces a beam whose power is officially secret but reportedly is several times greater than the temperature on the surface of the sun (about 5,500°C).

One of the earliest attempts in the development of laser weapons began during the late 1960s in the United States, at Kirtland Air Force Base in the deserts of New Mexico. The experimental high-energy laser weighed over 30 tons and required a crew of four to aim and fire it. The size of the weapon was massive, its huge fuel tanks and optics covering half an acre, the largest of its optics being a tracking telescope used to aim the beam. The first breakthrough came in November 1973 during trials against moving targets with the laser engaging a drone (an unmanned, remotely piloted aircraft) flying a 'racetrack' pattern around the test area. During the first test, the drone cut a corner and consequently caused a nearby steel weather tower to be aligned briefly between the aircraft and the laser tracking it. Since the tracking was effected by infra-red energy, the beam locked on to the top of the tower as a hotter target than the drone and melted the structure.

A second test brought more success. Another drone was launched and flown along a valley approximately 1.5 kilometres (nearly a mile) away from the base. At that distance, the 3.7-metre (12-foot) long aircraft was merely a small dot in the sky but the beam locked on to the heat emitted by the drone's engine and in just over a second burned a hole in the steel fuselage immediately below the wing. The fuel tank exploded and the drone broke up, the reusable engine unit descending on a small parachute and landing some 300 metres (985 feet) from the wreckage of the fuselage. This test had proved that the concept of a system reaching out at the speed of light and destroying a missile or aircraft was feasible. The problem in those early days, however, was the huge size and immensely high cost of such a weapon.

A high-energy laser comprises a glass tube filled with gas. The atoms that make up the gas are bombarded with electricity, 'exciting' them so that they emit light waves. At each end of the laser is a mirror that some light waves, travelling in harmony, bounce off and so head back into the gas, forcing other already excited atoms to release more light. As an increasing number of atoms are forced to release light, powerful light waves build up inside the gas and eventually burst through one of the mirrors. The amplified light waves are by then in perfect harmony, forming a laser beam. By injecting more energy to create an increased number of excited atoms, the beam's power is increased. As the energy input is stepped up, so the device that contains it must also grow.

Most early high-energy lasers grew so large that they never left the laboratories in which they were constructed. The team conducting experiments at Kirtland during the 1970s encountered the fundamental dilemma facing those attempting to build such weapons: the smaller the laser, the less power it possesses. However, the scientists at Kirtland not only were determined to overcome this problem but had also conceived a highly ambitious plan to put a high-energy laser on an airborne platform.

In 1973 an NKC-135A aircraft, one of fourteen USAF KC-135As permanently modified for special test projects, arrived at Kirtland Air Force Base where a high-energy laser was installed in it; the only indication of the aircraft's special role was a large 'hump' on its back. It was to serve as an airborne laser laboratory (ALL) containing a fully operational gas dynamic 500,000-watt laser built at a cost of 31.2 million dollars. On emission from the laser, the beam was relayed via a series of mirrors up to the telescope in the hump of the aircraft and fired from a turret with a power of 0.38 megawatts.

In October 1980 the ALL carried out its maiden flight; the laser system was so heavy that the aircraft could carry only enough fuel for four hours of flying, the KC-135's normal endurance being considerably more. There was some dangerous chemistry aboard the ALL. Hot carbon dioxide and nitrogen gases were blasted into a low pressure chamber where they expanded rapidly, releasing their energy as light, to form the beam. The fuels for the laser system, nitrogen and an oxidizer, were contained in liquid form in pressurized tanks. Not only were they also hazardous but helium gas, used to force the reactants from the storage tanks into the combustors, was stored in another tank at a pressure of 6,000 pounds per square inch (422 kilograms per square centimetre). The aircraft would have been destroyed in an instant if the tank had exploded through pressure.

Moreover, the laser beam itself proved to pose a hazard, this becoming apparent when burn marks appeared on panels inside the aircraft. Investigation of these revealed that what looked like sparks were in fact dust particles that had been caught in the path of the beam. The sides of the particles exposed to the beam vaporized, imparting them with momentum like miniature rocket engines. If the hurtling particles did not disintegrate, there was a risk of their striking the laser's sensitive optics. Furthermore, if one was ignited it could cause a sparkle effect, interfering with the beam sensors. The solutions to this problem were to ensure that the atmosphere within the laser compartment of the ALL was of clean-room air quality and to reduce the sensitivity of the beam sensors to the spectrum of the exploding dust while at the same time increasing their sensitivity to the laser radiation by the introduction of a narrow band filter.

After a few years of experiments, the ALL began a series of trials to discover whether it could intercept an AIM-9 Sidewinder air-to-air missile travelling at over 3,200 kilometres (2,000 miles) per hour. There was only one way to conduct such tests: the ALL itself would have to act as a target for the missile launched by an aircraft. In the event of the ALL being hit, it would be destroyed. There was only one safety precaution that could be taken: the missile was equipped with insufficient fuel to reach the ALL. However, there was a possibility that if the launching aircraft pitched upwards as it launched the Sidewinder, then there would be sufficient energy in the missile to arc over and head towards the ALL.

Just such a course of events did indeed take place on one occasion during the trials, and the members of the crew aboard the ALL could do nothing but watch as the missile headed directly towards the aircraft. It came close enough for them to see its fins but fortunately it ran out of fuel at the very last moment and dropped away, to the great relief of all concerned. The event was later recalled by Colonel John Otten, a member of the crew and the officer in charge of the test flights: 'Our chase airplane, the man who was flying with us and watching what was happening, disappeared. He was history. Our airplane was not getting out of the way of anything. We sat quietly and watched this missile come in toward us, and on the screen, you could see it coming toward you and it got big enough to where you could see the fins on it, you could actually see them start to wiggle. And then it dropped out and went away. Everybody was a little excited.'

It took three years before the ALL succeeded in destroying a missile, by which time the USAF was becoming impatient and time was running out. On 26 May 1983 it scored its first hit, the first of five in which the laser beam fractured the nose of the Sidewinder, breaking the homing lock and causing the missile to go off target. During the following autumn, the ALL succeeded in intercepting two US Navy BQM-34A Firebee drones representing sea-skimming cruise missiles. In each case the laser beam, from a range of over 1.5 kilometres (about a mile), struck the drones' flight control boxes and sent them diving into the sea.

While the system had proved capable of intercepting missiles, the trials had highlighted its limited range. Maximum effective range was achieved at night, while in daylight, with good weather, it was reduced to only a few miles beyond which atmospheric particles disrupted the laser beam. As a result of this limitation, the USAF decided the system was not an effective missile interceptor and in the mid-1980s the ALL was mothballed. It appeared that the dream of airborne laser air defence was over.

Elsewhere, however, even greater ambitions were being pursued. In the early 1980s, despite progress with arms control treaties, the threat of nuclear conflict was still very real. If a missile was launched from a Soviet silo or submarine, it was almost impossible to intercept during the flight to its target. But there were some scientists who still believed that there was a way of utilizing the power of the laser in space. Although it seemed very ambitious, there was a logical reason that led them to believe that a space-based laser would succeed where its airborne predecessor had failed. Whereas the latter has to shoot through the Earth's atmosphere, even when thin at high altitudes, a space-based laser has the advantage of being outside the atmosphere where there is nothing to penetrate. In essence, space is an ideal environment for a laser to operate.

From March 1983 onwards, President Ronald Reagan allocated up to 3 billion dollars per year to develop and build a space-based laser system designed to provide a defensive shield against nuclear ballistic missile attacks from within the Soviet Union. Popularly known as 'Star Wars', it was officially designated the Strategic Defence Initiative (SDI). The idea was to put a series of lasers on platforms in orbit around the Earth, at appropriate altitudes, which provided twenty-four-hour continuous coverage of the surface of the planet. If a missile was fired by a belligerent nation, it could be intercepted with a laser beam from a range of thousands of miles and destroyed as it broke through the clouds.

This sounded fine in theory but in practice, as far as most scientists were concerned, it was in the realms of science fiction. In order to intercept a missile in its boost phase, before it launches decoys, the laser would have to strike from a range of up to 8,000 kilometres (5,000 miles) which would require high-precision aiming. Furthermore, the accuracy of lasers had never been tested in space.

The only system designed to test the capabilities of a space-based laser is Alpha, developed by TRW Space & Electronics Group at its Capistrano Test Site. It is a system producing a beam approximately 30 centimetres (12 inches) in diameter and over a million watts in power which would have a range of 8,000 kilometres (5,000 miles) in the vacuum of space. This is generated by a mixture of deuterium, nitrogen trifluoride and helium producing fluorine, which is then burned with hydrogen in a mirrored chamber known as an optical resonator. This process creates hydrogen fluoride molecules which become excited before returning to a calm state, at which point they emit a cascade of photons that are amplified and converted into a beam. The beam-control system, comprising an assembly of optics and mirrors, magnifies the beam and directs it at its target.

Alpha was constructed during the 1980s but was not fired until 1989. By the beginning of the 1990s, by which time its total cost had reached 45 billion dollars, the SDI programme had not even produced a prototype. The Pentagon was swiftly becoming disenchanted and was beginning to think that the SDI scientists were on a wild goose chase. In 1993 the programme was shelved.

During the next five years, however, Alpha was fired twelve more times with a five-second full-power test firing taking place on 18 September 1996. Revival of the concept of ballistic missile defence took place that same year with the introduction of the Defend America Act which recognized the growing threat of ballistic missile proliferation and aimed to establish policy for the eventual development and deployment of a missile defence system covering the entire nation. During the following year, the United States and Russia agreed on a reinterpretation of the Anti-Ballistic

Missile Treaty of 1972, this resulting in the lifting of a ban on lasers and missile-based systems subject to certain limitations. Subsequently, approval was given for further development of Alpha, with a successful five-second test firing taking place in September 1996 and another during the following year. The latest known test-firing of Alpha took place in March 2000 when a six-second test resulted in a 25 per cent increase in power output and beam quality.

Under the auspices of the Ballistic Missile Defence Organization (BMDO), the revived concept, with a projected cost of 80 billion dollars, envisages up to twenty satellites, each carrying a deuterium-fluoride chemical SBL and weighing 35,000 kilograms (15,900 pounds), orbiting on platforms 800 to 1,300 kilometres (500 to 800 miles) above the Earth's surface and engaging missiles as they climb in boost phase to a point just outside the atmosphere. Each laser would possess sufficient capacity to engage up to 100 missiles at a range of up to 4,000 kilometres (2,500 miles), taking less than ten seconds to do so and only one second to switch between targets. If the programme keeps to schedule, a half-scale prototype, weighing 17,500 kilograms (38,500 pounds) and costing 1.5 billion dollars, will be launched into space via a Titan-4 missile between the years 2005 and 2008; an operational system comprising six high-energy space-borne lasers (SBL) is reportedly scheduled to be in operation by the end of 2010.

Meanwhile, prior to the revival of Star Wars, a new initiative to find a viable solution to the threat of Theatre Ballistic Missiles (TBMs) had been launched under the auspices of the Strategic Defence Initiative Organization (SDIO) which, through its previous involvement in the development of global missile defence systems, was already familiar with laser technology. In November 1991 the director of the SDIO initiated the first studies into the new concept of installing a high-energy laser on an airborne platform to intercept TBMs at long range. In August 1992, recommendations were given for the development of an Airborne Laser (ABL) prototype. Shortly afterwards, responsibility for the new development programme was transferred from the SDIO to the USAF. This time, the idea was to install a high-energy laser on a large aircraft for operation at very high altitudes where it would be unaffected by weather while still having time to intercept missiles during their boost stages.

The result was a chemical oxygen iodine laser (COIL) which produces light as a result of a chemical reaction: excited oxygen atoms are extracted from reacting hydrogen peroxide and chlorine and then pumped into iodine gas to generate a massive burst of energy which the iodine atoms release as an intense beam of laser light. Invented at the Phillips Laboratory at Kirtland Air Force Base in 1977, its principal advantage is that, unlike earlier lasers, it does not require a large power plant, being powered by a fuel comprising hydrogen peroxide and potassium hydroxide which are then combined with chlorine gas and water. Furthermore, its infra-red wavelength is only 1.315 microns, the shortest in the world for a high-powered laser. This wavelength travels through the atmosphere without any difficulty and has increased brightness and destructive strength when striking the target.

As described earlier, the principal limitation of the system carried on the ALL was one of range, and thus a high degree of effort was concentrated on overcoming this problem. Astronomers at Kirtland's Starfire Observatory, which tracks satellites and keeps them under surveillance, had already done so by developing an optical device that is capable of penetrating the atmosphere: a telescope, at the base of which is a mirror that bends. The device transmits a massive laser beam and any atmospheric distortion is viewed through the telescope as a shifting pattern of dots of light. A computer in the device calculates precise values for the shifts and these are relayed to tiny motors that push and pull at the mirror's surface, creating minuscule variations. The flexing in the mirror shifts the dots back to where they should be and the atmospheric distortion is reduced, improving the telescope's vision by 1,000 per cent. It was decided that a similar system would be incorporated in the new ABL laser: a flexible mirror increasing the power and range of the beam.

Meanwhile, consideration was being given to the type of platform that would carry the ABL. In September 1992 the aircraft manufacturer Boeing was awarded a contract to carry out an assessment on how well an existing large aircraft, such as a 707, 747 or even a B-52 bomber, would be suited to the role. In the event, a modified 747 proved to be the recommended airframe.

A year later a consortium of Boeing, TRW Space & Electronics Group and Lockheed Martin Missiles & Space, operating under the name Team ABL, submitted its proposals to the USAF ABL Program Office at Kirtland. With a strong history of successful management of large programmes and the development of weapon systems, Boeing offered extensive experience in integration of complex systems within aircraft through its work on programmes such as the AWACS. In addition to having designed and developed Alpha, TRW had considerable experience in the manufacture of lasers dating back to the late 1960s when it had developed and integrated the high-energy laser and beam control system aboard the ALL which had destroyed the Sidewinder missiles and Firebee drones in flight. Thereafter, during the early 1970s, TRW had also produced for the USAF the Mid-Infra-Red Advanced Chemical Laser (MIRACL), a megawatt-class continuous wave deuterium fluoride chemical laser. The third member of the consortium, Lockheed Martin, meanwhile had extensive experience in the development and manufacture of large-size optics and high-precision aiming systems for high-energy lasers.

In November 1996, after consideration of a competitive bid from another consortium, a 1.1 billion dollar Program Definition and Risk Reduction (PDRR) contract was awarded to Team ABL. The aim of this stage of the project was to develop a system that would prove conclusively that all the necessary technology was available. Based on a commercial 747-400F airframe, which was to be available for modification by early 1999, the prototype attack laser aircraft was to be flight tested by 2001. Designated YAL-1A, although still known as the ABL, it had to be able to demonstrate that it could intercept and destroy a missile in its 'boost' phase by the latter part of 2002. Boeing would be responsible for systems integration, airframe modification and development of battle management systems. The latter would include computers and software handling communications, intelligence, target detection and engagement. TRW would develop the laser while Lockheed Martin would be responsible for the beam and fire control systems. On successful completion of this initial contract, a further one worth approximately 4.5 billion dollars would be awarded for the engineering, manufacturing, development and production of a fleet of seven ABLs, the first three being completed by 2006 and the remaining four by 2008.

The internal configuration of the ABL has the megawatt-class high-energy laser, designed with aircraft safety and ease of field maintenance in mind, mounted in the rear half of the fuselage. Forward of that are situated two state-of-the-art diode-pumped solid-state lasers, the tracking illuminator laser (TILL) and the beacon illuminator laser (BILL), which acquire and track a target, and direct the main high-energy laser on to it. Dividing the laser compartment from the rest of the interior of the aircraft is a bulkhead equipped with an airlock, which provides controlled access to the rear of the aircraft. Forward of the bulkhead is the battle management compartment, containing modular consoles equipped with computer systems for target detection, identification, prioritization and nomination; surveillance and tracking; launch and impact point predictions; theatre interoperability; and common data/voice link communications to joint theatre assets. Mounted on the aircraft's distinctive 'hump' above the flight deck is the active laser ranger comprising a modified third-generation LANTIRN (Low Altitude Navigation and Targeting Infra-Red for Night) with a high-power gas dynamic CO_2 laser. This device acquires a target from the Infra-Red Search and Track (IRST) sensor cue, tracks it and directs the CO_2 laser for ranging, assisting in pinpointing the missile's launch and impact points.

Located forward of the battle management compartment is the beam/fire control system, which provides target acquisition and tracking, fire control engagement and sequencing, and aim point-and-kill assessment. The system also carries out high-energy laser beam wavefront control and atmospheric compensation, jitter control, alignment/beam-walk control, and beam containment for both the high-energy and illuminator lasers. It features calibration and diagnostics systems that provide autonomous real-time and post-mission analysis. Finally, on the nose of the aircraft, mounted in a turret assembly housed in a shell permitting rotation of 150° is a 150-centimetre (60-inch) telescope that focuses the laser beams on a target and collects returned images and signals. Incorporated in it is a flexible mirror, nearly 160 centimetres (62 inches) in diameter and just over 20 centimetres (8 inches) thick, equipped with 341 actuators that effect variations in the mirror's surface at the rate of approximately 1,000 per second.

The sequence of target engagement by the system would be as follows. The battle management systems provide target co-ordinates to the beam/fire control system that traverses the turret, centring the target in the acquisition sensor, acquiring the plume of the missile in the coarse track sensor. The TILL then acquires the body of the missile and actively tracks its nose. The beam control process begins after active tracking has been established by the firing of the BILL to establish the aiming point on the target for the high-energy laser. The resulting spot on the missile is imaged in the wavefront sensor and compared with the outgoing sample of the BILL. At this point, the high-energy laser is fired in a three- to five-second burst along the same path with similar wavefront correction, burning the skin of the missile's body until it ruptures and explodes.

The operational concept of the ABL is that of a rapidly deployable anti-TBM defence system which can deploy from the continental United States within a matter of hours.

Flown by crews of four and operating in pairs, ABL aircraft will fly orbits at high altitude over areas to be defended, scanning for the exhaust plumes of missiles. If missiles are launched, the ABL will detect them on passing through the clouds and destroy them, the resulting debris falling back on to enemy territory below. Given its size and lack of manoeuvrability, however, the ABL will be unable to fly over hostile territory because of its vulnerability to air and ground attack, and thus will have to engage targets at long range while flying at altitudes of around 12,000 metres (40,000 feet). The high-energy COIL laser will have a range of 300 to 600 kilometres (185 to 375 miles) but, as missiles would have to be intercepted while in the boost stage of their flight, it will only have between 40 and 100 seconds to engage and destroy its target. The ABL will be designed to possess a salvo engagement capability, capable of engaging multiple targets and carrying sufficient chemical laser fuel to destroy between twenty and forty missiles before needing to refuel.

Although capable of autonomous operations, the ABL will normally be a fully integrated element in a series of systems designed to provide missile defence in depth, destroying TBMs in the boost phase of launch.

Among other systems currently under development is the Airborne Tactical Laser (ATL), a lightweight high-energy laser designed specifically for tactical airborne or ground-based operations. Derived from the COIL system developed by TRW for the ABL, this new system is optimized for power levels of 100 to 500 kilowatts and is designed to operate at ground level with laser exhaust emissions being contained by a small sealed exhaust system.

Comprising the laser and an on-board optical sensor suite, the system will be configured in a roll-on/roll-off package for installation in aircraft, such as the AC-130 Spectre gunship, CH-47D Chinook or the V-22 Osprey, or ground vehicles. Intended for low-level below-cloud use against anti-ship or cruise missiles, the ALT will reportedly be capable of engaging targets at ranges of over 20 kilometres (12 miles) when used in the air-to-air or air-to-ground roles, and up to 10 kilometres (6 miles) in the ground-to-air role. It is envisaged that in the latter, a fully mobile ground-based system, contained in two vehicles, could be employed to counter short-range tactical rockets.

Equipped with different sensor and fire control systems, the ATL could also be used in an offensive role in special operations, carrying out precision strike tasks where pin-point accuracy and zero collateral damage are required. According to one report, the ATL laser will deliver sufficient power to place a beam on a 15-centimetre (6-inch) diameter target point and melt a metal object in a few seconds from a range of 8 kilometres (5 miles).

Also currently being developed is the tactical high energy laser (THEL), a joint programme being carried out by the US Army Space and Missile Defence Command (SMDC) and the Israeli Army under a 131-million-dollar contract funded by both countries. Designed as a ground-based short-range missile and rocket defence system, THEL comprises a pointer/tracker that will detect, track and target multiple rocket launches; a command, control, communications and intelligence (C3I) sub-system; and a deuterium fluoride chemical laser with a reported range of 5 kilometres (3 miles) and sufficient fuel capacity for up to sixty shots before refuelling.

The concept and effectiveness of such systems against short-range missiles and rockets was first demonstrated in

1996. On 9 February of that year, a test was carried out at the High Energy Laser Systems Test Facility at the US Army SMDC's missile test range at White Sands, New Mexico, in which a rocket was destroyed in mid-air. The laser used in the test was MIRACL, the megawatt-class chemical laser built in the 1970s by TRW. One of the most powerful lasers in the United States used for defence research purposes, MIRACL has been employed in several such tests, including one in October 1997 in which it successfully hit the USAF's small MSTI-3 satellite orbiting at an altitude of 418 kilometres (260 miles) above the Earth's surface. In other tests, MIRACL successfully shot down five Firebee drones and a Vandal supersonic missile. During the February 1996 test, however, it used only a small fraction of its power, equating to that of a ground-based mobile system of the type that THEL would comprise.

Two months after the test President Bill Clinton and Secretary of Defense William J. Perry attended a meeting with the Prime Minister of Israel, Shimon Peres, at which a commitment was made that the United States would help Israel in the development of a THEL system to assist in the defence of the north of Israel against short-range rockets, such as the Russian-manufactured 122-mm Katyusha, fired by the Lebanese terrorist group Hizbollah from southern Lebanon. On 18 July, the THEL demonstrator development was formally initiated by a memorandum of agreement between the governments of the two countries. The contract for the design, development and manufacture of a demonstrator system was awarded to TRW as prime contractor, the deadline for completion being the end of 1997, subsequently extended to March 1998. Despite problems over cost overruns and schedule delays, on 26 June 1999 the first test of the THEL laser took place successfully at TRW's Capistrano test facility in California.

In early 2000, it was announced that tests would be conducted at White Sands in April or May to determine whether THEL could intercept a single Katyusha rocket. If successful, further tests would be carried out against a salvo of rockets. If the tests were successful, the THEL demonstrator would be delivered to Israel. With withdrawal from southern Lebanon scheduled to take place by July and fearing an escalation in Hizbollah attacks on their positions in northern Israel, the Israelis were anxious to deploy THEL as soon as possible and thereafter develop a smaller, mobile version for integration into a missile defence system which would also include the Arrow missile interceptor. In June 2000, it was announced that a test firing of THEL had been carried out, resulting in the successful interception of a rocket.

A further ground-based theatre missile defence system was developed as a concept by the then West German consortium of Diehl and LFK under the auspices of a project commissioned by the Federal Ministry of Defence. Designated the High Energy Laser Experimental (HELEX), it was designed to provide mobile ground-to-air defence against low-level missiles and stand-off weapons launched by aircraft operating from Warsaw Pact bases in neighbouring Poland and Czechoslovakia. The distances from such bases were short and thus any warning times would have been limited. While West Germany already possessed a sophisticated air defence system based on aircraft, missiles and a chain of radar stations, this would have been exposed to heavy electronic countermeasures which would inevitably have degraded its performance. It was considered, however, that a laser air defence system would possibly provide an effective rapid response to any missile threat posed with little warning.

HELEX comprised a multi-megawatt gas dynamic carbon dioxide water-cooled laser, which was powered by burning a liquid consisting of a common hydrocarbon fuel, such as benzene, combined with a nitrogen compound oxidizer. The entire system, including fuel and coolant tanks, was mounted for mobility on a Leopard 2 tracked armoured chassis. Between 5 and 10 tons of laser fuel were carried, this being sufficient for up to approximately fifty shots. In order for the laser projector head and the passive target acquisition system to have a clear line of sight, they were mounted on an elevating arm that enabled them to be raised above surrounding trees or buildings.

Only mirrors suitable for the laser's wavelength of 10,600 nanometers could be used to direct and focus the beam; the use of transmission optics was reportedly excluded due to their fragility. The reflector in the projector head comprised a flexible concave mirror of over 1 metre (3 feet) in diameter. Its shape was altered by a large number of piezoelectric actuators that flexed it to the shape and axial angles needed to produce sufficient concentration of the beam at the required range, while also making the necessary adjustments to compensate for atmospheric turbulence and other factors. The information needed to control the surface of the mirror was furnished by the beam reflected by the target. For the interception of a fast-moving target flying at the speed of sound, the beam would have to be concentrated on the same spot for at least half a second.

HELEX's surveillance and target acquisition system was passive, possibly using satellite monitoring, and thus possessed the advantage of having no signature detectable by enemy electronic surveillance. Furthermore, its survivability in a battlefield environment was enhanced by its mobility, which allowed it to move quickly to a new location if necessary. It was estimated that one HELEX could have controlled an area against incursions by low-level aircraft or missiles up to a range, in good conditions, of nearly 10 kilometres (6 miles) although that figure could have been reduced by up to 50 per cent if the atmosphere in the area was heavily affected by smoke and pollution from battle, that being the principal limitation on the system.

The HELEX project was developed to the point where a small-scale version was tested in trials. It appears, however, that for reasons unknown it was not continued thereafter and was subsequently abandoned.

In addition to intercepting surface-to-surface missiles and rockets, lasers are now being developed for deployment against surface-to-air and air-to-air missiles.

A major threat facing aircraft operating in high risk areas in peacetime, or battlefield environments in time of war, is from man-portable infra-red (IR) guided heat-seeking surface-to-air missiles (SAM) such as the Russian SAM-7 Strela or the American FIM-92A Stinger. During the war in Afghanistan, Mujahideen guerrillas, equipped with Strelas and Stingers, shot down a total of 250 Soviet aircraft in ten years. During the Gulf War of 1991, 80 per cent of US losses of fixed-wing aircraft were reportedly from Iraqi IR-guided SAMs.

Until now, countermeasures against both man-portable SAMs and IR-guided air-to-air missiles have comprised IR decoy systems consisting of flares that are launched to divert missiles from aircraft as they take evasive action. These operate at a temperature higher than that of an aircraft engine and also emit more energy in particular bands of the electromagnetic spectrum. The current generations of IR-guided weapons are, however, equipped with seekers fitted with improved counter-countermeasure (CCM) capabilities, the more advanced types possessing the capability to seek the specific band closest to the heat signature of the aircraft and thus considerably decreasing the effectiveness of IR decoy systems.

Recent years, however, have seen development of systems that will not only provide early warning of attack but also enable aircraft to jam and ultimately destroy missiles. One example is 'Nemesis', formally designated the AN-AAQ-24 Directional Infra-Red Countermeasures (DIRCM). Jointly developed by Britain and the United States, it provides the capability of swift and accurate threat detection, tracking and countermeasures to defeat current and future generation IR-guided missile threats.

Warning systems have long been fitted to fixed-wing aircraft and helicopters to detect the approach of hostile missiles. Early IR-based detection models were, however, far from reliable as they suffered from high levels of false alarm rates. Later systems were designed to search for ultra-violet (UV) signatures that are emitted during missile motor burn. During the early 1980s, the two-colour IR concept assisted in reducing high false alarm rates, the situation being further improved with the development of quadrant UV warning systems. These suffered, however, from 'clutter' caused by UV transmissions from high-powered lighting in such areas as sports stadiums as well as the various types of UV emissions experienced during battlefield conditions. While helicopters are likely to encounter a missile during the period known as 'motor burn', fixed-wing aircraft will more probably be engaged at longer range after power burn-out (PBO), each situation presenting a differing signature in either the ultra-violet or infra-red bands or both.

To overcome such problems, Nemesis is equipped with two missile warning systems (MWS), the AAR-54 UV MWS and AAR-44 IR MWS. The AAR-54 will detect and locate the launch of a missile, tracking it while its rocket motor is burning up until the point of burnout. The AAR-44 thereafter continues to track the weapon in the supersonic, PBO stage of its flight. Each system covers a 90° conical sector while four additional sensors are needed to provide sufficient azimuth and elevation coverage. The MWS will classify potential targets as either clutter, non-threat missiles or hostile missiles and will provide accurate angle of arrival data to the system's fine track sensor (FTS), an imaging IR sensor which, housed in a turret assembly, tracks an incoming threat missile. Large rotary and fixed-wing aircraft, such as the CH-47 Chinook or C-130 Hercules transport, are fitted with turrets on either side of the fuselage. Thus a missile is tracked even through the nadir region and will be handed from jammer to jammer if it crosses into another sector during its homing phase.

Also housed in each turret are two 3-inch infra-red xenon or krypton flash lamps; those on smaller aircraft, such as the Lynx helicopter, are fitted with a single 2-inch lamp. As the missile is being tracked by the FTS, the lamps transmit modulated jamming energy at the missile's homing head, blinding it and causing the missile to drop away. Each turret assembly is also designed for optional installation of an air-cooled diode-pumped IR laser, operating in the mid-band region of the electro-magnetic spectrum which is eye-safe and can be used in populated areas. This will enable the system to cope with any missiles which may in the future be designed to overcome jamming by infra-red lamp systems.

During tests in 1998, the Nemesis DIRCM system proved highly successful. In June, live firing tests were conducted in the United States at the White Sands test range. Twelve IR-guided missiles were launched at an unmanned DIRCM-equipped helicopter, suspended from a cable and towed between two mountain peaks, and the system succeeded in defeating all of them.

At the time of writing, DIRCM is currently in early production and among the first to bring it into service will be the US Air Force Special Operations Command, which will fit sixty systems to its AC-130H/U Spectre gunships and MC-130E/H Combat Talon I/II special operations transports. In addition, a total of 131 systems will reportedly be fitted to fixed and rotary wing aircraft of the Royal Air Force and Royal Navy. A system for fast jets, designated Laser Infra-Red Counter Measures (LIRCM) and fitted only with a laser unit, is currently under development.

The concept of laser defence has recently been extended to countering anti-armour guided missiles by Northrop Grumman which has developed a system that will detect, track and jam such weapons. Designated the Directed Missile Countermeasure Device (DMCD) and designed to be mounted on an armoured vehicle, the system comprises four missile warning systems and a laser emitting 425 watts. It utilizes closed-loop tracking to maintain the laser beam on the missile and smart jamming techniques to maximize distances by which the latter will miss the vehicle. At the same time, it provides accurate data on the position and range of the missile launcher, its four MSWs picking up the signature of the launch. The system itself is fully stabilized so that the beam can be maintained on the missile even while the vehicle is mobile.

Lasers can also be used to detect and neutralize enemy electro-optical systems. One such system, developed for the US Army during the 1990s, was Stingray which was designed to detect and counter optical and electro-optical devices such as thermal imagers and rangefinders. This comprised a low-powered target acquisition laser and a high-powered device designed to neutralize optical and electro-optical targets. Developed for use on the Bradley M2 Infantry Fighting Vehicle, it was intended to enhance the vehicle's survivability on the battlefield. Two prototype systems were deployed on Bradleys in Kuwait during the Gulf War in 1991 but reportedly were not used during combat. A lightweight version of Stingray, designated Outrider, was subsequently developed for the US Marine Corps for mounting on the 0.50 calibre heavy machine-gun mount on the High-Mobility Multi-purpose Wheeled vehicle, popularly known as the 'Hummer'. Neither system, however, progressed beyond prototype stage and both development programmes were ultimately terminated.

The mid-1990s saw the development of man-portable laser systems for target acquisition and designation purposes with small arms and other weapons systems. Among these is the Target Location and Observation System (TLOS), a lightweight image-intensified day/night sight system using a low-powered infra-red laser to acquire direct-view electro-optical targets. Designed for use with either individual or crew-served weapons, the system comprises a laser projector, a third-generation image intensifier and two field-of-view objective lenses, and is capable of target acquisition and designation as well as night observation tasks. Weighing nearly 4.5 kilograms (9¾ pounds), TLOS has a detection range of 3,000 metres (9,840 feet) at night and 2,000 metres (6,560 feet) during the day.

Lasers have also featured in the development of non-lethal weapons. One such device is the SABER 203 which was developed by the Phillips Laboratory at Kirtland. Configured to the same shape and dimensions as a 40-mm grenade round and designed for use in the M203 launcher fitted under the barrel of the M16A2 rifle, the low-power device projected a beam intended to dazzle and disorientate a person. A number of SABER 203s were deployed with US Marine Corps units during Operation Restore Hope, which took place in Somalia from December 1992 to mid-1994, and it was found that when beams were projected on to torsos rather than into eyes, as the device had not been declared eye-safe, it had a dramatic effect on Somali gunmen who panicked and ran. Consequently, further development of the concept as a non-lethal weapon system was carried out under the name Hindering Adversaries with Less-than-lethal Technology (HALT) and the result was a device which can be mounted on an M-16 rifle or M4 carbine, with a hand-held version being planned for the future. In effect the device, which is eye-safe at point of emission from the projector, serves as a warning to belligerents, showing them that they are in the aiming point of a weapon and could be subject to lethal force if they persist in showing aggression.

The use of lasers for blinding is not permitted under the terms of the 1980 Certain Conventional Weapons Convention (CCWC), to which the United States and Russia are signatories. Its Amended Protocol IV, prohibits the use or transfer of laser weapons specifically designed to cause permanent blindness to unenhanced vision (the naked eye or the eye with corrective lenses). However, the potential of even low-powered lasers to cause long-lasting damage to vision or blindness is considerable. Aircrews have proved particularly susceptible to this threat, as was illustrated by an incident which took place in the United States on 4 April 1997. On that date, a Canadian Armed Forces CH-124 helicopter was tasked with carrying out surveillance on a Russian merchant vessel, the *Kapitan Man*, which was tracking a US Navy strategic ballistic missile submarine (SSBN), the USS *Ohio*, which was based at the US naval base at Bremerton in the state of Washington.

Aboard the aircraft, in addition to the pilot Captain Patrick Barnes, was Lieutenant Jack Daly, an officer of the Office of Naval Intelligence (ONI) serving at the Canadian Pacific Maritime Forces Command at Esquimalt, on Vancouver Island in the state of Victoria, as a member of the ONI's Foreign Intelligence Liaison Officer (FILO) Program. The role of the Program was to monitor the movements and activities of the *Kapitan Man* and a number of other spy ships, one Russian and others sailing under Chinese, Cypriot and Panamanian flags, which monitored the movements of nine US SSBNs operating out of their base at Bremerton. Four years previously, the *Kapitan Man* had been searched by the US authorities and was found to have anti-submarine warfare equipment on board; a subsequent search uncovered the presence of sonobuoys.

When the helicopter caught up with the *Kapitan Man*, she was one of three vessels in the Strait of Juan de Fuca, 8 kilometres (5 miles) north of the Port of Los Angeles, heading towards Puget Sound. The aircraft circled three times before passing over the ship, during which time Lieutenant Daly filmed it with a digital camera. While doing so, however, he was unaware that he was being 'lased' – attacked with a laser. On returning to Esquimalt, his camera was handed over to another member of the US Navy intelligence staff for downloading of the images, which clearly showed a red dot of bright light on the bridge of the *Kapitan Man*. By that time, Lieutenant Daly was suffering from a severe headache and irritation to his right eye. On the following day he was experiencing severe pain and discomfort, and examination of the eye revealed that the eyeball was swollen. That night Captain Barnes began to suffer similar problems.

A subsequent medical examination of Lieutenant Daly's right eye revealed four or five lesions on the retina, caused, specialists believed, by a 'repetitive pulsed laser'. While Captain Barnes appeared to be less injured, possibly because his helmet visor was lowered at the time, his eyes were also permanently damaged. Both men subsequently suffered constant pain and deteriorating vision. As a result of his injuries, Captain Barnes was grounded after the incident and told that he would never fly again.

As for the *Kapitan Man*, she was subsequently boarded by the US Coast Guard but the Russian crew refused to allow certain areas of the vessel to be searched. According to an article by journalist Bill Gertz, published in the *Washington Post*, the Russian embassy had been warned of the search beforehand by the State Department, giving time for the laser to be concealed. It appeared that the Clinton administration was prepared to allow the Russians to get away with committing a hostile act in order to avoid jeopardizing relations with them.

Another laser attack on aircrew took place on 24 October 1998 during peacekeeping missions in Bosnia near the town of Zenica, when the pilot and a crew member of a UH-60 Blackhawk helicopter came under attack from a laser on the ground. Both suffered minor burns to their corneas and had to be removed temporarily from flying duty. There were suspicions that the attack was carried out with a small hand-held device but such was the threat posed that US helicopter aircrews were thereafter issued with special protective glasses.

Similar incidents have also occurred in the United States, commercial airline crews being temporarily incapacitated by lasers. The pilot and co-pilot of a Boeing 737 taking off from Los Angeles were hit by a burst from a laser that lasted between five and ten seconds; the co-pilot suffering burns to his right eye and broken blood vessels. In another incident, a member of the crew of an aircraft landing at Phoenix, Arizona, was blinded by a laser, suffering after-images and loss of night vision for about one and a half hours afterwards. In November 1998, the National Air Intelligence Centre reported that lasers had been responsible for over fifty incidents of blinding of aircrew in the United States alone.

Lasers are seen by many as the weapons of the future. Proponents of the technology emphasize in particular their non-lethal potential in being able to disarm an enemy with minimum collateral damage and loss of life, able to be calibrated to cause either minimum damage or totally destroy targets far more precisely than conventional weapons. Furthermore, they have much to offer as defensive systems, from shooting down ballistic missiles to defeating SAMs targeted on aircraft or anti-armour guided weapons launched at armoured vehicles. Without a doubt, lasers will before long form an essential part of any modern arsenal.

TIMELINE

The following chronology shows the specifications of the key weapons that entered service between 1944 and 1990.

Entered Service	Country of Origin	Type	Designation	Warhead	Range
1944	Germany	Cruise	V1	900 kg (1,980 lb) high explosive	240 kms (5,000 miles)
1944	Germany	SRBM	V2	750 kg (1,600 lb) high explosive	320 kms (200 miles)
1954	USA	Rocket	MGR-1 Honest John	2, 20 or 40 kilotons	19 kms (12 miles)
1958	USSR	SLBM	SS-N-4 Sark	1.2–2 megatons	560 kms (350 miles)
1960	USA	SLBM	Polaris A-1	1 megaton	2,200 kms (1,400 miles)
1961	USSR	IRBM	SS-5 Skean	1 megaton	3,700 kms (2,300 miles)
1961	USA	ICBM	Titan I	4 megatons	8,000 kms (5,000 miles)
1961	USA	SLBM	Polaris A-2	3×200 kilotons	2,800 kms (1,750 miles)
1962	USA	ICBM	Minuteman I	1 megaton	8,000 kms (5,000 miles)
1962	USA	IRBM	MGM-31 Pershing 1a	400 kilotons	740 kms (460 miles)
1963	USSR	ICBM	SS-8 Sasin	5 megatons	11,260 kms (7,200 miles)
1963	USA	ICBM	Titan II	9 megatons	8,000 kms (5,000 miles)
1964	USA	SLBM	Polaris A-3	3×60 kiloton MIRV	4,720 kms (2,950 miles)
1965	USSR	ICBM	SS-9 Scarp	3×5 megaton MRV or 1×20–25 megaton	12,000 kms (7,500 miles)
1965	USA	ICBM	Minuteman II	1.2 megaton	13,000 kms (8,000 miles)
1965	USSR	Rocket	FROG-7	200 kilotons	65 kms (105 miles)

Entered Service	Country of Origin	Type	Designation	Warhead	Range
1965	USSR	SRBM	SS-1C Scud B	1 megaton	180 kms (110 miles)
1968	USSR	SLBM	SS-N-6 Sawfly	1 megaton	3,000 kms (1,860 miles)
1969	USSR	SRBM	SS-12 Scaleboard	500 kilotons	700–800 kms (435–500 miles)
1970	USA	ICBM	Minuteman III	3×170 kiloton MIRV	9,650 kms (6,000 miles)
1971	USA	SLBM	Poseidon C-3	10–14 40 kiloton MIRV	4,000–5,200 kms (2,485–3,230 miles)
1971	USA	SRBM	MGM-52 Lance	100 kilotons	120 kms (75 miles)
1976	USSR	IRBM	SS-20 Model 1	1 megaton	2,750 kms (1,700 miles)
1976	USSR	SLBM	SS-N-18 Model 1 Stingray	7×150 kiloton MIRV	6,500 kms (4,000 miles)
1978	USSR	SRBM	SS-21 Spider	100 kilotons	120 kms (75 miles)
1978	USSR	ICBM	SS-18 Satan (Model 2)	8×600 kilotons 1.6 megatons	9,250 kms (5,750 miles)
1979	USA	SLBM	Trident I C4	8×100 kilotons	6,800 kms (4,200 miles)
1984	USA	IRBM	Pershing II	5–50 kilotons	1,800 kms (1,120 miles)
1984	USSR	SLBM	SS-N-20 Sturgeon	10×100 kilotons	8,300 kms (5,160 miles)
1985	USSR	SRBM	SS-23 Scarab	200 kilotons	500 kms (310 miles)
1989	USA	SLBM	Trident II D5	8 or 14×375 kilotons	7,400–11,100 kms (4,560–6,900 miles)

GLOSSARY
of abbreviations

A&E	Audits and Examinations
ABL	Airborne Laser
ABM	Anti-Ballistic Missile Systems
AERE	Atomic Energy Research Establishment
AFSA	Armed Forces Security Agency
ALL	Airborne Laser Laboratory
AMEC	Arctic Military Environmental Co-operation
AP	Armour-piercing ammunition
APFSDS-T	Armour-piercing Fin-stabilized Discarding Sabot Tracer
ATACMS	Army Tactical Missile System
ATAS	Air-to-Air Stinger
ATL	Airborne Tactical Laser
AWACS	Airborne Warning and Control System
BAT	Brilliant Anti-Tank
BILL	Beacon Illuminator Laser
BKA	Federal Criminal Bureau
BMDO	Ballistic Missile Defence Organization
BND	German intelligence service (Bundesnachtrichtendienst)
C3I	Command, Control, Communications and Intelligence
CCM	Counter-Counter Measure
CCWC	Certain Conventional Weapons Convention
CEP	Circular Error Probability
CIA	Central Intelligence Agency
CISCO	Commonwealth of Independent States Co-operation
COIL	Chemical Oxygen Iodine Laser
CTR	Co-operative Threat Reduction (*aka* Nunn-Lugar) Program
DATEN	Department of Atomic Energy
DIRCM	Directional Infra-Red Counter Measures
DMCD	Directed Missile Countermeasure Device
DSMAC	Digital Scene Matching Area Correlator
DVO	Direct View Optic sight unit

FBI	Federal Bureau of Investigation
FCR	Fire Control Radar
FILO	Foreign Intelligence Liaison Officer
FLIR	Forward-looking Infra-red System
FTS	Fine Track Sensor
GAN	Russian State Nuclear Regulator
GCS	Ground Control Station
GMLRS	Guided Multi-Launch Rocket Systems
GPS	Global Positioning System
GRU	Main Intelligence Directorate of the General Staff
GSM	Ground Station Module
HA	Heavy Armour
HALT	Hindering Adversaries with Less-than-lethal Technology
HE	High Explosive ammunition
HEAT	High Explosive Anti-Tank ammunition
HEAT-MP-T	High Explosive Anti-Tank Multi-Purpose Tracer
HEDP	High Explosive Dual Purpose ammunition
HEI	High Explosive Incendiary ammunition
HELEX	High Energy Laser Experimental
HESH	High Explosive Squash Head ammunition
ICBM	Intercontinental Ballistic Missiles
IFF	Identification Friend or Foe
IHADSS	Integrated Helmet and Display Sighting System
INF	Intermediate-Range Nuclear Forces Treaty
IR	Infra-Red
IRBM	Intermediate Range Ballistic Missiles
IRST	Infra-Red Search and Track
JSTARS	Joint Surveillance Target Attack Radar System
KGB	Soviet security police
LANTIRN	Low Altitude Navigation and Targeting Infra-Red for Night
LIRCM	Laser Infra-Red Counter Measures
LKA	Bavaria's State Criminal Bureau
LRF/D	Laser Rangefinder/Designator
MAD	Mutual Assured Destruction
MADM	Medium Atomic Demolition Munitions
MaRV	Manoeuvring warhead

MBT	Main Battle Tanks
MCS	Master Control Station
Minatom	Russian Ministry of Nuclear Energy
MIRACL	Mid-Infra-Red Advanced Chemical Laser
MIRV	Multiple Independently Targeted Re-entry Vehicles
MLRS	Multi-Rocket Launch Rocket System
MOX	Mixed Uranium-plutonium Dioxide
MPC&A	Material Protection, Control and Accounting Program
MRBM	Medium Range Ballistic Missiles
MRV	Multiple Re-entry Vehicles
MWS	Missile Warning System
NASA	US National Aeronautical and Space Administration
NATO	North Atlantic Treaty Organisation
NBC	Nuclear Biological and Chemical warfare
NKVD	Soviet secret police
NSA	National Security Agency
ONI	Office of Naval Intelligence
PBO	Post-Burnout
PDRR	Program Definition and Risk Reduction
PNVS	Pilot Night Vision Sensor
PPS	Precise Positioning Service Programme
RAAF	Royal Australian Air Force
RAF	Royal Air Force
RAN	Royal Australian Navy
RARDE	Royal Armaments Research and Development Establishment
RFI	Radio Frequency Interferometer
RPVs/UAVs	Remotely Piloted Vehicles
RSTA	Reconnaissance, Surveillance and Target Acquisition
RVSN	Strategic Rocket Forces (Raketnyye Voyska Strategicheskogo Naznacheniya)
SADM	Special Atomic Demolition Munitions
SALT	Strategic Arms Limitation Talks
SAM	Surface-to-Air missile
SBL	Space-Borne Lasers
SCDL	Surveillance Control Data Link
SCS	Scientific Civil Service

SDI	Strategic Defence Initiative
SDIO	Strategic Defence Initiative Organization
SEAD	Suppression of Enemy Air Defences aircraft
SEP	System Enhancement Programme
SIS	Secret Intelligence Service
SLBM	Submarine-Launched Ballistic Missiles
SMDC	Space and Missile Defence Command
SPS	Standard Positioning Service
SRBM	Short Range Ballistic Missiles
SSBN	Strategic Ballistic Missile Submarines
START	Strategic Arms Reductions Talks
SVR	External Intelligence Agency (Sluzhba Vneshney Razvedki)
TADS	Target Acquisition Designation Sights system
TBM	Theatre Ballistic Missiles
TCU	Tracking Communication Unit
THEL	Tactical High Energy Laser
TI	Infra-Red Thermal Imager
TIALD	Thermal Imaging Airborne Laser Designator
TILL	Tracking Illuminator Laser
TLOS	Target Location and Observation System
TOW	Tube-launched, Optically-tracked, Wire-guided missile
TV	Television
UN	United Nations
USAF	United States Air Force
UV	Ultra-Violet

SELECTED BIBLIOGRAPHIES

Equinox: The Earth

Alvarez, Walter *T-Rex and the Crater of Doom* Penguin, 1998

Courtillot, Vincent *Evolutionary Catastrophes: The Science of Mass Extinction* Cambridge University Press 1999

Edwards, Katie and Rosen, Brian *From the Betginnning* The Natural History Museum 2000

Gould, Stephen Jay, with Benton, Michael and Stringer, Christopher *The Book of Life*, Hutchinson 1993

Hallam, A and Wignall, P.B. *Mass Extinctions and Their Aftermath* Oxford University Press 1997

Hynes, Gary *Mammoths, Mastodonts and Elephants: biology, behaviour and the fossil record* Cambridge University Press 1991

Lister, Adrian et al *Mammoths* Marshall Editions, London, 2000

MacDougall, Dougie *Still Waters Run Deep* Dougie MacDougall (Yellow Rock Cottage, Caol Ila, Port Askaig, Islay, Scotland) 1996

MacRae, Colin *Life Etched in Stone: Fossils of South Africa* Geological Society of South Africa 1999 (British Stockists, Natural History Museum, London)

Pernetta, John *Atlas of the Oceans* Phillips 1994

Ridley, Gordon, *Dive West Scotland* Underwater Publications Ltd, 1984

Sigurdsson, Haralder, Rymer, Hazel, et al *Encyclopaedia of Volcanoes* Academic Press 2000

Van Rose, Susanna and Mason, Roger *Earthquakes: Our trembling planet* British Geological Survey 1997

Ward, Peter *On Methuselah's Trail* Freeman 1992

Wilson, R.C.L. et al, *The Great Ice Age: climate change and life* Routledge/Open University 2000

Equinox: The Brain

Mind Readers

Baron-Cohen, Simon, and Patrick Bolton, *Autism*, Oxford Paperbacks, Oxford, 1993.

Frith, Uta, *Autism*, Blackwell Publishers, Oxford, 1989.

Rosner, Eleanor, and Sue R. Semel, *Williams' Syndrome*, Blackwell Publishers, Oxford, 1998.

Sacks, Oliver, *An Anthropologist on Mars*, Picador, London, 1995.

Scariano, Margaret M., *Emergence: Labelled Autistic – Temple Gradin*, Warner Books, New York, 1996.

Natural-born Genius

Devlin, Bernie (ed.) and Stephen E. Fienberg, *Intelligence, Genes, and Success: Scientists Respond to The Bell Curve*, Copernicus Books, New York, 1997.
Eysenck, Hans J., *Intelligence: The New Look*, Transaction Publishers, New Brunswick, 1998.
Gardner, Howard, Mindy Kornhaber and Warren Wake, *Intelligence: Multiple Perspectives*, Harcourt Publishers Ltd, New York, 1996.
Herrnstein, Richard J., and Charles Murray, *The Bell Curve: Intelligence and Class Structure in American Life*, The Free Press (a division of Simon and Schuster), New York, 1994.
Kinchloe, Joe L. and Shirley R. Steinberg (eds), and Aaron d'Gresson, *Measured Lies: The Bell Curve Examined*, St Martin's Press, New York, 1996.

Phantom Brains

Melzack, Ronald, and Patrick D. Wall, *The Challenge of Pain*, Penguin Books, London, 1993.
Ramachandran, V. S., and Sandra Blakeslee, *Phantoms in the Brain*, Fourth Estate Ltd, London, 1998 (paperback 1999).

Living Dangerously

Bernstein, Peter L., *Against the Gods: The Remarkable Story of Risk*, John Wiley and Sons, New York, 1998.
Tomlinson, Joe, *The Ultimate Encyclopedia of Extreme Sports*, Carlton Books, London, 1996.
Wilde, Gerald J. S., *Target Risk*, PDE Publications, Toronto, 1995.
Willis, Jim, Albert A. Okunade and William J. Willis, *Reporting on Risks*, Praeger Pub Text, Westport, Connecticut, 1997.
Zuckerman, Marvin (ed.), *Behavioral Expression and Biosocial Bases of Sensation Seeking*, Cambridge University Press, Cambridge, 1994.

Thin Air

Boukreev, Anatoli, and G. Watson Dewalt, *The Climb: Tragic Ambitions on Everest*, St Martin's Press, New York, 1998.
Krakauer, Jon, *Into Thin Air*, Random House, New York, 1998.
Lieberman, Philip, *Eve Spoke: Human Language and Human Evolution*, Macmillan Publishers Ltd, London, 1998.
Phillips, John L., *The Bends: Compressed Air in the History of Science, Diving and Engineering*, Yale University Press, Newhaven, Connecticut, 1998.

Lies and Delusions

Freeman, Walter J., *How Brains Make Up Their Minds*, Weidenfeld & Nicolson General, London, 1999.
Sacks, Oliver, *The Man Who Mistook His Wife for a Hat*,

HarperCollins, New York, 1986.
Weiskrantz, Lawrence, *Consciousness Lost and Found*, Oxford University Press, Oxford, 1998.

General

Cairns-Smith, A. G., *Secrets of the Mind: A Tale of Discovery and Mistaken Identity*, Springer Verlag New York Inc, New York, 1999.
Carter, Rita, and Christopher Frith, *Mapping the Mind*, Weidenfeld Illustrated, London, 1998.
Greenfield, Susan, *The Human Brain: A Guided Tour*, Weidenfield & Nicolson, London, 1997 (paperback 1998: Phoenix, a division of Orion Books).
Ridley, Matt, *Genome: The Autobiography of a Species in 23 Chapters*, Fourth Estate, London, 1999.

Equinox: Space

Apt, Jay, Michael Helfert, Justin Wilkinson and Roger Ressmeyer (ed.), *Orbit: NASA Astronauts Photograph the Earth*, National Geographic Society, Washington, 1996.
Begelman, Mitchell, and Martin Rees, *Gravity's Fatal Attraction: Black Holes in the Universe*, Scientific American Library, Chicago, 1996.
Charles, Philip, and Mark Wagner, *Exploring the X-ray Universe*, Cambridge University Press, Cambridge, 1995.
Couper, Heather, and Nigel Henbest, *Space Encyclopedia*, Dorling Kindersley, London, 1999.
Ferguson, Kitty, *Prisons of Light*, Cambridge University Press, Cambridge, 1996.
Henbest, Nigel, and Michael Marten, *The New Astronomy*, Cambridge University Press, Cambridge, 1996.
Hufbauer, Karl, *Exploring the Sun: Solar Science Since Galileo* (New Series in NASA History), Johns Hopkins University Press, New York, 1993.
Light, Michael, *Full Moon*, Jonathan Cape, London, 1999.
Padmanabhan, Thanu, *After the First Three Minutes*, Cambridge University Press, Cambridge, 1998.
Thorne, Kip, *Black Holes and Time Warps: Einstein's Outrageous Legacy*, W. W. Norton & Co., New York, 1994.
Turner, John, *Rocket and Spacecraft Propulsion*, John Wiley and Sons, London, 1999.
Wilson, Robert, *Astronomy Through the Ages*, UCL Press, London, 1997.

Equinox: Warfare

Andrew, Christopher and Vasili Mitrokhin. *The Mitrokhin Archive – The KGB in Europe and The West*. Allen Lane, London 1999.

Arnold, Lorna. *A Very Special Relationship – British Atomic Weapon Trials in Australia*. Her Majesty's Stationery Office, London 1987.

Bishop, Chris (ed.). *The Directory of Modern Military Weapons.* Greenwich Editions, London 1999.

Bonds, Ray (ed.) *The Soviet War Machine – An Encyclopedia of Russian Military Equipment and Strategy.* Salamander, London 1976.

Brooks, Geoffrey. *Hitler's Nuclear Weapons.* Leo Cooper, London 1992.

Bunn, Matthew. *The Next Wave – Urgently Needed New Steps to Control Warheads and Fissile Material.* Carnegie Endowment for International Peace, Washington 2000.

Clancy, Tom with General Chuck Horner. *Every Man a Tiger.* Sidgwick & Jackson, London 2000.

Cockburn, Andrew & Leslie. *One Point Safe – The Terrifying Threat of Russia's Unwanted Nuclear Arsenal.* Little Brown, London 1997.

Darwish, Adel & Gregory Alexander. *Unholy Babylon – The Secret History of Saddam's War.* Victor Gollancz, London 1991.

Dorril, Stephen. *MI6 – Fifty Years of Special Operations.* Fourth Estate, London 2000.

Gowing, Margaret. *Britain And Atomic Energy 1939–1945.* Macmillan, London 1964.

Shields, John M. & William C. Potter. *Dismantling the Cold War – US and NIS Perspectives on the Nunn-Lugar Cooperative Threat Reduction Program.* MIT Press, Cambridge, Mass. 1997.

Young, Peter. *The Machinery of War – An Illustrated History of Weapons.* Hart-Davis MacGibbon, London 1973.

INDEX

G

H

J

K

L

M

O

P

T